Hydraulics and Hydraulic Machines

Prof. Kamlesh Purohit
Head, Dept. of Mechanical Engineering,
Faculty of Engineering
MBM Engineering College,
JNV University, Jodhpur

Dr. S.P. Harsha
Associate Professor
Dept. of Mechanical & Industrial Engineering,
IIT, Roorkee

Dr. R.K. Purohit
Retd. Associate Professor
Dept. of Mechanical Engineering,
Faculty of Engineering
MBM Engineering College,
JNV University, Jodhpur

Published by:

Scientific Publishers (India)
5 A, New Pali Road, P.O. Box 91
Jodhpur 342 001 (India)

Branch Office
Scientific Publishers (India)
4806/24, Ansari Road, Daryaganj
New Delhi - 110 002 (India)

E-mail: info@scientificpub.com
Website: www.scientificpub.com

ISBN: 978-81-7233-829-9

Typeset by Rajesh Ojha
Printed in India

4CE3 : HYDRAULICS AND HYDRAULIC MACHINES

SYLLABUS

UNIT 1 — Dimensional Analysis & Models : Dynamical Similarity and Dimensional Homegeneity Model experiment, geometric, Kinematic and Dynamic similarity. Reynold's, froudes, Weber's, Euler and Mach numbers. Distorted river models and undistorted models, proper choice of scale ratios. Scale effect. Principle of dimensional analysis Rayleigh method, Buckingham theorem, applications of dimensional analysis to pipe Friction problems, resistance to motion of partially and fully submerged bodies and other simple problems. Ship model experiments.

UNIT 2 — Laminar Flow : Relation betweens shear & pressure gradient. Flow between plates & pipes. Equations for velocity distribution, pressure difference. **Turbulent Flow in pipes :** Theories of Turbulence, Nikuradse's Experiments. Hydrodynamically smooth & rough boundaries. Laminar, Sublayer, Equations of velocity distribution and friction coefficient. Stanton Diagram, Moody's diagram.

UNIT 3 — Flow through channels : Uniform, Non-Uniform and variable flow. Resistance equations of Chezy, Mannring and Bazin. Section factor for uniform flow. Most Efficient rectangular, triangular and trapezoidal sections. Equations of gradually varied flow in Prismatic channels. Limitation of its applicability and assumption made in its derivation. Specific emergy of flow. Critical depth in prismatic channels. Alternate depths. Rapid, critical and sub critical Flow Mild, steep and Critical Slopes. Classification of surface curves in prismatic channels and elementary computation

UNIT 4 — Rapidly varied flow: Hydraulic jump or standing wave in rectangular channels. Conjugate or sequent depths Losses in jump, location of jump. Broad crested weirs for channel flow: Measurement, velocity distribution in open channels, parshall flume. **Impact of free Jets :** Impact of a jet on a flat or a curved vane, moving and stationary vane, flow over radial vanes.

UNIT 5 — Centrifugal pumps and turbines : Volute and whirlpool chambers, Loses of head due to variation of discharge Manometric and Hydraulic efficiencies, Description of single and multistage pumps. Specific speed, characteristic curves. Model Test. Reaction and Impulse turbines, specific speed, Mixed flow turbines. Pelton wheel turbine, Francis turbine, propeller turbine and Kaplan turbine Efficiency, Characteristics of turbines. Basic principles of governing of turbines, Draft-tube, Selection of turbines, model tests.

Preface

It gives me immense pleasure in placing this book `Hydraulics and Hydraulic Machines' in the hands of B.Tech IV Semester students of Civil Engineering of Rajasthan Technical University (RTU).

The subject matter has been evolved from first principles and the treatment is then extended to considerable depth so that this single book covers the complete course of IV Semester of RTU.

The importance of illustrative examples cannot be under estimated, as they help in understanding the concepts of the subject clearly. Besides, they also follow the style of question set in RTU.

We are thankful to colleagues, friends and students who encouraged us to write this book.

We express our appreciation and gratefulness to our publisher Tanay Sharma for his most cooperative, painstaking attitude and untiring effort in bringing out this book in a very short period.

We have used several references and we are grateful to those learned authors.

Jodhpur

Prof. Kamlesh Purohit
Dr. S.P. Harsha
Dr. R.K. Purohit

Contents

1

Dimensional Analysis and Simlitude

1.1 INTRODUCTION

Dimensional analysis is one of the most important mathematical tools in the study of fluid mechanics. It is a mathematical technique which makes use of the study of dimensions as an aid to the solution of many engineering problems. The main advantage of a dimensional analysis of a problem is that it reduces the number of variables in the problem by combining dimensional variables to form nondimensional parameters. Dimensional analysis is particularly helpful in experimental work because it provides a guide to those things that significantly influence the phenomena; thus it indicates the direction in which experimental work should go.

Although many practical engineering problems involving fluid mechanics can be solved by using the equations and analytical procedures described in the preceeding chapters, there remain a large number of problems that rely on experimentally obtained data for their solution. In fact, it is probably fair to say that very few problems involving real fluids can be solved by analysis alone. But in the majority of more, complex fluid flow problems, especially those of turbulent flow, there are so many variables involved in the differential equation of fluid motion that a direct mathematical solution is simply out of the question. In these problems dimensional analysis can be used to great advantage in obtaining a functional relationship among the various variables involved in terms of nondimensional parameters.

When experimental testing of a full size prototype is either impossible or prohibitively expensive, the only feasible way of attacking the problem is through model testing in the laboratory. If we are to predict the prototype behaviour from measurements on the model, it is obvious that we cannot run just any test on any model. The model flow and the prototype flow must be related by known scaling laws. The laws of simlitude make it possible to predict the performance of the prototype, which means the full size device, from tests made with the model. It is not necessary that the same fluid be used for the model and its prototype. Neither is the model necessarily small than its prototype. Thus the flow in a Carburiettor might be studied in a very large model. And the flow of water at the entrance to a small centrifugal pump runner might be investigated by the flow of air at the entrance to a large model of the runner.

A few examples where models may be used are ships in towing basins, aeroplane in wind tunnels, hydraulic turbines, centrifugal pumps, spillways of dams, rivel channels, and the study of such phenomena as the action of waves and tides on beaches, soil erosion, and the transportation of sediment.

It should be emphasized that the model need not necessarily be different in size from is prototype. In fact, it may be the same device, the fact, it may be the same device, the variables in the case being the velocity and the physical properties of the fluid.

1.2 DIMENSIONS, DIMENSIONAL HOMOGENEITY AND UNITS

In fluid mechanics we will be dealing with a velocity of fluid characteristics. It is necessary to develop a system for describing these characteristics both qualitatively and quantitatively. The qualitative aspect serves to identify the nature, or type, of the characteristics (such as, length, time, velocity and stress). Whereas quantitative description requires both a number and a standard by which various quantities can be compared such standards are called units, and several systems of units are in common use. The qualitative description of physical quantities can be given in terms of basic dimensions such as mass, M, length, L and time T. Alternatively we could use force, F, L, and T as basic dimensions, since from Newton's second law

$$F = MLT^{-2} \quad \text{or} \quad \frac{FT^2}{ML} = 1 \tag{1.1}$$

where the symbol = is used to indicate dimensional equality.

Thus, in fluid mechanics, we can select [L] and [T] as fundamental dimensions together with either force [F] or mass [M], the remaining quality being regarded as a derived dimensions and expressed in terms of the three fundamental dimensions. In this book, we shall normally select the dimensions of mass [M] as a third fundamental dimension and treat force as a derived dimension, having the dimensions $[MLT^{-2}]$.

The dimensions of a number of common physical quantities are listed in Table 1.1.

Table 1.1 Dimensions associated with common physical quantities.

Quantity	MLT System
Length	L
Area	L^2
Volume	L^3
Time	T
Velocity	LT^{-1}
Acceleration	LT^{-2}
Force	MLT^{-2}
Mass	M
Density	ML^{-3}
Pressure and Stress	$ML^{-1}T^{-2}$
Energy and Work	$ML^{-1}T^{-2}$
Momentum	MLT^{-1}
Power	ML^2T^{-3}
Torque	$ML^{-2}T^{-2}$
Moment of a force	ML^2T^{-2}

RPM	T^{-1}
Angular Velocity	T^{-1}
Dynamic Viscosity	$ML^{-1}T^{-1}$
Kinematic Viscosity	L^2T^{-1}
Surface tension	MT^{-2}
Modulus of elasticity	$ML^{-1}T^{-2}$
Angle	$M^oL^oT^o$
Efficiency	$M^oL^oT^o$
Strain	$M^oL^oT^o$

The distinction between units and dimensions is that dimensions are properties that can be measured and units are the standard elements in terms of which these dimensions can be described quantitatively and assigned numerical values.

In dimensional analysis, we are concerned only with the nature of the factors involved in the situation and not with their numerical values. The notation adopted to indicate this is to enclose the name or symbol of the quantity in square brackets. Whenever a property in words or symbols, is enclosed in a square bracket it indicates that we are concerned with it only dimensionally and qualitatively, not quantitatively.

All theoretically derived equations are dimensionally homogeneous - that is, the dimensions of the left side of the equation must be the same as those on the right side, and all additive separate terms must have the same dimensions. We accept as a fundamental permise that all equations describing physical phenomena must be dimensionally homogeneous.

The principle of dimensional homogeneity requires that any equation which fully or correctly represents a possible physical situation must be dimensionally homogenous, so that each term has the same dimensional formula. It is therefore, possible to make a rapid check of any algebraic analysis of such a situation by substituting the dimensions of the quantities in each term and checking that, in the resultant dimensional equation, each fundamental dimension appears to the same power in each term.

Many engineering formulae in day-to-day use are committed to memory and, sometimes, there may be a doubt as to whether they have been recalled correctly. A quick dimensional check will establish the correct form immediately, although, of course, it will not indicate whether the values of any numerical constants are correct.

A dimensionally homogenous equation has the unique characteristics of being independent of the units chosen for measurements. For example, equation (1.1) holds true no matter what units of measurements are used.

Alternate forms of equation (1.1) also demonstrate that a dimensionally, homogeneous equation may be transferred to a non dimensional form because of the mutual dependence of fundamental dimensions. Although it is always possible to reduce a dimensionally homogeneous equation to a non dimensional form, the main difficulty in a complicated flow problem is in setting up the correct equation of motion. Therefore, a special mathematical method called dimensional analysis is required to determine the functional relationship among all the variables involved in any complex flow phenomena, in terms of non dimensional parameters.

1.3 DIMENSIONAL ANALYSIS

The fact that a complete physical equation must be dimensionally homogeneous and is, therefore, reducible to a functional equation among nondimensional parameters forms the basis for the theory of dimensional analysis. The following two methods of dimensional analysis are generally used in fluid mechanics. (1) The Rayleigh method and (2) The Buckingham π (pi) method.

Lord Rayleigh in 1899 gives a special form of relationship among the dimensionless groups, and has an inherent drawback that it does not provide any information regarding the number of dimensionless groups to be obtained as a result of dimensional analysis. It is for these reasons that it has become obsolete and is not favoured for use. The second method is the Buckingham method (developed by E. Buckingham in 1914) and is an important over the Rayleigh's method. Buckingham designated the dimensionless groups by the Greek capital letter π (pi). It is, therefore, often called as, Buckingham- π method. It offers an advantages over Rayleigh's method in letting us known, in advance of the analysis as to how many dimensionless groups are to be expected.

1.3.1 The Buckingham π - theorem

The Buckingham π-theorem states that if, in a dimensionally homogenous equation, there are n dimensional variables which are completely described by m fundamental dimensions (such as M, L, T etc.) they may be grouped into (n-m) π terms. Each π- term is a dimensionless parameter.

Buckingham used the symbol π to represent a dimensionless product, and this notation is commonly used. Although the π theorem is a simple one, its proof is not so simple and we will not include it here.

The π-theorem is based on the idea of dimensional homogeneity. Essentially we assume that for any physically meaningful equation involving n variables, such as

$$A_1 = f\left(A_2, A_3\ A_4, \text{----}\ A_n\right)$$

which can be transformed to

$$f_1\left(A_1, A_2, A_3\ A_4\ \text{----}\ A_n\right) = 0 \qquad (1.2)$$

The equation (1.2) is a dimensionally homogenous equation. It then follows that we can rearrange. the equation into a set of dimensionless products (pi terms) so that

$$\Phi\left(\pi_1, \pi_2, \text{----}\ \pi_{n-m}\right) = 0 \qquad (1.3)$$

or, $$\pi_1 = \Phi_1\left(\pi_2, \pi_3\ , \text{---}\ \pi_{n-m}\right) \qquad (1.4)$$

The required number of π-terms is fewer than the number of original variables by m, where m is determined by the minimum number of reference dimensions required to describe the original list of variables. Usually the references dimensions required to describe the variables will be the basic dimensions M, L and T or F, L and T. However, in some instances perhaps only two dimensions, such as L and T, are required, or may be just one, such as L. Also, in a few rare cases the variables may be described by some combination of basic dimensions, such as M/T^2 and L, and in this case m would be equal to two rather than three.

The theorem does not predict the functional form of Φ or Φ_1. The functional relation among the independent, dimensionless π parameters must be determined experimentally.

1.3.2 Selection of repeating variables

Three general rules guide the selection of repeating variables.

(1) In a dimensional analysis of any physical system the repeating variables must include among them all of the m fundamental dimensions. Each repeating variable must be dimensionally independent of the others. This means that the repeating variables cannot themselves be combined to form a dimensionless product. Although this is not an absolute requirement, mass density a characteristic velocity, and characteristic length need to be selected as repeating variables in order to produce common dimensionless groups.

(2) For any given problem we usually are interested in determining how one particular variables is influenced by the other variables. We would consider this variable to be the dependent variable, and we would want this to appear in only one π-term. Thus, do not choose the dependent variable as one of the repeating variables, since the repeating variables will generally appear in more than one π-term.

(3) For a fluid system, the most significant groups will result if the repeating variables are chosen so that one is geometrical characteristics, other is flow characteristics, and still other is fluid properties.

Variables with geometrical properties are, length, diameter, height, breadth, chord span etc.

Variables with flow characteristics are, velocity, angular velocity, rpm, acceleration, mass flow rate, flow rate, pressure, pressure gradient, force (drag, lift, shear) shear stress, torque and moment.

Variables with fluid properties are, Density, dynamic viscosity, kinematic viscosity. Specific weight, Elastic modulus, surface tension.

In ninety percent cases the choice of repeating variables are density, velocity and length (ρ, v and l).

1.3.3 Determining the π groups

We are looking for a method that will allow us to systematically form the π terms so that we are sure that they are dimensionless and independent and that we have the right number. Some experience admittedly is helping in compiling the list. Students, who do not have this experience often are troubled by the need to apply engineering judgment in an apparent massive dose. However, it is difficult to go wrong if a generous selection of parameters is made.

If you suspect that a phenomenon depends on a given parameter, include it. If your suspicion is correct, experiments will show that the parameter must be included to get consistent results. If the parameter is extraneous, an extra π parameter may result, but experiments will show that it may be eliminated from consideration. Therefore, do not be afraid to include all the parameters that you feel are important. With a little practice you will be able to readily complete a dimensional analysis for your problem.

The eight steps listed below are a recommended procedure for determining the π parameters.

Step 1. List all the variables that are involved in the problem. (Let n be the number of parameters).

Step 2. Express each of the variables in terms of basic dimensions. For the fluid mechanics problem the basic dimensions will be either, M, L and T or F, L and T. (Let m be the number of basic dimensional).

Step 3. Determine the required number of π terms.

The number of π terms is equal to n-m, where n is the number of variables in the problem (which is determined from step 1) and m is the reference dimensions required to describe these variables (which is determined from step 2). The reference dimensions usually correspond to the basic dimensions, and can be determined by an inspection of the dimensions of the variables obtained in step 2.

Step 4. Select a number of repeating variables, where the number required is equal to the number of reference dimensions. The repeating parameters chosen may appear in all the dimensions less groups obtained.

Step 5. Set up dimensional equation combining the repeated parameters, selected in step 4 with each of the other parameters in turn to form dimensionless groups.

Essentially each π term will be of the form $A_1^{a_i} A_2^{b_i} A_2^{c_i} A_i$ where A_i is one of the non repeating variables, A_1, A_2 and A_3 are the repeating variables and the exponents a_i, b_i and c_i are determined so that combination is dimensionless.

Step 6. Repeat Step 5 for each of the remaining non repeating variables.

Step 7. Check all the resulting π terms to make sure they are dimensionless.

Step 8. Express the final form as a relationship among the π terms, and think about what it means. Typically the final form can be written as

$$\pi_1 = \Phi\,(\pi_2, \pi_3, \pi_4 --- \pi_{n-m})$$

1.3.4 Some additional comments about dimensional analysis

In order to obtain the final expression in the desired manner the following useful suggestions may be noted.

(1) If a quantity is dimensionless (e.g. efficiency, angle etc), it is a π term without going through the above procedure.

For example

$$\pi = \eta, \pi = \theta$$

(2) If any two physical quantities have the same dimensions, their ratio will be one of the π term. For example (*l*/d) is dimensionless and hence it is a π term dimensionless and hence it is a π term which may be directly written without going through the above procedure.

(3) Any π term may be replaced by another π term obtained by adding or subtracting an absolute numeral from it.

(4) Any π term may be replaced by multiplying it by a numerical constant. For example, π1 may be replaced by $2\pi_1, \pi_1 / \sqrt{2}$ or so.

(5) Any π term may be replaced by multiplying it by another π term. For example π_1 may be replaced by ($\pi_1 \times \pi_2$) or π_1 and π_2 may be replaced by ($\pi_1 \times \pi_2$).

(6) Any π term may be replaced by any power of that term, including negative as well as fractional powers. For example, π_1 may be replaced by $1/\pi_1$ or π_2 may be replaced by π_2^2, π_3 may replaced by $\left(\frac{1}{\pi_3^{3/2}}\right)$ and π_2 can be formed by a combination of say π_3, π_4 and π_5 such as

$$\pi_2 = \frac{\pi_3\,\pi_4^2}{\sqrt{\pi_5}}$$

then π_2 is not an independent π term.

1.3.5 Uniqueness of π terms

There is not a unique set of π terms which arise from a dimensional analysis. However, the required number of π terms is fixed, and once a correct set is determined all other possible sets can be developed from this set by combinations of products of powers of the original set. For example

$$\pi_1 = \Phi\left(\pi_2, \pi_3\right) \tag{1.5}$$

We could always form a new set from this one by combining the π terms. For example we could form a new π term, π_3', by letting

$$\pi_3' = \pi^a,\ \pi_3^b$$

Where a and b are arbitrary exponents. Then the relationship could be expressed as

$$\pi_1 = \Phi_1\left(\pi_2, \pi_3'\right) \tag{1.6(a)}$$

or, $$\pi_3' = \Phi\ \left(\pi_2, \pi_3\right) \tag{1.6(b)}$$

All of these would be correct. It should be emphasized, however, that the required number of π terms cannot be reduced by this manipulation; only the form of the π terms is altered.

1.3.6 Limitations of dimensional analysis selection of variables - superfluous and omitted variables

All the fore going procedures deal with straight forward situations. There are some problems which do occur: In order to apply dimensional analysis we must first decide which variables are significant. If the problem is not understood well enough to make a good initial choice of variables, dimensional analysis will seldom provide clarification.

There is no simple procedure whereby the variables can be easily identified. Generally, one must rely on a good understanding of the phenomenon involved and the governing physical laws. If extraneous variables are included, then too many π terms appear in the final solution, and it may be difficult, time consuming, and expensive to eliminate these experimentally. If important variable are omitted, then an incorrect result will be obtain; and again, this may prove to be costly and difficult to ascertain. It is, therefore, imperative that sufficient time and attention be given to this first step in which the variables are determined.

Most engineering problems involve certain simplifying assumptions that have an influence on the variables to be considered. Usually we wish to keep the problem as simple as possible, perhaps even if some accuracy is sacrificed. A suitable balance between simplicity and accuracy is a desired goal. How accurate the solution must be depends on the objective of the study; that is, we may be only concerned with general trends and, therefore, some variables that are thought to have only a minor influence in the problem may be neglected for simplicity.

Since we wish to keep the number of variables to a minimum, it is important that all variables are independent. In a general term, if we have a problem in which the variables are

$$f(p, q, r, --- u, v, w, ---) = 0 \tag{1.7}$$

and it is known that there is an additional relationship among some of the variables, for example

$$P = f_1\ (u, v, w, -----) \tag{1.8}$$

then p is not required and can be omitted. Conversely, if it is known that the only way the variables u,v,w ------ enter the problem is through the relationship expressed by equation (1.8), then the variables u,v,w, can be replaced by the single variable p, thereby, reducing the number of variables.

It is thus evident that the method of dimensional analysis does not give any clue as to the correctness of the selection of the relevant variables of a particular phenomenon. It merely tells how the variables should be grouped so that from the experimental investigation it may be decided which of the variables are important ones. This is, however, the main limitation of the dimensional analysis.

1.4 MODELING AND SIMILITUDE

Models are widely used in fluid mechanics. A few examples where models may be used are airplanes in wind tunnels, ships in towing basins, spillways of dams, river channels, hydraulic turbines, centrifugal pumps, air and water pollution, harbours, dams, cars and trains, towing test on submarines. A recent application has arisen from the problems of air flow around buildings, which may be studied in wind tunnels and by examining smoke generation and propogation through building models Although the term "model", is used in many different contexts, the "engineering model" generally conforms to the following definition. A model is a representation of a physical system that may be used to predict the behaviour of the system in some desired respect. The physical system for which the predictions are to be made is called the prototype. Although mathematical or computer models may also conform to this definition our interest will be in physical models; that is models that resemble the prototype but are generally of a different size, may involve different fluids, and often operate under different conditions.

Usually a model is smaller than the prototype. Therefore, it is more easily handled in the laboratory and less expensive to construct and operate than a large prototype. Occasionally, if the prototype is very small, it may be advantageous to have a model that is larger than the prototype so that it can be more easily studied. For examples, the flow in a carburettor might be studied in a very large model, the flow of water at the entrance to a small centrifugal runner might be investigated by the flow of air, at the entrance to a large model of the runner and large models have been used to study the motion of red blood cells, which are approximately 8 μm in diameter.

Although studies of models do not always provide ready answers to all questions, the results often guide an engineer toward a better under standing of the problem and aid him in the proper design of the prototype. In fact, a model may be regarded as a mechanical analog for solving differential equations of fluid motion with a given set of rather complicated boundary condition; its main function is to obtain useful quantitative or qualitative information which can be safely transferred to its prototype. A thorough knowledge of principles of model similitude is therefore essential to the proper design, construction, and operation of the model as well as the interpretation of data obtained from model tests.

Silimarity of flow phenomena not only occurs between a prototype and its model but also may exist between various natural phenomena if certain laws of similarity are satisfied. Similarity thus becomes a means of correlating the apparently divergent results obtained from similar fluid phenomena and as such becomes a valuable tool of modern fluid mechanics; the application of the laws of similitude will be found to lead to more comprehensive solutions and, therefore, to a better understanding of fluid phenomena in general.

There are, in general three types of similarities to be established for complete similarity to exist between the model and its prototype. These are; geometrically, kinematically and dynamically similarity.

1.4.1 Geometric Similarity

One of the desirable features in model studies is that there be geometric similarity which means that the model and its prototype be identical in shape but differ only in size. The important consideration is that the flow patterns be geometrically similar. If the scale ratio is denoted by l_r, which means the ratio of the linear dimensions of the model to corresponding dimensions in the model.

Mathematically,

$$l_r = \frac{l_m}{l_p} \tag{1.9}$$

Subscript m and p refer to model and prototype respectively.

For true models there will be one scale ratio, and all lengths are fixed in accordance with this scale. The area ratio and volume ratios are written as l_{r2} and l_{r3} respectively. One of the most important factors in maintaining a complete geometric similarity between the two systems is the similarity in surface roughness. Failure to satisfy the geometric similarity in surface roughness results in a model with distorted scale. Therefore, complete geometric similarity is not always easy to attain.

1.4.2 Kinematic similarity

Kinematic similarity is the similarity of motions. To satisfy kinematic similarity, both the model and its prototype must undergo similar time rates of change of motions. Therefore the flow patterns in two systems must be geometrically similar (Fig 1.1). The corresponding velocity and acceleration ratios of two systems are then constant.

Velocity ratio $v_r = \frac{v_m}{v_p}$ (1.10)

and its value in terms of l_r will be determined by dynamic considerations as explained in the following section.

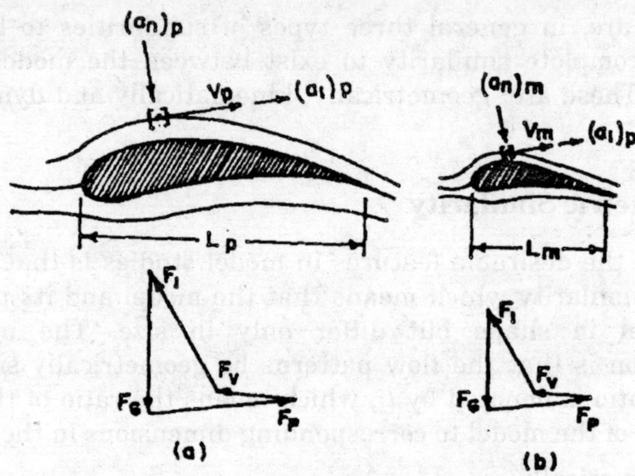

Fig. 1.1. (a) Prototype (b) Model.

As time T is dimensionally L/V, the time scale is

$$T_r = \frac{l_r}{v_r} \tag{1.11}$$

and in similar manner the acceleration scale is

$$a_r = \frac{a_m}{a_p} \tag{1.12}$$

$$\text{or,} \quad a_r = \frac{l_r}{T_r^2} = \frac{v_r^2}{l_r} \tag{1.13}$$

1.4.3 Dynamic similarity

Dynamic similarity means the similarity of forces. The model is said to be dynamically similar to its prototype if the ratios of all forces acting on corresponding fluid particles or corresponding boundary surface in the two systems are constant.

Forces that may act on a fluid element include those due to

(1) Viscous force F_V, (2) Gravity force F_G,

(3) Elastic force F_E, (4) Pressure force F_P, and

(5) Surface tension force F_S.

If the summation of forces on a fluid element does not add up to zero, the element will accelerate in accordance with Newton's law. Such an unbalanced force system can be transferred into a balanced system by adding an inertia force F_I that is equal and opposite to the resultant of the acting forces. Thus generally,

$$\Sigma F = F_V + F_G + F_E + F_P + F_S = \text{Resultant}$$

and $F_I = -\text{Resultant}$

Thus $F_V + F_G + F_E + F_P + F_S + F_I = 0$

For dynamic Similarity, we have

$$\frac{(F_I)_m}{(F_I)_p} = \frac{(F_V)_m}{(F_v)_p} = \frac{(F_G)_m}{(F_G)_p} = \frac{(F_E)_m}{(F_E)_p}$$

$$= \frac{(F_P)_m}{(F_P)_p} = \frac{(F_S)_m}{(F_S)_p} = F_r \qquad (1.14)$$

where, F_R is known on force ratio. The subscripts m and p refer to model and prototype.

1.4.4 Standard dimensionless numbers

Over the years, in the history of fluid mechanics a set of dimensionless number of dynamic similarity have come to be well known, and each is associated with the name of a prominent scientist or engineer, usually the one who pioneered its use.

Forces encountered in flowing fluids including those due to inertia, viscosity, gravity, elastic, pressure and surface tension. The ratio of any two forces will be dimensionless.

Inertia forces are important in most fluid mechanics problems. The ratio of the inertia force to each of the other forces listed above leads to five fundamental dimensionless numbers encountered in fluid mechanics.

1.4.4.1 Reynold's number (R_e)

Reynold's number is defined as the ratio of inertia force of a flowing fluid and the viscous force of the fluid. Therefore,

$$R_e = \frac{\text{Inertia force}}{\text{Viscous force}} = \frac{F_I}{F_V} \tag{1.15}$$

F_I = Mass × Acceleration of flowing fluid

= Density × volume × velocity/time

$$= \rho l^3 \left(\frac{u}{t}\right) = \rho l^2 \frac{l}{t} u = \rho l^2 \, u \, u = \rho l^2 u^2 \tag{1.16}$$

$$F_V = \textit{Shear stress} \times \textit{Area} = \tau A$$

$$= \mu \frac{du}{dy} A \quad = \mu \frac{u}{l} l^2 \quad = \mu \, u \, l$$

Substituting the value of F_I and F_V in equation (1.15)

$$R_e \; \frac{\rho u^2 \, l^2}{\mu \, u \, l} = \frac{\rho \, u \, l}{\mu} \qquad [1.17\text{ (a)}]$$

$$= \frac{u \, l}{\nu} \qquad [1.17\text{ (b)}]$$

where, l and u are a characteristic or typical length and velocity for the system.

For pipe flow,

$$R_e = \frac{ud}{\nu}$$

Where, d is the diameter of the pipe.

For flow between parallel plates

$$R_e = \frac{uh}{\nu}$$

Where, h is the distance between the two plates.

For boundary layer flow

$$R_e = \frac{ul}{\nu}$$

Where l is length from the leading edge.

1.4.4.2 Froude's number (F_R)

Froude's number is defined as the square root of the ratio of inertia force of flowing to the gravity force. Therefore,

$$F_r = \sqrt{\frac{Inertia\ \ force}{Gravity\ \ force}} = \sqrt{\frac{F_1}{F_G}} \tag{1.18}$$

$$F_1 = \rho\, u^2\, l^2 \qquad \text{[From equation (1.16)]}$$

F_G = mass × acceleration due to gravity $= \rho\, l^3\, g$

Substituting in equation (12.18)

$$F_r = \sqrt{\frac{\rho\, u^2\, l^2}{\rho\, l^3\, g}} = \frac{u}{\sqrt{lg}} \tag{1.19}$$

1.4.4.3 Mach's number (M)

Mach's number is the square root of the inertia forces to compressibility forces (Elastic forces) in the fluid. Therefore,

$$M = \sqrt{\frac{Inertia\ \ force}{Elastic\ \ force}} = \sqrt{\frac{F_1}{F_E}} \tag{1.20}$$

$$F_I = \rho\, u^2\, l^2 \qquad \text{[From equation (1.16)]}$$

F_E = Compressibility force

or Elastic force = Bulk modulus of elasticity × Area $= K\, l^2$

Substituting in equation (1.20)

$$M = \sqrt{\frac{\rho\, u^2\, l^2}{K\, l^2}} = \frac{u}{\sqrt{K / \rho}} = \frac{u}{c} \tag{1.21}$$

We know that the velocity of sound c is equal to $\sqrt{K / \rho}$.

1.4.4.4 Euler's number (E_u)

Euler's number is defined as the square root of the ratio of inertia force of flowing fluid to the pressure force. Therefore,

$$E_u = \sqrt{\frac{Inertia\ \ force}{Pressure\ \ force}} = \sqrt{\frac{F_I}{F_p}} \tag{1.22}$$

$F_I = \rho u^2 l^2$ [From equation (1.16)]

F_P = Pressure × area = p l^2

Substituting in equation (1.22)

$$E_u = \sqrt{\frac{\rho u^2 l^2}{p l^2}} = \frac{u}{\sqrt{p/\rho}} \quad (1.23)$$

1.4.4.5 Weber's number (W_b)

Weber number is defined as the squareroot of the ratio of the inertia force of the flowing fluid to the surface tension force. Therefore,

$$W_b = \sqrt{\frac{\text{Inertia force}}{\text{Surface tension force}}} = \sqrt{\frac{F_I}{F_S}} \quad (1.24)$$

$F_I = \rho u^2 l^2$ [From equation (1.16)]

F_S = Surface tension × length = σ_l

Substituting in equation (1.24)

$$W_b = \sqrt{\frac{\rho u^2 l^2}{\sigma l}} = \frac{u}{\sqrt{\frac{\sigma}{\rho l}}} \quad (1.25)$$

1.5 MODEL LAWS

The results obtained from the model tests may be transferred to the prototype by the use of model laws which may be developed from the principles of dynamic similarity.

The equation (1.14) can be expressed as

$$\left(\frac{F_I}{F_V}\right)_m = \left(\frac{F_I}{F_V}\right)_p \quad \text{or} \quad R_{e_m} = R_{e_p} \quad (1.26)$$

$$\left(\frac{F_I}{F_G}\right)_m = \left(\frac{F_I}{F_G}\right)_p \quad \text{or} \quad F_{r_m} = F_{r_p} \quad (1.27)$$

$$\left(\frac{F_I}{F_E}\right)_m = \left(\frac{F_I}{F_E}\right)_p \quad \text{or} \quad M_m = M_p \quad (1.28)$$

$$\left(\frac{F_I}{F_P}\right)_m = \left(\frac{F_I}{F_P}\right)_p \qquad or \quad E_{u_m} = E_{u_p} \tag{1.29}$$

$$\left(\frac{F_I}{F_S}\right)_m = \left(\frac{F_I}{F_S}\right)_p \qquad or \quad W_{b_m} = W_{b_p} \tag{1.30}$$

In the foregoing force ratio equations may be written in terms of the dimensionless numbers as indicated above.

It is almost impossible to obtain complete dynamic similarity between the model and its prototype (except models with unit scale ratio) in which all these five conditions are satisfied simultaneously because no two fluids are known to posses all the requisite properties to satisfy all these five conditions. Fortunately, in most engineering problems, only two or three forces may be predominant, whereas the others are relatively insignificant and can be safely neglected. For all practical purposes, therefore, a particular state of fluid motion can usually be simulated in a model by considering the predominant forces only.

1.5.1 Reynold's model law

If two systems, such as model and its prototype, are to be dynamically equivalent so far as inertia and viscous forces are concerned, they must both have the same value of Reynold's number (R_e). Thus for such a case, dynamic similarity is achieved when

$$R_{e_m} = R_{e_p}$$

or, $$\left(\frac{\rho u l}{\mu}\right)_m = \left(\frac{\rho u l}{\mu}\right)_p \tag{1.31}$$

or, $$\left(\frac{ul}{\nu}\right)_m = \left(\frac{ul}{\nu}\right)_p \tag{1.32}$$

or, $$\frac{u_r\, l_r}{\nu_r} = 1 \tag{1.33}$$

Where the various quantities with subscript r represent the corresponding scale ratios.

The scale ratios for various physical quantities based on Reynold's model law are obtained as follows

Using equation (1.33), we get

(a) Scale ratio for velocity

$$u_r = \left(\frac{\nu}{l}\right)_r$$

(b) Scale ratio for time

We know that,

$$u_r = \frac{l_r}{T_r}$$

or, $$T_r = \frac{l_r}{u_r}$$

(c) Scale ratio for acceleration

We know that,

$$a_r = \frac{u_r}{T_r}$$

or, $$a_r = \frac{u_r}{l_r / u_r} = \frac{u_r^2}{l_r} = \frac{\nu_r^2}{l_r^3}$$

(d) Scale ratio for discharge

We know that

$$Q_r = \rho_r \, u_r \, l_r^2$$

$$Q_r = \rho_r \frac{\nu_r}{l_r} l_r^2 = \rho_r \frac{\mu_r}{\rho_r} l_r = \mu_r \, l_r$$

(e) Scale ratio for force

$$F_r = \rho_r \, u_r^2 \, l_r^2 = \rho_r \frac{\nu_r^2}{l_r^2} l_r^2 = \rho_r \, \nu_r^2 = \rho_r \frac{\mu_r^2}{\rho_r^2} = \frac{\mu_r^2}{\rho_r}$$

The Reynold's model law is applicable to the following cases

(1) Submarine in the fully submerged condition

(2) Aircraft travelling at a Mach number much smaller than unity (i.e. submerged in air)

(3) Buildings and structures subjected to wind loadings.

(4) Testing of objects in a water tunnel or closed circuit wind tunnel.

(5) Flow through closed conduits.

(6) Flow around immersed bodies.

1.5.2 Froude's model law

If two systems, such as model and its prototype, are to be dynamically equivalent so far as inertia and gravity forces are concerned, they must both have the same value of Froudes number (Fr). Thus for such a case, dynamic similarity is achieved when

$$F_{r_m} = F_{r_p}$$

or,
$$\left(\frac{u}{\sqrt{l\,g}}\right)_m = \left(\frac{u}{\sqrt{l\,g}}\right)_p \tag{1.34}$$

or,
$$\frac{u_m}{\sqrt{l_m}} = \frac{u_p}{\sqrt{l_p}}$$

or,
$$u_r = \sqrt{l_r} \tag{1.35}$$

Scale ratio for various physical quantities based on Froude model law are obtained as follows :

(a) Scale ratio for discharge.

Since the velocity varies as $\sqrt{l_r}$ and the cross section are as l_r^2 it follows that

$$Q_r = \frac{Q_m}{Q_p} = v_r\, l_r^2 = \left(\sqrt{l_r}\right) l_r^2 = l_r^{5/2}$$

(b) Scale ratio for time

$$T_r = \frac{T_m}{T_p} = \frac{l_r}{u_r} = \sqrt{l_r}$$

(c) Scale ratio for acceleration

$$a_r = \frac{u_r}{T_r} = \frac{\sqrt{l_r}}{\sqrt{l_r}} = 1$$

(d) Scale ratio for force

$$F_r = \frac{F_m}{F_p} = \frac{\rho_m\, l_m^2\, u_m^2}{\rho_p\, l_p^2\, u_p^2}$$

If the same fluid is used for model and prototype, then

$$F_r = l_r^2\, u_r^2 \quad = l_r^2\, l_r \quad = l_r^3$$

(e) Scale ratio for work, energy, torque, moment

Moment = Force × perpendicular distance

$$M_r = \frac{M_m}{M_p} = \frac{(F \times l)_m}{(F \times l)_p} = \frac{l_m^4}{l_p^4} = l_r^4$$

(f) Scale ratio for pressure intensity

$$p = Force / \; Area = \frac{\rho\, u^2\, l^2}{l^2} = \rho\, u^2$$

$$P_r = \frac{p_m}{p_p} = \frac{\rho_m\, u_m^2}{\rho_p\, u_p^2}$$

Using the same fluid for model and prototype, then

$$P_r = u_r^2 = l_r$$

(g) Scale ratio for power

$$\text{Power} = \text{Rate of doing work} = \frac{Work}{Time} = \text{Force} \times \text{Velocity}$$

$$P_r = \frac{P_m}{P_p} = l_r^3\, u_r = l_r^3\, \sqrt{l_r} = l_r^{7/2}$$

The Froude's model law is applicable to the following cases.

(1) Flow with a free surface, e.g. spillways, sluices, channels. weirs, etc.

(2) Motion in river, ocean, etc where waves are likely to be formed.

(3) Hydraulic Jump and stiling basin.

(4) Flow of jets from on orifice or nozzie.

(5) Flow of one fluid past another fluid of different densities.

1.5.3 Mach model law

If two systems, such as model and its prototype, are to be dynamically equivalent so far as inertia and elastic (compressibility) forces are concerned, they must both have the same value of Mach number (M). Thus for such a case, dynamic similarity is achieved when

$$M_m = M_p$$

$$\left(\frac{u}{C}\right)_m = \left(\frac{u}{C}\right)_p \tag{1.36}$$

Where $C = \sqrt{K/\rho} = \sqrt{\gamma\ R\ T}$ (1.37)

For air $\gamma = 1.4$, $R = 287$ J/kg K and T in K

Therefore, $C = \sqrt{1.4 \times 287 \times T} = 20.25\sqrt{T}$ (1.38)

Mach model law is applicable to the following cases; e.g.

(1) Commonly used in compressible flow problems, particularly in the fields of gas dynamics and aerodynamics. (M > 0.3)

The effect of compressibility is considered only when the Mach number is greater than 0.3.

(2) Unsteady flow-Water hammer problems.

(3) Underwater testing of torpedoes.

(4) Aerodynamic testing.

1.5.4 Euler's model law

If two systems, such as model and its prototype, are to be dynamically equivalent so far as inertia and pressure forces are concerned, they must both have the same value of Euler number (Eu). Thus for such a case, dynamic similarity is achieved when

$$E_{u_m} = E_{u_p}$$

or,

$$\left(\frac{u}{\sqrt{p/\rho}}\right)_m = \left(\frac{u}{\sqrt{p/\rho}}\right)_p \tag{1.39}$$

Euler number would normally be used in problems in which pressure or the pressure difference between two points is an important variable.

For problems in which cavitation is of concern, the dimensionless group $(P_r - P_v) / \rho u^2 / 2$ is commonly used, where P_v is the vapour pressure and P_r is some reference pressure. Although this dimensionless group has the same form as the Euler number, it is generally referred as the cavitation number.

1.5.5 Weber's model low

If two systems, such as model and its prototype, are to be dynamically equivalent so far as inertia and surface tension forces are concerned, they must both have the same value of Weber number (W_b). Thus for such a case, dynamic similarity is achieved when

$$(W_b)_m = (W_b)_p$$

$$\left(\frac{u}{\sqrt{\sigma / \rho l}}\right)_m = \left(\frac{u}{\sqrt{\sigma / \rho l}}\right)_p \tag{1.40}$$

Weber number may be important in problems in which there is an interface between two fluids. Common examples of problems in which this parameter may be important include the flow of thin films of liquid, or in the formation of droplets or bubbles. Clearly, not all problems involving flows with an interface will require the inclusion of surface tension. More examples are (1) Flow over weirs involving very low heads, (2) Very thin sheet of liquid flowing over a surface, (3) Capillary waves in channels etc.

1.6 UNDISTORTED AND DISTORTED MODELS

A model is said to be perfect or undistorted when it is perfectly similar to its prototype. In undistorted models, the conditions of simlitude are completely satisfied. The model investigation results can be directly used for the prototype.

Models for which one or more of the similarity requirements are not satisfied are called distorted models.

Distorted models can arise for a variety of reasons.

(1) Geometric distortion, (2) Distortion of configuration (3) Material distortion and (4) Distortion of hydraulic quantities such as, discharge, velocity and time.

Distorted models can be successfully used, but the interpretation of results obtained is more difficult than the interpretation of results obtained with true models for which all similarity requirements are met. There are no general rules for handling distorted models, and essentially each problem must be considered on its own merits. The success of using distorted models depends to a large extent on the skill and experience of the investigator responsible for the design of the model, and in the interpretation of experimental data obtained from the model. Distorted models are widely used, especially in the models of rivers and open channels.

(a) River and harbour models

River and horbour engineering projects are costly undertaking and, as analytical techniques only provide approximate prediction of the likely effects of any river widening or harbour improvement, the use of models at an early stage in the design has many advantages. The problems arise when a suitable scale is to be chosen for the model; the adoption of a scale that will give reasonable channel depths will usually result in a model too large to be practical, while choosing a scale on area-cost criteria yields channel depths that are very small.

In order to provide a solution to these problems, distorted scaling is adopted (i.e. geometrical distortion), vertical scales of 1/100 and larger being typical while horizontal scales vary from 1/200 to 1/600. Since the velocity is largely determined by gravity forces, the vertical height of the model is used to calculate its froude number. When comparing Froude numbers of model and prototype, the Froude number of the prototype must also be based on its vertical height. Distortation of this sort is suitable if the overall discharge characteristics of a long length of river are being studied.

(b) Estuary Models

Estuary models may be constructed using the same general principles as outlined for river models; distorted scales are typically 1/50 to 1/150 vertical and 1/300 to 1/2500 horizontal. As the available data on tidal velocity distributions are likely to be sketchy, it is necessary to incorporate in the mode the whole tidal channel system, as well as a substantial portion of adjustment coast line.

The speed of propagation of the tide becomes an important design criteria as the tidal period governs the time available for result recording. The tidal period of the model, thus,

$$\frac{\text{Time in model}}{\text{Time in prototype}} = \frac{(\text{Distance/velocity})_{\text{model}}}{(\text{Distance/velocity})_{\text{prototype}}}$$

Now, as the Froude numbers are equal (as for river models) it follows:

$$\frac{(Time)_m}{(Time)_p} = \left(\frac{l_m}{l_p}\right)\left(\frac{u_p}{u_m}\right) = \left(\frac{l_m}{l_p}\right)\sqrt{(Depth)_p/(Depth)_m} \qquad (1.41)$$

The above equation refers to horizontal fluid velocity, the scale factor for vertical silt particle velocity is also of interest and may be derived in the same manner

$$\frac{\text{Rate of fall in model}}{\text{Rate of fall in prototype}} = \frac{\text{Depth Scale}}{\text{Time Scale}} \qquad (1.42)$$

Estuary models are now commonly used to investigate the effects of discharge of power station or industrial cooling water flows and, here, density variations may again have to be modelled, based on temperature. Other uses of estuary models include silting and erosion studies and the spread and deposit of effluent discharged into the sea. Generally, the role of the estuary model should be seen was as a method of comparing the attributes of various design solutions rather than as an accurate method of predicting the effects of one design.

1.7 SCALE EFFECT

It is not possible to predict the exact behaviours of the prototype by model testing. The behaviour of the prototype as predicted by two models with different scale ratios is generally not the same. This difference in the predicted behaviour of the prototype is attributed to difference in scale ratios and is known as scale effect.

Scale effect can be detected by testing a larger number of models with different scale ratios and the exact behaviour of prototype can then be predicted.

1.8 COMMENTS ON MODEL TESTING

While outlining the procedures involved in model we have tried not to that testing is a simple task that automatically gives results which are easy to interpret, accurate and complete. As in all experimental work, careful planning and execution are needed to obtain valid results. Models must be constructed carefully and accurately, and they must include sufficient detail in areas critical to the phenomenon being measured. Aerodynamic balances or other force measuring systems must be aligned carefully and calibrated correctly. Mounting methods must be devised that offer adequate rigidity and model motion, yet do to interfere with the phenomenon being measured.

Experimental facilities must be designated and constructed carefully. The quality of the flow in a wind tunnel must be documented. It should be as nearly uniform as possible, free from angularity, and with little swirl content. If they interfere with measurements boundary layers on tunnel walls must be removed by suction. Pressure gradients in a wind tunnel test section may cause erroneous readings due to horizontal buoyancy in a flow direction.

In the use of models it is essential that the fluid velocity should not be so low that laminar flow exists when the flow in the prototype is turbulent. Also, conditions in the model should not be such that surface tension is importent if such conditions do not exist in the prototype.

While model studies are very important and valuable it is necessary to exercise some judgment in transferring results from the model to the prototype. It is not always necessary or desirable that these various dimensionless ratios be adhered to in every case.

The roughness of a model should be scaled down in the same ratio as the other linear dimensions, which means that a small model should have surfaces that are much smoother than those in its prototype. But this requirement imposes a limit on the scale that can be used if true geometric similarity is to be achieved. However, in the case of a distorted model with a vertical scale larger than the horizontal scale, it may be necessary to make the model surface rough in order to simulate the flow conditions in the prototype.

Special facilities are needed for unusual conditions or for special test requirements, especially to achieved large Reynold's numbers. Many facilities are so large or specialized that they cannot be supported by university laboratories or private industry.

ILLUSTRATIVE EXAMPLES

1.1. Assuming that the resistance to motion, F, of a body moving in a compressible fluid is a function of a characteristic dimension, l, of the body, the velocity u of the body and fluid properties density, ρ, viscosity, μ, and bulk modulus, K; Find the expression for F in terms of these quantities using for F in terms of these quantities using Buckingham's π theorem.

$$F = \frac{1}{2}\left(\rho u^2 l^2\right)\Phi\left[\frac{\rho u l}{\mu}, \frac{u}{\sqrt{K/\rho}}\right]$$

Discuss the significance of the non dimensional groups obtained.

Solution

The relationship between the variables may be expressed as

$$F = \textit{function of } [u, \rho, l, \mu, K]$$

There are six variables in this problem, all of which may be expressed in terms of three primary dimensions M, L and T. Therefore, there will be 6-3 = 3 dimensional parameters or π - groups in the solution.

Applying the π theorem, it is first necessary to select three primary variables which between them include all the primary dimensions (M, L and T) involved. In this problem we may select, ρ, u and l as primary variables.

Each of the remaining variables is now taken in turn and used to form a dimensionless π groups as follows

$$\pi_1 = \rho^{a_1} u^{b_1} l^{c_1} F$$

$$\pi_2 = \rho^{a_2} u^{b_2} l^{c_2} \mu$$

$$\pi_3 = \rho^{a_3} u^{b_3} l^{c_3} K$$

Dimensional alaysis for π_1

$$\pi_1 = \rho^{a_1} u^{b_1} l^{c_1} F$$

$$\left[M^o L^o T^o\right] = \left[ML^{-3}\right]^{a_1} \left[LT^{-1}\right]^{b_1} \left[L\right]^{c_1} \left[MLT^{-2}\right]$$

Equating indices of M, L and T, we have

M : 0 = a_1 + 1

L : 0 = – 3 a_1 + b_1 + c_1 + 1

T : 0 = – b_1 – 2

Hence a_1 = –1, b_1 = –2 and c = –2

Therefore, $\pi_1 = \dfrac{F}{\rho u^2 l^2}$

Dimensional analysis for π_2

$$\pi_2 = \rho^{a_2} \ u^{b_2} \ l^{c_2} \ \ \mu$$

$[M^o L^o T^o] = [ML^{-3}]^{a2} \ [LT^{-1}]^{b2} \ [L]^{c2} \ [ML^{-1} \ T^{-1}]$

Equating indices

M : 0 = a_2 +1

L : 0 = –3a_2 + b_2 + c_{2-1}

T : 0 = – b_2 – 1

Hence a_2 = – 1, b_2 = – 1, c_2 =–1

Therefore, $\pi_2 = \dfrac{\mu}{\rho u l}$

Dimensional analysis for π_3

$$\pi_3 = \rho^{a_3} \ u^{b_3} \ l^{c_3} \ \ K$$

$[M^o L^o T^o] = \left[ML^{-3}\right]^{a_3} \left[LT^{-1}\right]^{b_3} \left[L\right]^{c_3} \ [ML^{-1} \ T^{-2}]$

Equating indices

M : 0 = a_3 +1

L : 0 = –3a_3 + b_3 + c_{3-1}

T : 0 = – b_3 – 2

Hence a_3 = – 1, b_3 =–2, c_3 = 0

Therefore, $\pi_3 = \dfrac{K}{\rho u^2}$

Dimensional analysis does not generate any numerical coefficients. In the π_2 term, the numerical constant 1/2, which normally multiplies $(\rho v^2\ l^2)$ to give it physical significance. It is certainly permissible to multiply any dimensionless group by a dimensionless constant to give it physical significance. Therefore,

$$\pi_1 = 2\ \pi_1 = \frac{F}{\frac{1}{2}\left(\rho\, u^2\, l^2\right)}$$

and
$$\pi_2 = \frac{1}{\pi_2} = \frac{\rho\, u\, l}{\mu}$$

$$\pi_3 = \sqrt{\frac{1}{\pi_3}} = \frac{u}{\sqrt{K/\rho}}$$

The Solution may be stated in the form

$$\pi_1' = \Phi\left(\pi_2'\ \pi_3'\right)$$

or,
$$\frac{F}{\frac{1}{2}\rho\, u^2\, l^2} = \Phi\left(\frac{\rho\, u\, l}{\mu}, \frac{u}{\sqrt{K/\rho}}\right)$$

or,
$$F = \frac{1}{2}\left(\rho\, u^2\, l^2\right)\Phi\left[\frac{\rho\, u\, l}{\mu}, \frac{u}{\sqrt{K/\rho}}\right]$$

The group $\dfrac{F}{\frac{1}{2}\rho\, u^2\, l^2}$ is sometimes referred to as the drag coefficient C_D.

The group $\dfrac{\rho\, u\, l}{\mu}$ is the Reynold's number, R_e, which is a criterion of dynamic similarity when viscous and inertia forces are significant.

The group $\dfrac{u}{\sqrt{K/\rho}}$ is the Mach number, M, which is a criterion of dynamic similarity when elastic and inertia forces are significant.

1.2. Show by the method of dimensions that for a screw propeller, the relation between the thrust F, torque T, diameter D, speed of travel u, speed of rotation N, mass density ρ and dynamic viscosity μ may be put in the form

$$F = \rho u^2 D^2 \Phi \left[\frac{\rho u^2 D^3}{T}, \frac{DN}{u}, \frac{\rho u D}{\mu} \right]$$

Solution

Let F = function of (ρ, u, D, T, N, μ)

n = number of variables = 7

m = number of primary variables = 3

(here M, L and T)

π groups = n – m = 7 –3 = 4

Selecting ρ, u and D as a repeated variables, then

$$\pi_1 = \rho^{a_1} u^{b_1} D^{c_1} F$$

$$\pi_2 = \rho^{a_2} u^{b_2} D^{c_2} T$$

$$\pi_3 = \rho^{a_3} u^{b_3} D^{c_3} N$$

$$\pi_4 = \rho^{a_4} u^{b_4} D^{c_4} \mu$$

For π_1

$$\pi_1 = \rho^{a_1} u^{b_1} D^{c_1} F$$

$$\left[M^o L^o T^o\right] = \left[ML^{-3}\right]^{a_1} \left[LT^{-1}\right]^{b_1} \left[L\right]^{c_1} \left[MLT^{-2}\right]$$

Equating indices

$$M : 0 = a_1 + 1$$

$$L : 0 = -3 a_1 + b_1 + c_1 + 1$$

$$T : 0 = -b_1 - 2$$

Which gives $a_1 = -1, b_1 = -2, c_1 = -2$

Therefore, $\pi_1 = \dfrac{F}{\rho u^2 D^2}$

For π_2

$$\pi_2 = \rho^{a_2} u^{b_2} D^{c_2} T$$

$$\left[M^o L^o T^o\right] = \left[ML^{-3}\right]^{a_2} \left[LT^{-1}\right]^{b_2} \left[L\right]^{c_2} \left[ML^2T^{-2}\right]$$

Equating indices

$$M : 0 = a_2 + 1$$

$$L : 0 = -3\,a_2 + b_2 + c_2 + 2$$

$$T : 0 = -b_2 - 2$$

Which gives $a_2 = -1, b_2 = -2, c_2 = -3$

Therefore, $\pi_2 = \dfrac{F}{\rho\, u^2\, D^3}$

or, $\pi_2' = \dfrac{1}{\pi_2} = \dfrac{\rho\, u^2\, D^3}{T}$

For π_3

$$\pi_3 = \rho^{a_3}\, u^{b_3}\, D^{c_3}\, N$$

$$\left[M^o L^o T^o\right] = \left[ML^{-3}\right]^{a_3} \left[LT^{-1}\right]^{b_3} \left[L\right]^{c_3} \left[T^{-1}\right]$$

Equating indices

$$M : 0 = a_3$$

$$L : 0 = -3\,a_3 + b_3 + c_3$$

$$T : 0 = -b_3 - 1$$

Which gives $a_3 = 0, b_3 = -1, c_3 = +1$

Therefore, $\pi_3 = \dfrac{ND}{\mu}$

For π_4

$$\pi_4 = \rho^{a_4}\, u^{b_4}\, D^{c_4}\, \mu$$

$$\left[M^o L^o T^o\right] = \left[ML^{-3}\right]^{a_4} \left[LT^{-1}\right]^{b_4} \left[L\right]^{c_4} \left[ML^{-1}\, T^{-1}\right]$$

Equating indice

$$M : 0 = a_4 + 1$$

$$L : 0 = -3\,a_4 + b_4 + c_4 - 1$$

$T : 0 = -b_4 - 1$

Which gives $a_4 = -1, b_4 = -1, and\ c_4 = -1$

Therefore $\pi_4 = \dfrac{\mu}{\rho u D}$

$$\pi_4' = \frac{1}{\pi_4} = \frac{\rho u D}{\mu}$$

The functional relation can be expressed as

$$\pi_1 = \Phi\left(\pi_2', \pi_3, \pi_4'\right)$$

$$\frac{F}{\rho u^2 D^2} = \Phi\left[\frac{\rho u^2 D^3}{T}, \frac{DN}{u}, \frac{\rho u D}{\mu}\right]$$

$$F = \rho u^2 D^2 \Phi\left[\frac{\rho u^2 D^3}{T}, \frac{DN}{u}, \frac{\rho u D}{\mu}\right]$$

1.3. Derive on the basis of Buckingham's π theorem to present the thrust developed by a propeller.

$$F = \rho u^2 D^2 \Phi\left[\frac{\omega D}{u}, R_e, M\right]$$

Where, F = thrust developed, ω = angular velocity

u = Speed of advance D = diameter

μ = dynamic viscosity ρ = mass density

and c = speed of sound

Solution

Let F = function of (ρ, u, D, ω, μ, c)

n = number of variables = 7

m = number of primary variables = 3

π groups = 7 – 3 = 4

Choosing ρ, u nd D as a repeated variables, then

$$\pi_1 = \rho^{a_1} u^{b_1} D^{c_1} F$$

$$\pi_2 = \rho^{a_2} u^{b_2} D^{c_2} \omega$$

$$\pi_3 = \rho^{a_3}\ u^{b_3}\ D^{c_3}\ \mu$$

$$\pi_4 = \rho^{a_4}\ u^{b_4}\ D^{c_4}\ c$$

For π_1

$$\pi_1 = \rho^{a_1}\ u^{b_1}\ D^{c_1}\ F$$

$$\left[M^o\ L^o\ T^o\right] = \left[ML^{-3}\right]^{a_1}\ \left[LT^{-1}\right]^{b_1}\ [L]^{c_1}\ \left[MLT^{-2}\right]$$

Equating indices

$$M : 0 = a_1 + 1$$

$$L : 0 = -3a_1 + b_1 + c_1 + 1$$

$$T : 0 = -b_1 - 2$$

Which gives $a_1 = -1, b_1 = -2, c_1 = -2$

therefore, $\pi_1 = \dfrac{F}{\rho u^2 D^2}$

For π_2

$$\pi_2 = \rho^{a_2}\ u^{b_2}\ D^{c_2}\ \omega$$

$$\left[M^o\ L^o\ T^o\right] = \left[ML^{-3}\right]^{a_2}\ \left[LT^{-1}\right]^{b_2}\ [L]^{c_2}\ \left[T^{-1}\right]$$

Equating indices

$$M : 0 = a_2$$

$$L : 0 = -3a_2 + b_2 + c_2$$

$$T : 0 = -b_2 - 1$$

Which gives $a_2 = -0, b_2 = -1, c_2 = 1$

therefore, $\pi_2 = \dfrac{\omega D}{u}$

For π_3

$$\pi_3 = \rho^{a_3}\ u^{b_3}\ D^{c_3}\ \mu$$

$$\left[M^o\ L^o\ T^o\right] = \left[ML^{-3}\right]^{a_3}\ \left[LT^{-1}\right]^{b_3}\ [L]^{c_3}\ \left[ML^{-1}T^{-1}\right]$$

Equating indices

$$M : 0 = a_3 + 1$$

$$L : 0 = -\ 3a_3 + \ b_3 \ + c_3 - 1$$

$$T \ : 0 \ = -\ b_3 \ - 1$$

Which gives $a_3 = -1, b_3 = -1, c_3 = -1$

therefore, $\pi_3 = \dfrac{\mu}{\rho\, u\, D}$

$$\pi_3' = \frac{1}{\pi_3} = \frac{\rho\, u\, D}{\mu} = \text{the Reynold's number } R_e$$

π_4 can be calculated by inspection

$$\pi_4 = \frac{c}{u}$$

$$\pi_4' = \frac{1}{\pi_4} = \frac{u}{c} = \text{ the Mach number M}$$

Thus the functional relation can be expressed as

$$\pi_1 = \Phi\left(\pi_2, \pi_3', \pi_4'\right)$$

$$\frac{F}{\rho u^2 D^2} = \Phi\left[\frac{\omega\, D}{u}, \frac{\rho u\, D}{\mu}, \frac{u}{c}\right]$$

$$F = \rho\, u^2\, D^2\, \Phi\left[\frac{\omega D}{u}, R_e, M\right]$$

1.4. Using Buckingham's π theorem, show that velocity through a circular orifice is given by

$$V = \sqrt{2gH}\ \Phi\left[\frac{H}{D,}\ \frac{VH}{\nu}\right]$$

Where H is the head causing flow, D is the diameter of the orifice, υ is the kinematic viscosity and g is the acceleration due to gravity.

Solution

Let V = function of (H, D, g, ν)

There are a total number of variables n = 5 Number of primary dimensions (Here L and T), m = 2. Dimensionless π terms = n-m = 5-2 = 3

Assuming H and g as repeating variables, then

$$\pi_1 = H^{a_1} g^{b_1} V$$

$$\pi_2 = H^{a_2} g^{b_2} D$$

$$\pi_3 = H^{a_3} g^{b_3} \upsilon$$

Dimensional analysis for π_1

$$\pi_1 = H^{a_1} g^{b_1} V$$

$$\left[L^o T^o\right] = [L]^{a_1} \left[LT^{-2}\right]^{b_1} \left[LT^{-1}\right]$$

Equating indices

$$L: 0 = a_1 + b_1 + 1$$

$$T: 0 = -2\, b_1 - 1$$

Hence $a_1 = -\dfrac{1}{2},\ b_1 = -\dfrac{1}{2}$ *and*

$$\pi_1 = \frac{V}{\sqrt{gH}} \text{ or } \pi_1' = \frac{\pi_1}{\sqrt{2}} = \frac{V}{\sqrt{2gH}}$$

Dimensional analysis for π_2

$$\pi_2 = H^{a_2} g^{b_2} D$$

Applying the above technique or by inspection

$$\pi_2 = \frac{D}{H} \text{ or } \pi_2' = \frac{1}{\pi_2} = \frac{H}{D}$$

Dimensional analysis for π_3

$$\pi_3 = H^{a_3} g^{b_3} \upsilon$$

$$\left[L^o T^o\right] = [L]^{a_3} \left[LT^{-2}\right]^{b_3} \left[L^2 T^{-1}\right]$$

Equating indices

$$L: 0 = a_3 + b_3 + 2$$

$$T: 0 = -2\, b_3 - 1$$

Hence $a_3 = -\dfrac{3}{2},\ b_3 = -\dfrac{1}{2}$ *and*

$$\pi_3 = \frac{v}{H^{3/2} g^{1/2}} = \frac{v}{\sqrt{gH}\,H}$$

or, $$\frac{\pi'}{3} = \pi_1 / \pi_3 = \frac{V}{\sqrt{gH}} \times \frac{\sqrt{gH}\,H}{v} = \frac{VH}{v}$$

Thus the functional relationship can be expressed as

$$\pi_1' = \Phi\left(\pi_2', \pi_3'\right)$$

$$\frac{V}{\sqrt{2gH}} = \Phi\left(\frac{H}{D}, \frac{VH}{v}\right)$$

or, $$V = \sqrt{2gH}\;\Phi\left(\frac{H}{D}, \frac{VH}{v}\right)$$

1.5. **The efficiency η of a fan depends on the kinematic viscosity υ of the fluid, the angular velocity W, diameter D of the rotor and the discharge Q. Express η in terms of dimensionless parameter using Buckingham's π theorem.**

$$\eta = \Phi\left[\frac{\omega D^2}{v}, \frac{Q}{\omega D^3}\right]$$

Solution

The relationship between the variables may be expressed as

$$\eta = \text{function of } [v, \omega, D, Q]$$

There are five variables in this problem, all of which may be expressed in terms of two primary dimensions L and T. Therefore, there will be 5-2 = 3 dimensional parameters or π groups in the solution.

Repeating variables from the expression to be proved, are W and D.

Each of the remaining variables is now taken in turn and used to form a dimensionless π groups as follows

$$\pi_1 = \omega^{a_1} D^{b_1} \eta$$

$$\pi_2 = \omega^{a_2} D^{b_2} v$$

$$\pi_3 = \omega^{a_3} D^{b_3} Q$$

Dimensional analysis for π_1

$$\pi_1 = \omega^{a_1} D^{b_1} \eta$$

$$\left[L^o T^o\right] = \left[T^1\right]^{a_1} [L]^{b_1} \left[L^o T^o\right]$$

Equating indices of L and T, we have

L : 0 = b_1 + 0

T : 0 = $-a_1$ + 0

Hence $a_1 = b_1 = 0$ and

$$\pi 1 = \eta$$

The above value of π_1 can also be written by inspection.

Dimensional analysis for π_2

$$\pi_2 = \omega^{a_2} D^{b_2} \upsilon$$

$$\left[L^o T^o\right] = \left[T^{-1}\right]^{a_2} [L]^{b_2} \left[L^2 T^{-1}\right]$$

Equating indices

L : 0 = b_2 + 0

T : 0 = $- a_2 - 1$

Hence $a_2 = -1, b_2 = -2$ *and*

$$\pi_2 = \frac{v}{\omega D^2}$$

Dimensional analysis for π_3

$$\pi_3 = \omega^{a_3} D^{b_3} Q$$

$$\left[L^o T^o\right] = \left[T^{-1}\right]^{a_3} [L]^{b_3} \left[L^3 T^{-1}\right]$$

Equating indices

L : 0 = b_3 + 3

T : 0 = $-a_3 - 1$

Hence $a_3 = -1, b_3 = -3$ *and*

$$\pi_3 = \frac{Q}{\omega D^3}$$

As per given in the question.

From the π_2 term

$$\pi_2' = \frac{1}{\pi_2} = \frac{\omega D^2}{v}$$

Thus the functional relationship can be expressed as

$$\pi_1 = \Phi\left[\pi_2, \pi_3\right]$$

$$\eta = \Phi\left[\frac{\omega D^2}{\upsilon}, \frac{Q}{\omega D^3}\right]$$

1.6. Using Buckingham's π theorem show that the discharge, Q, of a centrifugal pump can be expressed as

$$Q = ND^3\ \Phi\left[\frac{ND}{\sqrt{gh}}, \frac{ND^2}{\nu}\right]$$

Where, N = Speed of the pump in rpm,

D = diameter of the impeller

ρ = Kinematic viscosity

g = acceleration due to gravity

Solution

Let Q = function of (N, D, g, H, ν)

n = number of variables = 6

m = number of primary variables = 2

(Here L and T)

π groups = n – m = 6 – 2 = 4

Selecting N and D as a repeated variables, then

$$\pi_1 = N^{a_1} D^{b_1} Q$$

$$\pi_2 = N^{a_2} D^{b_2} g$$

$$\pi_3 = N^{a_3} D^{b_3} H$$

$$\pi_4 = N^{a_4} D^{b_4} \nu$$

π_3 can be obtained by inspection

$$\pi_3 = \frac{H}{D}$$

For π_1

$$\pi_1 = N^{a_1} D^{b_1} Q$$

$$\left[L^o T^o\right] = \left[T^{-1}\right]^{a_1} [L]^{b_1} \left[L^3 T^{-1}\right]$$

Equating indices

$$L : 0 = b_1 + 3$$

$$T : 0 = -a_1 - 1$$

Which gives $a_1 = -1, b_1 = -3$

Therefore, $\pi_1 = \dfrac{Q}{ND^3}$

For π_2

$$\pi_2 = N^{a_2} D^{b_2} g$$

$$\left[L^o T^o\right] = \left[T^{-1}\right]^{a_2} [L]^{b_2} \left[LT^{-2}\right]$$

Equating indices

$$L : 0 = b_2 + 1$$

$$T : 0 = -a_2 - 2$$

Which gives $a_2 = -2, b_2 = -1$

Therefore, $\pi_2 = \dfrac{g}{N^2 D}$

For π_4

$$\pi_4 = N^{a_4} D^{b_4} \nu$$

$$\left[L^o T^o\right] = \left[T^{-1}\right]^{a_3} [L]^{b_3} \left[L^2 T^{-1}\right]$$

Equating indices

$$L : 0 = b_3 + 2$$

$$T : 0 = -a_3 - 1$$

Which gives $a_3 = -1, b_3 = -2$

Therefore, $\pi_4 = \dfrac{\nu}{ND^2}$

According to question

$$\pi_2' = \frac{1}{\sqrt{\pi_2}} \times \frac{1}{\sqrt{\pi_3}} = \frac{N\sqrt{D}}{\sqrt{g}} \times \frac{\sqrt{D}}{\sqrt{H}} = \frac{ND}{\sqrt{gH}}$$

$$\pi_4' = \frac{1}{\pi_4} = \frac{ND^2}{\nu}$$

Thus the functional relation can be expressed as

$$\pi_1 = \Phi\left(\pi_2', \pi_4'\right)$$

$$Q = ND^3\ \Phi\left[\frac{ND}{\sqrt{gH}}, \frac{ND^2}{\nu}\right]$$

1.7. The general formula for the discharge Q over a V - notch is

$$Q = H^2\sqrt{gH}\ \Phi\left[\frac{H\sqrt{gH}}{\upsilon}, \theta\right]$$

Where, H = head over apex, υ = Kinematic viscosity, g = acceleration due to gravity and θ = notch angle.

Solution

Q = function of (H, g, υ, θ)

n = number of variables = 5

m = number of primary dimensions = 2

(here, L and T)

n – m = π groups = 5 – 2 = 3

Selecting g and H as repeating variables, then

$$\pi_1 = g^{a_1} H^{b_1} Q$$

$$\pi_2 = g^{a_2} H^{b_2} \upsilon$$

$$\pi_3 = g^{a_3} H^{b_3} \theta$$

For π_1 term

$$\pi_1 = g^{a_1} H^{b_1} Q$$

$$\left[L^o T^o\right] = \left[LT^{-2}\right]^{a_1} \left[L\right]^{b_1} \left[L^3 T^{-1}\right]$$

Equating indices

$$L : 0 = a_1 + b_1 + 3$$

$$T : 0 = -2a_1 - 1$$

Hence $a_1 = 1/2,\ b_1 = -5/2$ *and*

$$\pi_1 = \frac{Q}{g^{1/2} H^{5/2}} = \frac{Q}{H^2 \sqrt{gH}}$$

For π_2 term

$$\pi_2 = g^{a_2} H^{b_2} v$$

$$\left[L^o T^o\right] = \left[LT^{-2}\right]^{a_2} \left[L\right]^{b_2} \left[L^2 T^{-1}\right]$$

Equating indices

$$L : 0 = a_2 + b_2 + 2$$

$$T : 0 = -2a_1 - 1$$

Hence $a_2 = -\frac{1}{2},\ b_2 = -\frac{3}{2}$ *and*

$$\pi_2 = \frac{v}{g^{1/2} H^{3/2}} = \frac{v}{H^2 \sqrt{gH}}$$

$$\pi_2' = \frac{1}{\pi_2} = \frac{H\sqrt{gH}}{v}$$

For π_3 term

$$\pi_3 = g^{a_3} H^{b_3} \theta$$

Applying the above technique or by inspection

$$\pi_3 = \theta$$

Thus the functional relationship can be expressed as

$$\pi_1 = \Phi\left(\pi_2' \ \pi_3\right)$$

$$\frac{Q}{H^2 \sqrt{gH}} = \Phi\left(H^2 \frac{\sqrt{gH}}{v}, \theta\right)$$

or, $$Q = H^2 \sqrt{gH}\ \Phi\left(H \frac{\sqrt{gH}}{\nu}, \theta\right)$$

1.8. **A submarine-launched missile, 1 m diameter by 5 m long, is to be studied in a water tunnel to determine the loads acting on it during its underwater launch. The maximum speed during this initial part of the missile's flight is 10 m/s. Calculate the mean water tunnel flow velocity if a 1/20 scale model is to be employed and dynamic similarity is to be achieved. The kinematic viscosity for model and prototype is some.**

Solution

In this problem, Reynold's model law is used.

$$R_{e_m} = R_e$$

$$\frac{u_m l_m}{\nu_m} = \frac{u_p l_p}{\nu_p}$$

$$u_m = u_p \frac{l_p}{l_m} = 10 \times 20 = 200 \text{ m/s} \qquad \textbf{Ans}$$

1.9. **An airship of 3 m diameter and 20 m length is to be studied in a wind tunnel. The air ship speed range to be investigated is at the docking end of its range, a maximum of 2 m/s. Calculate the mean model wind tunnel speed, if the model is made to 1/10 scale. Assume the same air pressure and temperature for model and prototype**

Solution

In the wind tunnel, object is completely submerged hence Reynold's model law is applicable.

$$R_{e_m} = R_{e_p}$$

$$\frac{\rho_m u_m l_m}{\mu_m} = \frac{\rho_p u_p l_p}{\mu_p}$$

For same air pressure and temperature

$$u_m l_m = u_p l_p$$

or, $$u_m = u_p \frac{l_p}{l_m} = 2 \times 10 = 200 \text{ m/s}$$ Ans

1.10. A 1 : 16 model of a submarine is to be tested in a towing tank containing salt water. If the submarine moves at 5 m/s at what velocity should the model be towed for dynamic similarity? Assume kinematic viscosity and density for model and prototype are same. Also determine the drag force ratio between model and its prototype.

Solution

For the submarine, Reynold's number will be used. Therefore,

$$R_{e_m} = R_{e_p}$$

$$\frac{u_m l_m}{v_m} = \frac{u_p l_p}{v_p}$$

or, $$\frac{u_m}{u_p} = \frac{l_p}{l_m} = \frac{1}{l_r}$$

or, $$u_m = \frac{u_p}{l_r} = 5 \times 16 = 80 = 80 \text{ m/s}$$ Ans

and $$\frac{F_{D_m}}{\rho_m u_m^2 l_m^2} = \frac{F_{D_p}}{\rho_p u_p^2 l_p^2}$$

or, $$\frac{F_{D_m}}{F_{D_p}} = F_{D_r} = \left(\frac{u_m}{u_p}\right)^2 \left(\frac{l_m}{l_p}\right)^2 \left(\frac{\rho_m}{\rho_p}\right)$$

$$= \left(\frac{l_p}{l_m}\right)^2 \times \left(\frac{l_m}{l_p}\right)^2 = 1$$ Ans

1.11. A ship 300 m long moves in sea water, whose relative density is 1.03. A 1 : 100 model of this ship is to be tested in wind tunnel. The velocity of air in the wind tunnel around the model is 30 m/s and the resistance of the model is 600 N. Determine the velocity of ship in sea water and also the resistance of the ship in sea water. Take pressure and temperature as 1.01325 bar and 15°C and kinematic viscosity of air and sea water as 0.018 Stokes and 0.012 stokes.

Solution

$$p = \rho R T$$

$$1.01325 \times 10^5 = \rho \times 287 \times 288$$

$$\rho = 1.2258 \text{ kg/m}^3$$

In the wind tunnel the ship is completely submerged hence Reynold's model law is applicable.

$$R_{e_m} = R_{e_p}$$

$$\frac{u_m l_m}{\nu_m} = \frac{u_p l_p}{\nu_p}$$

$$u_p = u_m \times \frac{l_m}{l_p} \times \frac{\nu_p}{\nu_m} = 30 \times \frac{1}{100} \times \frac{0.012}{0.018} = 0.2\ m/s \qquad \text{Ans}$$

Now, $$\frac{F_m}{\rho_m u_m^2 l_m^2} = \frac{F_p}{\rho_p u_p^2 l_p^2}$$

$$F_p = F_m \times \frac{\rho_p}{\rho_m} \times \left(\frac{u_p}{u_m}\right)^2 \times \left(\frac{l_p}{l_m}\right)^2 = 600 \times \frac{1030}{1.2258} \times \left(\frac{0.2}{30}\right)^2 \times (100)^2$$

$$= 224071.3548 \text{ N} = 224.071 \text{ kN} \qquad \text{Ans.}$$

1.12. The performance of a ship was predicted by taking its model and prototype ship was 400 m and its model was 10 m long. The kinematic viscosity of air is 13 times that of water. The velocity of air around the model in the wind tunnel was measured as 39 m/s. The pressure and temperature of air is 1.01325 bar and 293 K. The experiments showed that the resistance of the model was ship 500 N. Find (a) the velocity of actual ship in water, (b) the resistance of the actual ship in water. Take density of sea water as 1030 kg/m³.

Solution

$$p = \rho RT$$

$$1.01325 \times 10^5 = \rho \times 287 \times 293$$

$$\rho = 1.2049\ kg/m^3$$

In the wind tunnel, the ship is completely submerged, hence the Reynold's model law is applicable.

$$R_{e_m} = R_{e_p}$$

$$\frac{u_m l_m}{v_m} = \frac{u_p l_p}{v_p}$$

$$u_p = u_m \left(\frac{v_p}{v_m}\right)\left(\frac{l_m}{l_p}\right) = 39\left(\frac{1}{13}\right)\left(\frac{10}{400}\right) = 0.075\ m/s$$

Now, $$\frac{F_p}{\rho_p u_p^2 l_p^2} = \frac{F_m}{\rho_m u_m^2 l_m^2}$$

or, $$F_p = F_m \times \frac{\rho_p}{\rho_m} \times \left(\frac{u_p}{u_m}\right)^2 \times \left(\frac{l_p}{l_m}\right)^2$$

$$= 500 \times \frac{1030}{1.2049} \times \left(\frac{0.075}{39}\right)^2 \times (40)^2$$

$$= 2529.12\ \text{N} = 2.529\ \text{kN}$$ Ans

1.13. A 1/5th scale model of a torpedo is tested in a wind tunnel using air at 23 bars and 20°C, with a speed of 48 m/s and a viscosity of 184×10^{-6} poise. The resistance is found to be 76 N. Calculate the corresponding speed of the full scale torpedo, and the power required to drive it in sea water of density 1025 kg/m³ of which the viscosity 164×10^{-4} poise. Take for air R = 287 J/kgK.

Solution

In the wind tunnel object is completely submerged hence Reynold's model law will be used.

$$p_m = \rho_m R T$$

$$23 \times 10^5 = \rho_m \times 287 \times 293$$

$$\rho_m = 27.35\ kg/m^3$$

Now, $$R_{e_m} = R_{e_p}$$

$$\frac{\rho_m\, u_m\, l_m}{\mu_m} = \frac{\rho_p\, u_p\, l_p}{\mu_p}$$

$$\frac{u_p}{u_m} = \frac{\rho_m}{\rho_p} \times \frac{\mu_p}{\mu_m} \times \frac{l_m}{l_p} \quad = \frac{27.55}{1025} \times \frac{164 \times 10^{-4}}{184 \times 10^{-6}} \times \frac{1}{5} = 0.4756$$

$$u_p = u_m \times 0.4756 = 48 \times 0.4756 = 22.832 \text{ m/s}$$

Now, $$\frac{F_p}{\rho_p\, u_p^2\, l_p^2} = \frac{F_m}{\rho_m\, u_m^2\, l_m^2}$$

$$\frac{F_p}{F_m} = \frac{\rho_p}{\rho_m} \times \left(\frac{u_p}{u_m}\right)^2 \times \left(\frac{l_p}{l_m}\right)^2$$

$$= \frac{1025}{27.35} \times (0.4756)^2 \times (5)^2 \quad = 211.9289$$

$$F_p = F_m \times 211.9289 \quad = 76 \times 211.9289$$

$$= 16106.5983 \text{ N} \quad = 16.1 \text{ kN}$$

Power $P = F_p\, u_p = 16.1 \times 22.832 = 367.745 \text{ kW}$ Ans

1.14. An airplane wing of 1 m chord is to move at 40 m/s in air. A model of 80 mm chord is to be tested in a wind tunnel with air velocity at 48 m/s. For air temperature of 20°C in each case, what should be the pressure in the wind tunnel? Take $\nu_{20°C} = 14.86 \times 10^{-6}\ m^2/s$

Solution

Airplane is completely submerged in air therefore, Reynold's number is used. Equating Reynold's number of model and prototype.

$$R_{e_m} = R_{e_p}$$

$$\frac{u_m\, l_m}{\nu_m} = \frac{u_p\, l_p}{\nu_p}$$

$$\frac{48 \times 0.08}{\nu_m} = \frac{40 \times 1}{14.86 \times 10^{-6}}$$

$$\nu_m = 1.426 \times 10^{-6}\ m^2/s$$

Therefore, $\nu_r = \frac{\nu_m}{\nu_p} = \frac{14.26\times10^{-6}}{14.86\times10^{-6}} = 0.096 = 1:10.416$

We know that $\upsilon_r = \frac{\mu_r}{\rho_r}$

Absolute viscosity is not affected by pressure change but density increase with pressure

Therefore, $\rho_r = \frac{\mu_r}{\nu_p} = 10.416$

$$p_m = \rho_m R T_m$$

$$p_p = \rho_p R T_p$$

or, $\frac{P_m}{P_p} = \frac{\rho_m}{\rho_p} = \rho_r = \frac{1}{1/10.416} = 10.416$

Assuming P_p is one atmosphere ie the standard condition therefore the pressure in the wind tunnel i.e. P_m is 10.416 atm. Ans

1.15. An aircraft is to fly in air at 0°C and 75 percent of sea level standard pressure at a speed of 100 m/s. A 1 : 15 scale model is to be tested in a wind tunnel at a maximum speed of 100 m/s in order to avoid compressibility effects. The air in the wind tunnel is at 25°C, What should be the pressure in the wind tunnel for dynamically similar flow as for the prototype? Give answer in terms of sea level atmospheres. Take $\mu_m = 1.8\times10^{-6}$ *Pas and* $\mu_p = 1.7\times10^{-6}$ *Pas.*

Solution

$\frac{l_m}{l_p} = 1:15,\ p_m = \rho_m\ RT_m$

$$\rho_m = \frac{p_m}{RT_m}$$

and $\rho_p = \frac{p_p}{RT_p}$

Equating R_{e_m} *and* R_{e_p} we get

$$R_{e_m} = R_{e_p}$$

$$\frac{\rho_m\, u_m\, l_m}{\mu_m} = \frac{\rho_p\, u_p\, l_p}{\mu_p}$$

or, $$\frac{p_m\, u_m\, l_m}{RT_m\, \mu_m} = \frac{p_p\, u_p\, l_p}{RT_p\, \mu_p}$$

or, $$\frac{l_m\, p_m}{\mu_m\, T_m} = \frac{l_p\, p_p}{\mu_p\, T_p}$$

or, $$p_m = p_p\left(\frac{l_p}{l_m}\right)\left(\frac{T_m}{T_p}\right)\left(\frac{\mu_m}{\mu_p}\right) = p_p(15)\left(\frac{298}{273}\right)\left(\frac{1.8}{1.7}\right)$$

$$= \mathrm{p_p} \times 17.336 \quad = 0.75 \times 17.336 \quad = 13 \text{ sea level atmosphere.}$$

1.16. The measured resistance in fresh water of a 2.5 m ship model moving at 2 m/s was 45N. (a) What would be the velocity of the 40 m prototype? (b) What force would be required to drive the prototype at this speed in salt water? Take density of sea water as 1030 kg/m³.

Solution

In this problem gravity forces are prodominantly used; hence Froude model law is applicable.

(a) $F_{rm} = F_{rp}$

$$\frac{u_m}{\sqrt{l_m\, g}} = \frac{u_p}{\sqrt{l_p\, g}}$$

or, $$\frac{u_m}{u_p} = \sqrt{l_r} = \sqrt{\frac{2.5}{40}} = \sqrt{1/16}$$

or, $$u_p = 2 \times \sqrt{16} = 8\ m/s$$ **Ans**

(b) $$\frac{F_m}{\rho_m\, u_m^2\, l_m^2} = \frac{F_p}{\rho_p\, u_p^2\, l_p^2}$$

or, $$F_p = F_m\left(\frac{\rho_p}{\rho_m}\right)\left(\frac{u_p}{u_m}\right)^2\left(\frac{l_p}{l_m}\right)^2 = 45\left(\frac{1030}{1000}\right)(16)\,(16)^2$$

$$= 189849.6 \text{ N} = 189.8496 \text{ kN}$$

$\therefore \quad Fp = 189.8496 \text{ kN}$ Ans

1.17. Oil of kinematic viscosity 4.645×10^{-5} m²/s is to be used in a prototype in which both viscous and gravity forces dominate. A model scale of 1 : 5 is also desired. What viscosity of model liquid is necessary to make both the Froudes number and the Reynolds number same in model and prototype?

Solution

Using the scale ratios of Froude and Reynold's number

According to Froudes law

$$F_{r_m} = F_{r_p}$$

$$\frac{u_m}{\sqrt{l_m g}} = \frac{u_p}{\sqrt{l_p g}}$$

or, $$\frac{u_m}{u_p} = \sqrt{l_r}$$

According to Reynold's law

$$R_{e_m} = R_{e_p}$$

$$\frac{U_m l_m}{\upsilon_m} = \frac{u_p l_p}{\upsilon_p}$$

or, $$\frac{u_m}{u_p} = \frac{l_p}{l_m} \times \frac{\upsilon_m}{\upsilon_p}$$

Equating (1) and (2)

$$\sqrt{l_r} = \frac{1}{l_r} \times \upsilon_r$$

or, $$\upsilon_p = l_r^{3/2}$$

or, $$\upsilon_m = \upsilon_p \times l_r^{3/2} = 4.645 \times 10^{-5} \times \left(\frac{1}{5}\right)^{3/2}$$

$0.41546 \times 10^{-5} \; m^2/s$ Ans

1.18. A Ship model, scale 1 : 50, has a wave resistance of 30 N at its design speed, calculate the prototype wave resistance.

Solution

In this problem Froude number will be used. Equating Froude number of model and prototype.

$$F_{r_m} = F_{r_p}$$

$$\frac{u_m}{\sqrt{l_m\, g}} = \frac{u_p}{\sqrt{l_p\, g}}$$

or, $$\frac{u_m}{u_p} = u_r = \sqrt{\frac{l_m}{l_p}} = \sqrt{l_r} = \sqrt{1/50} \qquad (1)$$

and $$\frac{F_{\omega_p}}{\rho_p\, u_p^2\, l_p^2} = \frac{F_{\omega_m}}{\rho_m\, u_m^2\, l_m^2}$$

or, $$F\omega_p = F_{\omega_m}\left(\frac{\rho_p}{\rho_m}\right)\left(\frac{u_p}{u_m}\right)^2\left(\frac{l_p}{l_m}\right)^2 = \frac{F_{\omega_m}}{\left(\sqrt{l_r}\right)^2\, l_r^2} = \frac{F_{\omega_m}}{l_r^3}$$

$$= \frac{30}{(1/50)^3} = 30\,(50)^3 = 3750 \times 10^3 \text{ N} = 3750 \text{ kN} \quad \text{Ans}$$

1.19. If the scale ratio between a model spillway and its prototype is 1 : 49 what velocity, discharge and power ratio should apply between model and prototype? If the model discharge is 1 m³/s, find the prototype discharge.

Solution

For the Spillway, Froude number is used. Equating Froude number of model and prototype

$$F_{r_m} = F_{r_p}$$

$$\frac{u_m}{\sqrt{l_m\, g}} = \frac{u_p}{\sqrt{l_p\, g}}$$

or, $$\frac{u_m}{u_p} = u_r = \sqrt{\frac{l_m}{l_p}} = \sqrt{l_r} \qquad \text{Ans}$$

The discharge Q = area × velocity

therefore, $Q_r = A_r\, u_r = l_r^2 \sqrt{l_r} = l_r^{5/2}$ Ans

The power $P = \rho\, Q\, g\, h_f$

therefore, $P_r = Q_r\, h_r = l_r^{5/2}\, l_r = l_r^{7/2}$ Ans

Now, $$\frac{Q_m}{Q_p} = l_r^{5/2} = \left(\frac{1}{49}\right)^{5/2} = 5.9499 \times 10^{-5}$$

therefore, $Q_p = 1 \times 10^5 / 5.9499 = 16807\ m^3/s$ Ans

1.20. Water is flowing in an open channel at a depth of 1.5 m with a velocity of 7.5 m/s. At a particular location a hydraulic jump is formed and the depth increases to 2.2 m. Another channel is built where a similar jump is formed. If the flow depth in the new dynamically similar channel is 6 m, estimate the flow velocity and the height of jump.

Solution

For the open channel and Hydraulic jump, Froude number will be used. Equating Froude number for hydraulic jump for both conditions.

$$\left(\frac{u}{\sqrt{l\,g}}\right)_{at\ 1.5\ m} = \left(\frac{u}{\sqrt{l\,g}}\right)_{at\ 6m}$$

Here, $l_r = \dfrac{1.5}{6} = \dfrac{1}{4}$

therefore, $u_r = \sqrt{l_r} = \sqrt{\dfrac{1}{4}} = \dfrac{1}{2}$

or, $$\frac{u\ at\ 1.5\ m\ depth}{u\ at\ 6m\ depth} = \frac{1}{2}$$

$\therefore$ $u\ at\ 6\ m\ depth = 2 \times 7.5 = 15\ m/s$ Ans

and $\dfrac{h_m}{h_p} = l_r = \dfrac{1}{4}$

$h_p = 2 \times 2.2 = 8.8\ m$ Ans

1.21. A model weir built 1/16th full scale is found to discharge 0.5 m³/min when the head over it is 75 mm. Assuming dyna-

mically similar conditions, find (a) the corresponding head over the full scale weir (b) the corresponding discharge over the latter.

Solution

In the weir problem, Froude model law will be used.

$$F_{r_m} = F_{r_p}$$

$$\frac{u_m}{\sqrt{l_m g}} = \frac{u_p}{\sqrt{l_p g}}$$

$$\frac{u_m}{u_p} = \sqrt{l_r} = \sqrt{1/16} = \frac{1}{4}$$

Therefore, $\frac{Q_m}{Q_p} = A_r\, u_r \quad = l_r^2 \sqrt{l_r} = l_r^{5/2}$

$\therefore \quad Q_p = \frac{Q_m}{l_r^{5/2}} = 0.5 \times (16)^{5/2} = 0.5 \times 1024 \quad = 512$ m³/min **Ans**

and

$$\frac{H_m}{H_p} = l_r$$

or, $\quad H_p = \frac{H_m}{l_r} = \frac{75}{1/16} = 1200\ mm$ **Ans**

1.22. **A 1 : 50 scaled model of a proposed power station tail race is to be used to predict prototype flow. If the design load rejection by pass flow is 1200 m³/s what water flow rate should be used on the model?**

Solution

For the power station tail race level the Froude model law will be used. Therefore,

$$F_{r_m} = F_{r_p}$$

$$\frac{u_m}{\sqrt{l_m g}} = \frac{u_p}{\sqrt{l_p g}}$$

$$u_r = \sqrt{l_r}$$

Therefore, $Q_r = A_r\, u_r \; = l_r^2 \sqrt{l_r} = l_r^{5/2}$

$$Q_m = Q_p \, (l_r)^{5/2} = 1200 \left(\frac{1}{50}\right)^{5/2}$$

$$= 0.06788 \, m^3/s = 244.376 \text{ m}^3\text{/hr}$$ Ans

1.23. 1 in 20 model of a stiling basin the height of the hydraulic jump in the model is observed to be 200 mm. What is the height of the hydraulic jump in the prototype? If the energy dissipated in the model is 75 Watt, What is the corresponding value in prototype?

Solution

For the hydraulic jump, Froude model law will be used.

$$\frac{H_m}{H_p} = l_r = \frac{1}{20}$$

Therefore, $H_p = H_m \times 200 = 200 \times 20 = 4000 \; mm = 4 \; meters$ Ans

$$F_{r_m} = F_{r_p}$$

or, $u_r = \sqrt{l_r}$

and $Q_r = l_r^{\,5/2}$

Therefore, $P_r = Q_r \, l_r \; = l_r^{7/2}$

$$\frac{P_m}{P_p} = \left(\frac{1}{20}\right)^{7/2}$$

or, $P_p = P_m \times (20)^{7/2} = 75 \, (20)^{3.5}$

$$= 2683281.573 \text{ Watt} = 2.68 \text{ MW}$$ Ans

EXERCISES

A. THEORY

1.1. What is dimensional analysis? state its uses.

1.2. Distinguish between dimensions and units.

1.3. What are fundamental and derived units?

1.4. Explain the principle of dimensional homogeneity.

1.5. What is dimensional homogeneity? What is meant by homogeneous equation? What is a rational equation?

1.6. Name the two methods of dimensional analysis and explain anyone.

1.7. What are the drawbacks of Rayleigh method of dimensional analysis.

1.8. Distinguish between Rayleigh and Buckingham's dimensional analysis methods.

1.9. What are the advantages of Buckingham's π method over the Rayleigh method.

1.10. Describe Buckingham's π theorem to formulate a dimensional homogenous equation between various physical quantities effecting a certain phenomenon.

1.11. How is the choice of repeating variables made in the case of Buckingham's π method.

1.12. What do you mean by repeating variables? How are the repeating variables selected for dimensional analysis?

1.13. Write a recommended procedure for determining the π parameters.

1.14. Write the useful comments about dimensional analysis.

1.15. To obtain the final expression in the desired manner, what are the useful suggestions.

1.16. What is meant by uniqueness of π terms?

1.17. What are limitations of dimensional analysis?

1.18. What do you mean by superfluous and omitted variables?

1.19. What is model analysis and its uses?

1.20. Describe the advantages of model testing.

1.21. What are the three types of similarities used in model analysis? explain.

1.22. What are standard dimensionless numbers? Explain all the numbers.

1.23. Define and explain Reynold's number, Froude's number and Mach number. What are their significances for fluid flow problems?

1.24. What is meant by geometric, kinematic and dynamic similarities? Are these similarities truly attainable? If not why?

1.25. Reynold's and Froude numbers represent ratio of forces influencing fluid motions. Define these force ratios and, hence, determine the conventional expression for each ratio. Illustrate the distinctive applications of these dimensionless numbers to engineering practice.

1.26. What are model laws? Explain, Reynold's model law and its applications.

1.27. Explain Mach's model law and it's applications.

1.28. Explain Euler's model law and its applications.

1.29. Explain Weber model law and its applications.

1.30. Explain model tests of system dependent on Reynold's and Froude numbers.

1.31. Write the sequential steps on system dependent on Reynold's and Froude, numbers.

1.32. Explain model tests of system dependent on Reynold's and Mach numbers.

1.33. Explain the model analysis of turbomachines.

1.34. What do you mean by undistorted and distorted models?

1.35. Explain river and harbour models.

1.36. Explain estuary models.

1.37. Explain scale effect in model testing. How is it found?

1.38. How is scale effect in a model testing minimised?

1.39. What are scale ratios of a distorted model?

1.40. What are distorted models? What are the reasons of constructing such models for river and harbour engineering projects?

1.41. What are the advantages and disadvantages of distorted models?

1.42. What are the limitations under which a hydraulic model could be tested?

1.43. Prove that the scale ratio for a distorted model is given by

(a) Times ratio, $T_r = \frac{l_x}{\sqrt{l_y}}$

(b) Discharge ratio, $Q_r = l_x\, l_y^{3/2}$

(c) Manning's rugovisty constant ratio, $n_r = l_y^{2/3} / \sqrt{l_x}$

(d) Slope ratio, $S_r = {}^{l_y}\!/_{l_x}$

1.44. River models are often distorted and have their surfaces roughened artificially. Account for these practices.

1.45. A tests on ski-jump spillway are to be made on a geometrically similar model with a scale ratio of l_r. Determine the corresponding scales of velocity, discharge pressure, force, power, acceleration and time. In what respects, if any, may the flow pattern of the full size spillway be expected to differ from that of a model.

1.46. How will you determine the total drag of a partially submerged object (ship)?

1.47. Derive scale ratios for distorted model for its vertical cross sectional area, velocity of flow, discharge, time of flow, bed slope and the rugosity coefficient in Manning's formula if l_y is the vertical scale and l_r is the horizontal scale.

1.48. Describe how is the total resistance of a cargo vessel, influenced simultaneously by Reynold's and Froude number, determined.

1.49. Describe how the drag force of a supersonic aircraft influenced simultaneously by Reynold's and Froude number, determined.

1.50. Define dimensionless head and power coefficient.

B. UNSOLVED PROBLEMS

(a) Buckingam's π theorem

1.1. The friction torque T of a disc of diameter D rotating at a speed N, fluid of viscosity μ and density ρ in a turbulent flow is given by using π theorem.

$$T = \rho D^5 N^2 \Phi\left[\frac{\mu}{\rho N D^2}\right]$$

1.2. The efficiency η of a fan depends on the density ρ, the dynamic viscosity μ of the liquid, the angular velocity ω, diameter D of the rotar and the discharge Q. Express η in terms of dimensionless parameters by using π theorem.

$$\eta = \Phi\left[\frac{\mu}{\rho\,\omega D^2}, \frac{Q}{D^3\,\omega}\right]$$

1.3. The resisting force F of a supersonic plane during flight can be considered as dependent upon the length of the aircraft l, velocity V, air viscosity μ, air density ρ and bulk modulus of air K. Express the functional relationship between these variables and the resisting force by using π theorem.

$$\frac{F}{\rho V^2 l^2} = \Phi\left[\frac{\mu}{\rho V l}, \frac{K}{\rho V^2}\right]$$

1.4. Using Buckingham's π theorem, show that velocity through a circular orifice is given by

$$V = \sqrt{2gH}\,\Phi\left[\frac{D}{H}, \frac{\mu}{\rho VH}\right]$$

Where H is the head causing flow, D is the diameter of the orifice, μ is the coefficient of viscosity, ρ is the mass density and g is the acceleration due to gravity.

1.5. The discharge Q flowing over a V (vee) notch weir with notch of angle θ is some function of the elevation h, of upstream liquid surface above the bottom of the notch, the velocity of approach V, to the weir and gravity g. Using Buckingham's π theorem and adopting g and h as the repeated variables, determine the form of the discharge equation.

$$Q = g^{1/2}\, h^{5/2}\, \Phi\left[\frac{V}{\sqrt{gh}}, \theta\right]$$

1.6. Using Buckingham's π theorem show that the discharge of a centrifugal pump can be expressed as

$$Q = N\,D^3\,\Phi\left[\frac{gH}{N^2\,D^2}, \frac{\mu}{\rho\,N\,D^2}\right]$$

Where N = speed of the pump in rpm, D = diameter of the impeller, g = acceleration due to gravity, H = manometric head; μ = dynamic viscosity; ρ = fluid density.

1.7. Show that the power P developed by a hydraulic turbine is given by using π theorem.

$$P = \rho\,D^5\,N^3\,\Phi\left[\frac{N^2\,D^2}{gH}\right]$$

Where ρ is the mass density of the fluid, N the speed of the rotation, D the diameter of the rotor and H is the variable head.

1.8. Prove the formula using Buckingham's π theorem.

$$\frac{Q}{g^{1/2}\,H^{5/2}} = \Phi\left[\frac{\rho\,D\sqrt{gH}}{\mu}, \frac{D}{H}, \frac{\sigma}{\rho\,gH^2}, \frac{K_s}{H}\right]$$

Where Q = Volume rate of flow

D = Diameter H = Head

σ = Surface tension K_s = Surface roughness

g = Acceleration due to gravity

ρ = Mass density and μ = Dynamic viscosity

1.9. Assuming that for turbulent flow in rough pipe the resistance per unit area of solid boundary τ is dependent upon the viscosity μ, the density ρ, the velocity V of the fluid, the diameter D of the pipe and size of roughness K, show by using Buckingham's π theorem.

$$\frac{\tau}{\rho\,V^2} = \Phi\left[\frac{V\,D\,\rho}{\mu}, \frac{K}{D}\right]$$

1.10. Show by using Buckingham's π theorem, that one form of the law govering the flow of liquids in pipes is

$$\frac{S}{\rho\,V^2} = \Phi\left[\frac{\rho\,VD}{\mu}\right]$$

Where S is the frictional force per unit wetted area $\left[S = \frac{F}{\pi D l}\right]$ and other variables have their usual significance.

1.11. Show by using Buckingham's π theorem, that the formula for flow over a V notch is given by

$$Q = g^{1/2}\ H^{5/2}\ \Phi\left[\frac{g^{1/2}\ H^{3/2}}{\nu}\right]$$

Where

Q is discharge

H is head

g acceleration due to gravity

ν is the kinematic viscosity.

(b) Model Analysis

1.1. The drag of an airfoil at zero angle of attack is a function of the flow density, viscosity, and velocity in addition to a length parameter. A 1 : 10 scale model of an airfoil was tested in a wind tunnel at a Reynold's number of 5.5×10^6 based on chord length. Test conditions in the wind tunnel air stream were 10 atm absolute pressure and 15°C. The prototype airfoil has a chord length 2m, and it is to be flown in air at standard conditions. Determine the velocity at which the wind tunnel model was tested, and the corresponding prototype velocity. [V_m = 39.2 m/s, = V_p]

1.2. An airship to operate at 20 m/s in air at standard conditions. A model is constructed to 1 : 20 scale and tested in wind tunnel at the same air temperature to determine drag. (a) What criterion should be considered to obtain dynamic similarity? (b) If the model is tested at 20 m/s, what pressure should be used in the wind tunnel? (c) If the model drag force is 250 N, what will be the drag of the prototype? [2.02 MPa, 5 kN]

1.3. A 1 : 5 scale model of a torpedo is to be tested in a wind tunnel to determine the drag force. The prototype operates in water, has a diameter of 533 mm, and is 6.7 m long. The desired operating speed of the prototype is 28 m/s. To avoid compressibility effects in the wind tunnel the maximum speed is limited to 110 m/s. However, the pressure in the wind tunnel can be varied while holding the temperature constant at 20°C. At what minimum pressure should the wind tunnel be operated to achieve a dynamically similar test? At dynamically similar test conditions the drag force on the model is measured as 618 N. Evaluate the drag force to be expected on the full scale torpedo [1.93 MPa abs, 43.7 kN].

1.4. The moment exerted on a submarine by its rudder is to be studied with a 1 : 20 scale model in a water tunnel. If the torque measured on the water model is 5 Nm for a tunnel velocity of 15 m/s, What are the coresponding torque and speed for the prototype? [T_p = 100 Nm, V_p = 0.75 m/s]

1.5. A torpedo shaped object 900 mm diameter is to move in air at 60 m/s and its drag is to be estimated from tests in water on a half scale model. Determine the necessary speed of the model and the drag of the full scale object if that of the model is 1140 N. Take densities : air 1.2 kg/m^3 & water 1000 kg/m^3. Viscoities of air 1.86×10^{-5} Pas and water 1.01×10^{-3} Pas [7.82 m/s, 322 N]

1.6. An aircraft is to fly at a height of 9 km (where the temperature and pressure are -45°C and 30.2 kPa respectively) at 400 m/s. A 1 : 20 scale model is tested in a pressurized wind tunnel in which the air is at 15°C. For complete dynamic similarity what pressure and velocity should be used in the wind tunnel? For air at T Kelvin, $\mu \times T^{3/2} / (T + 117)$ [821 kPa, 450 m/s]

1.7. For model and Prototype, show that, when gravity and inertia are the only influences, the ratio of flows Q is equal to the ratio of the length dimension to the five halves power.

1.8. A 1 : 50 scale model of a proposed dam is to be used in a design study. If the design value of the discharge over the spillway at flood stage is 15000 m^3/s, what water flow rate should be established in the model to simulate this flow? Also, if a velocity of 1.2 m/s is measured at a point in the model, what is the velocity at a corresponding point in the model, what is the velocity at a corresponding point in the prototype flow? [0.8485 m^3/s, 8.485 m/s].

1.9. If the scale ratio between a model spillway and its prototype is 1 : 25 what velocity and discharge ratio should apply between model and prototype? If the prototype discharge is 3000 m^3/s What is the model discharge [1/5, 1/3125, 0.96 m^3/s].

1.10. A model of a proposed dam spillway was constructed to a scale of 1 : 15. The design flood discharge over the spillway is 1000 m^3/s. What discharge should be provided in the model? What is the velocity in the prototype corresponding with a velocity of 1.5 m/s in the model at the corresponding point? [V_p = 7.5 m/s, Q_m = 0.32 m^3/s].

2

Laminar Viscous Flow

2.1 INTRODUCTION

When the Euler equations were developed it was assumed that there were no frictional forces acting upon the fluid element. In any fluid such forces must exist and the Euler equations cannot give an accurate description of the behaviour of the fluid especially if its viscosity is large.

There are many problems of fluid flow where the viscosity should be taken into account. Even a fluid with small viscosity may cause considerable shear stress on an immersed body.

More complex laminar viscous flows are multidimensional and multi-directional. The normal stresses are no longer simply the pressure but also involve the linear strain rates and viscosity. The normal stresses are different in different directions. More than one component of the shear stress is present, and the expressions for the shear stresses involve sums of two velocity gradients. These more complex viscous flows are analyzed using a set of general momentum differential equations called the Navier-Stoke equations, which contain complete expressions for the normal and shear stresses. Except for the simplest flow situations, the exact solution of the Navier-Stokes equations is so formidable that numerical integration using a high speed digital computer become necessary. Before the advent of computers, however, many approximate analytical solutions of the Navier-Stokes equations were developed. They involved some very sophisticated mathematics coupled with great insight into the physics of the flow.

This chapter briefly introduces the general differential equations of motion for a laminar viscous flow, the Navier-Stokes equations.

2.2. HAGEN-POISEUILLE FLOW

Probably the best known exact solution to the Navier-Stokes equation is straight circular tube of constant cross section. This type of flow is commonly called Hagen-Poiseuille flow. It is named in honour of J.L. Poiseuille (1799-1869), a French Physician and G.H.L. Hagen (1797-1884), a German hydraulic engineer. Poiseuille was interested in blood flow through capillaries and deduced experimentally the resistance laws, for laminar flow through circular tubes. Hagen's investigation of flow in tubes was also experimental. It was actually after the work of Hagen and Poiseuille that the theoretical results presented in this section were determined, but their names are commonly associated with the solution of this problem.

The derivation of the Hagen-Poiseuille equation for laminar flow in straight round pipes is based on the following two assumptions :

(1) the viscous property of fluid following Newton's law of viscosity, that is $\tau = \mu \dfrac{du}{dr}$ and

(2) there is no relative motion between fluid particles and solid boundaries, that is, no slip of fluid particles at the solid boundary.

Consider the flow through a horizontal circular tube of radius R as is shown in Fig. 2.1.

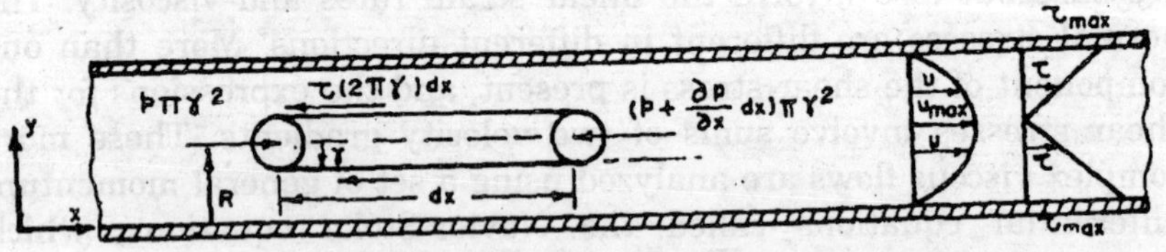

Fig. 2.1. Laminar flow in circular pipe-Hagen Poiseuille flow.

The momentum equation for the flow of fluid through the chosen free body (in the absence of grawitational force) is obtained from the conditions of equilibrium in the x direction which requires that shearing force acting on the circumferential area of the cylinder be equal to the pressure force acting on the cross section of the cylinder. At any radial position, the equilibrium condition can be expressed by

Pressure force = Shear force

$$p\,\pi\,r^2 - \left(p + \frac{\partial p}{\partial x}\,dx\right)\pi\,r^2 = \tau 2\pi\,r\,dx$$

or, $$p\,\pi\,r^2 - p\,\pi\,r^2 - \frac{\partial p}{\partial x}\,dx\;\pi\,r^2 = \tau 2\pi\,r\,dx$$

or, $$-\frac{\partial p}{\partial x}\,dx\;\pi r^2 = \tau\,2\,\pi\,r\,dx$$

or, $$\left(-\frac{\partial p}{\partial x}\right)\frac{r}{2} = \tau$$

or, $$\tau = \left(-\frac{\partial p}{\partial x}\right)\frac{r}{2} \qquad (2.1)$$

In accordance with the first assumption $\tau = \mu \frac{du}{dy}$. Since $y = R - r$ it follows that $dy = -dr$, Newton's law of viscosity then becomes

$$\tau = -\mu\frac{du}{dr} \qquad (2.2)$$

By Combining equation (2.1) and (2.2)

$$-\mu\frac{du}{dr} = -\frac{\partial p}{\partial x}\frac{r}{2}$$

$$du = \frac{r}{2\mu}\left(\frac{\partial p}{\partial x}\right)dr \qquad (2.3)$$

Integrating with respect to r

$$u = \left(\frac{\partial p}{\partial x}\right)\frac{r^2}{4\mu} + A \qquad (2.4)$$

Where A is a constant of integration which may be evaluated with reference to a suitable boundary condition. The approximate boundary condition in this case is $u = 0$ when $r = R$. Putting the boundary condition in the equation (2.4).

$$0 = \left(\frac{\partial p}{\partial x}\right)\frac{R^2}{4\mu} + A$$

$$A = \left(-\frac{\partial p}{\partial x}\right)\frac{R^2}{4\mu} \quad (2.5)$$

Substituting into equation (2.4)

$$u = \left(\frac{\partial p}{\partial x}\right)\frac{r^2}{4\mu} - \left(\frac{\partial p}{\partial x}\right)\frac{R^2}{4\mu}$$

$$\frac{1}{4\mu}\left(-\frac{\partial p}{\partial x}\right)\left(R^2 - r^2\right) \quad (2.6)$$

Thus, at any cross section the velocity distribution is parabolic.

The maximum velocity u_{max} occurs at the centre of the tube. The maximum velocity is obtained by putting r = o in equation (2.6)

$$u_{max} = \left(-\frac{\partial p}{\partial x}\right)\frac{R^2}{4\mu} \quad (2.7)$$

To obtain a relationship between the volume rate of flow, Q, passing through the tube and the pressure gradient, we consider the flow through the differential, washer shaped ring of Fig. 2.2. Since u is constant on this ring the volume rate of flow through the differential area is given by

$$dQ = u\left(2\pi r\right) dr$$

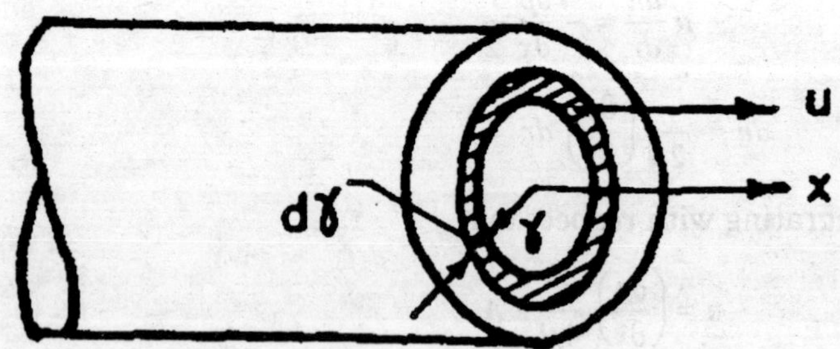

Fig. 2.2. Flow through differential annular ring.

and therefore

$$Q = 2\pi \int_0^R u\, r\, dr$$

Substituting the value of u from equation (2.7)

$$Q = 2\pi \int_o^R \frac{1}{4\mu}\left(-\frac{\partial p}{\partial x}\right)\left(R^2 - r^2\right) r\, dr$$

$$= \frac{2\pi}{4\mu}\left(-\frac{\partial p}{\partial x}\right)\int_o^R \left(R^2\, r - r^3\right) dr$$

$$= \frac{\pi}{2\mu}\left(-\frac{\partial p}{\partial x}\right)\left[\frac{R^2 r^2}{2} - \frac{r^4}{4}\right]_o^R = \frac{\pi}{2\mu}\left(-\frac{\partial p}{\partial x}\right)\left[\frac{R^4}{2} - \frac{R^4}{4}\right]$$

$$= \frac{\pi R^4}{2\mu}\left(-\frac{\partial p}{\partial x}\right)\left[\frac{1}{2} - \frac{1}{4}\right] = \frac{\pi R^4}{8\mu}\left(-\frac{\partial p}{\partial x}\right) \tag{2.8}$$

The average velocity can be obtained by using the relation

$$Q = \text{Area} \times \text{Average velocity} = \pi\, R^2\, u_{av}$$

or, $$u_{av} = \frac{Q}{\pi R^2}$$

Substituting the value of Q from equation (2.8)

$$u_{av} = \frac{1}{\pi R^2}\left[\frac{\pi R^4}{8\mu}\left(-\frac{\partial p}{\partial x}\right)\right]$$

$$u_{av} = \frac{R^2}{8\mu}\left(-\frac{\partial p}{\partial x}\right) \tag{2.9}$$

or, $$u_{av} = \frac{1}{2}\left[\frac{R^2}{4\mu}\left(-\frac{\partial p}{\partial x}\right)\right] = \frac{1}{2}\left(u_{\max}\right)$$ [From equation (2.7)]

$$u_{av} = \frac{u_{\max}}{2}$$

or, $$u_{max} = 2\, u_{av} \tag{2.10}$$

The velocity distribution can be written in terms of u_{max} as

$$\frac{u_{av}}{u_{\max}} = 1 - \left(\frac{r}{R}\right)^2 \tag{2.11}$$

As was true for the similar case of flow between parallel plates (some times referred to as plane Poiseuille flow).

The relationship can be expressed in terms of pressure drop, dp, which occurs over a length $l = x_2 - x_1$, along the tube, since

$$\frac{dp}{l} = -\frac{\partial p}{\partial x} \tag{2.12}$$

Substituting in equation (2.8)

$$Q = \frac{\pi R^4 \, dp}{8 \mu l} \tag{2.13}$$

For a given pressure drop per unit length, the volume rate of flow is inversely proportional to the viscosity and proportional to the tube radius to the fourth power. Equation (2.13) is commonly called Poiseuille's law.

Substituting the equation (2.12) into equation (2.9) gives

$$u_{av} = \frac{R^2}{8 \mu} \frac{dp}{l}$$

or, $$dp = \frac{8 \mu u_{av} l}{R^2}$$

$$dp = \frac{32 \mu u_{av} l}{D^2} \tag{2.14}$$

Dividing by ρg in the equation (2.14) gives

$$\frac{dp}{\rho g} = \frac{32 \mu u_{av} l}{\rho g D^2}$$

or, $$h_f = \frac{32 \mu u_{av} l}{\rho g D^2} \tag{2.15}$$

We know that the head loss due to friction is given by

$$h_f = \frac{\lambda l u_{av}^2}{2g D}$$

therefore,

$$\frac{\lambda l u_{av}^2}{2g D} = \frac{32 \mu u_{av} l}{\rho g D^2}$$

or, $$\lambda = \frac{64 \mu}{\rho u_{av} D}$$

$$\lambda = \frac{64}{R_e} \tag{2.16}$$

The shear stress distribution can be obtained very easily by using either equation (2.1) or equation (2.2)

$$\tau = \left(-\frac{\partial p}{\partial x}\right)\frac{r}{2} = \frac{dp}{l}\left(\frac{r}{2}\right) = \frac{8\,\mu\,u_{av}\,l}{R^2\,l}\left(\frac{r}{2}\right)$$

$$= \frac{4\,\mu\,u_{av}\,r}{R^2} \tag{2.17}$$

The generalized shear distribution is obtained by using equation (2.17). The maximum shear stress at the wall of the pipe is obtained by putting r = R, therefore,

$$\tau = \tau_{\max} = \frac{4\,\mu\,u_{av}\,R}{R^2} = \frac{4\,\mu\,u_{av}}{R} = \frac{8\,\mu\,u_{av}}{D} \tag{2.18}$$

$$= \frac{8\mu}{D}\left(\frac{u_{\max}}{2}\right) = \frac{4\mu u_{\max}}{D}$$

$$\text{or,} \qquad \frac{\tau_{\max}}{\tau_{\max}} = \frac{4\,\mu}{D} \tag{2.19}$$

Numerous experiments performed to substantiate the theoretical results show that the theory and experiment are in agreement for the laminar flow of Newtonian fluids in circular tube or pipes. As long as the Reynold's number has a value less than the critical Reynold's number, the laminar flow can be maintained. According to experiments the critical Reynold's number is approximately equal to 2300. For $R_e > R_{e_{critical}}$ the flow pattern is entirely different and becomes turbulent.

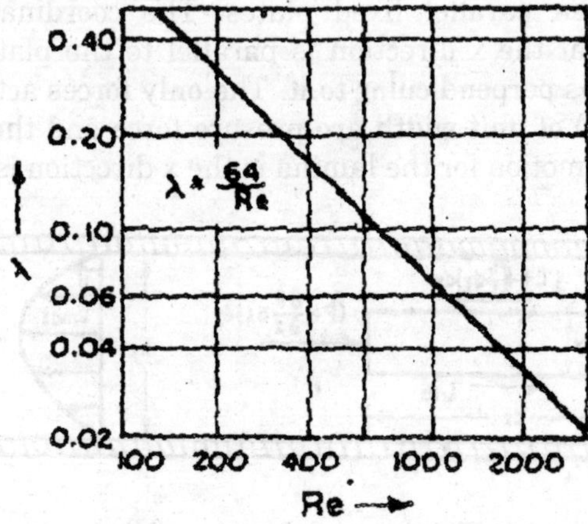

Fig. 2.3. Resistance coefficient λ v/s Reynold's number Re.

The laminar equation for pressure loss in pipes, equation (2.16), is in excellent agreement with experimental results for the laminar range, as seen from Fig. 2.3 which reproduces experimental points measured by G. Hagen.

The Navier-Stokes equation in cylindrical coordinates subject to the following assumptions steady flow, axially symmetric flow, uniform flow, flow parallel to pipe wall and continuity equation are used to simplify the above derivations for velocity distribution

$$v_z = \frac{1}{4\,\mu}\left(\frac{\partial p}{\partial z}\right) r^2 + A \ln\ r + B$$

with the proper boundary conditions at r = 0, it follows that A = 0 at the centre of the tube. At the wall (r = R) the velocity must be zero so that

$$B = \frac{1}{4\,\mu}\left(-\frac{\partial p}{\partial z}\right) R^2$$

and the velocity distribution becomes

$$v_z = \frac{1}{4\,\mu}\left(-\frac{\partial p}{\partial z}\right)\left(R^2 - r^2\right)$$

This is same equation (2.6) which we obtained earlier.

2.3. PLANE POISEUILLE FLOW

Consider the steady flow as shown in Fig. 2.4. which is assumed to occur between parallel fixed plates. The coordinate system is defined such that the x direction is parallel to the plate orientation and y direction is perpendicular to it. The only forces acting on a thin lamina (dx x dy) of unit width are presure force and the shear force. The equation of motion for the lamina is the x direction is as follows:

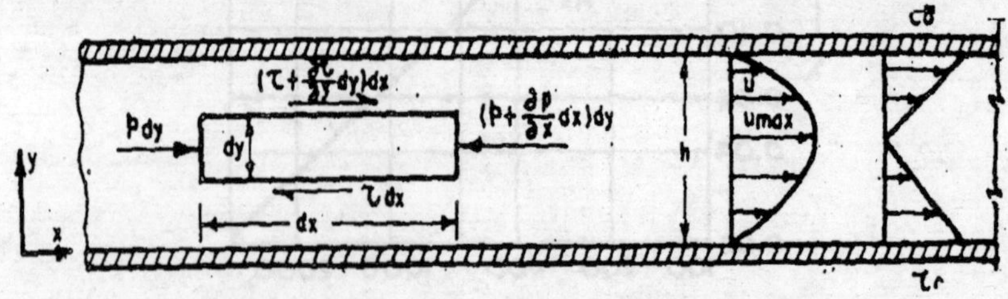

Fig. 2.4. Flow between fixed parallel plates.

Pressure force = Shear force

$$pdy - \left(p + \frac{\partial p}{\partial x}\, dx\right) dy = \tau\, d\, x - \left(\tau + \frac{\partial \tau}{\partial y} dy\right) dx$$

or, $$pdy - pdy - \frac{\partial p}{\partial x}\, dx\, dy = \tau\, dx - \tau\, dx - \frac{\partial \tau}{\partial y}\, dx\, dy$$

or, $$-\frac{\partial p}{\partial x}\, dx\, dy = -\frac{\partial \tau}{\partial y}\, dx\, dy$$

or, $$\frac{\partial \tau}{\partial y} = \frac{\partial p}{\partial x} \qquad (2.20)$$

When Newton's law of viscosity $\tau = \mu \dfrac{du}{dy}$ is substituted in the equation (2.20)

$$\frac{\partial}{\partial y}\left(\mu \frac{\partial u}{\partial y}\right) = \frac{\partial p}{\partial x}$$

or, $$\mu \frac{\partial^2 u}{\partial y^2} = \frac{\partial p}{\partial x} \qquad (2.21)$$

Integrating twice the equation (2.21) with respect to y we get

$$u = \frac{1}{2\,\mu}\left(\frac{\partial p}{\partial x}\right) y^2 + Ay + B \qquad (2.22)$$

The two constants A and B can be determined from the boundary conditions. In this case

$$u = 0 \quad at \quad y = 0 \qquad (i)$$

and $$u = 0 \quad at \quad y = h$$

The boundary conditions are used in the basic governing equation (2.22)

$$u = \frac{1}{2\,\mu}\left(\frac{\partial p}{\partial x}\right) y^2 + Ay + B$$

From first boundary condition

$$0 = 0 + 0 + B$$

Therefore $B = 0$

From second boundary condition

$$0 = \frac{1}{2\mu}\left(\frac{\partial p}{\partial x}\right)h^2 + Ah + 0$$

$$Ah = \frac{1}{2\mu}\left(-\frac{\partial p}{\partial x}\right)h^2$$

$$A = \frac{h}{2\mu}\left(-\frac{\partial p}{\partial x}\right)$$

Substituting the values of the constant A and B in equation (2.22)

$$u = \frac{1}{2\mu}\left(\frac{\partial p}{\partial x}\right)y^2 - \frac{hy}{2\mu}\left(\frac{\partial p}{\partial x}\right)$$

$$= \frac{1}{2\mu}\left(\frac{\partial p}{\partial x}\right)\left[y^2 - hy\right]$$

$$= \frac{1}{2\mu}\left(-\frac{\partial p}{\partial x}\right)\left[hy - y^2\right] \qquad (2.23)$$

The velocity distribution curve for the laminar flow between parallel flat plates is a parabola. The negative sign for the pressure gradient indicates that there is a drop in pressure in the direction of flow.

The maximum velocity which occur at the mid point between the plates, can be obtained by putting $y = \frac{h}{2}$ in equation (2.23).

$$u_{max} = \frac{h^2}{8\mu}\left(-\frac{\partial p}{\partial x}\right) \qquad (2.24)$$

The average velocity can be calculated as follows

$$u_{av} = \frac{1}{h}\int_o^h u\,dy \qquad = \frac{1}{h}\int_o^h \frac{1}{2\mu}\left(-\frac{\partial p}{\partial x}\right)\left(hy - y^2\right)dy$$

$$= \frac{1}{2\mu h}\left(-\frac{\partial p}{\partial x}\right)\int_o^h \left(hy - y^2\right)dy = \frac{1}{2\mu h}\left(-\frac{\partial p}{\partial x}\right)\left[\frac{hy^2}{2} - \frac{y^3}{3}\right]_o^h$$

$$= \frac{1}{2\mu h}\left(-\frac{\partial p}{\partial x}\right)\left[\frac{h^3}{2} - \frac{h^3}{3}\right] = \frac{h^3}{2\mu h}\left(-\frac{\partial p}{\partial x}\right)\left[\frac{1}{2} - \frac{1}{3}\right]$$

$$= \frac{h^2}{12\mu}\left(-\frac{\partial p}{\partial x}\right) \tag{2.25}$$

The discharge Q of the laminar flow can be obtained very easily

$$Q = h \times \text{width of the plate} \times u_{av}$$

$$q = \frac{Q}{\textit{width of the plate}} = h\,u_{av} = \frac{h^3}{12\mu}\left(-\frac{\partial p}{\partial x}\right)$$

or, $$q = \frac{h^3}{12\mu}\left(-\frac{\partial p}{\partial x}\right) \tag{2.26}$$

The relation between the average velocity and maximum velocity can be obtained by combining the equation (2.24) and (2.25)

$$u_{av} = \frac{8}{12} \times \frac{h^2}{8\mu}\left(-\frac{\partial p}{\partial x}\right)$$

or, $$u_{av} = \frac{2}{3}\,u_{\max}$$

or, $$u_{\max} = \frac{3}{2}\,u_{av} \tag{2.27}$$

The pressure drop can be obtained from the equation (2.25)

$$\left(-\frac{\partial p}{\partial x}\right) = \frac{12\mu\,u_{av}}{h^2}$$

or, $$\partial p = -\frac{12\,\mu\,u_{av}}{h^2}$$

Integrating between the points 1 and 2

$$p_1 - p_2 = \frac{-12\,\mu\,u_{av}}{h^2}\,(x_1 - x_2)$$

$$\Delta p = \frac{12\,\mu\,u_{av}}{h^2}\,(x_2 - x_1)$$

$$\Delta p = \frac{12\,\mu\,u_{av}\,l}{h^2} \tag{2.28}$$

The loss of head can be written as

or, $$h_f = \frac{\Delta p}{\rho g} = \frac{12\,\mu\,u_{av}\,l}{\rho\,g\,h^2} \tag{2.29}$$

The shear stress distribution in the flowing fluid across any section of the parallel plates may be determined by using the Newton's law of viscosity

$$\tau = \mu\frac{du}{dy} = \mu\left[\frac{1}{2\mu}\left(-\frac{\partial p}{\partial x}\right)[h-2y]\right] = \frac{1}{2}\left(-\frac{\partial p}{\partial x}\right)(h-2y)$$

$$= (-\partial p/\partial x)[h/2 - y] \tag{2.30}$$

The maximum shear stress or at the wall can be determined by putting y = 0 in the equation (2.30)

$$\tau_o = \tau_{max} = \left(-\frac{\partial p}{\partial x}\right)\frac{h}{2} \tag{2.31}$$

or, $$\tau_o = \frac{12\,\mu\,u_{av}}{h^2} \times \frac{h}{2} = \frac{6\,\mu\,u_{av}}{h} \tag{2.32}$$

$$= \frac{6\,\mu}{h} \times \frac{2}{3}u_{max} = \frac{4\,\mu\,u_{max}}{h} \tag{2.33}$$

The shearing stress is zero at the centre line of the parallel flat plates $\left(y = \frac{h}{2}\right)$ and increases linearly with the distance from the centre line, attaining a maximum value at the boundary or at plate wall (y = o and y = h). The linear variation of shearing stress is shown in Fig. 2.4.

The local friction coefficient c_f is defined by

$$(\tau)_{y=0} = \frac{1}{2}\rho\,u_{av}^2\;C_f$$

$$C_f = \frac{(\tau)_{y=0}}{\frac{1}{2}\rho u_{av}^2} = \frac{6\,\mu\,u_{av}}{h} / \frac{1}{2}\rho\,u_{av}^2$$ [Using equation 2.16]

$$= \frac{12\,v}{u_{av}\,h} = \frac{12}{R_e}$$

Where, $R_e = \frac{u_{av}\, h}{\nu}$ is the Reynold's number of the flow based on the average velocity and the channel height.

The flow will be laminar if the Reynold's number, $R_e = \frac{\rho\, u\, h}{\mu}$ remains below about 1400. For flow with larger Reynold's number the flow becomes turbulent and the preceding analysis is not valid since the flow field is complex, three dimensional and unsteady.

2.4 COUTTE FLOW

Another simple parallel plate flow can be developed by fixing one plate and letting the other plate move with a constant velocity U, as is illustrated in Fig. 2.5.

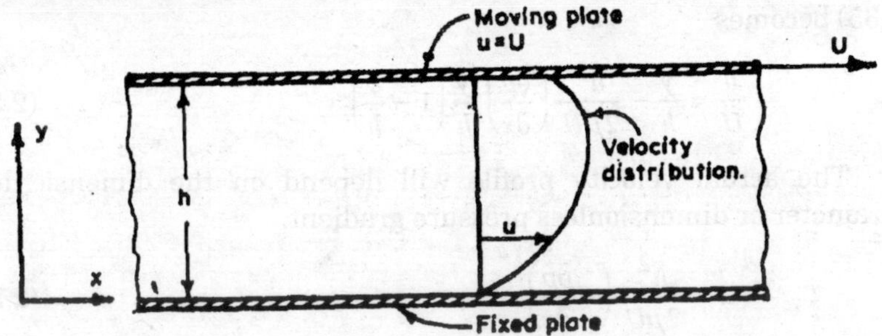

Fig. 2.5. Coutte flow

The Navier-Stoke's equations reduce to the same form as those in the preceding section and the solution for the velocity distribution is given by the equation (2.22). However, for the moving plate problem the boundary conditions are different. The origin of the coordinate system at the bottom plate and designate the distance between the two plates as h (Fig. 2.5). The two constants A and B in equation (2.22) can be determined from the boundary conditions

$$u = 0 \text{ at } y = 0$$

and $\quad u = U \text{ at } y = h$

From first boundary condition

$$0 = 0 + 0 + B$$

therefore, $B = 0$

From second boundary condition

$$U = \frac{1}{2\mu}\left(\frac{\partial p}{\partial x}\right)h^2 + Ah$$

$$A = \frac{U}{h} - \frac{h}{2\mu}\left(\frac{\partial p}{\partial x}\right) \qquad (2.34)$$

Substituting in equation (2.22)

$$u = \frac{1}{2\mu}\left(\frac{\partial p}{\partial x}\right)y^2 + U\frac{y}{h} - \frac{hy}{2\mu}\left(\frac{\partial p}{\partial x}\right)$$

$$u = \frac{Uy}{h} - \frac{h^2}{2\mu}\left(\frac{\partial p}{\partial x}\right)\frac{y}{h}\left(1 - \frac{y}{h}\right) \qquad (2.35)$$

Equation (2.19) is the equation for the generalized Coutte flow.

In non dimensional form the velocity distribution in equation (2.35) becomes

$$\frac{u}{U} = \frac{y}{h} - \frac{h^2}{2\mu U}\left(\frac{\partial p}{\partial x}\right)\frac{y}{h}\left(1 - \frac{y}{h}\right) \qquad (2.36)$$

The actual velocity profile will depend on the dimensionless parameter or dimensionless pressure gradient.

$$\alpha = \frac{h^2}{2\mu U}\left(-\frac{\partial p}{\partial x}\right) \qquad (2.37)$$

The dimensionless velocity is plotted in Fig. 2.6 as a function of the distance from the stationary plate for various dimensionless pressure gradient α. In particular, one can observe that, when the pressure gradient along the passage is zero, the velocity distribution between the two plate is linear. From the velocity distribution it can also be seen that, for positive dimensionless pressure gradient $\alpha > 1$, there is a reverse flow in the vicinity of the stationary wall.

The simplest type of Coutte flow is one for which the pressure gradient is zero, that is, the fluid motion is caused by the fluid being dragged along by the moving boundary. In this case, with $\frac{\partial p}{\partial x} = 0$, equation (2.38) simply reduces to

$$u = U\frac{y}{h} \qquad (2.38)$$

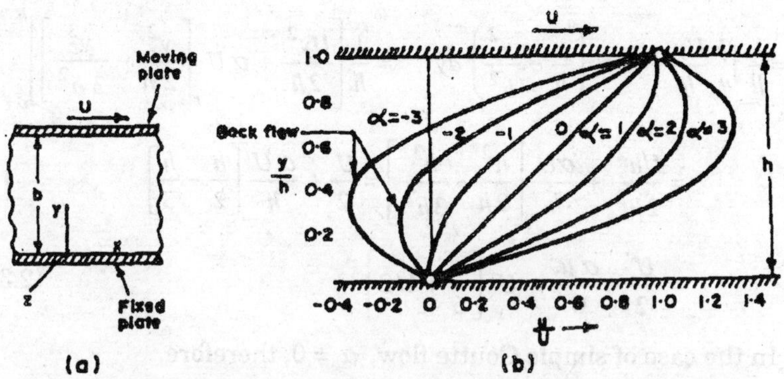

Fig. 2.6. Velocity distribution as a function of parameter α

Which indicates that the velocity varies linearly between the two plates as shown in Fig. 2.7 for α = 0. This situation would be approximated by the flow between closely spaced concentric cylinder in which one cylinder is fixed and the other cylinder rotates with a constant angular velocity w.

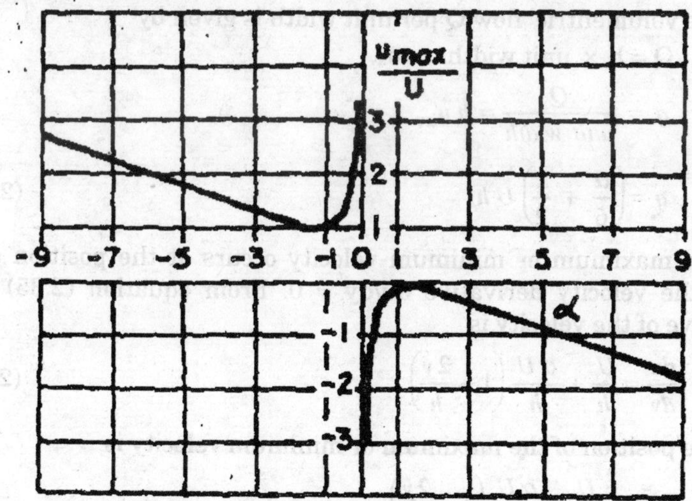

Fig. 2.7. Value of maximum velocity as a function of the dimensionless pressure gradient.

The average velocity distribution of the Coutte flow between two parallel straight walls can be calculated by the expression.

$$u_{av} = \frac{1}{h}\int_o^h u \, dy$$

$$u_{av} = \frac{1}{h}\int_o^h \frac{Uy}{h} + \alpha U\left(\frac{y}{h} - \frac{y^2}{h^2}\right) dy \quad = \frac{1}{h}\left[\frac{Uy^2}{2h} + \alpha U\left[\frac{y^2}{2\,h} - \frac{y^3}{3\,h^2}\right]\right]_o^h$$

$$= \frac{Uh^2}{2h^2} + \frac{\alpha U}{h}\left[\frac{h^2}{2h} - \frac{h^3}{3h^2}\right] = \frac{U}{2} + \frac{\alpha U}{h}\left[\frac{h}{2} - \frac{h}{3}\right]$$

$$= \frac{U}{2} + \frac{\alpha u}{6} = U\left[\frac{\alpha}{6} + 1/2\right] \tag{2.39}$$

In the case of simple Coutte flow, $\alpha = 0$, therefore,

$$u_{av} = \frac{U}{2} \tag{2.40}$$

When $\alpha = -3$ the average velocity becomes zero. This means that there is actually no flow across the passage between plates because the influence of the adverse pressure gradient in balance with the dragging action of the viscous force.

The volumentric flow Q per unit width is given by

$$Q = h \times \text{unit width} \times u_{av}$$

$$q = \frac{Q}{unit\ width} = h\,u_{av}$$

$$q = \left(\frac{\alpha}{6} + \frac{1}{2}\right) U\,h \tag{2.41}$$

The maximum or minimum velocity occurs at the position of y where the velocity derivative du/dy = 0. From equation (2.35) the derivative of the velocity is

$$\frac{du}{dy} = \frac{U}{h} + \frac{\alpha U}{h}\left(1 - \frac{2y}{h}\right) \tag{2.42}$$

The position of the maximum or minimum velocity is

$$0 = \frac{U}{h} + \frac{\alpha U}{h}\left(1 - \frac{2y}{h}\right)$$

or, $$\frac{U}{h} + \frac{\alpha U}{h} = \frac{\alpha U}{h}\left(\frac{2y}{h}\right)$$

$$1 + \alpha = \alpha\left(\frac{2y}{h}\right)$$

or, $$\frac{y}{h} = \frac{1}{2\alpha} + \frac{1}{2} \tag{2.43}$$

It is interesting to note that the maximum velocity for α = 1 occurs at $\frac{y}{h}$ = 1. The minimum velocity for α = –1 at $\frac{y}{h}$ = 0.

This implies that for α = –1 the velocity gradient at the stationary wall is zero and it becomes negative (reverse flow) for any value of α less than –1.

Equation (2.43) breaks down when $-1 < \alpha < 1$ because the maximum and minimum values of y/h have already been reached at α = 1 and α = –1, respectively.

The values of minimum and maximum velocities can be obtained by substituting the value of y from equation (2.43) into equation (2.36).

$$u_{min} = \frac{U}{4}(1+\alpha)^2 / \alpha \quad for\ \alpha \leq -1 \tag{2.44}$$

$$u_{max} = \frac{U}{4}\frac{(1+\alpha)^2}{\alpha} \quad for\ \alpha \geq +1$$

For the purpose of comparison, this equation is plotted in Fig. 2.7.

From equation (2.42) the shearing stress distribution in a Coutte flow can be obtained, and it is

$$\tau = \mu\frac{du}{dy} = \mu\frac{U}{h} + \frac{\mu U \alpha}{h}\left(1 - \frac{2y}{h}\right) \tag{2.45}$$

In dimensionless form, this shear stress distribution becomes.

$$\frac{\tau h}{\mu U} = 1 + \alpha(1 - 2y/h)$$

For a simple Coutte flow when $\alpha = 0$ the shearing stress remains constant across passage between plates.

At the position $\frac{y}{h} = \frac{1}{2}$, the second term on the right hand side of equation (2.45) becomes zero; hence the shearing stress is independent of α at the centre of the passage. The shearing stress at the stationary wall is positive for $\alpha > -1$ and negative for $\alpha < -1$. Since the velocity gradient at the stationary wall is zero for $\alpha = -1$, the value of the shearing stress is zero for this condition.

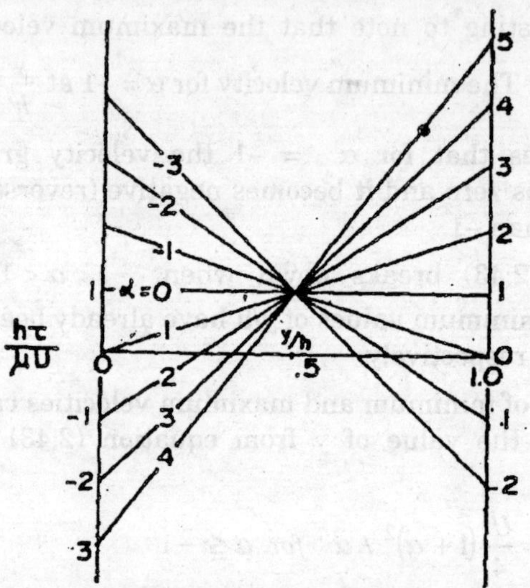

Fig. 2.8. Pressure distribution of the dimensionless pressure gradient.

ILLUSTRATIVE EXAMPLES

2.1. Oil of relative density 0.9 and kinematic viscosity of 330 cS. is pumped over a distance of 1500 meter through a 75 mm diameter tube at a rate of 25000 kg/hr. Determine whether the flow is laminar and calculate the pumping power required, assuming 70 percent mechanical efficiency. Also calculate the shear stress at the tube walls.

Solution

$$\nu = \frac{\mu}{S\rho}$$

or, $\mu = \nu\, S\, \rho = 330 \times 10^{-6} \times 0.9 \times 10^{3} = 0.297$ kg/ms

$$m = s\ \rho\ Q$$

$$\frac{25000}{3600} = 0.9 \times 1000 \times Q$$

$$Q = 7.716 \times 10^{-3}\ m^3/s$$

$$Q = \frac{\pi}{4} D^2\ u_{av}$$

$$7.716 \times 10^{-3} = \frac{\pi}{4} (0.075)^2\ u_{av}$$

$$u_{av} = 1.7465\ m/s$$

$$R_e = \frac{S\ \rho\ u_{av}\ D}{\mu} = \frac{0.9 \times 1000 \times 1.7465\ \times .075}{0.297}$$

= 396.9318 Less than 2000

Therefore the flow is laminar

$$h_f = \frac{32\ \mu\ u_{av}\ l}{s \rho\ g\ D^2} = \frac{32 \times 0.29 \times 1.7465\ \times 1500}{0.9 \times 1000 \times 9.81 \times (0.075)^2}$$

= 501.3398 m

$$P = \frac{S\ \rho\ g\ Q\ h_f}{\eta}$$

$$= \frac{0.9 \times 1000\ \times 9.81 \times 7.716\ \times 10^{-3}\ \times 501.3398}{0.7}$$

= 48790.798 = 48.79 kW

$$\tau = \frac{4\ \mu\ u_{av}\ r}{R^2}$$

The maximum shear stress or at wall shear stress is obtained by putting r = R

$$\tau_o = \frac{4\ \mu\ u_{av}\ R}{R^2} = \frac{4\ \mu\ u_{av}}{R} = \frac{8\ \mu\ u_{av}}{D}$$

$$= \frac{8 \times 0.297 \times 1.7465}{0.075} = 55.329\ N/m^2$$ Ans

2.2. **900 liters/min of oil are pumped through a pipe 600 m long and 100 mm diameter. The oil has an absolute viscosity of 1 Poise and a relative density of 0.928. Find the Reynold's number and friction head.**

If a pump having an efficiency of 65 percent is used to over come the friction, also to elevate the oil through 20 m, determine the power required to drive the pump.

Solution

Discharge through pipe,

$$Q = a \times u$$

$$\frac{900 \times 10^{-3}}{60} = \frac{\pi}{4}(0.1)^2 \times u$$

$$u = 1.9098 \text{ m/s}$$

$$S = \frac{\rho_{oil}}{1000}$$

$$\therefore \rho_{oil} = 0.928 \times 1000 = 928\ kg/m^3$$

$$\therefore R_e = \frac{\rho\, ud}{\mu} = \frac{928 \times 1.9098 \times 0.1}{1/10} = 1772.29$$

The flow is laminar

Therefore, $$\lambda = \frac{64}{R_e} = \frac{64}{1772.29} = 0.036$$

Friction head,

$$h_f = \frac{\lambda\, l\, u^2}{2g\, d} = \frac{0.036 \times 600 \times (1.9098)^2}{2 \times 9.81 \times 0.1} = 40.278 \text{ m} \qquad \text{Ans}$$

Total Pumping head $H = 40.278 + 20 = 60.278$ m

Therefore,

Power to drive pump $= \dfrac{S\rho Q\, g\, H}{\eta_p}$ η_p is the efficiency of pump.

$$= \frac{0.928 \times 10^3 \times 15 \times 10^{-3} \times 9.81 \times 60.278}{0.65} = \frac{8231.328}{0.65}$$

$$= 12663.58 \text{ Watt} = 12.663 \text{ kW} \qquad \text{Ans}$$

2.3. A 100 mm diameter smooth pipe carries Olive oil horizontally at an average velocity of 1.4 m/s. What is the pressure gradient when the oil temperature is (a) 20°C and (b) 40°C (c) Explain why the pressure gradient is changed so little for the warmer, less viscous oil.

Take

$$\nu_{20^\circ c} = 8.3 \times 10^{-5}\ m^2/s \qquad and \qquad \nu_{40^\circ c} = 3.5 \times 10^{-5}\ m^2/s$$

Density is 918 kg/m³

Solution

(a) $R_e = \frac{uD}{\nu} = \frac{1.4 \times 0.1}{8.3 \times 10^{-5}} = 1686.7469 \approx 1687$ (Laminar flow)

$$\lambda = \frac{64}{R_e} = \frac{64}{1687} = 0.03793$$

$$\frac{dp}{l} = \frac{\lambda}{D} \times \frac{\rho u^2}{2} = \frac{0.03793}{0.1} \times \frac{918 \times (1.4)^2}{2} = 341.2979 \text{ Pa/m}$$

(b) $R_e = \frac{uD}{\nu} = \frac{1.4 \times 0.1}{3.5 \times 10^{-5}} = 4000$ (Trubulent flow)

From the chart λ = 0.04

$$\frac{dp}{l} = \frac{\lambda}{D} \times \frac{\rho u^2}{2} = \frac{0.04}{0.1} \times \frac{918 \times (1.4)^2}{2} = 359.856 \text{ Pa/m}$$

(c) The pressure gradient is changed so little because the flow changes from laminar to turbulent for the warmer, less viscous oil.

2.4. Crude oil of dynamic viscosity 1.5 poise and relative density 0.9 flows through a 20 mm diameter vertical pipe. Two pressure gauges have been fixed at 20 m apart. The pressure gauges fixed at higher level reads 200 kPa and that at the lower level reads 6 bar. Find the direction and the rate of flow through the pipe.

Solution

$$\rho = 0.9 \times 1000 = 900 \; kg/m^3$$

$$\mu = 1.5 \; poise = \frac{1.5}{10} \; Ns/m^2 = 0.15 \frac{Ns}{m^2}$$

$$p_1 = 6 \times 10^5 \; N/m^2 = 6 \times 10^5 \; N/m^2$$

$$p_2 = 200 \times 10^3 \; N/m^2 = 6 \times 10^5 \; N/m^2$$

Z = 20 m

Now,

$$p_1 - \rho g z - p_2 = \frac{32 \, \mu \, u_{av} \, l}{D^2}$$

$$(p_1 - p_2) - \rho g z = \frac{32 \, \mu \, u_{av} \, l}{D^2}$$

$$(6-2)\times 10^5 - 900 \times 9.81 \times 20 = \frac{32 \times 0.15\, u_{av} \times 20}{(0.02)^2}$$

$$400000 - 176580 = 240000\, u_{av}$$

$$223420 = 240000\, u_{av}$$

$$u_{av} = 0.9309 \text{ m/s}$$

$$Q = \frac{\pi}{4} D^2\, u_{av} \quad = \frac{\pi}{4}(0.02)^2 \times 0.9309 = 2.9245 \times 10^{-4}\ m^3/s$$

$$= 0.29245 \text{ liters/s}$$ Ans

2.5. Glycerin (density is 1250 kg/m³ and dynamic viscosity is 1 kg/ms) flows through a 100 mm diameter pipe 60 m long and at an upward slope of 10° at a flow rate of 25 liters per second. At the lower inlet the pressure is 8 bar gauge.

(a) What is the Reynold's number of the flow ?

(b) Why is the pressure drop from viscous shear? Also calculate head loss due to friction.

(c) What is the average wall shear stress?

(d) What is the pressure at the higher, downstream end of the pipe?

Solution

$$Q = \frac{\pi}{4} D^2 \times u$$

$$0.025 = \frac{\pi}{4}(0.1)^2 \times u$$

$$u = 3.188 \text{ m/s}$$

$$R_e = \frac{\rho\, u\, D}{\mu} = \frac{1250 \times 3.183 \times 0.1}{1}$$

$$= 397.875 \approx 398$$ Ans

(b) $$dp = \lambda\, \frac{l}{D}\, \frac{\rho u^2}{2} = \frac{64}{R_e} \times \frac{60}{0.1} \times 1250 \times \frac{(3.183)^2}{2}$$

$$= 610944.0603 \text{ Pa} \approx 6.1 \text{ bar}$$ Ans

$$h_f = \frac{dp}{\rho g} = 6.1 \times 10^5 / 1250 \times 9.81 = 49.745\ m$$ Ans

(c) $\tau_o = \frac{dp\ R}{2\ l} = \frac{61 \times 10^5 \times 0.05}{2 \times 60} = 254.167\ N/m^2$

(d) $\frac{p_1}{\rho g} = \frac{p_2}{\rho g} + dz + h_f$

or, $\frac{p_2}{\rho g} = \frac{P_1}{\rho g} - dz - h_f$

$$\frac{P_2}{\rho g} = \frac{8 \times 10^5}{1250 \times 9.81} - 60 \sin 10° - 49.745$$

$$= 65.2395 - 10.418 - 49.745 = 5.075 \text{ m}$$

Therefore,

$P_2 = 5.075 \times 1250 \times 9.81 = 62238.35\ Pa = 62.2$ kPa Ans

2.6. An oil (relative density 0.8) a kinematic viscosity of $6 \times 10^{-5}\ m^2/s$ flows through a 50 mm diameter tube at an average velocity of 0.12 m/s. Calculate the pressure drop in 65 m of tube and the velocity 10 mm from the tube centre.

Solution

$$\mu = \nu\,\rho = 6 \times 10^{-5} \times 800 = 0.048 \text{ kg/ms}$$

$$R_e = \frac{u_{av}\ D}{\nu} = \frac{0.12 \times 0.05}{6 \times 10^5} = 100$$

Hence the flow is laminar

Now pressure drop,

$$\Delta p = \frac{32\ \mu\ u_{av}\ l}{D^2} = \frac{32 \times 0.048 \times 0.12 \times 65}{(0.05)^5}$$

$= 4792.32$ N/m² $= 4.792$ k Pa Ans

The velocity at 10 mm from the tube centre i.e. r = 10 mm

$$u = \left(-\frac{\partial p}{\partial x}\right) \frac{1}{4\mu}\left(R_e - r^2\right)$$

$$= \frac{4792.32}{65} \times \frac{1}{4 \times 0.048}\left[(0.025)^2 - (.01)^2\right]$$

$$= \frac{4792.32 \times 5.25 \times 10^{-4}}{65 \times 4 \times 0.048}$$

$= 0.2016$ m/s **Ans**

2.7. Linseed oil at 25°C flows in a 25 mm diameter horizontal pipe. (a) What is the maximum average velocity for which the flow may be considered laminar? (b) What is the pressure drop in 50 m of pipe at that flow? (c) What is the wall shear stress? Take S = 0.93, $\mu = 3.31 \times 10^{-2}$ kg/ms.

Solution

$$\rho = 0.93 \times 1000 = 930\ kg/m^3$$

$$\nu = \mu/\rho = \frac{3.31 \times 10^{-2}}{930} = 3.559 \times 10^{-5}\ m^2/s$$

The Reynold's number is 2000 for calculating maximum average velocity for laminar flow,

$$R_e = \frac{u_{av}\ D}{\nu}$$

$$2000 = \frac{u_{av} \times 0.025}{3.559 \times 10^{-5}}$$

$u_{av} = 2.8473$ m/s **Ans**

(b) $$dp = \frac{32\ \mu\ u_{av}\ l}{D^2} = \frac{32 \times 3.31 \times 10^{-2} \times 2.8473 \times 50}{(0.025)^2}$$

$= 241268.8$ N/m² $= 2.412$ bar **Ans**

(c) $$\tau_o = \frac{8\ \mu\ u_{av}}{D} = \frac{8 \times 3.31 \times 10^{-2} \times 2.8473}{0.025}$$

$= 30.1586$ N/m² **Ans**

2.8. Fluid of density 1260 kg/m³ and viscosity 0.9×10^{-3} Ns/m² passes between two infinite plates 20 mm separation. If the flow rate is 0.5 liters/s unit width calculate the pressure drop/ unit length if both plates are stationary.

Solution

$Q = A \times u_{av}$

Q / unit width = h u_{av}

$0.5 = 0.02 \times u_{av}$

$u_{av} = 0.025$ m/s

$$R_e = \frac{\rho\, u_{av}\, h}{\mu} = \frac{1260 \times 0.025 \times 0.02}{0.9 \times 10^{-3}} = 700$$

The flow is laminar

$$\frac{dp}{dx} = \frac{12\, \mu\, u_{av}}{h^2}$$

$$= \frac{12 \times 0.9 \times 10^{-3} \times 0.025}{(0.02)^2} = 0.675 \text{ N/m}^2\text{/m} \qquad \text{Ans}$$

2.9. Oil of specific gravity 0.90 and coefficient of viscosity 100 cp (centi poise) flows between two fixed parallel plates kept 12 mm apart. If the average velocity is 1.5 m/s Calculate, (a) the maximum velocity, (b) plate shear stress, (c) velocity and shear stress at a distance of 3 mm from one of the plates and (d) head loss in a distance of 30 m.

Solution

For two fixed parallel plates

(a) $$u_{max} = \frac{3}{2} u_{av} = \frac{3}{2} \times 1.5 = 2.25 \text{ m/s} \qquad \text{Ans}$$

(b) $$u_{av} = \left(-\frac{dp}{dx}\right) \frac{h^2}{12\mu}$$

$$1.5 = \left(-\frac{dp}{dx}\right) \times \frac{\left(\frac{12}{1000}\right)^2}{12 \times 0.1}$$

$$\left(-\frac{dp}{dx}\right) = \frac{1.5 \times 12 \times 0.1 \times (1000)^2}{(12)^2} = 12500$$

$$\tau_o = \left(-\frac{dp}{dx}\right) \frac{h}{2} = 12500 \times \frac{12}{1000 \times 2} = 75 \text{ N/m}^2 \qquad \text{Ans}$$

(c) $$(u)_y = \frac{1}{2\,\mu} \left(-\frac{dp}{dx}\right) \left(hy - y^2\right)$$

$$= \frac{1}{2 \times 0.1}(12500)\left[\frac{12}{1000} \times \frac{3}{1000} - \left(\frac{3}{1000}\right)^2\right]$$

$$= \frac{12500 \times 2.7 \times 10^{-5}}{2 \times 0.1} = 1.69 \text{ m/s}$$ **Ans**

$$(\tau)_y = \left(-\frac{dp}{dx}\right)\left(\frac{h}{2} - y\right)$$

$$= 12500\left(\frac{12}{2 \times 1000} - \frac{3}{1000}\right) = 37.5 \text{ N/m}^2$$ **Ans**

(d) head loss

$$h_f = \frac{12\, \mu\, u_{av}\, l}{\rho\, g\, h^2}$$

$$= \frac{12 \times 0.1 \times 1.5 \times 30}{(0.9 \times 1000 \times 9.81)\left(\frac{12}{1000}\right)^2} = 42.47\ m$$ **Ans**

2.10. **Kerosene oil flows upwards through inclined parallel plates at the rate of 2 liters/s per meter width. The distance between the parallel plates is 10 mm and inclination of plates is 30° with the horizontal. Determine the difference of pressure between two sections 10 m apart. Take S = 0.82 and μ = 1.92 × 10⁻³ Ns/m² for kerosene oil.**

Solution

$$u = \frac{Q}{h} = \frac{2 \times 10^{-3}}{10 \times 10^{-3}} = 0.2\, m/s$$

$$h_f = \frac{12\, \mu\, u\, l}{\rho\, g\, h^2} = \frac{12 \times 1.92 \times 10^{-3} \times 0.2 \times 10}{820 \times 9.81 \times \left(10 \times 10^{-3}\right)^2}$$

$$= \frac{0.04608}{0.80442} = 0.05728\ m$$

But $$h_f = \left(\frac{p_1}{\rho g} + z_1\right) - \left(\frac{p_2}{\rho g} + z_2\right)$$

$$0.05728 = \frac{p_1 - p_2}{\rho g} + (z_1 + z_2)$$

$$\frac{p_1 - p_2}{\rho g} = 0.05728 + 10 \sin 30° = 0.05728 + 5 = 5.05728$$

$$p_1 \; p_2 = (\rho g)(5.05728) = 40681.77 \text{ N/m}^2 = 40.68 \text{ k Pa}$$ Ans

2.11. **Show that the rate of change of pressure at any point along laminar flow between parallel surfaces is given by**

$$\frac{dp}{dx} = \mu\left(\frac{d^2u}{dy^2}\right)$$

Where μ is the viscosity of the fluid and *u* is viscosity at the distance *y* from one of the surface.

Water at 20°C flows between two large parallel plates at a distance 2 mm apart. If the average velocity is 0.15 m/s. Find the maximum velocity, pressure drop, wall shearing stress and the frictional coefficient. The dynamic viscosity at 20°C is given by 1.02 × 10^-3^ kg/ms.

Solution: For theory see text.

The maximum velocity is given by

$$u_{\max} = \frac{3}{2} u_{average} = \frac{3}{2} \times 0.15 = 0.225 \text{ m/s}$$ Ans

The pressure drop

$$\frac{dp}{dx} = 12 \mu \frac{u_{av}}{h^2}$$

$$= \frac{12 \times 1.02 \times 10^{-3} \times 0.15}{\left(2 \times 10^{-3}\right)^2} = 459 \text{ N/m}^2$$ Ans

The wall shearing stress

$$\tau_o = -\frac{h}{2}\left(\frac{dp}{dx}\right) = -\frac{2}{2} \times 10^{-3}(-459) = -0.459 \text{ N/m}^2$$ Ans

$$R_e = \frac{\rho u h}{\mu} = \frac{1000 \times .15 \times 2 \times 10^{-3}}{1.02 \times 10^{-3}} = 294.11$$

The frictional coefficient

$$\lambda = 12 / R_e = \frac{12}{294.11} = 0.408 \qquad \text{Ans}$$

EXERCISES

A. THEORY

2.1. Show that for laminar flow in a circular pipe head loss is given by

$$h_f = \frac{32 \mu\, u\, l}{S \rho\, g\, D^2}$$

2.2. Prove that for a steady laminar flow through a circular pipe the velocity distribution across the section is parabolic and the average velocity is half the maximum local velocity.

2.3. Show that the volumetric flow through a circular pipe is given by the expression $\frac{\pi}{8\,\mu} R^4 \left(-\frac{dp}{dx}\right)$ where R is the radius of the pipe, μ = dynamic velocity.

2.4. Derive an expression for velocity distribution and pressure drop in the fully developed flows region of the flow through a circular pipe.

2.5. Derive an expression for velocity distribution for laminar flow in a circular pipe. Show that the maximum velocity is twice the average velocity.

2.6. For a laminar flow in a circular pipe, find out the relation between friction coefficient, λ and the Reynold's number of flow.

2.7. Show from first principles that the loss of pressure due to laminar flow in a horizontal circular pipe is given by

$$dp = \frac{32\, \mu\, u\, l}{D^2}$$

and hence that under these conditions the friction coefficient λ in the expression.

$$\frac{\lambda\, l\, u^2}{2g\, D} \text{ is } \frac{64}{\text{Reynold's number}}$$

2.8. **Derive an expression for velocity distribution for laminar flow occuring between two parallel plates.**

2.9. **For laminar flow in a pipe, show that the shear stress at a radial distance r from the centre line of the pipe is given by**

$$\tau = \left(-\frac{dp}{dx}\right)\frac{r}{2}$$

2.10. **Using Hagen-Poiseuille equation show that**

$$\frac{\tau_{max}}{u_{max}} = \frac{4\mu}{D}$$

2.11. **Using Hagen-Poiseuille equation show that**

$$u_{max} = \left(-\frac{dp}{dx}\right)\frac{R^2}{4\,\mu}$$

2.12. **Using Hagen-Poiseuille equation show that**

$$\frac{u_{av}}{u_{max}} = 1 - \left(\frac{r}{R}\right)^2$$

2.13. **Using Hagen-Poiseuille equation show that**

$$Q = \frac{\pi\, R^4\, dp}{8\,\mu l}$$

2.14. **Laminar flow takes place between two stationary plates parallel to each other at a distance h apart from each other where h/2 is the distance from the centre line to the plate.**

Show that the velocity profile for the flowing fluid is given by

$$u = \frac{h^2}{8\,\mu}\left(-\frac{\partial p}{\partial x}\right)\left[1 - 4\,(y/h)^2\right]$$

Also show that

$$\tau = 4\,\mu\,\frac{u_{max}}{h}$$

2.15. **For a plane Poiseuille flow, when both the plates are stationary, show that, the velocity u is given by**

$$u = \frac{h^2}{8\,\mu}\left(-\frac{\partial p}{\partial x}\right)\left[1 - 4\,(y/h)^2\right]$$

2.16. Show that for a plane Poiseuille flow the maximum velocity is given by

$$u_{max} = \frac{h^2}{8\,\mu}\left(\frac{\partial p}{\partial x}\right)$$

where h/2 is the distance from the centre line to the plate. Also show that the friction coefficient

$$\lambda = \frac{12}{R_e}$$

where R_e is the Reynold's number

2.17. Show that for a plane Poiseuille flow that pressure drop across length l is given by

$$p_2 - p_1 = \frac{12\mu\, u_{av}\, l}{h^2}$$

B. UNSOLVED PROBLEMS

2.1. Estimate the power required in kW per 1 km length of a pipe line to over come the viscous resistance to the flow of crude oil (sp. gravity = 0.85) having dynamic viscosity equal to 0.1 Ns/m^2 through horizontal 75 mm diameter pipe at the rate of 5 liters/s. [3.21 kW]

2.2. Calculate the loss of head due to friction and the power required to maintain flow in a horizontal circular pipe of 40 mm diameter and 750 m long when water (μ = 1.14 × 10^{-3} Ns/m^2) flow at a rate of 4 liter/min. [R_e = 1856, h_f = 92.4 × 10^{-3} m, P = 0.0605 Watts]

2.3. The centre line velocity is measured as 3m/s in a 0.1 m diameter tube. If the fluid flowing has a density of 1260 kg/m^3 and a viscosity of 0.9 Ns/m^2 determine whether the flow is laminar and calculate the pressure gradient necessary. [R_e = 210, 4.3 $kN/m^2/m$]

2.4. Oil of specific gravity 0.9 and viscosity 1 poise pumped through a 50 mm diameter pipe at the rate of 280 liters/min. Show that the flow is stream line and hence estimate the power required to pump the oil up through a pipe line 75 m long which rises 7.5 m over this length. [R_e = 1071, P = 1.404 kW]

2.5. What power will be required per one kilometer length of pipe line to overcome the viscous resistance to the flow of crude oil (S.G. 0.85) of viscosity 1.9 poises through a horizontal 100 mm diameter pipe at the rate of 650 liters/hr. [2.5 Watt]

2.6. Oil of viscosity 0.5 poise flows through a 20 mm diameters pipe. The velocity along the centre line of the pipe is 0.4 m/s.

(a) The pressure drop per unit length of pipe

(b) The velocity at a distance d/4 from the wall of the pipe.

(c) The discharge

(d) The shear stress in the oil at the pipe wall.

[7.85 kPa, 0.3 m/s, 62.8 cm^3/s.3.92 kPa]

2.7. Oil of dynamic viscosity 0.07 kg/ms flows through a 20 mm diameter pipe with a mean velocity of 0.5 m/s. Calculate the pressure drop which occurs over a length of 50 m of pipe. [1.4 bar]

2.8. The velocity along the centre line of the Hagen-Poiseuille flow in a 100 mm diameter pipe is 2 m/s. If the viscosity of the fluid is 0.07 kg/ms and its relative density is 0.92, Calculate (a) the volumetric flow in m^3/s, (b) the shearing stress of the fluid at the pipe wall, (c) the local frictional coefficient. [7.85×10^{-3} m^3/s, 2.2 Pa, 0.0487]

2.9. Calculate the diameter of a vertical pipe needed for flow of a liquid at a Reynold's number of 1200 when the pressure remains constant throughout the pipe. Take kinematic viscosity of the fluid as 1.92×10^{-3} m^2/s. [D = 243 mm]

2.10. A lubricating oil of relative density 0.92 and dynamic viscosity 0.085 Ns/m^2 is to be pumped at a rate of 2400 liters/min with energy gradient in laminar flow not exceeding 0.04. Determine the least diameter of the pipe to satisfy the above requirements. [275 mm]

2.11. An oil of coefficient of viscosity 1.5 poise and specific gravity 0.9 flows through a 30 mm vertical pipe. Two pressure gauges are fixed 20 m apart. The gauge A fixed at the top records 200 kPa and the gauge B fixed at the bottom records 500 kPa. Find the direction of flow and the rate of flow. [Upwards, R_e = 139, Q = 0.5468 liters/s]

2.12. An oil of density 917 kg/m^3 is being pumped in a 150 mm diameter pipe. The discharge is measured as 14.2 liters/s. The drop in pressure in a stretch of 800 mm of pipe line, both ends of which are at th same elevation, is measured as 95 kPa. Estimate the absolute viscosity of the oil. [R_e = 1059, μ = 0.1041 kg/ms]

2.13. In a steady laminar flow between parallel plates, h = 2 mm, $\mu = 5 \times 10^{-2}$ Ns/m^2 and $\frac{\partial p}{\partial x} = -10$ kPa. Determine q, maximum shear stress (τ_o) and maximum velocity (u_{max}). [q = 0.1333 liter/s m, τ_o = 10 Pa, u_{max} =0.1 m/s]

2.14. Kerosene oil flows upwards through inclined parallel plates at the rate of 2 liters/s per meter width. The distance between the parallel plates is 10 mm and the inclination of plates is 20° with the horizontal. Determine the difference of pressure between two sections 10 m apart. Take density and viscosity of kerosene oil as 800 kg/m^3 and 2×10^{-3} Ns/m^2

2.15. A flow of 60 liters/s per meter width of glycerine of relative density 1.25 and dynamic viscosity 1.5 kg/ms takes place between two parallel plates having a gap of 25 mm between them. Calculate the (a) maximum velocity (b) boundary shear stress and (c) energy gradient [3.6 m/s, 864 Pa, h_f/l = 5.648]

2.16. Water flows between two very large parallel flat plates 20 mm apart. If the average velocity of the water is 0.15 m/s, what is the shearing stress at 5 mm and 10 mm from the lower plate. [$\tau_{5\,mm}$ = 0.0225 Pa, $\tau_{10\,mm}$ = 0]

2.17. Oil flows between parallel plates, one of which is at rest and other moves with a velocity U. If the pressure is decreasing in the direction of flow at a rate of 5 N/m^3, dynamic viscosity is 0.05 Ns/m^2. The space between two plates is 40 mm and volumetric flow per unit width is 0.02 m^2/s. What is the velocity. [u_{av} = 0.5 m/s, U = 1.16 m/s].

3

Turbulent flow through Pipes

3.1 INTRODUCTION

In the chapter eleven i.e. in the last chapter various properties of fully developed laminar pipe flow were discussed. Since turbulent pipe flow is actually more likely to occur than laminar flow in practical situations, it is necessary to obtain similar information for turbulent pipe flow. However, turbulent flow is a very complex process.

The turbulent flow is characterized by rapid and continuous mixing of the fluid in a rather chaotic manner throughout the pipe. In the turbulent flow the Reynold's number is large (more than 4000), the inertia forces, which dominate in the flow, tend to diffuse the fluid particles, hence causing the mixing of the fluid.

The discussion so far has hinged upon only the physical phenomenon of the disorderlines and random irregularities of turbulent flow; yet we do not have a clear cut definition of turbulence.

In his book Turbulence, Hinze suggests, "Turbulent fluid motion is an irregular" condition of flow in which the various quantities show a random variation with time and space coordinates, so that statistically distinct average values can be discerned.

According to the definition suggested by Taylor and von Karman, turbulence can be generated by fluid flow past solid surfaces or by the flow of layers of fluids at different velocities past or over one another. The definition above indicates that there are two distinct types of turbulence.

(1) Turbulence generated by the various effect due to the pressure of a solid wall is designated by wall turbulence;

(2) Turbulence, in the absence of a wall, generated by the flow of layers of fluids at different velocities is called free turbulence.

Turbulent flow through pipes and past bodies are examples of wall turbulence, and turbulent jet mixing regions and wakes fall into the category of free turbulence.

Numerous persons have devoted considerable effort in attempting to understand the variety of baffling aspects of turbulence. Although a considerable amount of knowledge about the topic has been developed, the field of turbulent flow still remains the least understood area of fluid mechanics.

Fortunately, engineers can evade such troublesome details by time averaging values of the flow. Time averaging produces statistical correlations that involve fluctuating velocities, pressure and temperatures. These fluctuating properties are the unknown variable that appear in the governing equation of turbulent motion. Since no general solution of these fluctuating properties are known to exist, we are forced to model them in terms of properties we do know.

In this chapter, we shall be investigating internal turbulent flows or elementary topics of wall turbulence. The simplest example is a fully developed turbulent flow in a circular pipe. This is a flow that is independent, in the time mean, of the stream wise position along the pipe, that is independent of either an inlet or outlet state and that has conditions statistically identical at each axial position.

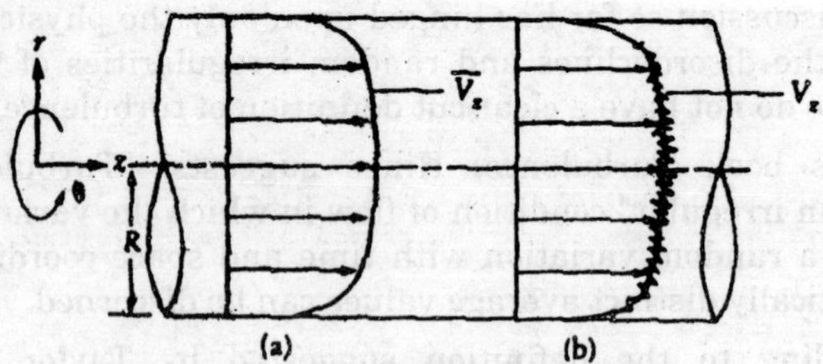

Fig. 3.1. Turbulent flow in a tube.

Fig. 3.1 shows turbulent flow in a tube. If the velocity at each point in the cross section is measured with the help of pitot static tube, the velocity profile in Fig. 3.1(a) results. If the velocity is instead measured with a more sensitive device, such as hot wise anemometer, the profile in Fig. 3.1(b) is obtained. Transition from laminar shown in Fig. 3.2, Fig. 3.3 shows shear stress and velocity distribution in a pipe.

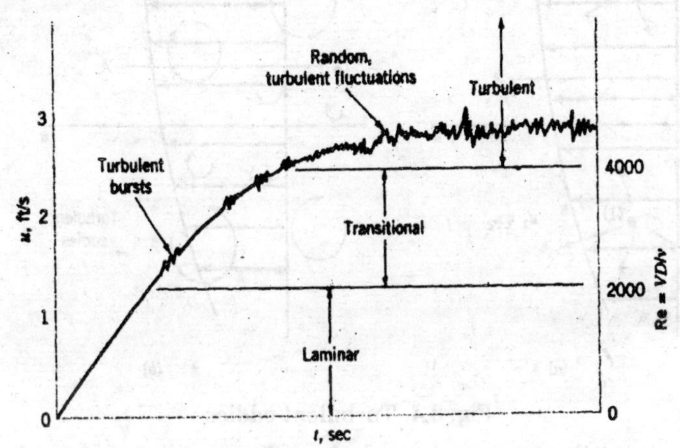

Fig. 3.2. Transition from Laminar flow.

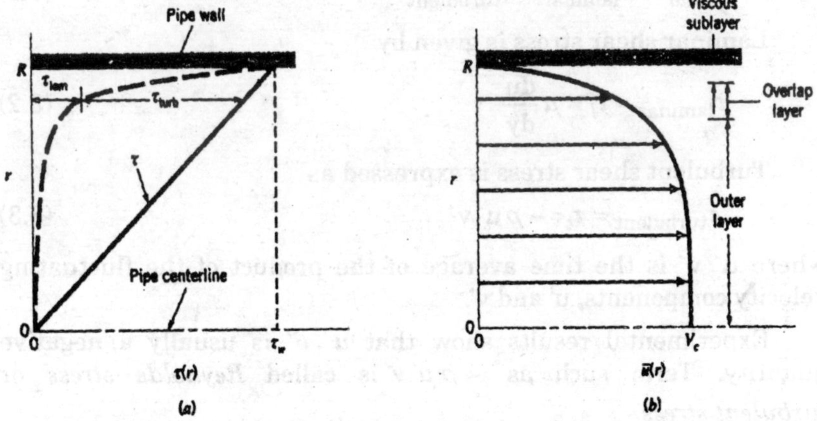

Fig. 3.3. Velocity and shear stress distribution in a turbulent flow near a wall.

3.2 TURBULENT SHEAR STRESS

Turbulent flow is characterized by random and rapid fluctuations of swirling regions of fluid, called eddies, throughout the flow. These fluctuations provide an additional mechanism for momentum and energy transfer (Fig. 3.4).

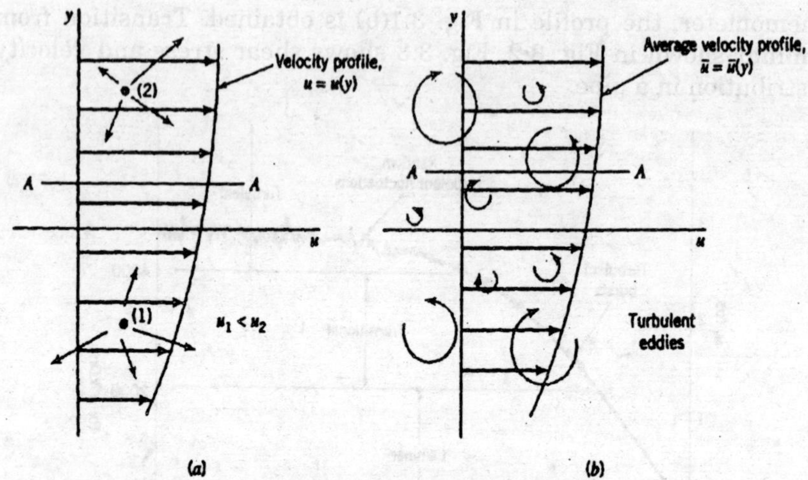

Fig. 3.4. Turbulent eddies

The total shear stress in turbulent flow can be expressed as

$$\tau_{total} = \tau_{laminar} + \tau_{turbulent} \quad (3.1)$$

Laminar shear stress is given by

$$\tau_{laminar} = \tau_l = \mu \frac{du}{dy} \quad (3.2)$$

Turbulent shear stress is expressed as

$$\tau_{turbulent} = \tau_t = -\rho u' v' \quad (3.3)$$

where $u' v'$ is the time average of the product of the fluctuating velocity components, u' and v'.

Experimental results show that $u' v'$ is usually a negative quantity. Term such as $-\rho \mu' v'$ is called *Reynolds stress* or *turbulent stress.*

3.3 BOUSSINES EDDY VISCOSITY

In analogy with the coefficient of viscosity, Boussines eddy viscosity defined the mixing coefficient, μ_t for the Reynolds stress in turbulent flow by putting

$$\tau_{turbulent} = -\rho u' v' = \mu_t \frac{d\bar{u}}{dy} \quad (3.4)$$

where, μ_t is the eddy viscosity or turbulent viscosity or apparent or virtual viscosity, which accounts for momentum transport by turbulent eddies. Then the total shear stress can be expressed as

$$\tau_{total} = (\mu + \mu_t)\frac{d\bar{u}}{dy} \quad (3.5(a))$$

$$= \rho(v + v_t)\frac{d\bar{u}}{dy} \quad (3.5(b))$$

where $v_t = \frac{\mu_t}{\rho}$ is the **Kinematic eddy viscosity** or kinematic turbulent viscosity.

3.4. PRANDTL'S MIXING LENGTH THEORY

Prandtls' mixing length theory is based on the concept of momentum exchange in fluids. Let us imagine that two adjacent layers of fluid in a turbulent flow moving at different velocities. When fluid particles move from one layer to the other, there is an exchange of momentum between the two layers. This transverse exchange of momentum produces a shear force between the two layers (Fig. 3.5).

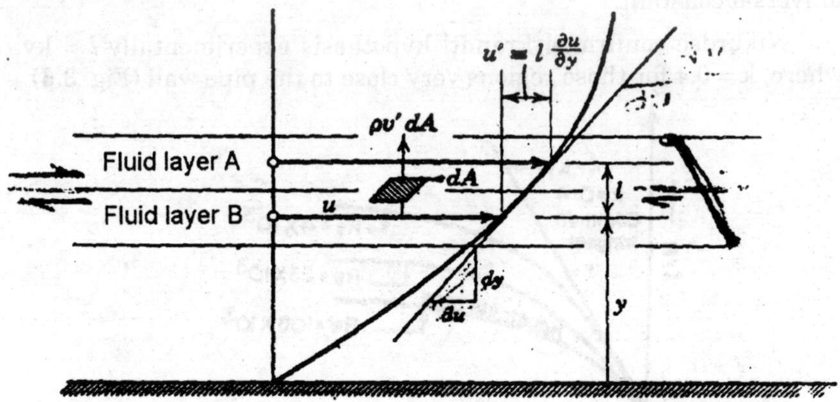

Fig. 3.5. Prandtl conceptional model of a two dimensional parallel turbulent flow.

Prandtl introduced the concept of mixing length. He postulated that the average distance a fluid particle had to travel before its velocity changed to that of the new region was related to the average velocity gradient, and so in a shear flow

$$u' = l\frac{d\bar{u}}{dy}$$

Prandtl assumed v' to be of the some order of magnitude as u' that is u' = v'.

$$v' = l\ \frac{d\bar{u}}{dy}$$

Substituting the values u' and v' in equation (3.3)

$$\rho u' v' = \rho l^2 \left(\frac{d\bar{u}}{dy}\right)^2 \quad (3.6)$$

The equation (3.6) is known as **Prandtl's mixing length** hypothesis.

The l can be calculated from the experimental data. For pipes, l to be a linear function of the distance y from the pipe wall.

$$l' \alpha\, y$$

or $\quad l = ky \quad (3.7)$

where K is the constant of proportionality and is called the universal constant.

Nikurdse confirmed Prandtl hypothesis experimentally l = ky, where, k = 0.4 for those regions very close to the pipe wall (Fig. 3.6).

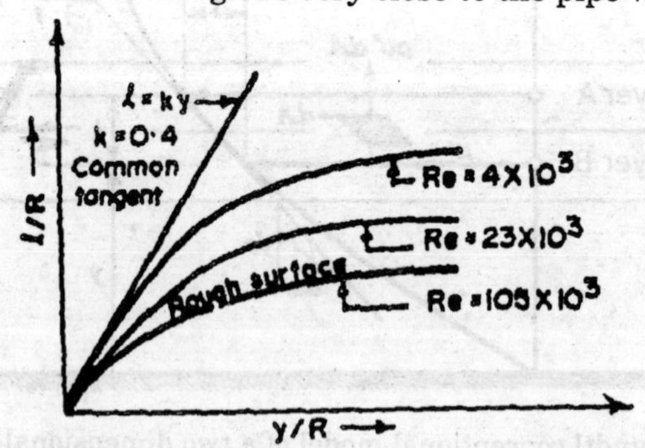

Fig. 3.6. Curve showing l/R versus v/R for turbulent pipe flow (not to scale)

3.5 SHEAR VELOCITY OR FRICTION VELOCITY

At any radial position, the equilibrium condition for pipe flow (Hagen-Poiseuille flow) can be expressed by

Pressure force = Shear force

$$p\,\pi r^2 - \left(p + \frac{\partial p}{\partial x}dx\right)\pi r^2 = \tau\, 2\pi r\, dx$$

Solving we get,

$$\tau = \left(-\frac{\partial p}{\partial x}\right)\frac{r}{2} \tag{3.8}$$

$$\tau = \left(\frac{p_1 - p_2}{x_2 - x_1}\right)\frac{r}{2} = \left(\frac{p_1 - p_2}{L}\right)\frac{r}{2}$$

$$= \left(\frac{p_1 - p_2}{\rho g}\right)\left(\frac{r}{2L}\right)(\rho g) = h_f\left(\frac{r}{2L}\right)(\rho g)$$

when $r = R$, $\tau = \tau o$

$$\tau_o = h_f\left(\frac{\rho g R}{2L}\right)$$

or, $$\tau_o = h_f\left(\frac{\rho g D}{4L}\right)$$

or, $$h_f = \frac{4L\tau_o}{\rho g D} \tag{3.9}$$

We know that, head lost due to friction can be written as

$$h_f = \frac{\lambda L u_{av}^2}{2gD} \tag{3.10}$$

where $\lambda = 4f$

Equating (3.9) and (3.10)

$$\frac{4L\tau_o}{\rho g D} = \frac{\lambda L u_{av}^2}{2gD}$$

or, $$\tau_o = \frac{\lambda \rho u_{av}^2}{8}$$

or, $$\frac{\tau_o}{\rho} = \left(\frac{\lambda}{8}\right)u_{av}^2$$

or, $$\sqrt{\frac{\tau_o}{\rho}} = \sqrt{\frac{\lambda}{8}}\,u_{av}$$

or, $$u_* = \sqrt{\frac{\lambda}{8}}\, u_{av} \tag{3.11}$$

$u_* = \sqrt{\frac{\tau_o}{\rho}}$ is termed as **shear velocity or friction velocity.**

It is to be noted that u_* is not an actual velocity of the fluid-it is merely a quantity that has dimensions of velocity. u_* is constant for any given case of turbulent pipe flow.

3.6 PRANDTL'S UNIVERSAL VELOCITY DISTRIBUTION EQUATION FOR TURBULENT PIPE FLOW

The following derivation is for a general case of turbulent pipe flow and therefore, apply equally well to turbulent flow in both smooth and rough pipes. The mechanism of turbulent motion in the central region of the pipe must necessarily be the same, regardless of the surface conditions at the pipe wall.

According to Prandtl's mixing length theory, the turbulent shearing stress between any two layers of fluid is given by

$$\tau_t = \rho \left(l \frac{du}{dy} \right)^2$$

Prandtl further assumed that near the pipe wall.

(a) $\tau_t = \tau_o$, that is, the turbulent shearing stress remaining approximately constant.

(b) The mixing length l has a linear relationship with y, the distance from the boundary (i.e. l = ky).

Thus

$$\tau_o = \rho k^2 y^2 \left(\frac{du}{dy} \right)^2$$

or, $$\left(\frac{du}{dy} \right)^2 = \frac{\tau_o}{\rho K^2 y^2}$$

or, $$\frac{du}{dy} = \frac{\sqrt{\tau_o}}{\rho} \times \frac{1}{k\,y}$$

or, $$\frac{du}{dy} = \frac{u_*}{k\,y}$$

or, $$du = \frac{u_*}{k\,y}\,dy$$

By Integrating,

$$\int du = \int \frac{u_*}{k\,y}\,dy$$

or, $$u = \frac{u_*}{k}\,ln\,y + C \qquad (3.12)$$

The constant of integration C is determined from the boundary conditions at the centre of the pipe.

At the pipe axis (i.e. at centre of the pipe)

$$u = u_{max} \text{ and } y = R, \text{ so that}$$

$$u_{max} = \frac{u_*}{k}\,ln\;R + C$$

$$C = u_{max} - \frac{u_*}{k}\,ln\;R$$

By substituting the expression for C into equation (3.12) and noting that K = 0.4 (suggested by Nikuradse)

$$u = \frac{u_*}{k}\,ln\;y + v_{max} - \frac{u_*}{k}\,ln\,R$$

or, $$u = \frac{u_*}{k}\,ln\;\frac{y}{R} + u_{max}$$

or, $$u = \frac{u_*}{0.4}\,ln\,\frac{y}{R} + u_{max}$$

or, $$u = 2.5\,u_*\;ln\;\frac{y}{R} + u_{max} \qquad (3.13)$$

The equation (3.13) is known as **Prandtl's universal velocity distribution equation for turbulent flow in both smooth and rough pipes.**

The equation (3.13) can also be expressed in a non dimensional form.

$$\frac{u_{max} - u}{u_*} = 2.5\, ln\, \frac{R}{y} \tag{3.14}$$

and $$\frac{u_{max} - u}{u_*} = 5.75\, \log_{10} \frac{R}{y} \tag{3.15}$$

The term (u_{max} – u) is known as `velocity defect'. As depicted in Fig. 3.7 the equation governs the relationship between u_{max} and u. Unless either the maximum or wall velocity is known it does not of itself provide sufficient information to enable the actual velocities across the section to be determined. The logarthmic form of the equation does, however, indicate that the velocity gradient is much steeper adjacent to the walls than in the central portion of the pipe.

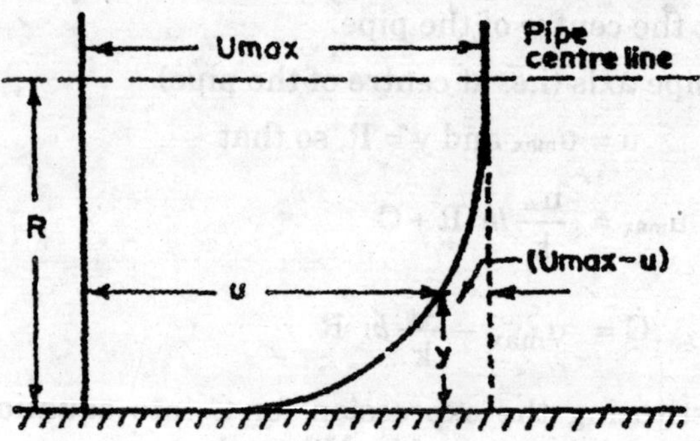

Fig. 3.7. Velocity deficiency distribution

3.7 HYDRODYNAMICALLY SMOOTH AND ROUGH BOUNDARIES

The flow conditions in the vicinity of the pipe wall depend on the relative magnitude of the laminar sublayer thickness δ' and the pipe wall roughness Ks as shown in Fig. 3.8. If the Reynolds number of flow is of such a numerical value that Ks is smaller than δ', the wall roughness are submerged within the laminar sublayer and, therefore, do not interfere with the formation of the laminar sublayer and buffer zone. The velocity distribution would be exactly the same as that in a smooth pipe. The pipe is in fact considered to be hydraulically smooth. When Ks is larger than δ', the wall roughness protrude beyond the laminar sub-layer and create additional turbulence in the flow. As a result, the relative roughness R/Ks may be regarded as a similarity parameter for turbulent flow in rough pipes because geometrically similar rough surfaces will result in dynamically similar turbulent flow patterns. Hence the resulting velocity distribution equation for turbulent flow in rough pipes is a function of the relative roughness R/Ks and the Reynolds number Re of flow.

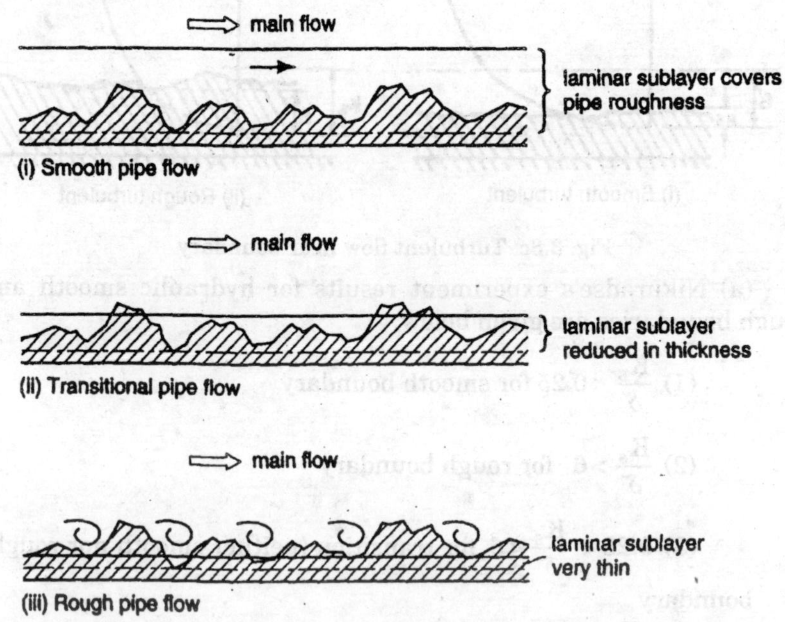

Fig. 3.8(a). Smooth and Rough boundaries with reference to Laminar sub layer thickness

Fig. 3.8(a) shows with reference to laminar sublayer, Fig. 3.8(b) shows the structure of the turbulent boundary layer and 12.8(c) shows the difference between rough and smooth boundary.

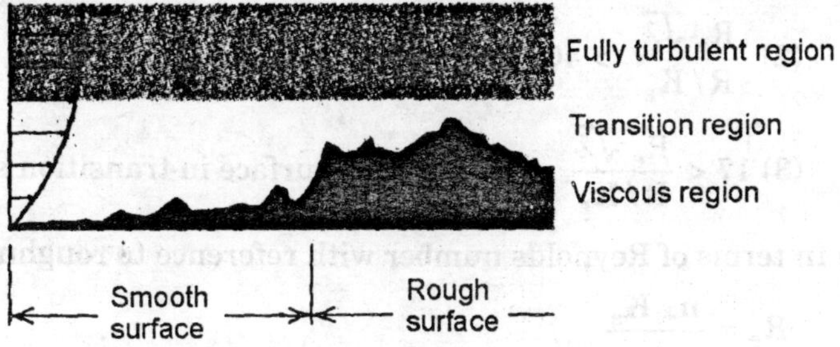

Fig. 3.8(b). Structure of turbulent boundary layer.

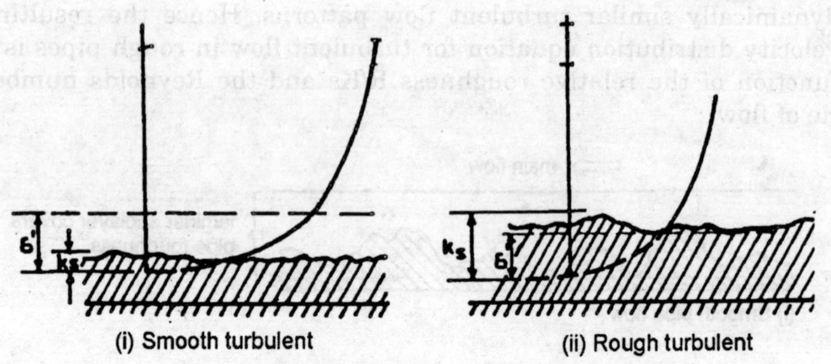

Fig. 3.8c. Turbulent flow near boundary

(a) Nikuradse's experiment results for hydraulic smooth and rough boundaries are given below.

(1) $\frac{K_s}{\delta'} < 0.25$ for smooth boundary

(2) $\frac{K_s}{\delta'} > 6$ for rough boundary

(3) $0.25 < \frac{K_s}{\delta'} < 6$ for transition (neither smooth nor rough) boundary

(b) In terms of Reynolds roughness number.

Reynolds roughness number is written as $\frac{R_e\sqrt{\lambda}}{R/K_s}$

(1) $\frac{R_e\sqrt{\lambda}}{R/K_s} < 17$ for smooth pipe surface.

(2) $\frac{R_e\sqrt{\lambda}}{R/K_s} > 400$ for rough pipe surface.

(3) $17 < \frac{R_e\sqrt{\lambda}}{R/K_s} < 400$ the pipe surface in transition stage.

(c) In terms of Reynolds number with reference to roughness.

$$R_e = \frac{u_* K_s}{v}$$

(1) $\frac{u_* K_s}{v} < 4$ for smooth boundary

(2) $\frac{u_* K_s}{v} > 100$ for rough boundary

(3) $4 < \frac{u_* K_s}{v} < 100$ for transition boundary.

3.8 VELOCITY DISTRIBUTION FOR TURBULENT FLOW IN SMOOTH AND ROUGH PIPES-PRANDTL KARMAN VELOCITY DISTRIBUTION EQUATION

Prandtl's equation (equation 3.14) gives a general non-dimensional form for the logarthmic velocity distribution in the central region of the pipe where the turbulent flow is completely developed. This equation, however, is not applicable to the points near the pipe wall because the velocity distribution in the laminar sub layer is entirely different. At y = 0, equation (3.12) gives the value of u equal to – ∞ as (minus infinity). This is absurd and is due to the fact that equation that equation (3.12) is not applicable at y = 0, At a certain distance y′ from the wall, the velocity is zero (Fig. 3.9).

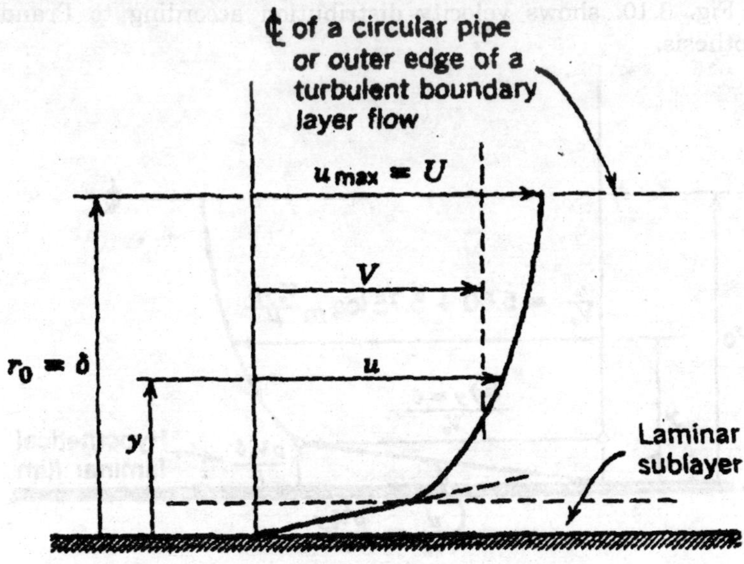

Fig. 3.9. Velocity profile near a pipe wall
(vertical scale greatly exaggerated)

The constant C can be obtained by putting y = y′ and u = 0 in the equation (3.12).

$$0 = \frac{u_*}{k} \ln y' + C$$

$$C = -\frac{u_*}{k} \ln y'$$

Substituting the value of C in equation (3.12)

$$u = \frac{u_*}{k} \ln y - \frac{u_*}{k} \ln y' = \frac{u_*}{k} \ln \frac{y}{y'}$$

Putting k = 0.4

$$u = 2.5\, u_* \ln \frac{y}{y'}$$

or, $$u = 2.5 \times 2.3\, u_* \log_{10} \frac{y}{y'}$$

$$= 5.75\, u_* \log_{10}\left(\frac{y}{y'}\right) \qquad (3.16)$$

Fig. 3.10. shows velocity distribution according to Prandtl's hypothesis.

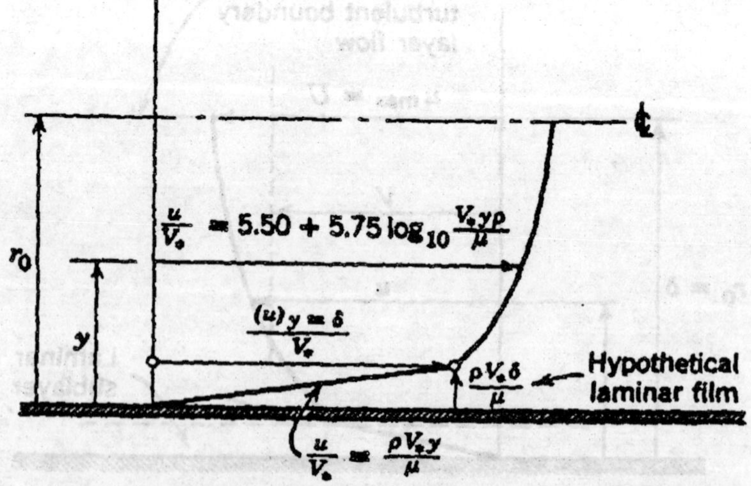

Fig. 3.10. Velocity distribution according to Prandtl's hypothesis.

3.8.1 Velocity distribution in smooth pipes

The logarthmic velocity distribution in not applicable to the region in the vicnity of the pipe wall. The logarthmic velocity distribution applies only to turbulent flow in the central region of the pipe, and it cannot be applied to regions close to the pipe boundary.

In the vicnity of pipe wall, there is a laminar sublayer in which the velocity distribution is parabolic. Therefore, it can be assumed that the flow is laminar with a constant velocity gradient. Fig. 3.11 and velocity distribution in a turbulent boundary layer is shown in Fig. 3.12.

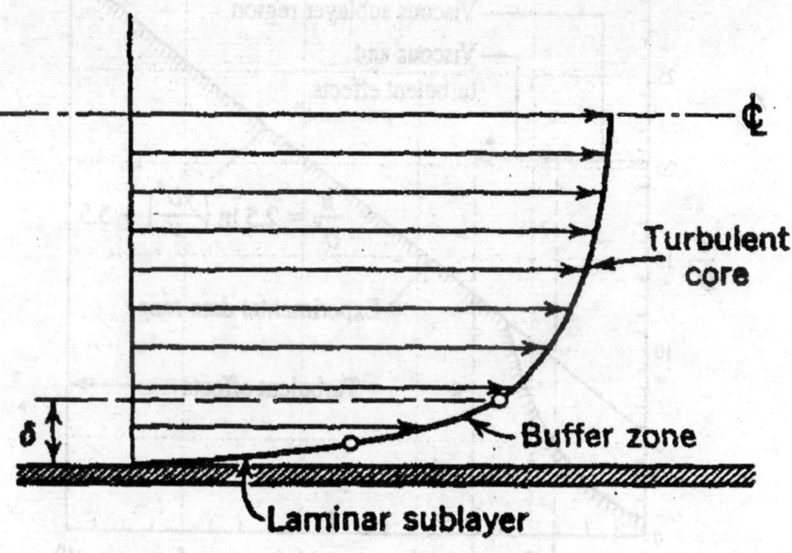

Fig. 3.11. Velocity distribution in a turbulent boundary layer.

That is,

$$\left(\frac{du}{dy}\right)_{\delta'-0} = \frac{y}{f'} = \frac{\tau_o}{\mu}$$

or, $$\frac{u}{y'} = \frac{\tau_o}{\rho} \times \frac{\rho}{\mu}$$

or, $$\frac{u}{y'} = \frac{u_*^2}{v}$$

or, $$\frac{u}{u_*} = \frac{u_* \, y'}{v} \tag{3.17}$$

or, expressed in a more general non dimensional form,

$$\frac{u}{u_*} = \frac{u_* \, \delta'}{\nu} = \text{Constant} \tag{3.18}$$

The terms $\frac{u_* \, y'}{\nu}$ and $\frac{u_* \, \delta'}{\nu}$ on the right hand side of equations (3.17) and (3.18) are dimensionally equivalent to a form of Reynold's number.

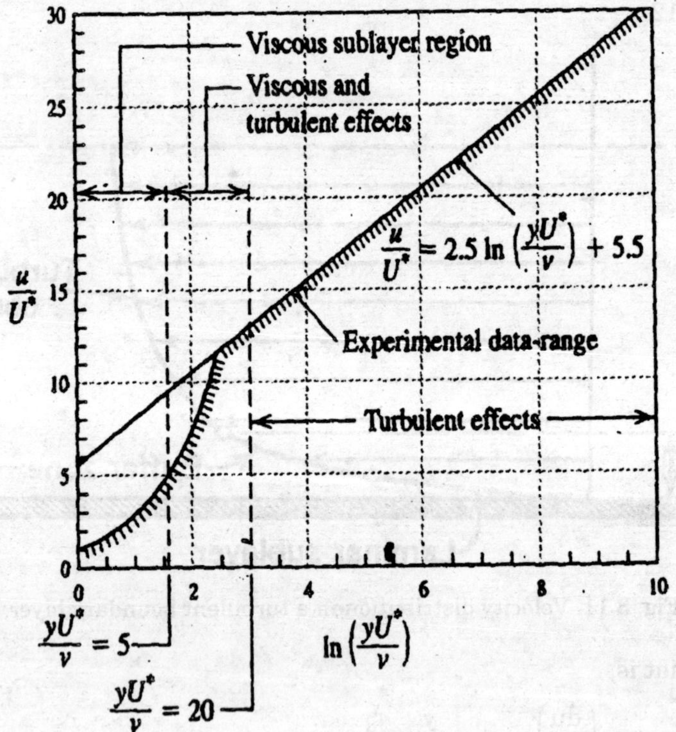

Fig. 3.12. Graph of experimental data between verses $ln\left(\frac{\rho u}{\nu}\right)\frac{u}{u_*}$

Since the velocity distribution is assumed to be linear within the hypothetical laminar film of thickness δ', the non dimensional equations (3.17) and (3.18) are significant in that

(1) the $\frac{u}{u_*}$ is a linear function of same modified Reynold's number $u_* y' / \nu$;

(2) the modified Reynold's number $\frac{u_* \delta'}{\nu}$ and has a constant numerical value of approximately 11.6 for all turbulent flow in smooth pipes of any size.

Nikuradse's experimental studies of turbulent flow in a smooth pipes have shown that for turbulent flow in smooth pipes of any size of the value of the parameter $\left(\frac{u_* y}{\nu}\right)$ is approximately 11.6 and for $y = y'$ it is approximately 0.108. That means

$$\frac{u_* \delta'}{\nu} = 11.6 \tag{3.19}$$

and
$$\frac{u_* y'}{\nu} = 0.108 \tag{3.20}$$

Solving (3.19) and (3.20), we get

$$y' = \frac{\delta'}{107} \tag{3.21}$$

Substituting the value of y' in equation (3.16), we get

$$\frac{u}{u_*} = 5.75 \quad \log_{10} \frac{y}{\frac{0.108\,\nu}{u_*}}$$

$$\frac{u}{u_*} = 5.75 \log_{10} \left(\frac{u_* y}{0.108\,\nu}\right)$$

$$= 5.75 \log_{10} \frac{u_* y}{\nu} + 5.75 \log_{10} \left(\frac{1}{0.108}\right)$$

$$= 5.75 \log_{10} \frac{u_* y}{\nu} + 5.75 \log_{10} (9.259)$$

$$= 5.75 \log_{10} \frac{u_* y}{\nu} + 5.5$$

$$\frac{u}{u_*} = 5.5 + 5.75 \log_{10} \frac{u_* y}{\nu} \tag{3.22}$$

The equation (3.22) is known as the **Karman-Prandtl equation** for velocity distribution in turbulent flow near hydrodynamically **smooth boundaries.**

3.8.2 Velocity distribution in rough pipes

In the case of hydrodynamically rough boundaries, the height Ks, of the roughness elements is large in comparison with the laminar sublayer (Fig. 3.8c), create additional turbulence in the flow. In this case laminar sub layer is destroyed and hence one would except y' to be proportional to the average height of protrusions Ks.

Nikuradses experimental data on velocity distribution measurements for turbulent flow in artifically roughned pipes with Ks >> δ, gave

$$y' = \frac{K_s}{30} \tag{3.23}$$

Substituting the value of y′ in equation (3.20), we get.

$$\frac{u}{u_*} = 5.75 \log_{10} \frac{y}{K_s/30} = 5.75 \log_{10}\left(\frac{y}{K_s}\right) + 5.75 \log_{10} 30$$

$$= 5.75 \log_{10} y/K_s + 8.5 \tag{3.24}$$

This is the **Karman-Prandtl equation** for velocity distribution in turbulent flow near **rough boundaries**.

3.9 VELOCITY DISTRIBUTION FOR TURBULENT FLOW IN TERMS OF AVERAGE VELOCITY

In the last article (3.8), velocity distribution equation for turbulent flow in smooth (equation 3.22) and rough (equation 3.24) pipes were derived separately. However, these two equations can be transformed to identical form if both are expressed in terms of the average velocity of flow in pipe.

By definition, the average velocity u_{av}, is obtained by

$$u_{av} = \frac{Q}{A} = \frac{\text{Total discharge}}{\text{Area of the pipe}}$$

$$= \frac{Q}{\pi R^2} = \frac{\int_0^R u(2\pi r)\,dr}{\pi R^2} \qquad (3.25)$$

Consider an elementary ring of thickness dr at a radial distance *r* from the centre of the pipe at a distance *y* from the pipe boundary as shown in Fig. 3.13.

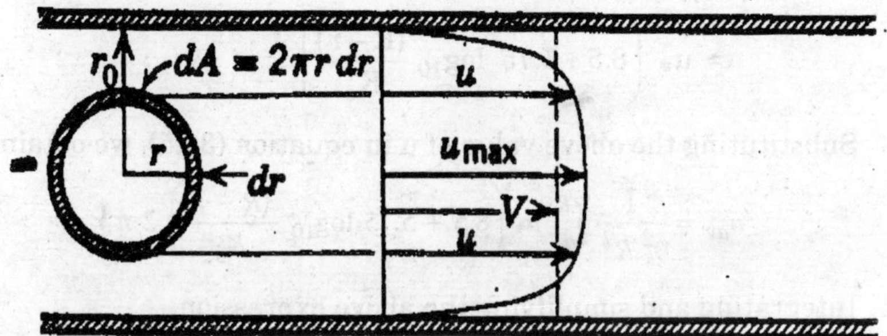

Fig. 3.13. Circular pipe with elemental ring.

Thus $\quad y = R - r$

Where R is the radius of the pipe.

3.9.1 Turbulent flow in smooth pipes

In a turbulent flow in smooth pipes, equation (3.22) gives the velocity u as

$$u = u_* \left(5.5 + 7.5 \log_{10} \frac{u_* y}{\nu}\right)$$

$$u = u_* \left(5.5 + 7.5 \log_{10} \frac{u_* (R - r)}{\nu}\right)$$

Substituting the above value of u in equation (3.25), we obtain

$$u_{av} = \frac{1}{\pi R^2} \int_0^R u_* \left(5.5 + 5.75 \log_{10} \frac{u_* (R - r)}{\upsilon}\right) 2\pi r\,dr$$

Integrating and simplifying the above expression

$$\frac{u_{av}}{u_*} = 1.75 + 5.75 \log_{10} \frac{u_* R}{\nu} \qquad (3.26)$$

3.9.2 Turbulent flow in rough pipes

In a turbulent flow in rough pipes, equation (3.24) gives the velocity u as

$$u = u_* \left(8.5 + 5.75 \log_{10} \frac{y}{K_s}\right)$$

$$= u_* \left(8.5 + 5.75 \log_{10} \frac{(R-r)}{K_s}\right)$$

Substituting the above value of u in equation (3.25), we obtain

$$u_{av} = \frac{1}{\pi R^2} \int_o^R u_* \left(8.5 + 5.75 \log_{10} \frac{(R-r)}{K_s}\right) 2\,\pi\, r\, dr$$

Integrating and simplifying the above expression

$$\frac{u_{av}}{u_*} = 4.75 + 5.75 \log_{10} \frac{R}{K_s} \qquad (3.27)$$

3.9.3 Difference between point velocity and average velocity for smooth and rough pipes

(a) Smooth pipes

The point velocity is given by the equation (3.22)

$$\frac{u}{u_*} = 5.5 + 5.75 \log_{10} \frac{u_* y}{\nu}$$

or,
$$\frac{u}{u_*} = 5.5 + 5.75 \log_{10} \frac{u_* (R-r)}{\upsilon}$$

and the average velocity is given by the equation (3.26)

$$\frac{u_{av}}{u_*} = 1.75 + 5.75 \log_{10} \frac{u_* R}{\upsilon}$$

Difference of the above two velocities is given by

$$\frac{u}{u_*} - \frac{u_{av}}{u_*} = 5.5 + 5.75 \log_{10} \frac{u_* (R-r)}{\upsilon} - \left[1.75 + 5.75 \log_{10} \frac{u_* R}{\upsilon}\right]$$

$$\frac{u - u_{av}}{u_*} = (5.5 - 1.75) + 5.75 \left[\log_{10} \frac{u_* (R-r)}{\upsilon} - \log_{10} \frac{u_* R}{\upsilon}\right]$$

$$= 3.75 + 5.75 \log_{10} \frac{u_* (R - r)}{\upsilon} \times \frac{\upsilon}{u_* R}$$

$$= 3.75 + 5.75 \log_{10} \left(\frac{R - r}{R} \right) = 3.75 + 5.75 \log_{10} \frac{y}{R} \qquad (3.28)$$

(b) Rough pipes

The point velocity is given by the equation

$$\frac{u}{u_*} = 8.5 + 5.75 \log_{10} \left(\frac{y}{K_s} \right)$$

and the average velocity is given by the equation

$$\frac{u_{av}}{u_*} = 4.75 + 5.75 \log_{10} \frac{R}{K_s}$$

Difference of the above two velocities is given by

$$\frac{u - u_{av}}{u_*} = (8.5 - 4.75) + 5.75 \left[\log_{10} \frac{y}{K_s} - \log_{10} \frac{R}{K_s} \right]$$

$$= 3.75 + 5.75 \log_{10} \frac{y}{K_s} \times \frac{k_s}{R} = 3.75 + 5.75 \log_{10} \frac{y}{R} \qquad (3.29)$$

Equation (3.28) and (3.29) are identical and independent of kinematic viscosity υ and roughness factor K_s.

Thus the velocity distribution for the turbulent core is seen to be identical for both smooth and rough pipes because the mechanism of turbulent motion of fluid in the region outside the boundary layer is essentially independent of the conditions at the boundary surface. Therefore, when equation (3.28) and (3.29) are used to plot a velocity distribution curve in rough pipes, the pipe wall roughness k need not be known. Thus the two equations (3.28) and (3.29) have a definite advantage over the special equation (3.24), which are for rough pipe only.

3.10 TURBULENT PIPE COEFFICIENT

The ratio of average velocity to maximum velocity (u_{av} / u_{max}) which occurs at the centre of the pipe is sometimes called the pipe coefficient or pipe factor. For laminar flow in round pipes the pipe coefficient is always equal to 0.5. The expression for the turbulent pipe coefficient may be derived from equation (3.28) or (3.29)

$$\frac{u - u_{av}}{u_*} = 3.75 + 5.75 \log_{10} \frac{y}{R}$$

From equation (3.11)

$$u_* = u_{av} \sqrt{\frac{\lambda}{8}} \qquad \text{[where } \lambda = 4f\text{]}$$

Substituting in above equation

$$\frac{u - u_{av}}{u_{av} \sqrt{\frac{\lambda}{8}}} = 3.75 + 5.75 \log_{10} \frac{y}{R}$$

or, $$\frac{u}{u_{av}} - 1 = \sqrt{\frac{\lambda}{8}} \left[3.75 + 5.75 \log_{10} \frac{y}{R} \right]$$

$$\frac{u}{u_{av}} = 1 + \sqrt{\lambda} \left(1.33 + 2 \log_{10} \frac{y}{R} \right) \qquad (3.30)$$

Which indicates the velocity distribution for turbulent flow to be dependent on λ, regardless of whether the pipe wall is smooth or rough. However, better agreement with Nikuradse's experimental data occurs when the constant 1.33 and 2 are changed to 1.43 and 2.15, respectively, equation (3.30) then becomes

$$\frac{u}{u_{av}} = 1 + \sqrt{\lambda} \left(1.43 + 2.15 \log_{10} \frac{y}{R} \right) \qquad (3.31)$$

By Substituting the conditions at the centre of the pipe, that is $u = u_{max}$ at $y = R$, into equation (3.31), the turbulent pipe coefficient

$$\frac{u_{max}}{u_{av}} = 1 + 1.43 \sqrt{\lambda} \qquad (3.32)$$

or, $$\frac{u_{av}}{u_{max}} = \frac{1}{1 + 1.43 \sqrt{\lambda}} \qquad (3.33)$$

The pipe coefficient equation (3.33) is seen to be a function of coefficient of friction or friction factor ($\lambda = 4$ f) alone.

3.11 THE CHRONOLOGICAL DEVELOPMENT OF TURBULENT PIPE FLOW THEORIES

Table 3.1 lists the names of the main contributors, and their contributions, to turbulent pipe flow theories in chronological order.

Table 3.1 The chronological development of turbulent pipe flow theories

S.N.	Year	Name	Contribution.
1.	1913	Blasius	Friction factor equation for smooth pipes.
2.	1914	Stanton & Pannell	Experimental values of the friction factor for smooth pipes.
3.	1930	Nikuradse	Experimental values of the friction factor for artifically rough pipes.
4.	1930s	Prandtl & Von Karman	Equation for rough and smooth friction factors.
5.	1937-1939	Colebrook & White	Experimental values of the friction factor for commercial pipes & transition formula.
6.	1944	L.F. Moody	The Moody diagram for commercial pipes
7.	1958	Ackers	The hydraulic research station charts & tables for the design of pipes and charts.
8.	1975	Barr	Direct solution of the Colebrook-White equation.
9.	1976	J W Murdock	No iteration is needed for sufficient accurany D in most problems.
10.	1976	Swamee & Jain	Simple and more accurate explicit equation for friction factor for flow in commercial pipes.
11.	1983	S.E. Haaland	The best yet produced. It combines reas-onable simplicity with acceptable accu-racy (with in about 1.5%) and is useful if an algebraic expression is required.

The Colebrook–White transition formula represent the culmination of all the previous work, and can be applied to any fluid in any pipe operating under turbulent flow conditions. The later contributions of Moody, Ackers and Barr etc are mainly concerned with the practical application of Colebrook–White equation.

In the initial research, the original investigators persumed that the friction factor (λ or 4f) was constant. This was subsequently found to be incorrect. Equations relating λ or 4f to both the Reynold's number and the pipe roughness were developed later.

3.11.1 Smooth pipes and Blasius equation

At the beginning of the 20th century, Blasius came to the important conclusion that there were two pipes of pipe friction in turbulent flow. The first type be associated with smooth pipes where the viscosity effects predominate so that the friction factor is dependent solely on the Reynold's number [$\lambda = \Phi\left(R_e\right)$] the second type was relevant to rough pipes where the viscosity and roughness effects influence the flow and friction factor is dependent both on the Reynold's number and a parameter of relative roughness.

On the basis of his own researches and of the experimental data supplied by Saph and Schoder, Blasius deduced the following expression for the friction in smooth pipes.

$$\lambda = \frac{0.316}{\left(R_e\right)^{0.25}} \tag{3.34}$$

3.11.2 Stanton and Pannell

Stanton and Pannell working at the National Physical laboratory, investigated in considerable detail the head loss in smooth brass pipes conveying both air and water. On a dimensionless graph of λ (actually τ_o/u_{av}^2) against R_e the plotted points were found to be compactly situated along a single curve. The curve in the turbulent zone agreed quite well with the Blasius formula for values of R_e up to 10^5, but after this point there was increasing divergence, indicating that the power by which u should be raised was not constant but increased with R_e. Confirmation of this characteristic was provided by the experimental results of other investigators.

3.11.3 Nikuradse experimental results using artificially rough pipes

The next major experimental contributions were made by Nikuradse at Gottingen in the period (1933).

Nikuradse made a major contribution to the theory of pipe flow by objectively differentiating between smooth and rough turbulence in pipes. He carried out a painstaking series of experiments to determine both the friction factor and the velocity distributions at various Reynold's numbers up to 3×10^6. In these experiments, pipes were artificially roughened by sticking uniform sand grains on to smooth pipes. He defined the relative roughness (K_s/D) as the ratio of the sand grain size to the pipe diameter. By using pipes of different diameter and sand grains of different size, he produced a set of experimental results of λ and R_e for a six values of K_s/D, ranging from $K_s/D = \frac{1}{30}$ *to* $K_s / D = 1/1014$. The test data produced the series of smooth curves shown in Fig. 3.14. These curves were found to be defined not by D but by the ratio K_s/D.

He plotted his results as log λ against log R_e for each value of K_s/D, as shown in Fig. 3.14. This figure shows that these are five regions of flow as follows.

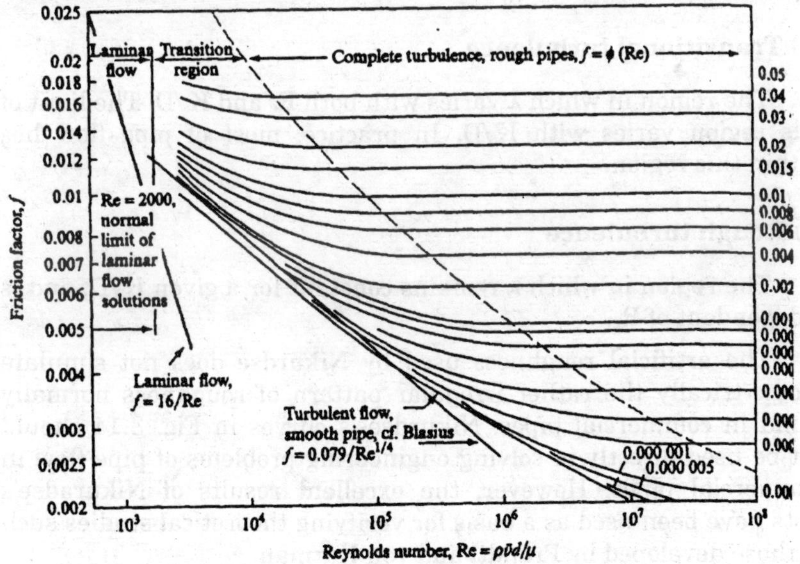

Fig. 3.14. Nikurdse experimental results.

(a) Laminar flow

The laminar portion is a straight line and is unaffected by the degree of roughness. In this region the relative roughness has no influence on the friction factor λ.

Hagen-Poiseuille equation equated with Darcy-Weisbach equation gives

$$\lambda = \frac{64}{R_e} \tag{3.35}$$

Darcy-Weisbach equation may also be used for laminar flow, provided that λ is evaluated by equation (3.35).

(b) Transition from Laminar to Turbulent flow

An unstable region between R_e = 2000 and 4000. Fortunately, pipe flow normally lies outside this region.

(c) Smooth turbulence

The limiting line of turbulent flow, approached by all values of relative roughness as R_e decreases.

(d) Transitional turbulence

The region in which λ varies with both R_e and K_s/D. The limit of this region varies with K_s/D. In practice, most of pipe flow lies within this region.

(e) Rough turbulence

The region in which λ remains constant for a given K_s/D, and is independent of R_e.

The artificial roughness used by Nikurdse does not simulate geome-trically the rather irregular pattern of roughness normally found in commercial pipes. Nikurdses's curves in Fig. 3.14 should not be used directly in solving engineering problems of pipe flow in commercial pipes. However, the excellent results of Nikuradse's tests have been used as a basis for verifying theoretical studies such as those developed by Prandtl and von Karman.

3.11.4 The smooth and rough laws of Prandtl and von Karman

The publication of Nikuradse's experimental results (particularly his velocity distribution measurements) were used by Prandtl and Von Karman to support their own work on turbulent boundary layers. By combining their theories of turbulent boundary layer flows with the experimental results, they derived the semi empirical smooth and rough laws.

(a) Universal resistance law for turbulent pipe flow

The fact that both the velocity distribution and the friction loss in turbulent pipe flow are related to the shearing stress τo at the pipe wall as shown in preceding articles in the equation (3.11)

$$u_* = \sqrt{\frac{\lambda}{8}}\, u_{av}$$

or, $$\lambda = \frac{8\, u_*^2}{u_{av}^2} \qquad (3.36)$$

Which may be used in the derivation of universal resistance laws for turbulent flow in smooth and rough pipes.

For turbulent flow in smooth pipes, the average velocity u_{av} of flow can be expressed in a non dimensional form shown in equation (3.26).

$$\frac{u_{av}}{u_*} = 1.75 + 5.75 \log_{10} \frac{u_*\, R}{\upsilon}$$

Combining equation (3.36) and (3.37) yields

$$\frac{u_{av}}{u_{av}\, \sqrt{\lambda/8}} = 1.75 + 5.75 \log_{10} \frac{u_{av}\, \sqrt{\lambda/8}\, R}{\upsilon} \qquad (3.37)$$

Which may be simplified and rearranged in the following form

$$\frac{1}{\sqrt{\lambda}} = -0.91 + 2.03 \log_{10} \frac{u_{av}\, D\sqrt{\lambda}}{\upsilon}$$

or, $$\frac{1}{\sqrt{\lambda}} = -\,0.91 + 2.03 \log_{10} \left(R_e\, \sqrt{\lambda}\right) \qquad (3.38)$$

By plotting Nikurdse's experimental data of turbulent flow in smooth pipes in Fig. 3.15 with $\frac{1}{\sqrt{\lambda}}$ *versus* $\log_{10}\left(R_e\, \sqrt{\lambda}\right)$ equation (3.38) is found to confirm with experimental results after the constants are changed slightly, the equation then becomes

$$\frac{1}{\sqrt{\lambda}} = -\,0.8 + 2 \log_{10} \left(R_e\, \sqrt{\lambda}\right) \qquad (3.39)$$

This is Prandtl's universal resistance equation for turbulent flow in smooth pipes. The universal equation (3.39) may be safely used to predict the value of λ for any arbitrarily large Reynold's number of flow beyond the range of available experimental data.

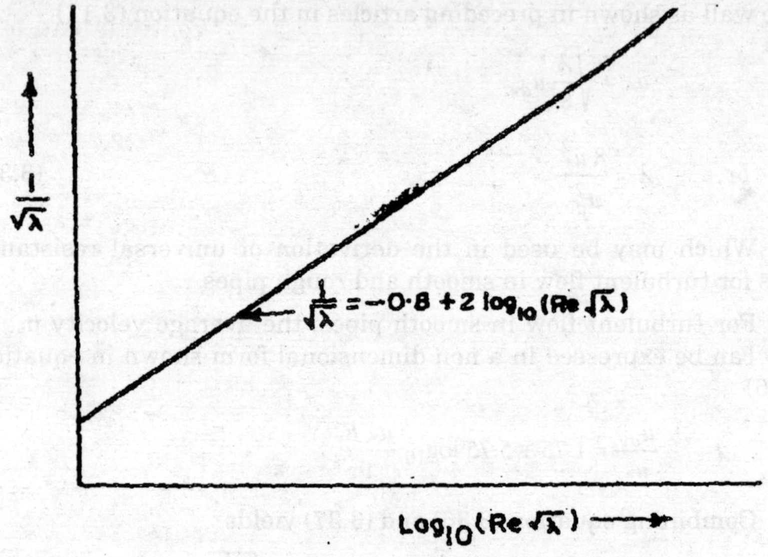

Fig. 3.15. Experimental verification of universal resistance law for turbulent flow in smooth pipe.

The equation (3.39) is simplified in a form given below.

$$\frac{1}{\sqrt{\lambda}} - \log \lambda = 2 \log_{10} R_e - 0.8$$

With the help of above equation λ can be calculated by trial and error method.

The value of λ can alternatively be obtained by Nikurdse

$$\lambda = 0.0032 + \frac{0.221}{(R_e)^{0.237}} \tag{3.40}$$

and from Dr. A.K. Jain's equation.

$$\frac{1}{\sqrt{\lambda}} = 1.8 \log_{10} R_e - 1.5186 \tag{3.41}$$

Alternatively the two equations 3.40 and 3.41 give approximately the same value of λ as obtained by the equation (3.39).

The equation (3.39) can also be written in a new form as

$$\frac{1}{\sqrt{\lambda}} = 2 \log_{10}\left(\frac{R_e \sqrt{\lambda}}{2.51}\right) \quad (3.42)$$

Since the wall shear stress is governed by the viscous force, that is $\tau = \mu$ (du/dy) at the wall, the thickness of the viscous sublayer must be an important parameter in relation to friction effects. The dimensionless ratio δ'/D from equation

$$\frac{\delta'}{D} = \frac{11.6\, \upsilon}{u_* D} = \frac{11.6\, \upsilon}{u\sqrt{\lambda/8}\, D} = \frac{32.8}{R_e \sqrt{\lambda}}$$

From the above equation the decrease of sub layer thickness with increasing Reynold's number is readily seen. Furthermore, writing the above equation in the form

$$R_e \sqrt{\lambda} = \frac{32.8}{(\delta'/D)}$$

and substituting it in the equation (3.39) and (3.42) we get

$$\frac{1}{\sqrt{\lambda}} = 2 \log_{10}\left(\frac{32.8}{\delta'/D}\right) - 0.8 \quad (3.43)$$

and $$\frac{1}{\sqrt{\lambda}} = 2 \log_{10}\left(\frac{32.8/(\delta'/D)}{2.51}\right) = 2 \log_{10}\left(\frac{13.067}{\delta'/D}\right) \quad (3.44)$$

For turbulent flow in completely rough pipes, a similar procedure may be followed.

$$\frac{u_{av}}{u_*} = 4.75 + 5.75 \log_{10}\frac{R}{K_s} \quad (3.45)$$

Substituting $u_* = \sqrt{\frac{\lambda}{8}}\, u_{av}$ in equation (3.45) we get

$$\frac{1}{\sqrt{\lambda}} = 1.68 + 2.03 \log_{10}\frac{R}{K_s} \quad (3.46)$$

Again a comparison with Nikurdse's experimental data on turbulent flow in completely rough pipes as plotted in Fig. 3.16 shows the slight modification of constant in equation (3.46) will improve the agreement with test results. The modified universal resistance equation for turbulent flow in completely rough pipes is

$$\frac{1}{\sqrt{\lambda}} = 1.74 + 2 \log_{10} \frac{R}{K_s} \tag{3.47}$$

The equation (3.47) can also be written in a new form as

$$\frac{1}{\sqrt{\lambda}} = 2 \log_{10} \frac{3.7\,D}{K_s} \tag{3.48}$$

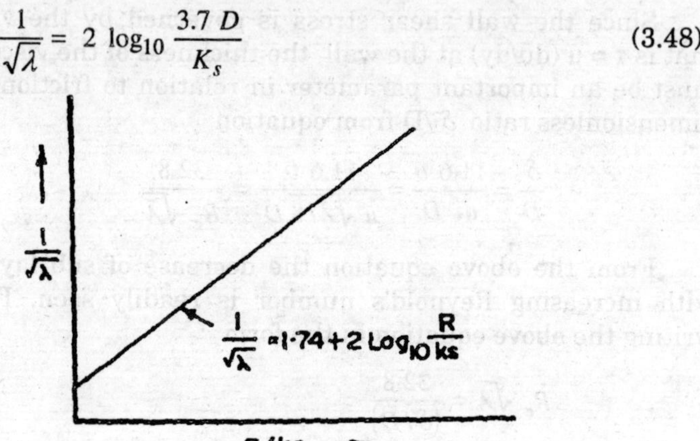

Fig. 3.16. Experimental verification of universal resistance law for turbulent flow in completely rough pipes.

3.10.5 The Colebrook-White transition formula

The experimental work of Nikurdse and the theoretical work of van Karman and Prandtl provided the frame work for a theory of pipe friction. However, these results were not to direct use to engineers because they applied only to artificially roughned pipes. Commercial pipes have roughness which is uneven both in size and spacing, and do not, therefore, necessarily correspond to the pipes used in Nikurdse's experiment.

Colebrook and White (1938 and 1939) therefore carried out tests at Imperial College with pipes that had been roughened with sand grains which were mixed in size. They discovered that in the turbulent transition region the $\lambda - R_e$ curves exhibited a gradual change from smooth to rough turbulent in constrast to Nikurdse's `S' shaped curves for uniform roughness, size and spacing. Colebrook then went on to determine the effective roughness size of many commercial pipes.

He was thus able to publish a list of K_s values applicable to commercial pipes.

Colebrook and White found that by judiciously combining the Prandtl and von Karman smooth and rough laws in a single expression which would represent the entire turbulent flow range of the λ - R_e diagram. The expression is

$$\frac{1}{\sqrt{\lambda}} = -2\log_{10}\left(\frac{2.51}{R_e\sqrt{\lambda}} + \frac{K_s}{3.7\,D}\right) \tag{3.49}$$

For $K_s \to 0$ this equation (3.49) transform into equation (3.48), valid for hydraulically smooth pipes.

For $R_e \to \infty$, it transforms into equation (3.48) for the completely rough regime.

The Colebrook and White equation with its asymptotes, are shown in Figs. 3.17 & 3.18

The coordinates of Fig. 3.17 are chosen to be $1/\sqrt{\lambda} - 2\log_{10}\dfrac{3.7\,D}{K_s}$ and Reynold's roughness number $R_e = \dfrac{u_* K_s}{\upsilon} = R_e\sqrt{\dfrac{\lambda}{8}}\dfrac{K_s}{D}$, so that the smooth pipe equation (3.42) appears as a straight line with a slope of 2, and the rough pipe equation (3.48) plots as a horizontal line.

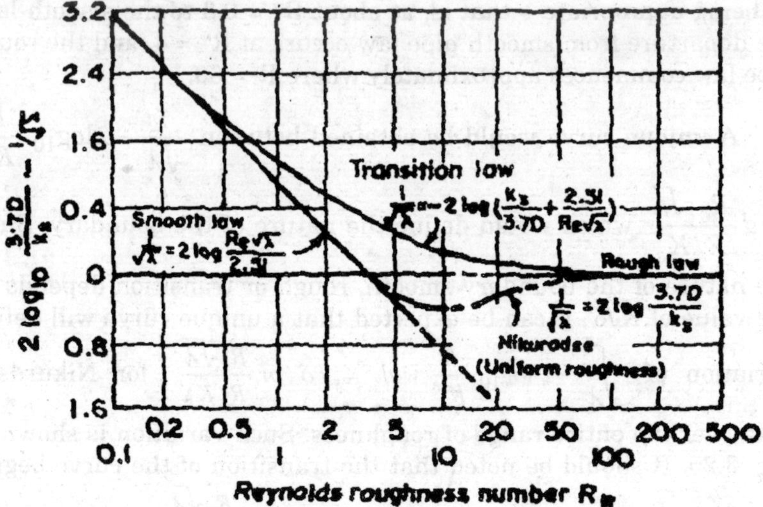

Fig. 3.17. Comparison of transitition zone curves for commercial and uniform roughness.

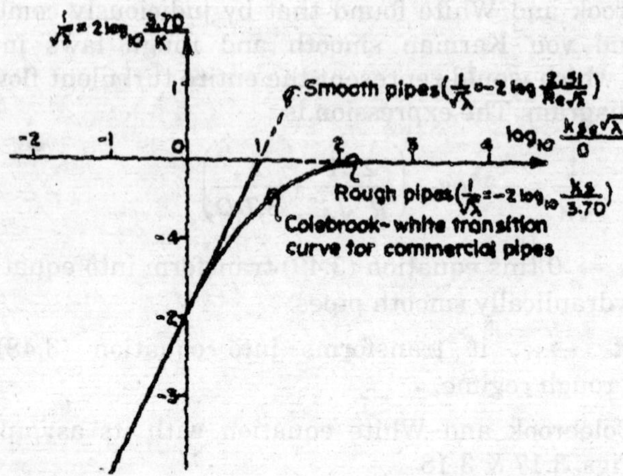

Fig. 3.18. Colebrook-White curve for commercial pipes.

The transition curve, shown in Figs. 3.17 and 3.18, was found to be in reasonable agreement with the experimental data. An essential feature of the expression is that at low values of R_e, the first term in the brackets predominate and the smooth law is approached; whilst at high values of R_e the second term predominates and the rough law is approached. The curve is asymtotic at approximately those point which the experimental results indicate as being appropriate – that is, at about R* = 0.3 to the smooth law; the departure from smooth pipe law occurs at R* = 4, and the rough pipe law commences approximately where R* = 60.

A unique curve would be obtained between $\frac{1}{\sqrt{\lambda}} - 2\log_{10}\frac{R}{K_2}$ *and* $\frac{R_e\sqrt{\lambda}}{R/K_s}$ which would define the nature of the boundary. Since the nature of the boundary smooth, rough or transition depends on the value of K_s/δ', it can be expected that a unique curve will define variation of $\frac{1}{\sqrt{\lambda}} - 2\log_{10}\frac{R}{K_s}$ *with* K_s/δ', *or* $\frac{R_e\sqrt{\lambda}}{R/K_s}$ for Nikurdse's data over the entire range of roughness. Such variation is shown in Fig. 3.25. It should be noted that the transition of the curve begins when the ratio K_s/δ' is approximately 0.25 or $\frac{R_e\sqrt{\lambda}}{R/K_s}$ is about 17 and it ends when K_s/δ is about 6 or $\frac{R_e\sqrt{\lambda}}{R/K_s} \approx 400$. Therefore, a pipe will act as a smooth pipe if K_s/δ is less than 0.25 and will act as a rough when K_s/δ is greater than 6.

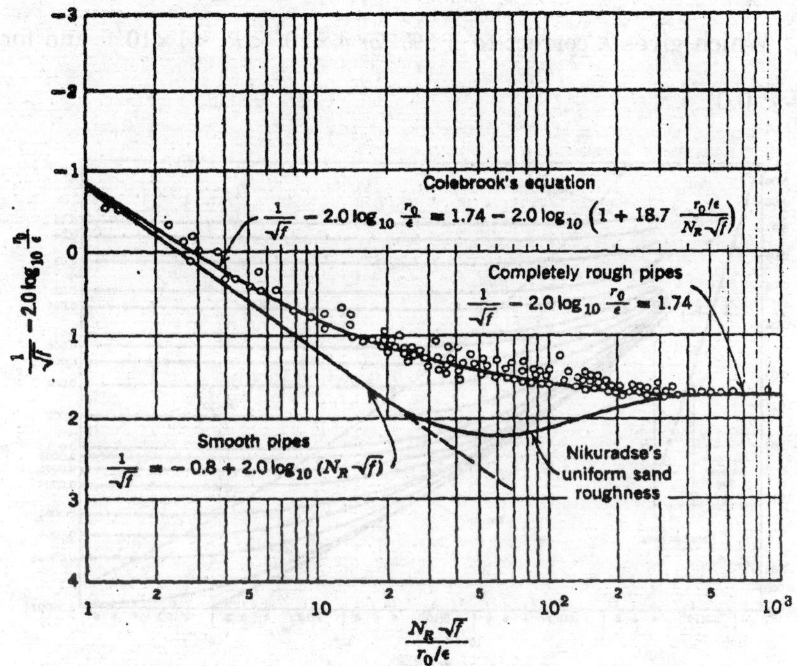

Fig. 3.19. Nikurdse data over the entire range of roughness

Equation (3.49) was not at first used very widely by engineers, mainly because it was not expressed directly in terms of the standard engineering variables of diameter, discharge and hydraulic gradient. In addition, the equation is implicit and requires a trial and error solution. In the 1940s, slide rules and logarithm tables were the main computational aids of the engineer, since pocket calculators and computers were not then available. So these objections to the use of the Colebrook-White equation were not unreasonable.

3.11.6 Moody diagram for commercial pipes

The first attempt to make engineering calculations easier was made by Moody. He produced a $\lambda - R_e$ plot based on equation (3.49) for commercial pipes, as shown in Fig. 3.20 which is now known as the Moody diagram. He also presented an explicit formula for λ.

$$\lambda = 0.0055\left[1+\left(\frac{2000\,K_s}{D}+\frac{10^6}{R_e}\right)^{1/3}\right] \qquad (3.50)$$

Which gives λ correct to $\pm 5\%$ *for* $4\times10^3 < R_e < 1\times10^7$ and for $\frac{K_s}{D} < 0.01$.

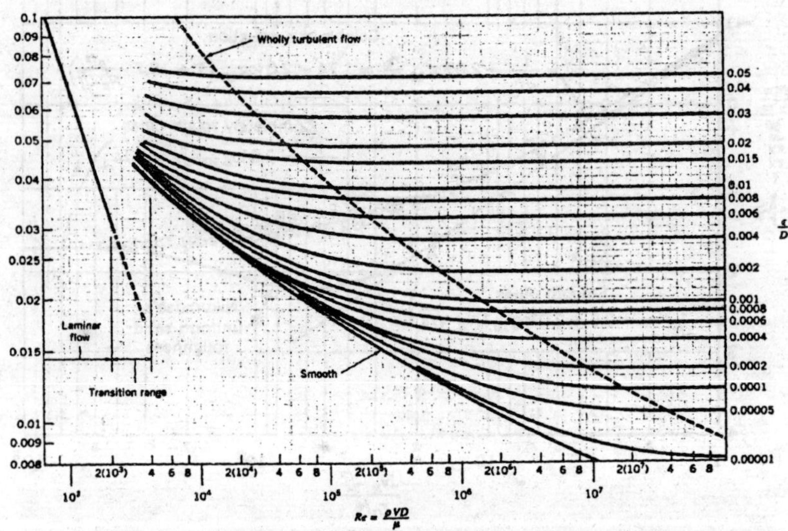

Fig. 3.20. The Moody diagram

Several important observations can be made from these curves.

(1) For a given Reynolds number of flow, as the relative roughness D/K_s is increased, the friction factor λ decreases.

(2) For a given relative roughness D/K_s, the friction factor λ decreases with increasing Reynold's number until the zone of complete turbulence is reached.

(3) Within the zone of complete turbulence, the Reynolds number has no effect on the friction factor λ.

(4) As the relative roughness D/K_s increases, the value of the R_e at which the zone of complete turbulence begins also increases.

Fig. 3.21 shows a chart for fully developed flow in circular pipes Fig. 3.22, 11.23 and 11.24 shows moody diagrams for type 1, type 2 and type 3 pipe flow problems.

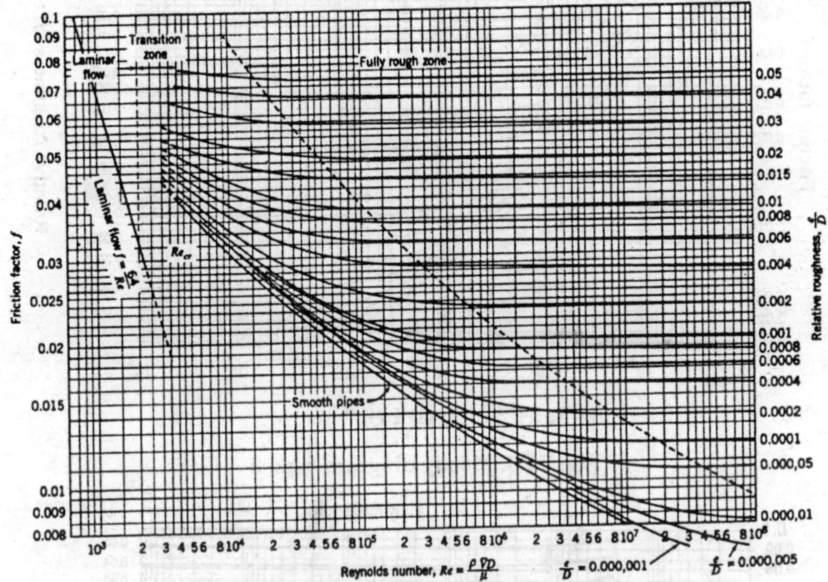

Fig. 3.21. Chart for fully developed flow in a circular pipe.

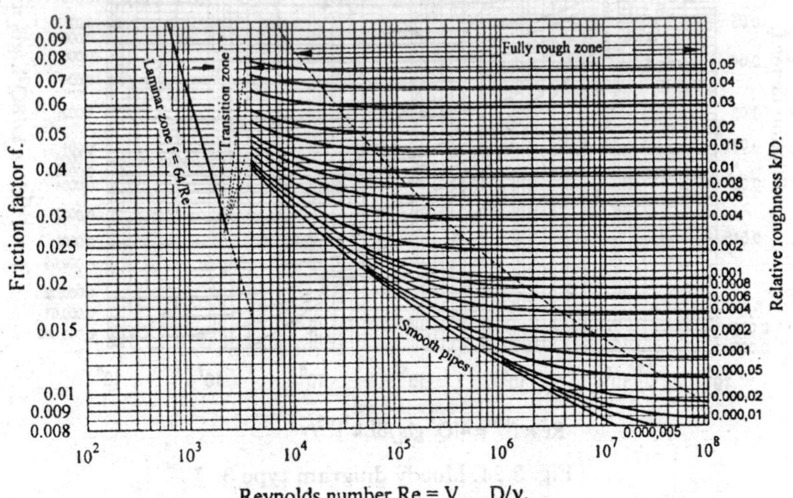

Fig. 3.22. Moody diagram type 1

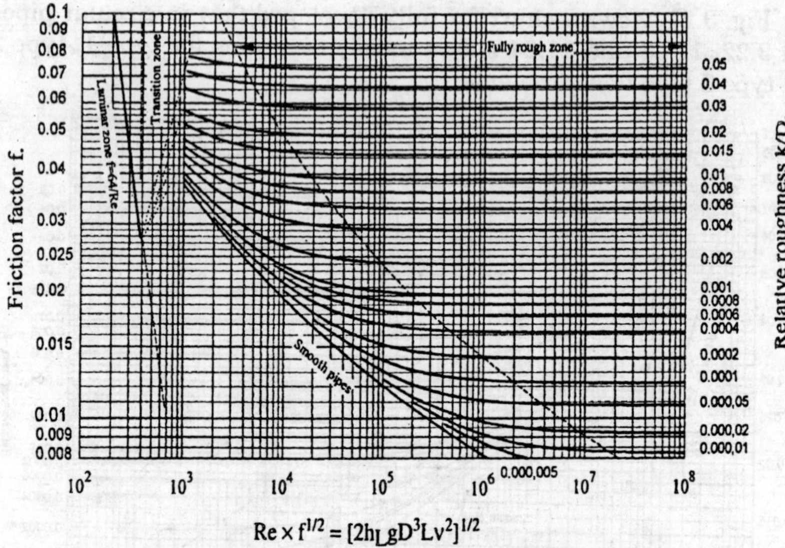

Fig. 3.23. Moody diagram type 2

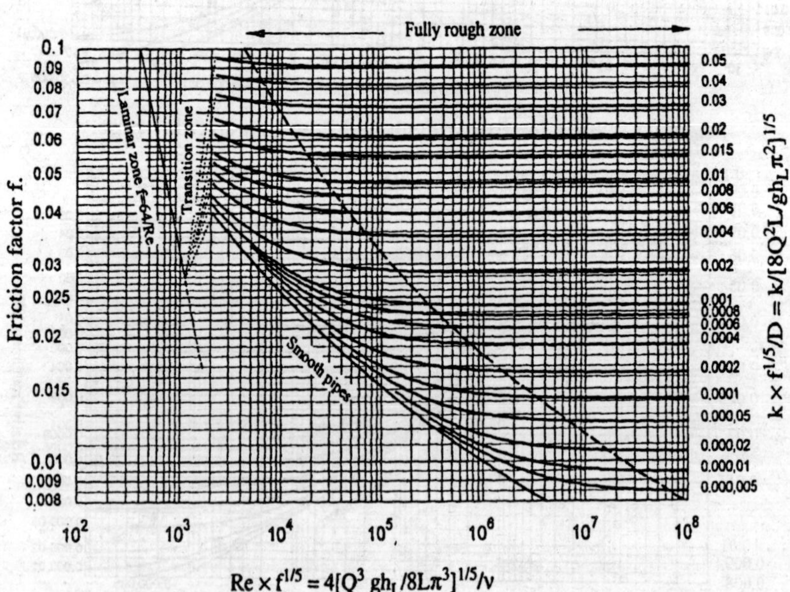

Fig. 3.24. Moody diagram type 3

Fig. 3.25 is a simplified sketch of Moody's diagram in which the various zones are identified.

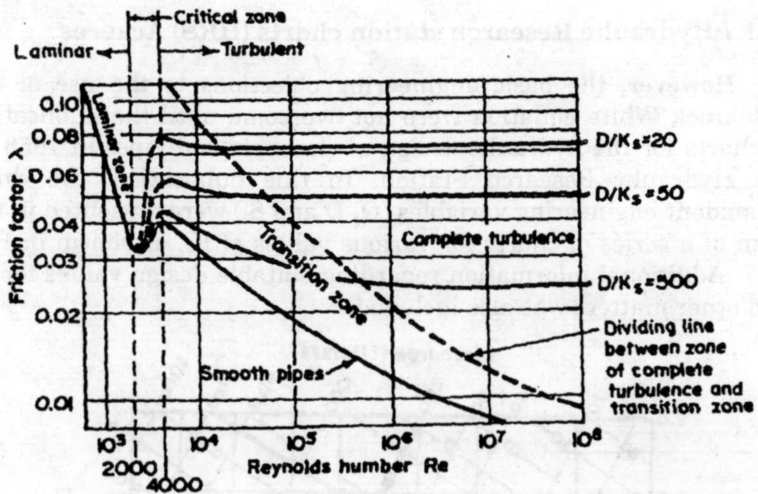

Fig. 3.25. Explanation of parts of Moody diagram.

Fig. 3.26 shows relative surface roughness for various pipes.

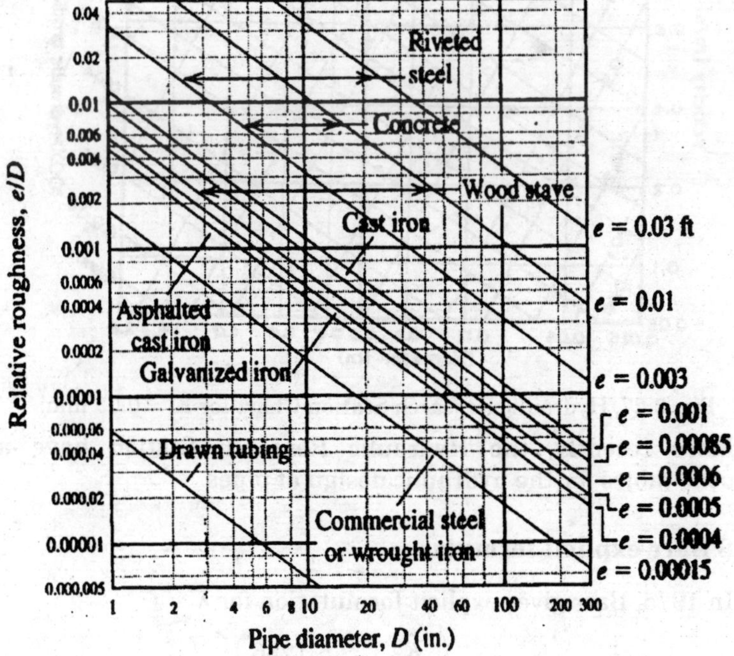

Fig. 3.26. Relative surface roughness for various pipes.

3.11.7 Hydraulic Research station charts (HRS) Ackeres

However, the basic engineering objections to the use of the Colebrook White equation were not overcome until the publication of charts for the hydraulic design of channels and pipes in 1958 by the Hydraulic Research Station. In this publication, the three dependent engineering variables (Q, D and S_f) were presented in the form of a series of charts for various values of K_s, as shown in Fig. 3.27. Additional information regarding suitable design values for K_s and other matters was also included.

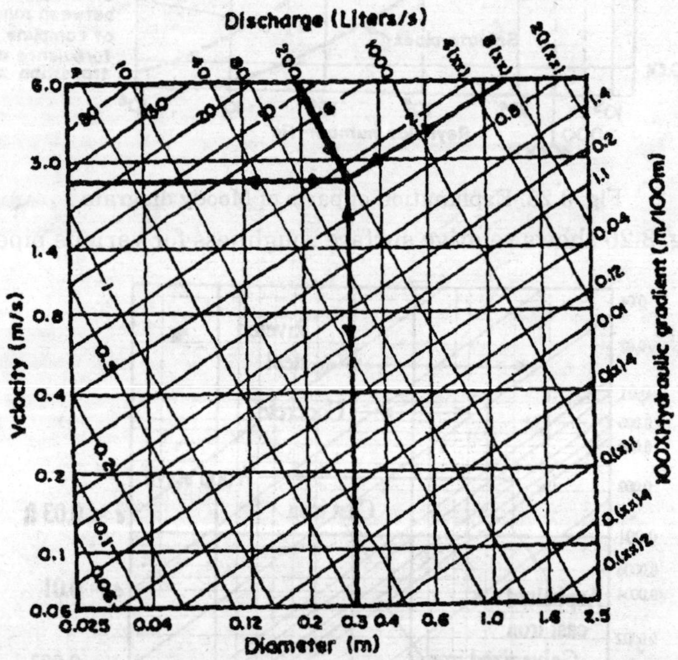

Fig. 3.27. Hydraulic Research Station Chart for K_s = 0.03 mm.

More recently, the Hydraulic Research Station have also produced tables for the hydraulic design of pipes.

3.11.8 Barr explicit formula

In 1975, Barr gives explicit formulation for λ

$$\frac{1}{\sqrt{\lambda}} = -2\log_{10}\left(\frac{K_s}{3.7\,D} + \frac{5.1286}{R_e^{0.99}}\right) \tag{3.51}$$

In this formula the smooth law component $\left(\frac{2.51}{R_e\sqrt{\lambda}}\right)$ has been replaced by an approx-imation $\left(5.1286/R_e^{0.99}\right)$. For $R_e > 10^5$ this provides a solution for S_f (h_f/L) to an accuracy better than ± 1 %.

3.11.9 Murdock formula

In 1976, J.W. Murdock solves the Colebrook-White equation. First, an approximate value for λ can be calculated from

$$\lambda_a = 0.0055\left[1+\left(\frac{2\times10^4}{D/K_s}+\frac{10^6}{R_e}\right)^{1/3}\right] \tag{3.52}$$

Then, we can use a modified form of equation (3.49) to find the actual λ as follows

$$1/\sqrt{\lambda} = -2\log_{10}\left[\frac{1}{3.7\,D/K_s}+\frac{2.51}{R_e\sqrt{\lambda_a}}\right] \tag{3.53}$$

Since λ_a is quite close to the actual value of λ, no iteration is needed for sufficient accuracy in most problems.

3.11.10 Swamee and Jain's explicit equation

Dr. A.K. Jain has obtained a simple and more accurate explicit equation for friction factor λ for flow in commercial pipes. The equation has a logarithmic form.

$$\frac{1}{\sqrt{\lambda}} = 1.14 - 2\log_{10}\left(\frac{K_s}{D}+\frac{21.25}{R_e^{0.9}}\right) \tag{3.54}$$

It is valid for $5000 \le R_e \le 10^8$ and $10^{-6} < K_s/D < 10^{-2}$. This equation yields λ within 1% of the equation (3.49).

3.11.11 S.E. Haaland's formula

The best yet produced is probably that by S.E. Haalland.

$$\frac{1}{\sqrt{\lambda}} = -1.8\log_{10}\left[\frac{6.9}{R_e}+\left(\frac{K_s}{3.71\,D}\right)^{1.11}\right] \tag{3.55}$$

It combines reasonable simplicity with acceptable (within about 1.5%) and is useful if an algebraic expression is required.

3.12 NON CIRCULAR PIPES

For non circular pipes it is common practice to use an effective diameter termed the hydraulic diameter D_h which is defined as

$$D_h = 4\left(\frac{A}{P}\right) = 4 \times hydraulic\ radius \qquad (3.56)$$

For regular noncircular shapes of pipes, the friction factors for Turbulent flow can be predicted from Moody's diagram or, HRS chart. The equations already developed for circular pipes may also be used by applying the concept of hydraulic radius. D_h is used in Reynold's equation as $R_e = \dfrac{\rho u D_h}{\mu}$ and in the Darcy's-Weisbach equation as

$$h_f = \frac{\lambda l u^2}{2 g D_h} \qquad (3.57)$$

The above method using Moody's diagram and Reynolds number computed with the help of hydraulic radius gives satisfactory results.

3.13 ROUGHNESS OF PIPES WITH AGE (OLD PIPES)

When the pipes become old, the friction increases with the rust and sedimentation accumulated on the pipe wall.

The value of K_s given are for new and clean pipes. Colebrook-White gave the following equation for the determination of roughness of old pipes.

$$K_s = K_o + \propto t \qquad (3.58)$$

where, $\propto$ is a coefficient, t is number of years of use of the pipe and K_o is the initial value of K_s.

For pipelines in service for a long period, say 20 years or more, the absolute roughness of old concrete or cement lined steel pipes may increase three times over that of new pipes; for cast-iron pipes the increase may be upto twenty times and for wrought iron and steel pipes it may reach even forty times.

ILLUSTRATIVE EXAMPLES

3.1. **Oil (specific gravity S = 0.85) is flowing in a 150 m long smooth pipe line at the rate of 80 liters per second. The loss of head in the pipe line is 67 m. If the kinematic viscosity of the oil is 0.1 stoke. Calculate diameter of the pipe.**

Solution

$$Q = \frac{\pi}{4} d^2 u$$

$$80 \times 10^{-3} = \frac{\pi}{4} d^2 u$$

$$u = \frac{80 \times 10^{-3} \times 4}{\pi d^2} = \frac{101.859 \times 10^{-3}}{d^2}$$

$$R_e = \frac{ud}{\upsilon} = \frac{101.859 \times d^{1/4}}{0.1 \times 10^{-4} \times d^2} = \frac{10185.916}{d} \qquad (1)$$

Using Blasius formula

$$\lambda = \frac{0.3164}{(R_e)^{1/4}} = \frac{0.3164 \times d^{1/4}}{(10185.916)^{0.25}} = 0.03149\ d^{1/4}$$

$$h_f = \frac{\lambda\, l\, u^2}{2g\, d}$$

$$67 = \frac{0.03149\ d^{1/4} \times 150 \times (101.859 \times 10^{-3})^2}{2 \times 9.81 \times d^5}$$

$$67 = \frac{2.4978 \times 10^{-3}}{d^{19/4}}$$

$$d^{19/4} = 3.728 \times 10^{-5}$$

$$d = (3.728 \times 10^{-5})^{4/19} = 0.11687 \text{ m} = 116.8 \text{ mm} \qquad \text{Ans}$$

Check

$$R_e = \frac{10185.916}{0.11687} = 87154.5856 \quad \text{Less than } 10^5$$

3.2. A smooth pipe carries 7 liters of water per second at 20°C (υ = 1 centistoke) with a head loss of 75 mm per 10 m of length. Determine the pipe diameter.

Solution

$$Q = \frac{\pi}{4} d^2 u$$

$$7 \times 10^{-3} = \frac{\pi}{4} d^2 u$$

$$u = \frac{8.915 \times 10^{-3}}{d^2}$$

$$R_e = \frac{ud}{\upsilon} = \frac{8.915 \times 10^{-3} \times d}{d^2 \times 10^{-6}} = \frac{8.915 \times 10^{-3}}{d} \quad (1)$$

$$h_f = \frac{\lambda l u^2}{2gd} = \frac{\lambda l}{d} \times \left(\frac{8.915 \times 10^{-3}}{d^2}\right)^2 \times \frac{1}{2 \times 9.81}$$

$$\frac{75}{1000} = \frac{\lambda}{d^5} \times 40.5 \times 10^{-6}$$

$$d^5 = \frac{40.5 \times 10^{-6} \times 1000}{75} \lambda = 54 \times 10^{-5} \quad \lambda$$

or, $$\lambda = \frac{d^5}{54 \times 10^{-5}}$$

Using the Blasius formula

$$\lambda = \frac{0.3164}{(R_e)^{1/4}} = \frac{0.3164}{\left(\frac{8.915 \times 10^3}{d}\right)^{1/4}}$$

or, $$\frac{d^5}{54 \times 10^{-5}} = \frac{0.3164\, d^{1/4}}{(8.915 \times 10^3)^{1/4}}$$

or, $$d^{19/4} = \frac{54 \times 10^{-5} \times 0.3164}{9.716} = 1.7583 \times 10^{-5}$$

$$d = 0.100392 \text{ m} = 100.3 \text{ mm}$$

Therefore $R_e = \frac{8.915 \times 10^{+3}}{0.100392} = 88801.89$ Less than 10^5

3.3. Determine (a) the shear stress at the walls of a 300 mm diameter pipe when water flowing causes a measured lost head of 15 mm in 300 m of pipe length, (b) the shear stress 50 mm from the centre line of the pipe, (c) shear velocity, (d) the average velocity for a value of 0.05 and (e) the ratio of u/u*.

Solution

(a) $$\tau_o = \frac{\rho g h_f R}{2l}$$

$$= \frac{1000 \times 9.81 \times 15 \times 0.15}{2 \times 300} = 36.7875 \text{ N/m}^2$$ Ans

(b) The shear stress varies linearly from centre line to wall

$$\frac{\tau}{r} = \frac{\tau_o}{R}$$

$$\tau = \frac{\tau_o r}{R} = \frac{36.7875 \times 50}{150} = 12.2625 \ N/m^2$$ Ans

(c) $$u_* = \sqrt{\frac{\tau_o}{\rho}} = \sqrt{\frac{36.7875}{1000}} = 0.1918 \ m/s$$ Ans

(d) $$h_f = \frac{\lambda l u^2}{2g \ d}$$

$$15 = \frac{0.05 \times 300 \times u_{av}^2}{2 \times 9.81 \times 0.3}$$

$$u_{av}^2 = 5.886$$

Therefore,

$$u_{av} = 2.426 \ m/s$$

The average velocity = 2.426 m/s Ans

(e) $$\tau_o = \frac{\mu u}{y}$$

or, $$\tau_o = \frac{\rho \upsilon u}{y}$$

or, $$\frac{\tau_o}{\rho} = \frac{\upsilon u}{y}$$

or, $$u_*^2 = \frac{\upsilon u}{y}$$

or, $$\frac{u}{u_*^2} = \frac{y}{\upsilon}$$

or, $$\frac{u}{u_*} = \left(\frac{y}{\upsilon}\right) u_*$$ Ans

3.4. **When 0.28 m³/s of water flow in a 300 mm pipeline, 63 kW are lost in friction in 300 mm of pipe. Calculate head loss, friction factor, friction velocity, and shear stress at the pipe wall.**

Solution

$$P = \rho\ g\ Q\ h_f$$

$$63 \times 1000 = 1000 \times 9.81 \times 0.28 \times h_f$$

$$h_f = \frac{63}{9.81 \times 0.28} = 22.93\ m$$ Ans

$$h_f = \frac{\lambda\ l\ u^2}{2g\ d}$$

$$22.93 = \frac{\lambda\ l\ Q^2}{12.1\ d^5} = \frac{\lambda \times 300 \times (0.28)^2}{12.1 \times (.3)^5}$$

λ= 0.0286 Ans

$$Q = \frac{\pi}{4}(d)^2\ u$$

$$0.28 = \frac{\pi}{4}(0.3)^2\ u$$

u = 3.96 m/s

$$u_* = u\sqrt{\frac{\lambda}{8}} = (3.96)\sqrt{\frac{0.0286}{8}} = 0.237$$ Ans

$$\tau_o = \rho\ u_*^2 = 1000 \times (0.237)^2 = 56.169 = 56.169\ \text{Pa}$$ Ans

3.5. **Fluid of density 1030 kg/m³ and kinematic viscosity 1.86 × 10⁻⁵ m²/s flows parallel to a very smooth plane surface. The velocities at 75 mm and 4.4 mm from the wall are measured and found to be 0.3 m/s and 0.08 m/s, respectively. Using both of these measurements calculate the shearing stress on the surface.**

Solution

The locations and magnitudes of velocities suggest that one is in the turbulent region, the other in the viscous sub layer.

We know that

$$\frac{u}{u_*} = 5.5 + 5.75 \log_{10}\left(\frac{u_* y}{\upsilon}\right)$$

$$\frac{0.3}{u_*} = 5.5 + 5.75 \log_{10}\left(\frac{u_* \times 0.075}{1.86 \times 10^{-5}}\right)$$

Solving u* by trial and error method.

$$u^* = 0.018 \text{ m/s}$$

In the laminar film

$$\tau_o = \rho\, u_*^2 = \frac{\upsilon \rho\, u}{y}$$

$$\text{Therefore, } u_*^2 = \frac{1.86 \times 10^{-5} \times 0.08}{0.0044} = 3.38 \times 10^{-4}$$

$$u_* = 0.01838 \quad \text{as before}$$

Which seems to confirm the original assumption.

For friction confirmation calculate δ'

We know that

$$\frac{u_*\, \delta'}{\upsilon} = 11.6$$

$$\text{or,} \quad \frac{0.018\, \delta'}{1.86 \times 10^{-5}} = 11.6$$

$$\delta' = 0.01198$$

Since $\delta' > y$

or, $0.01198 > 0.0044$,

u = 0.08 m/s is the viscous sub layer

Therefore,

$$\tau_o = \rho u_*^2 = 1030 \times (0.018)^2 = 0.33372 \text{ N/m}^2$$ Ans

3.6. A pipe 200 mm diameter conveys water. Measurements indicate that velocity at the pipe centre and that at a distance 50 mm from the pipe centre are 2.5 m/s and 2 m/s respectively. The flow in the pipe is turbulent, calculate the shear velocity and the wall shear stress.

Solution

Using the velocity deficiency equation

$$\frac{u_{max} - u}{u_*} = 5.75 \log_{10}\left(\frac{R}{y}\right)$$

$$\frac{2.5 - 2}{u_*} = 5.75 \log_{10}\left(\frac{100}{50}\right)$$

$$\frac{0.5}{u_*} = 1.73092$$

$$u_* = 0.288 \ m/s$$ Ans

$$u_* = \sqrt{\frac{\tau_o}{\rho}}$$

$$\tau_o = \rho u_*^2 = 1000 \times (0.288)^2 = 83.442 \text{ N/m}^2$$ Ans

3.7. Crude oil of relative density 0.86 and kinematic viscosity 1.3 $\times 10^{-5}$ m²/s is discharged through a 300 mm diameter pipe line at the rate of 100 liters/s. Determine (a) the centre line velocity and (b) the velocity just above the laminar sub layer.

Solution

$$Q = \frac{\pi}{4} d^2 u$$

$$100 \times 10^{-3} = \frac{\pi}{4} (0.3)^2 \times u$$

u = 1.4147 m/s

$$R_e = \frac{ud}{\upsilon} = \frac{1.4147 \times 0.3}{1.3 \times 10^{-5}} = 32646.923$$

The value of R_e number comes out to be less than 10^5. The Blasius formula can be used for finding the friction factor

$$\lambda = \frac{0.3164}{(R_e)^{1/4}} = \frac{0.3164}{(32646.923)^{0.25}} = \frac{0.3164}{13.44} = 0.0235$$

$$\sqrt{\lambda/8} = \sqrt{\frac{0.0235}{8}} = 0.054198$$

$$u_* = u_{av}\sqrt{\lambda/8} = 1.4147 \times 0.054198 = 0.0766$$

For finding the value of u_{max}

$$\frac{u_{max}}{u_{av}} = 1 + 1.43\sqrt{\lambda} = 1 + 1.43\sqrt{0.0235}$$

$$u_{max} = 1.4147\,(1 + 0.2192)$$

$$u_{max} = 1.7248\ m/s$$ Ans

(b) Hence the velocity just above the laminar sub layer

$$\tau_o = \mu\frac{u}{\delta} = \frac{\upsilon\rho u}{\delta}$$

Therefore,

$$u = \frac{\tau_o\,\delta}{\rho\,\upsilon} = \frac{\tau_o}{\rho} \times \frac{1}{\upsilon} \times \frac{11.6\,\upsilon}{u_*} = \frac{\tau_o}{\rho} \times \frac{11.6}{\sqrt{\tau_o/\rho}} = \sqrt{\tau_o/\rho} \times 11.6$$

$$= u_* \times 11.6 = 0.0766 \times 11.6 = 0.89\ m/s$$ Ans

3.8. A 100 mm smooth pipe line carries 38 liters/s of water at 20°C. Calculate (a) the friction factor, (b) maximum velocity, (c) shear velocity, (d) shear stress at the wall (e) the thickness of the laminar sub layer. Take ν = 1 cs.

Solution

$$Q = \frac{\pi}{4} d^2 u$$

$$38 \times 10^{-3} = \frac{\pi}{4}(0.1)^2 \times u$$

u = 4.838 m/s

$$R_e = \frac{ud}{\upsilon} = \frac{4.838 \times 0.1}{1 \times 10^{-6}} = 483831.027 = 4.838 \times 10^5$$

$$\log_{10} R_e = \log_{10}\left(4.838 \times 10^5\right) = 5.6846$$

The value of R_e comes out to greater than 10^5 therefore, Blausius formula cannot be used.

(a) For turbulent flow in smooth pipe

By Nikurdse

(1) $$\lambda = 0.0032 + \frac{0.221}{(R_e)^{0.237}} = 0.0032 + \frac{0.221}{\left(4.838 \times 10^5\right)^{0.237}}$$

$$= 0.0032 + 9.934 \times 10^{-3} = 0.013$$

(b) $$u_* = u_{av}\sqrt{\lambda/8} = 4.838\sqrt{\frac{0.013}{8}} = 0.195 \text{ m/s}$$ Ans

(c) $$\frac{u_{max}}{u_{av}} = 1 + 1.43\sqrt{\lambda}$$

$$= 1 + 1.43\sqrt{0.013} = 1 + 0.163 = 1.163$$

$$u_{max} = 4.838 \times 1.163 = 5.6268 \text{ m/s}$$ Ans

(d) $$u_* = \sqrt{\frac{\tau_o}{\rho}}$$

$$\tau_o = \rho u_*^2 = 1000 \times (0.195)^2 = 38.025 \text{ N/m}^2$$ Ans

(e) $$\delta' = 11.6\frac{\upsilon}{u_*} = \frac{11.6 \times 1 \times 10^{-6}}{0.195}$$

$$= 5.9487 \times 10^{-5} \text{ m} = 0.05948 \text{ mm}$$ Ans

3.9. **A 300 mm diameter pipe line carries 500 liters/s of water at 20°C. Determine the maximum permissible height of roughness so that the pipe may still be called smooth. Take ν = 1 cs.**

Solution

$$Q = \frac{\pi}{4} d^2 u$$

$$500 \times 10^{-3} = \frac{\pi}{4}(0.3)^2 \times u$$

u = 7.0753 m/s

$$R_e = \frac{ud}{\upsilon} = \frac{7.0735 \times 10^{-6}}{1 \times 10^{-6}} = 2.11 \times 10^6$$

The R_e comes out to be greater than 10^5 therefore, Blausius formula cannot be used.

For turbulent flow in smooth pipes

By Nikurdse

$$\lambda = 0.0032 + \frac{0.221}{R_e^{0.237}}$$

$$= 0.0032 + \frac{0.221}{2.11 \times 10^{6\ 0.237}} = 0.0105$$

$$\therefore u_* = u\sqrt{\lambda/8} = 7.0735\sqrt{\frac{0.0105}{8}} = 0.2562 \text{ m/s}$$

Thickness of the laminar sub layer

$$\delta' = \frac{11.6\ \upsilon}{u_*} = \frac{11.6 \times 1 \times 10^{-6}}{0.25626}$$

$$= 4.526619 \times 10^{-5} \text{ m} = 0.04526 \text{ mm}$$

Limiting value of height of roughness so that the pipe behaves as smooth

$$K_s = \frac{\delta'}{4} = \frac{0.04526}{4} = 0.0113 \text{ mm}$$ Ans

3.10. Kerosene (S = 0.8 and μ = 1.90 × 10^{-3} kg/ms) is to flow in a 200 mm diameter pipe at a velocity of 3 m/s with a minimum loss of energy Determine the permissible height of surface roughness of the pipe wall.

Solution

$$R_e = \frac{\rho\ u\ d}{\nu} = \frac{800 \times 3 \times 0.2}{1.9 \times 10^{-3}} = 252631.5789 = 2.526 \times 10^5$$

The R_e number comes out to be greater than 10^5. Therefore, Blausius formula cannot be used. At the permissible height of roughness in the pipe the flow will be in smooth turbulent regime.

By Nikurdse method

$$\lambda = 0.0032 + \frac{0.221}{(R_e)^{0.237}} = 0.0032 + \frac{0.221}{\left(2.526 \times 10^5\right)^{0.237}}$$

$$= 0.0032 + \frac{0.221}{19.0717} = 0.0032 + 0.01158 \quad = 0.01478$$

Since the pipe wall behave as smooth, at the limiting value

$$\frac{R_e \sqrt{\lambda}}{(R_o / K_s)} = 17$$

$$\frac{2.56 \times 10^5 \sqrt{0.01478}}{(R_o / K_s)} = 17$$

$$\frac{R_o}{K_s} = \frac{30733.95}{17} = 1807.879$$

$$K_s = 5.5313 \times 10^{-5} \; = 0.00553 \text{ mm}$$ Ans

3.11. The wall shear stress and hence the pressure gradient may be estimated for fully developed turbulent flow from velocity measurements. Water flows at $u_{av} = 0.86\ u_{max} = 2.1$ m/s in a 150 mm diameter pipe.

(a) What is the wall shear stress?

(b) What is the pressure drop in 30 m of pipe?

Solution

We know that

$$\frac{u_{max}}{u_{av}} = 1 + 1.43 \sqrt{\lambda}$$

$$\frac{u_{max}}{0.86 u_{max}} = 1 + 1.43 \sqrt{\lambda}$$

$$\frac{1}{0.86} = 1 + 1.43 \sqrt{\lambda}$$

$$1.16279 = 1 + 1.43\sqrt{\lambda}$$

$$\sqrt{\lambda} = \frac{0.16279}{1.43} = 0.113839$$

$$\sqrt{\lambda} = 0.012959$$

We know that

(a) $$u_* = \sqrt{\frac{\tau_o}{\rho}} = u_{av}\sqrt{\lambda/8}$$

$$\tau_o = \frac{\rho u_{av}^2 \lambda}{8} = \frac{1000 \times (2.1)^2 \times 0.012959}{8}$$

$= 7.1439$ N/m² **Ans**

(b) $$\frac{\Delta p}{\rho} = \frac{\lambda l u^2}{2d}$$

or, $$\Delta p = \lambda\left(\frac{l}{d}\right)\frac{\rho u^2}{2} = 0.012959 \times \frac{30}{0.15} \times 1000 \times \frac{(2.1)^2}{2}$$

$= 5714.919 = 5.714$ k Pa **Ans**

3.12. In a fully turbulent flow in a 150 mm diameter pipe the centre line velocity is 2.5 m/s and the local velocity at mid radius is 2.28 m/s. Find the discharge and the height of the roughness projections.

Solution

For both smooth and rough boundaries, the maximum velocity u_{max} is at y = R and is given by

$$\frac{u - u_{max}}{u_*} = 5.75 \log_{10}\frac{y}{R}$$

or, $$\frac{u_{max} - u}{u_*} = 5.75 \log_{10}\frac{R}{y} \qquad (1)$$

and in terms of u_{max}, u_{av} and u, are related as

$$\frac{u_{max} - u_{av}}{u_*} = 3.75 \qquad (2)$$

Using equation (1)

$$\frac{2.5-2.28}{u_*} = 5.75 \log_{10}\left(\frac{R}{R/2}\right)$$

$$\frac{0.22}{u_*} = 1.73092$$

$$u_* = 0.12709\ m/s$$

Using equation (2)

$$\frac{2.5-u_{av}}{0.127} = 3.75$$

$$2.5 - u_{av} = 0.4766$$

$$u_{av} = 2.023375\ m/s$$

The discharge $Q = \frac{\pi}{4} d^2\ u_{av}$

$$= \frac{\pi}{4}(0.17)^2 \times 2.023375$$

$$= 0.03575\ \text{m}^3\text{/s} = 35.75\ \text{liters/s}$$ Ans

For rough turbulent flow

$$\frac{u_{max}}{u_*} = 5.75 \log_{10}\frac{R}{K_s} + 8.5$$

or, $$\frac{2.5}{0.127} - 8.5 = 5.75 \log_{10}\frac{R}{K_s}$$

or, $$\frac{19.685-8.5}{5.75} = \log_{10}\frac{R}{K_s}$$

or, $$\frac{11.185}{5.75} = \log_{10}\frac{R}{K_s}$$

$$\log_{10}\frac{R}{K_s} = 1.945$$

$$\frac{R}{K_s} = 88.15$$

$$\frac{0.075}{K_s} = 88.15$$

$$K_s = 8.508 \times 10^{-4}\ m = 0.8508\ \text{mm}$$ Ans

3.13. A 200 mm diameter pipe line has equivalent sand grain roughness magnitude of 0.2 mm. At what velocity of flow will the roughness

(a) Cause the flow to depart from smooth pipe regime. The transition from the smooth pipe regime begins at Reynold's roughness number $\frac{R_e\sqrt{\lambda}}{R/K_s} = 17$ and

(b) Cause the flow to become fully rough turbulent. Fully rough flow begins at Reynold's roughness number

$$\frac{R_e\sqrt{\lambda}}{R/K_s} = 400 \ \textit{Take}\ \nu = 1\ cs.$$

Solution

(a) $$\frac{R_e\sqrt{\lambda}}{R/K_s} = 17$$

$$R_e\sqrt{\lambda} = 17 \times \left(\frac{0.1}{0.0002}\right) = 8500 \qquad (1)$$

We know that for smooth pipe flow

$$\frac{1}{\sqrt{\lambda}} = 2\log_{10} R_e\sqrt{\lambda} - 0.8 = 2\log_{10} 8500 - 0.8$$

$$= 7.8588 - 0.8 = 7.0588$$

Using equation (1)

$$R_e\sqrt{\lambda} \times \frac{1}{\sqrt{\lambda}} = 8500 \times 7.0588$$

$$R_e = 8500 \times 7.0588 = 60000 = 6 \times 10^4$$

$$R_e = \frac{u\,d}{\nu}$$

$$6 \times 10^4 = \frac{u \times 0.2}{1 \times 10^{-6}}$$

u = 0.3 m/s

Upper limit velocity for smooth pipe regime is 0.3 m/s<T>Ans

(b) $$\frac{R_e \sqrt{\lambda}}{R / K_s} = 400$$

$$R_e \sqrt{\lambda} = 400 \times \frac{0.1}{0.0002} = 2 \times 10^5$$

We know that for rough turbulent pipe flow

$$\frac{1}{\sqrt{\lambda}} = 2 \log_{10} \frac{R}{K_s} + 1.74$$

$$\frac{1}{\sqrt{\lambda}} = 2 \log_{10} \left(\frac{0.1}{0.0002}\right) + 1.74 = 5.3979 + 1.74 = 7.1379$$

Therefore,

$$R_e \sqrt{\lambda} \times \frac{1}{\sqrt{\lambda}} = 2 \times 10^5 \times 7.1379 = 1427588 = 1.427 \times 10^6$$

$$R_e = \frac{U d}{\upsilon}$$

$$1.427 \times 10^6 = \frac{u \times 0.2}{1 \times 10^{-6}}$$

u = 7.1379 m/s Ans

The flow will be at fully rough turbulent at 7.1379 m/s

3.14. A 100 mm rough pipe line with roughness 0.12 mm carries 38 liters/s of water at 20°C (ν = 1 cs) Calculate (a) friction factor, (b) shear velocity, (c) shear stress at the wall and (d) maximum velocity

Solution

$$Q = \frac{\pi}{4} d^2 u$$

$$38 \times 10^{-3} = \frac{\pi}{4}(0.1)^2 \times u$$

u = 4.838 m/s

$$R_e = \frac{ud}{\nu} = \frac{4.838 \times 0.1}{1 \times 10^{-6}} = 483831.027 = 4.838 \times 10^5$$

The value of R_e comes out to be greater than 10^5 therefore, Blausius formula cannot be used.

(a) For turbulent flow in rough pipe

$$\frac{1}{\sqrt{\lambda}} = 1.74 + 2\log_{10}\frac{R}{K_s}$$

$$\frac{1}{\sqrt{\lambda}} = 1.74 + 2\log_{10}\frac{0.05}{0.12/1000} = 1.74 + 2 \times 2.61978$$

$$= 1.74 + 5.2395 = 6.9795$$

$$\sqrt{\lambda} = 0.143275$$

$$\lambda = 0.0205$$ Ans

(b) $$u_* = u\sqrt{\lambda/8} = 4.838\sqrt{\frac{0.0205}{8}} = 0.243 \text{ m/s}$$

(c) $$\tau_o = \rho u_*^2 = 1000 \times (0.243)^2 = 59.1 \ N/m^2$$

(d) $$\frac{u_{max}}{u_{av}} = 1 + 1.43\sqrt{\lambda} = 1 + 1.43\sqrt{0.0205}$$

$$= 1 + 0.2047 = 1.2047$$

$$u_{max} = 4.838 \times 1.2047 = 5.8285 \text{ m/s}$$ Ans

3.15. Mean point velocities measured with the help of a pitot tube at mid point and quarter point of a 200 mm diameter pipe were found to be 1.5 m/s and 1.35 m/s respectively. If the flow in the pipe is turbulent, determine discharge, friction factor and average height of roughness projections.

Solution

We know that

$$\frac{u_{max} - u}{u_*} = 5.75 \log_{10} (R/y)$$

$$\frac{1.5 - 1.35}{u_*} = 5.75 \log_{10} \left(\frac{100}{50}\right)$$

$$\frac{0.15}{u_*} = 5.75 \log_{10} 2$$

$$\frac{0.15}{u_*} = 1.7309$$

$$u_* = 0.0866\ m/s$$

We know a general formula

$$\frac{u - u_{av}}{u_*} = 3.75 + 5.75 \log_{10} \frac{y}{R}$$

When $u = u_{max}$ at $y = R$

Therefore,

$$\frac{u_{max} - u_{av}}{u_*} = 3.75$$

$$\frac{1.5 - u_{av}}{0.0866} = 3.75$$

$$u_{av} = 1.175\ m/s$$

$$Q = \frac{\pi}{4} d^2\ u_{av} \qquad = \frac{\pi}{4} (0.2)^2 \times 1.175 = 0.0369\ \text{m}^3\text{/s} \qquad \text{Ans}$$

Also, $u_* = u_{av} \sqrt{\lambda/8}$

$$0.0866 = 1.175\ \sqrt{\lambda/8}$$

$$0.0737 = \sqrt{\lambda/8}$$

$$\lambda/8 = 5.432 \times 10^{-3}$$

$$\lambda = 0.04345 \qquad \text{Ans}$$

Therefore,

$$\frac{1}{\sqrt{\lambda}} = 2\log_{10}\left(\frac{R}{K_s}\right) + 1.74$$

$$\frac{1}{\sqrt{0.04345}} = 2\log_{10}\left(\frac{R}{K_s}\right) + 1.75$$

$$\log_{10}\frac{R}{K_s} = \frac{4.79 - 1.75}{2} = 1.5235$$

$$\frac{R}{K_s} = 33.383$$

$$K_s = \frac{0.1}{33.383} = 2.9955 \times 10^{-3}\ m = 2.9955\ mm \qquad \text{Ans}$$

3.16. In a fully turbulent rough pipe flow occur in a 150 mm diameter pipe carrying oil of density 900 kg/m³. If at a distance of 25 mm from the wall the velocity is 1.5 m/s and the velocity gradient is 7 per second; calculate (a) the boundary shear stress (b) equivalent sand grain roughness, (c) discharge (d) friction factor.

Solution

In a fully turbulent rough pipe flow

$$\frac{u}{u_*} = 5.75\log_{10}\frac{y}{K_s} + 8.5 \qquad (1)$$

$$u = u_*\left[2.5\,In\frac{y}{k_s} + 8.5\right]$$

Differentiating with respect to y

$$\frac{du}{dy} = u_*\left[\frac{2.5}{y}\right]$$

$$7 = \frac{2.5\,u_*}{0.025}$$

$$u_* = 0.07\ \text{m/s}$$

$$\tau_o = \rho u_*^2, \quad = 900 \times (0.07)^2 = 4.41\ \text{N/m}^2 \qquad \text{Ans}$$

(b) u = 1.5 m/s, $u_* = 0.07$, y = 0.025

Substituting in equation (1)

$$\frac{1.5}{0.07} = 5.75 \log_{10} \frac{0.025}{K_s} + 8.5$$

$$\log_{10} \frac{0.025}{K_s} = 2.248$$

$$\frac{0.025}{K_s} = 177.193$$

$$K_s = 1.41 \times 10^{-4}\ m = 0.141 \text{ mm}$$ Ans

(c) For a turbulent pipe flow

$$\frac{u_{av}}{u_*} = 5.75 \log_{10} \frac{R}{K_s} + 4.75$$

$$\frac{u_{av}}{0.07} = 5.75 \log_{10} \left(\frac{0.075}{1.41 \times 10^{-4}} \right) + 4.75$$

$$= 15.67 + 4.75 = 20.422$$

u_{av} = 1.4295 m/s

$$Q = \frac{\pi}{4} d^2\, u_{av} = \frac{\pi}{4} (0.15)^2 \times 1.4295$$

$$= 0.02526 \text{ m}^3\text{/s} = 25.26 \text{ liters/s}$$ Ans

(d) $u_* = u_{av} \sqrt{\lambda/8}$

$$\sqrt{\lambda/8} = 0.048968$$

$$\sqrt{\lambda/8} = 0.048968$$

or, $\lambda = 0.01918$ Ans

3.17. A liquid a relative density 0.9 and dynamic viscosity 6.5 × 10^{-4} kg/ms flows through a pipe of diameter 150 mm at the rate of 60 liters/s. If the loss of head in 100 m length of pipe is 5 m, determine whether the pipe is rough or smooth.

Solution

Discharge $Q = \frac{\pi}{4} d^2 u$

$$60\times10^{-3} = \frac{\pi}{4}(0.15)^3\, u$$

$$u = 3.395 \text{ m/s}$$

$$R_e = \frac{\rho\, u\, d}{\mu} = \frac{900\times 3.395\times 0.15}{6.5\times10^{-4}} = 7.05\times10^5$$

$$h_f = \frac{\lambda\, l\, u^2}{2g\, d} \qquad 5 = \frac{\lambda\times100\times(3.395)^2}{2\times9.81\times0.15}$$

$$\lambda = 0.01276$$

Let us assume that the pipe is hydrodynamically rough, then

$$\frac{1}{\sqrt{\lambda}} = 2\log_{10}\left(\frac{R}{K_s}\right) + 1.74$$

$$\frac{1}{\sqrt{0.01276}} - 1.74 = 2\log_{10}\frac{R}{K_s}$$

$$2\log_{10}\frac{R}{K_s} = 8.85 - 1.74$$

$$\log_{10}\frac{R}{K_s} = \frac{7.11}{2}$$

$$\log_{10}\frac{R}{K_s} = 3.555$$

$$\frac{R}{K_s} = 3590.598$$

Thus

$$\frac{R_e\sqrt{\lambda}}{R/K_s} = \frac{7.05\times10^5\times\sqrt{0.01276}}{3590.598} = 196.346\times\sqrt{0.01276} = 22.179$$

Since the parameter $\left[\dfrac{R_e\sqrt{\lambda}}{R/K_s}\right]$ is more than 17, and less than 400, the boundary acts in transition. Thus the assumption made for the pipe to be rough is not correct. Hence the pipe is in the transition range.

EXERCISES

A. THEORY

3.1. What is turbulence? Define (a) intensity of turbulence, (b) Reynold's stress. Explain Prandtl mixing length hypothesis and derive the relationship between the mixing length and shear stress.

3.2. For turbulent flow explain how momentum may be transferred between adjoining layers in the fluid which are moving with different velocities and obtain an expression for the resulting shear stress in terms of velocity gradient.

3.3. Define "eddy viscosity" as given by Boussinesq. What is its numerical value in comparison with the viscosity of fluid?

3.4. Define shear velocity or friction velocity for turbulent flow. Give its value in terms of shearing stress.

3.5. Describe Prandtl mixing length theory for finding the shear stress in a turbulent flow in circular pipes. What is universal constant used in this theory?

3.6. Define mixing length concept of Prandtl. Show it by means of sketch. How does the mixing length vary with the distance from the wall in turbulent flow?

3.7. Describe von-karman similarly theory. Where it is used?

3.8. Describe experimental work for pipe friction giving Darcys, Blasius and Nikuradse results.

3.9. State the universal velocity distribution laws for turbulent flow. Derive Prandtl universal velocity distribution equation.

3.10. What is the difference between smooth and rough pipes? Describe with the help of a diagram.

3.11. Describe laminar sub layer in case of turbulent flow in circular pipes. How does it change in case of smooth and rough pipes? What is true thickness of laminar sub layer? How does it differ from thickness of normal laminar sub layer?

3.12. Describe universal resistance laws of turbulent flow in smooth and rough pipes. Derive equations in each case. How is the single relation applicable for both smooth and rough pipes calculated?

3.13. Describe how the roughness of the pipe wall affects the frictional resistance to flow in pipes.

3.14. Explain the difference between wall turbulence and free turbulent shear flow.

3.15. Explain the Reynold's concept of momentum exchange, and derive the following equation for Reynold's shear stress,

$$\tau = -\rho \overline{u'\ v'}$$

3.16. Define eddy and molecular viscosity

3.17. Show that shear velocity is given by the expression

$$u_* = u_{av} \sqrt{\lambda/8}$$

3.18. Describe Taylor's vorticity transport theory. Where is it used?

3.19. Name the semiempirical theories of turbulence and explain any one.

3.20. Describe Reynolds's Time Average concept.

3.21. Show that shear velocity is given by the expression

$$u_* = \sqrt{\frac{\tau_o}{\rho}}$$

3.22. Derive the Prandtl's universal velocity distribution equation for turbulent flow in a non dimensional form.

$$\frac{u_{max} - u}{u_*} = 5.75 \log_{10} \frac{R}{y}$$

Explain why this relationship does not apply close to the pipe wall.

3.23. Derive the Prandtl-Karaman velocity distribution equation

$$u = 5.75 \log_{10} \left(y / y'\right)$$

3.24. Show that for turbulent flow in a smooth pipe

$$\frac{u}{u_*} = 5.5 + 5.75 \log_{10} \frac{u_* y}{\nu}$$

where u = velocity at radius y,

u_* = shear velocity

ν = kinematic viscosity

3.25. Assuming that

$$\frac{u}{u_*} = 5.5 + 5.75 \log_{10} \frac{u_* y}{\nu}$$

Show that for turbulent flow in a smooth pipe

$$\frac{1}{\sqrt{\lambda}} = -0.8 + 2 \log_{10} R_e \sqrt{\lambda}$$

3.26. Show that for turbulent flow in a rough pipe

$$\frac{u}{u_*} = 8.5 + 5.75 \log_{10} \left(\frac{y}{K_s}\right)$$

Where K_s = average height of protrusions

3.27. Assuming that

$$\frac{u}{u_*} = 8.5 + 5.75 \log_{10} (y / K_s)$$

Show that for turbulent flow in a rough pipe

$$\frac{1}{\sqrt{\lambda}} = 1.74 + 2 \log_{10} \frac{R}{K_s}$$

3.28. Show that the velocity distribution for turbulent flow in terms of average velocity for smooth pipes

$$\frac{u_{av}}{u_*} = 1.75 + 5.75 \log_{10} \frac{u_* R}{\upsilon}$$

and hence deduce the formula

$$\frac{u - u_{av}}{u_*} = 3.75 + 5.75 \log_{10} y/R$$

3.29. Show that the velocity distribution for turbulent flow in terms of average velocity for rough pipes

$$\frac{u_{av}}{u_*} = 4.75 + 5.75 \log_{10} \frac{R}{K_s}$$

and hence deduce the formula

$$\frac{u - u_{av}}{u_*} = 3.75 + 5.75 \log_{10} \frac{y}{R}$$

3.30. Show that the average velocity in pipe flow can be related to the centre line (maximum) velocity and the pipe friction factor by

$$\frac{u_{max}}{u_{av}} = 1 + 1.43\sqrt{\lambda}$$

3.31. Derive an expression for the turbulent pipe coefficient as

$$\frac{u}{u_{av}} = 1 + \sqrt{\lambda}\left(1.43 + 2.15 \log_{10} \frac{y}{R}\right)$$

3.32. Plot a curve of the ratio of the average to centre line velocity $\left(\frac{u_{av}}{u_{max}}\right)$ for fully developed turbulent flow in a smooth pipe as a function of the Reynold's number (R_e) of the flow. Use linear scale for $\frac{u_{av}}{u_{max}}$ as ordinate and a log scale for R_e as abscissa.

3.33. What is Colebrook-White formula? where is it used?

3.34. Describe some empirical relationships for finding the friction factor.

3.35. Describe Nikurdse experiments and their significance.

3.36. Describe the results of Nikurdse's experiments on the effect of the relative roughness R/K_s on the value of the resistance coefficient λ for turbulent flow in circular pipes.

3.37. Describe Moody's chart. What are the basis to draw such a diagram? State its uses.

3.38. Show how the smooth pipe and rough pipe formula may be combined to give an approximate relation for λ in terms of the relative roughness K_s/d and the pipe Reynold's number.

3.39. What do you mean by R* = Reynold's roughness number. Give the values of R* for smooth, departure from smooth pipe law and rough pipe law.

3.40. Draw a curve between K_s/δ' and $R_e\sqrt{\lambda}/R\,K_s$ and show on the curve the nature of the boundary.

3.41. Draw Colebrook-White transition curve for commercial pipes and explain its significance.

3.42. Draw a curve between $\frac{1}{\sqrt{\lambda}} - 2\log_{10}\frac{3.7\,d}{K_s}$ and Reynold's roughness number (R_*) and explain it.

3.43. Draw a simplified sketch of Moody's diagram and show the various zones.

3.44. Draw a hydraulic research station charts (HRS) for any value of K_s and discuss it.

3.45. Write down the various explicit equation for friction factor λ for flow in commercial pipes.

(a) Moody's formula; (b) Barr's explicit formula; (c) Murdock formula; (d) Swamee and Jain's explicit formula; (e) SE Haaland's formula

3.46. How is the head loss due to friction in commercial pipe determined? In what way the old pipes will be different from the new pipes.

3.47. What do you understand by roughening of pipes with age? How is roughness of an old pipes determined?

3.48. "The friction factor λ becomes independent of Reynold's number at very high Reynold's number" state the significance attached to this statement.

3.49. Draw Stanton and Pannell curve and discuss it.

3.50. What is Blasius equation and explain its significance.

B. UNSOLVED PROBLEMS

3.1. Fully developed flow in a 300 mm C.I. pipe is at an average velocity of 1.8 m/s. What is the centre line velocity for (a) fuel oil at 10°C (υ = 0.002 m²/s) and (b) crude oil at 20°C (ν = 5.6×10^{-6} m²/s) ? [(a) 3.6 m/s, (b) 2.16 m/s]

3.2. If the velocities in a 600 mm pipe carrying oil are 4.5 m/s and 4.2 m/s on the centre line and at a radial distance of 100 mm from the axis, calculate the discharge in the pipe. [0.958 m³/s]

3.3. A pipe line carrying water has average height of irregularities projecting from the surface of the boundary of the pipe as 0.15 mm. What type of boundary is it ? The shear stress developed is 4.9 Pa. The kinematic viscosity of water is 1 cs.

$$[R_e = \frac{u_* K_s}{\upsilon} = 10.5, \text{ transition state}]$$

3.4. A pipe of 250 mm diameter carries air (ρ = 1.22 kg/m^3, υ = 1.5 × 10^{-5} m^2/s) at an average velocity of 8 m/s. The equivalent sand grain roughness of the pipe is 0.5 mm. Calculate the friction factor λ by assuming the flow to be fully rough-turbulent. What is the shear stress at the boundary? [λ = 0.0234, τ_o = 0.2285 Pa]

3.5. Determine the wall shearing stress in a pipe diameter 100 mm, which carries water. The velocities at the pipe centre and 30 mm, from the pipe centre are 2 m/s and 1.5 m/s respectively. [47.676 N/m^2]

3.6. Water (at 20°C) flows in a 75 mm diameter smooth brass pipeline. Wall shear stress is 3.68 N/m^2. Calculate the thickness of the viscous sub layer, the friction factor, the mean velocity and flow rate, the centreline velocity, the shear stress and velocity 25 mm from the pipe centreline, and the head lost in 1000 m of this pipe. [u_* = 0.061 m/s, δ' = 0.19 mm, u_{av} =1.29 m/s, u_{max} =1.54 m/s u_{25} = 1.35 m/s, h_f = 20.4 m]

3.7. For the conditions of the problem 12.6, viz, u_{av} = 1.29 m/s, d= 75 mm, ρ = 998 kg/m^3 and ν = 1 cs. Calculate the friction factor, wall shear stress, centreline velocity, and the velocity 25 mm from the pipe centre line, using the seventh root law. [R_e=96750 λ=0.018, u_{max}=1.58 m/s, u_{25} =1.35 m/s, τ_o =3.7 Pa]

3.8. The mean velocity in a 300 mm pipe line is 3 m/s. The relative roughness of the pipe is 0.002 and the kinematic viscosity of water 9 × 10^{-7} m^2/s. Determine the friction factor, centre line velocity, velocity 50 mm from the pipe wall, and the head lost in 300 m of this pipe under the assumption that the pipe is rough. [λ = 0.0234, u_* = 0.162 m/s, u_{max} = 3.62 m/s, u_{50} = 3.17 m/s, h_l = 10.7 m of water]

3.9. What is the pressure drop in 15 m of 6 mm tubing for a horizontal flow of hydraulic fluid (μ = 0.014 kg/ms and S = 0.848) at 50°C at a velocity of 2 m/s? What is the wall shear stress? [37.3 Pa]

3.10. What is the flow rate for water at 15°C in a commercial steel pipe, 250 mm diameter, when the head loss in 300 mm of pipe is 5 m ? Take K_s = 0.000045 m, and ν = 1.40 × 10^{-6} m^2/s [144 liters/s]

3.11. Water flows at the rate of 91 liters/s over a distance of 500 m in a horizontal commercial steel pipe with a pressure drop not to exceed 825 kPa. What is the minimum size pipe which can be used? Take ν = 1 cs. [0.152 m]

3.12. Water at 40°C flows in a 75 mm pipe at Reynold's number 8 $\times 10^4$. If the pipe contains a uniform sand roughness of grain size 0.15 mm diameter, how much head loss is to be expected in 300 m of the pipe? How much head loss would be expected if this pipe were smooth? [2.1 m, 1.88 m]

3.13. Calculate the loss of head and the pressure drop when air at an absolute pressure of 101.3 kPa and 15°C flows through 600 m of a 450 mm by 300 mm smooth rectangular duct with a mean velocity of 3 m/s. [R_n = 0.09 m, R_e = 73950, λ = 0.019, h_l = 14.5 m of air, δp = 174 Pa]

3.14. A smooth pipe of diameter 80 mm and 800 m long carries water at the rate of 480 liters/min. Calculate the loss of head, wall shearing stress, centreline velocity, velocity and shear stress at 30 mm form the pipe wall. Also calculate the thickness of laminar sub layer. Take ν = 0.015 stokes. [23.92 m, 5.860 Pa, 1.88 m/s, 1.825 m/s, 0.2274 mm]

3.15. For turbulent flow in a pipe of diameter 30 mm, find the discharge when the centreline velocity is 2 m/s and the velocity at a point 100 mm from the centre as measured by pitot tube is 1.6 m/s. [102.7 liters/s].

4

Open Channel Flow

4.1 INTRODUCTION

Liquids are transported from one location to another using natural or constructed conveyance structures and flow passages. These flow passages may have cross sections that are open or closed at the top. The structures with open tops are referred to as open channels and those with closed tops are referred to closed conduits.

Open channel flow involves the flow of a liquid in a channel or conduit that is not completely filled. These exists a free surface between the flowing fluid (usually water) and fluid above it (usually the atmosphere). The forces causing flow are due to gravity, and the forces retarding flow are due to viscous shear along the channel bed. Thus both the Froude number and the Reynold's number define the state of the flow.

Open channel flows are different from a pipe flows. The flow in a pipe takes place under pressure. Whereas the fluid flow in open channels is characterised primarily by the exposure of a free surface to atmospheric pressure. Water flowing in open channels is acted upon by all the forces that affect pipe flow with addition of gravitational and surface tension forces that are the direct consequence of the free surface.

Water flows in a channel will always be subjected to some resistance. To overcome the resistance and to cause the flow of water in a channel, it is constructed with its bottom slopping towards the direction of flow, so that a component of the weight of the flowing water in the direction of flow is developed which causes the flow of water in the channel.

Despite the similarily between the two kinds of flow, it is much more difficult to solve problems of flow in open channels than in pressure pipes. Flow conditions in open channels are complicated by the fact that the position of the free surface is likely to change with respect to time and space and also by the fact that the depth of flow, the discharge, and the slopes of the channel bottom and of the free surface are independent.

The treatment of open channel flow is somewhat more empirical than the pipe flow. The empirical method is the best available at present and, if cautiously applied, can yield results of practical value.

The flow in a closed conduit is not necessarily pipe flow. It must be class-ified as open channel flow if it has a free surface. The storm sewer, for example, which is a closed conduit, is generally designed for open channel flow because the flow in the sewer is expected to maintain a free surface most of the time.

The channels may be classified as natural channel or artificial channel. A natural channel is the one which has irregular sections of varying shapes, which is developed in a natural way. The artificial channel is the one which is built artificially for carrying water for various purposes. Flow in rivers, flumes, flow of rain water in the gutters of our houses, drainage, ditches, sewers, and gutters along roads, flow in chutes, partly filled culverts etc. are example of flow in open channels.

A prismatic channel has a constant cross section and constant bed slope, though the flow cross section may vary in the direction of flow because of changes in the depth.

This chapter is devoted to flow through channels (prismatic) whose geometry is stable and whose beds are non erodible.

4.2 FLOW CLASSIFICATION

In the study of pipe flow we found that there are many ways to classify a flow–laminar, turbulent, stcady, uniform and so on. For open channel flow the existence of a free surface allows additional types of flow. The extra degree of freedom because the fluid is free to select its free surface location and configuration (because it does not completely fill a pipe) allows important phenomenon in open channel flow that cannot occur in pipe flow. Some of the classifications of the flows are described below.

4.2.1 Steady uniform flow

Steady uniform flow, in which the depth is constant, both with time and distance. This constitutes the fundamental type of flow in open channel in which the gravity forces in equilibrium with the resistance forces.

4.2.2 Steady non uniform flow

Steady non uniform, in which the depth varies with distance, but not with time. The flow may be either

(a) Gradually varying flow (GVF)

In this type of flow, the depth changes slowly with distance along the channel or

(b) Rapidly varying flow (RVF)

In this type of flow, the depth changes considerably over a relatively short distance.

4.2.3 Unsteady flow

Unsteady flow in which the depth varies with both time and distance (Unsteady uniform flow is very rare). This is the most complex flow type, requiring the solution of energy, momentum and friction equations through time.

The various flow types are all shown in Fig 4.1.

(a) Steady uniform flow (b) Unsteady uniform flow (c) Steady varied flow (d) Unsteady flow.

4.2.4 Laminar and turbulent flows

As for any flow geometry, open channel flow may be laminar, transitional, or turbulent depending on various conditions involved. Which type of flow occurs depends on the Reynold's number, $R_e = \dfrac{\rho\, v\, R_h}{\mu}$, where R_h is the hydraulic radius of the channel.

A general rule is that open channel flow is laminar if $R_e \leq 500$, turbulent if $R_e \geq 2000$, in between these two limits, the flow is in a transitional state.

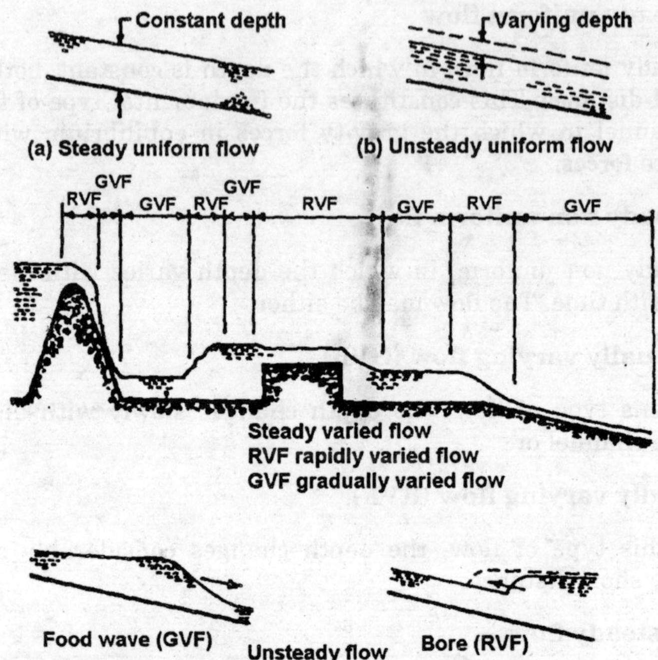

Fig. 4.1. Classification of open channel flow

The values of these dividing Reynold's numbers are only approximate a precise knowledge of the channel geometry is necessary to obtain specific values. Since many open channel flows involve water (which has a fairly small viscosity) and have relatively large characteristic lengths, it is uncommon to have laminar open channel flows.

4.2.5 Subcritical, critical and supercriticial flows

Open channel flows involve a free surface that can deform from its undisturbed relatively flat configuration to form waves. Such waves moves across the surface at speeds that depend on their size and properties of the channel. The character of an open channel flow may depend strongly on how fast the fluid is flowing relative to how fast a typical wave moves relative to the fluid. The dimensionless parameter that describes this behaviour is termed the Froude number, $F_r = \dfrac{V}{\sqrt{gl}}$

$F_r < 1$ Subcriticial (tranquil) flow

$F_r = 1$ Critial flow

$F_r > 1$ Supercritical (rapid) flow.

A combined effect of viscosity and gravity may produce any one of four regimes of flow in an open channel, namely.

(1) Subcritical-laminar flow, = $F_r < 1$ and $R_e \leq 500$

(2) Supercritical flow, $F_r > 1$ and $R_e \leq 500$.

(3) Subcritical turbulent flow, $F_r < 1$, $R_e \geq 2000$

(4) Supercritical, turbulent flow, $F_r > 1$, $R_e \geq 2000$.

The depth velocity relationship for the four regimes in a wide open channel can be shown by a logarthmic plot (Fig 4.2).

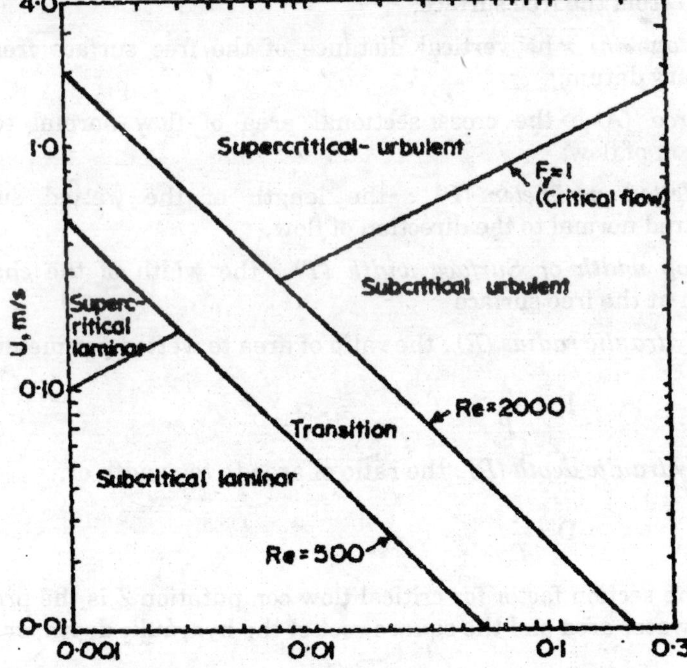

Fig. 4.2. Classification of flow in a wide rectangular channel.

4.3 GEOMETRIC ELEMENTS OF CHANNEL SECTION

Geometric elements are properties of a channel section that can be defined entirely by the geometry of the section and the depth of flow. These elements are very important and are used extensively in flow computations.

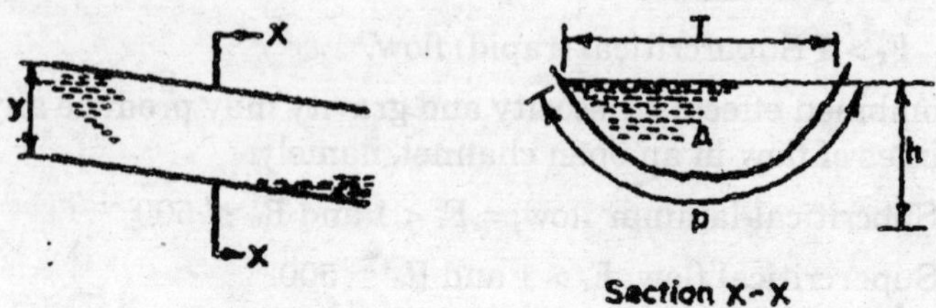

Fig. 4.3. Definition sketch of geometric channel properties

The commonly used geometric properties are shown in Fig 4.3 and defined as follows.

Depth (y) : the vertical distance of the lowest point of a channel section from the free surface;

Stage (h) : he vertical distance of the free surface from an arbitrary datum;

Area (A) : the cross sectional area of flow normal to the direction of flow;

Wetted perimeter (P) : the length of the wetted surface measured normal to the direction of flow,

Top width or Surface width (T) : the width of the channel section at the free surface

Hydraulic radius (R) : the ratio of area to wetted perimeter or,

$$R = \frac{A}{P} \tag{4.1}$$

Hydraulic depth (D) : the ratio of area to top width or,

$$D = \frac{A}{T} \tag{4.2}$$

The section factor for critical flow computation Z is the product of the water area and the square root of the hydraulic depth, or

$$Z = A\sqrt{D} = A\sqrt{\frac{A}{T}} \tag{4.3}$$

The section factor for uniform flow computation $AR^{2/3}$ is the product of the wetted area and the hydraulic radius to two thirds power.

4.4 VELOCITY DISTRIBUTION IN A CHANNEL SECTION

The point velocity in an open channel varies continuously across the cross section because of friction along the boundary. However, the velocity distribution is not axisymmetric due to the presence of the free surface. One might expect to find the maximum velocity at the free surface, where the shear stress is negligible, but the maximum velocity normally occurs below the free surface and the fluid velocity is zero on the wetted perimeter, where a wall shear stress is developed. This shear stress is seldom uniform along the wetted perimeter. Typically variations are shown in Fig 4.4 for various channel shapes. The velocity distribution in a channel section depends on the various factors such as the shape of the section, the roughness of the channel and the presence of bends in the channel alignment.

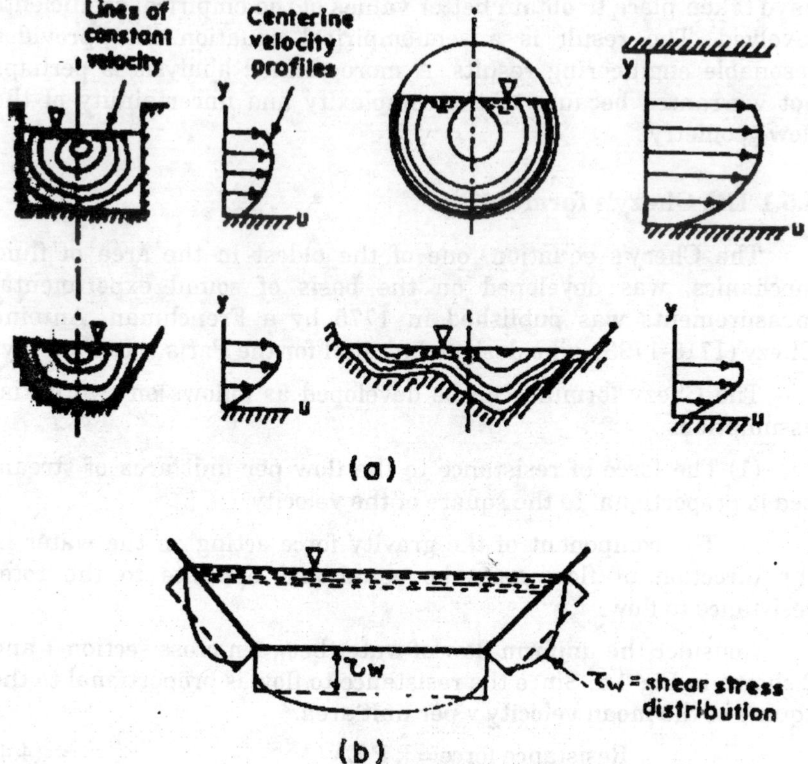

Fig. 4.4. Velocity distribution in open channel

4.5 UNIFORM CHANNEL FLOW

In the uniform channel flow, the conditions at all cross sections are identical and the consideration of a single cross section will be representative of the whole.

The death and the average velocity remain constant, and thus the channel bed, water surface, and the total head line are parallel ($S_o = S_w = S_f = S$). This type of flow rarely occurs in a natural stream and exists mainly in relatively long man mad channels. Even though uniform flow is not common, its treat-ment is important because many non uniform flows approach a uniform flow condition, and uniform flow criteria are used in analyzing many nonuniform flows.

The basic equations used to determine the uniform flowrate in open channels were derived many years ago. Continual refinements have taken place to obtain better values of the empirical coefficients involved. The result is a semiempirical equation that provides resonable engineering results. A more refined analysis is perhaps not warranted because of the complexity and uncertainlity of the flow geometry.

4.5.1 The Chezy's formula

The Chezy's equation, one of the oldest in the area of fluid mechanics, was developed on the basis of sound experimental measurements was published in 1775 by a Frenchman, Antoine Chezy (1718-1798), who designed a canal for the Paris water supply.

The Chezy formula can be developed as follows on two initial assumptions.

(1) The force of resistance to the flow per unit area of stream bed is proportional to the square of the velocity.

(2) The component of the gravity force acting on the water in the direction of flow must be equal and opposite to the total resistance to flow.

Consider the uniform flow of water between cross section 1 and 2 shown in Fig 4.5. Since the resistance to flow is proportional to the square of the mean velocity v per unit area.

$$\text{Resistance force} = k\, P\, L\, v^2 \qquad (4.4)$$

in which K = a friction coefficient

P = the length of the wetted perimeter in cross section

L = the length of channel between sections 1 and 2.

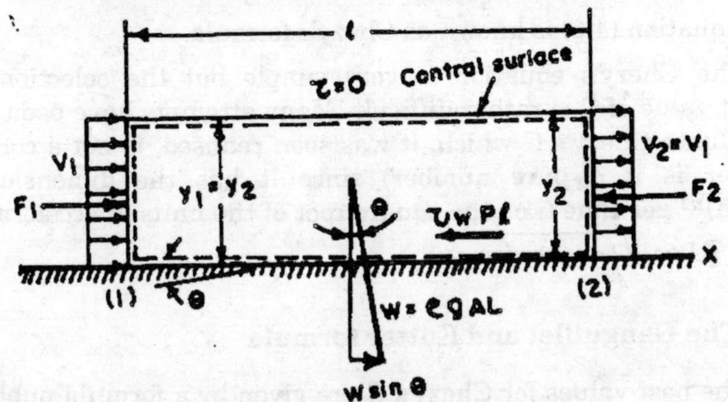

Fig. 4.5. Uniform flow in open channel

If the channel bed slopes at an angle θ to the horizontal, then sinθ = h/L, h being the fall in level over the length L. The slope of the channel bed S is defined as

$$S = \sin\theta = h/L$$

The water weight components causing the flow will be

$$W = \rho\, g\, AL \sin\theta = \rho g\; AL\; S \qquad (4.5)$$

in which

A = cross section area of the channel flow

ρ = density of water.

Equating equation (4.5) to (4.4) gives

$$\rho\, g\; ALS = k\; PL\, v^2$$

$$\text{or,} \qquad v^2 = \left(\frac{\rho\, g}{k}\right)\left(\frac{A}{P}\right) S$$

$$v^2 = \left(\frac{\rho\, g}{k}\right)(R\; S)$$

$$v = \sqrt{\frac{\rho g}{k}}\; \sqrt{RS} \qquad (4.6)$$

$$v = C\; \sqrt{RS}$$

in which $C = \sqrt{\frac{\rho g}{k}}$ = Chezy's constant

Equation (4.6) is known as Chezy's formula.

The Chezy's equation is very simple but the selection of a correct value of C is rather difficult. Many attempts have been made to evaluate Chezy's C which, it was soon realised, is not a constant (neither is it a pure number) since it has the dimensions of $(\text{length})^{1/2}$ per time (i.e., the square root of the units of acceleration). or $L^{1/2}\,T^{-1}$ or $\sqrt{Acceleration}$.

4.5.2 The Ganguillet and Kutter formula

The best values for Chezy's C are given by a formula published in 1869 by two Swiss engineers Ganguillet and Kutter in which C is expressed in terms of the bed slope S, the hydraulic radius R and a coefficient of roughness n.

$$C = \frac{23 + \dfrac{0.00155}{S} + \dfrac{1}{n}}{1 + \left(23 + \dfrac{0.00155}{S}\right)\dfrac{n}{\sqrt{R}}} \tag{4.7}$$

The coefficient n in this formula is known as Kutter's n.

4.5.3 The Manning's formula

In 1889, Robert Manning (1816-1897), an Irish engineer, developed the following somewhat modified equation for open channel flow to more accurately describe the R dependence.

$$v = \frac{1}{n} R^{2/3}\, S^{1/2} \tag{4.8}$$

Equation (4.8) is termed the Manning equation, and the parameter n is the Manning resistance coefficient. Its value is dependent on the surface material of the channel's wetted perimeter and is obtained from experiments. It is not dimensionless, having the units of $s/m^{1/3}$.

The value of chezy's C is obtained by comparing the Mannings and chezy's formula

$$C = \frac{1}{n} R^{1/6} \tag{4.9}$$

The equation (4.9) provides an important relationship between chezy's C and Manning's n.

The Chezy's C is recognised to be a function of the hydraulic radius and also of a roughness coefficient n. This roughness coefficient n is for practical purposes identical with kutter's n and the same values are normally used.

The Manning formula has found favour with engineers for a long time because of its simplicity and the satisfactory results that it gives. The Ganguillet and Kutter formula is seldom used because of its complexity, except in tabulated or graphical form.

The values of Manning's n are given in Table 4.1

Table 4.1 Values of the Manning coefficient n. (Source Munson etc)

	Wetted Perimeter	n
A	***Natural Channels***	
	Clean and Straight	0.030
	Sluggish with deep pools	0.040
	Major rivers	0.035
B	***Floodplains***	
	Pasture, farmland	0.035
	Light brush	0.050
	Heavy brush	0.075
	Trees	0.15
C	***Excavated earth channels***	
	Clean	0.022
	Gravelly	0.025
	Weedy	0.030
	Stony, Cobbles	0.035
D	***Artificially lined channels***	
	Glass	0.010
	Brass	0.011
	Steel smooth	0.012

Steel, painted	0.014
Steel, riveted	0.015
Cast iron	0.013
Concrete, finished	0.012
Concrete, unfinished	0.014
Planed Wood	0.012
Clay tile	0.014
Brick Work	0.015
Asphalt	0.016
Corrugated metal	0.022
Rubble masonary	0.025

4.6 MOST ECONOMICAL SECTION OF CHANNEL

The engineer faced with the problem of designing a channel cross section must keep two requirements in mind. His design must fulfil the function for which it is needed and it must not cost more than is essential to construct and maintain. The cost of construct is roughtly proportional to the cross sectional area of channel and the problem of maintenance is related to the stability of the sides if the channel is cut through erodible material.

If the excavation cost is assumed to be proportional to the cross sectional area A then for the most economical design it is necessary to determine the shape which will require the minimum cross sectional are a for a given discharge. the determination of this most economical shape, called the best hydraulic section.

Manning's equation indicates that for a given channel slope, roughness, and cross section a shape which has a minimum perimeter will convey the maximum discharge. Or a given flow will require a minimum cross section to convey it when the perimeter is also a minimum.

Minimizing the perimeter gives the most efficient cross section, the semicircle being the best. Fabricated channels or flumes may be made seimicirculars, rectangular, or trapezoidal in cross section. In any case, the most efficient section is one for which the hydraulic radius is one half the depth. This condition is automatically met for a semi circle. For a rectangular channel, the width should equal twice the depth (b = 2y). For a trapezoidal channel, a half hexagon is most efficient.

Criteria for the design of an open channel for uniform flow depend on whether the channel is nonerodible, erodible, or grass lined. Important criteria in the design of non erodible channels are lining costs (which depend on the channel shape as well as the lining material), minimum flow velocities in order to avoid deposition of sediment which might be carried in the stream, and free board requirements. In erodible channels, the maximum flow velocities are limited by the probability of bed scouring. Grass lining are used in erodible channels to prevent bed erosion, and such a channel may be designed on the basis of stability of the bed or for maximum capacity.

4.6.1 Rectangular channel section

This is frequently used for small channels constructed in concrete. For a rectangular channel section having bottom width b and depth of flow y as shown in Fig 4.6.

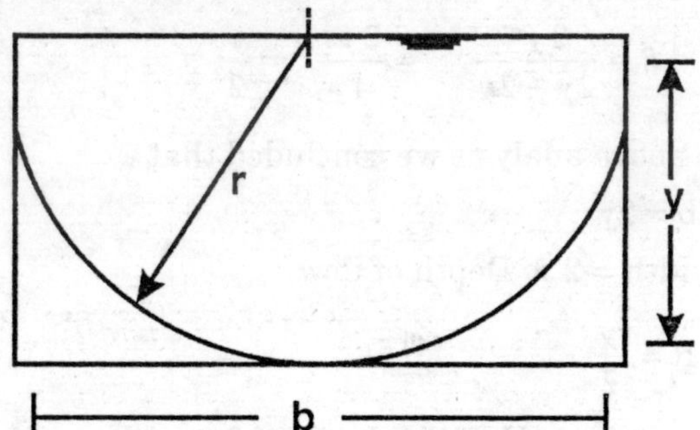

Fig. 4.6. Rectangular channel section

$$\text{Area } A = b\,y \tag{4.10}$$

$$\text{Wetted Perimeter } P = b + 2y \tag{4.11}$$

From (4.10),

$$b = \frac{A}{y}$$

Substituting the value of b in equation (4.11)

$$P = \frac{A}{y} + 2y \tag{4.12}$$

Assuming a fixed value for A and differentiating equation (4.12) with respect to y and equated to zero for obtaining the condition of minimum P.

$$\frac{dp}{dy} = -\frac{A}{y^2} + 2 = 0$$

or, $A = 2y^2$

or, $by = 2y^2$ [From equation (4.10)]

or, $b = 2y$ (4.13)

The equation (4.13) gives b = 2y which is a rectangle that can be circumscribed around a semicircle (Fig 4.6)

In this case, hydraulic radius R is given by

$$R = \frac{A}{P} = \frac{by}{b + 2y} \quad \text{[From equation (4.10) and (4.11)]}$$

but b = 2y from equation (4.13)

therefore,

$$R = \frac{2y^2}{2y + 2y} = \frac{2y^2}{4y} = \frac{y}{2} \quad (4.14)$$

From the above analysis we concluded that

(1) $b = 2y$

Bottom width = 2 × Depth of flow.

(2) $R = \frac{y}{2}$

Hydraulic radius = Half the depth of flow.

In wide channels the vaule of R approaches y. For example when b = 10 y, $R = \frac{20y^2}{22y} = 0.91\,y$.

4.6.2 Trapezoidal channel section

Fig 4.7 shows a trapezoidal channel section whose sides slope at a gradient of 1 vertical to z horizontal. The bottom width is b and depth of flow is y and the length of the sloping side $y\sqrt{1 + z^2}$

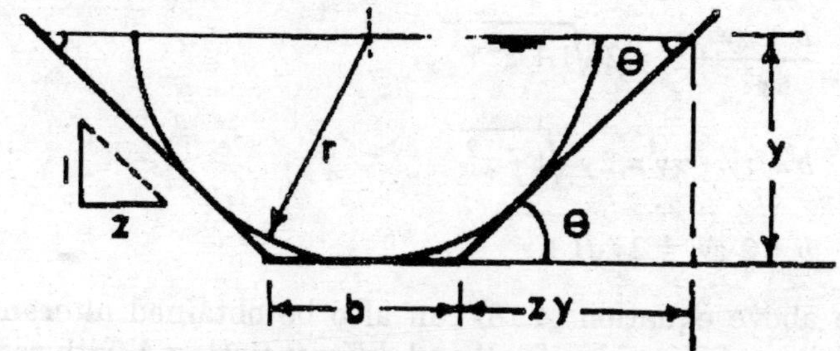

Fig. 4.7. Trapezoidal channel section

Area A = Area of trapezium.

$$= \frac{\text{Sum of the parallel sides}}{2} \times \text{Height}$$

$$= \left[\frac{b + (b + 2\,zy)}{2}\right] y = \left[\frac{2b + 2\,zy}{2}\right] y = (b + zy)\, y \qquad (4.15)$$

Wetted perimeter,

$$p = b + 2y\sqrt{1 + z^2} \qquad (4.16)$$

From equation (4.15)

$$b = \frac{A}{y} - zy$$

Substituting in equation (4.16)

$$p = \frac{A}{y} - zy + 2\,y\sqrt{1 + z^2} \qquad (4.17)$$

Assuming a fixed value for A and differentiating equation (4.17) with respect to y and equated to zero for the condition of minimum p.

$$\frac{dp}{dy} = -\frac{A}{y^2} - z + 2\sqrt{1 + z^2} = 0$$

or, $$\frac{A}{y^2} + z = 2\sqrt{1 + z^2}$$

or, $$\frac{(b + zy)\,y}{y^2} + z = 2\sqrt{1 + z^2}$$ [From equation (4.15)]

$$\text{or,} \quad \frac{b + zy}{y} + z = 2\sqrt{1 + z^2}$$

$$\text{or,} \quad b + zy + zy = 2y\sqrt{1 + z^2}$$

$$\text{or,} \quad b + 2\,zy = 2y\sqrt{1 + z^2} \tag{4.18}$$

The above equation (4.18) can also be obtained alteranatively by assuming a fixed value for P and differentiating A with respect to y and equated to zero. i.e. $\frac{dA}{dy} = 0.$

Equation (4.18) is the required condition for a trapezoidal channel section to be most economical. It shows that for a trapezoidal channel section to be most economical.

Top width = 2 × Lenth of the slopping sides.

In this case, the hydraulic radius R is given by

$$R = \frac{A}{P} = \frac{(b + zy)\,y}{b + 2y\sqrt{1 + z^2}}$$

$$= \frac{(b + zy)\,y}{b + b + 2\,zy} \quad \text{[From equation 4.18]}$$

$$= \frac{(b + zy)\,y}{2\,(b + zy)} = \frac{y}{2} \tag{4.19}$$

Now assuming that the side slope z is a variable and that A and y have fixed values. Equating $\frac{dp}{dz}$ to zero will now give the minimum value of p.

Differentiating equation (4.17) with respect to z and equated to zero.

$$\frac{dp}{dz} = 0 - y + \frac{2y}{2\sqrt{1 + z^2}}(2z) = 0$$

$$\text{or,} \quad \frac{2\,zy}{\sqrt{1 + z^2}} = y$$

$$\text{or,} \quad 2z = \sqrt{1 + z^2}$$

Squaring on both the sides

or, $4z^2 = 1 + z^2$

or, $3z^2 = 1$

or, $z^2 = \frac{1}{3}$

or, $z = \frac{1}{\sqrt{3}}$ (4.20)

therefore, $\tan\theta = \frac{1}{z} = \frac{1}{1/\sqrt{3}} = \sqrt{3}$

$$\theta = \tan^{-1}\left(\sqrt{3}\right) = 60^\circ \qquad (4.21)$$

Therefore, the most economical section which has θ = 60° will also have its top width equal to twice the slopping sides. It will therefore have the form of half a regular hexagon or three adjacent equilateral triangles of side 2y/3.

Let Fig. 4.8 shows the trapezoidal channel of the most economical section.

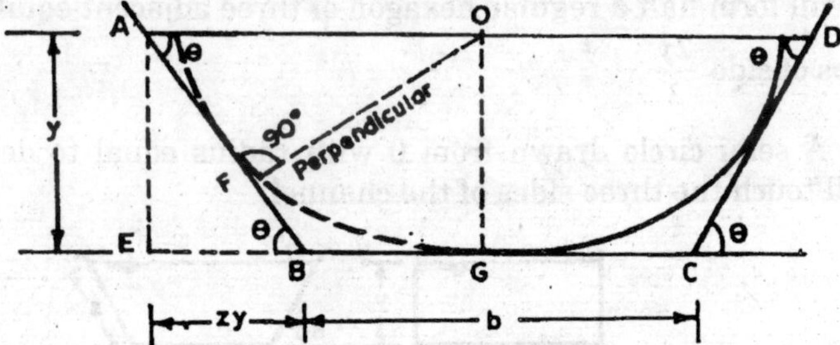

Fig. 4.8. Trapezoidal most economical section

From the triangle, OAF

$$\sin\theta = \frac{OF}{AO}$$

or, OF = AO sin θ (4.22)

From the right angled triangle AEB

$$\sin\theta = \frac{AE}{AB} = \frac{y}{y\sqrt{1+z^2}} = \frac{1}{\sqrt{1+z^2}}$$

Substituting in equation (4.22)

$$OF = AO \frac{1}{\sqrt{1+z^2}} = \left(\frac{b+2nz}{2}\right)\frac{1}{\sqrt{1+z^2}}$$

$$= \frac{b+2nz}{2\sqrt{1+z^2}} = \frac{2y\sqrt{1+z^2}}{2\sqrt{1+z^2}}$$

$$= y = \text{Depth of flow} \tag{4.23}$$

From the above analysis we concluded that

(1) $b + 2\,zy = 2y\sqrt{1+z^2}$

Top Width = 2 × Length of the slopping sides

(2) $R = \frac{y}{2}$

$$\textit{Hydraulic radius} = \frac{\textit{Depth of flow}}{2}$$

(3) $\theta = 60°$.

It will form half a regular hexagon of three adjacent equilateral triangles of side $\frac{2y}{3}$

(4) A semi circle drawn from 0 with radius equal to depth of flow will touch the three sides of the channel.

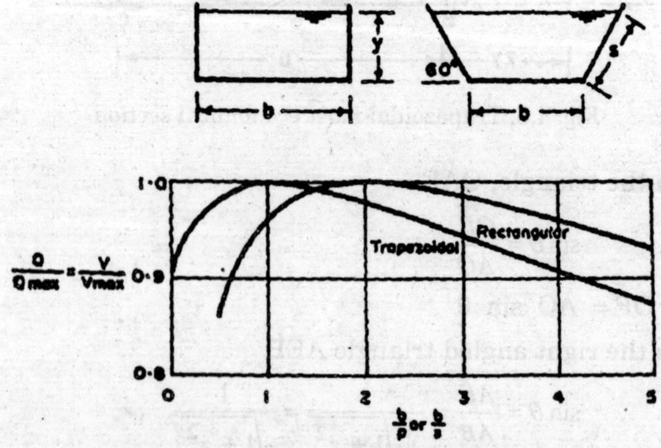

Fig. 4.9. Effect of channel geometry on capacity

Fig. 4.9 Compares the flow rates of various rectangular and trapezoidal channels of constant cross section in terms of the (maximum) flow rate for the most efficient geometry. It is seen that the ratio of bottom width to side dimensions may vary from 0 to 4 for the trapezoidal channel, and from 0.6 to more than 5 for the rectangular channel with less than a 10 per cent decrease in flow rate for a given area.

4.6.3 Triangular channel section

For the triangular channel section put b = 0 in the trapezoidal section (Fig. 4.10). The equation (4.15) and (4.16) reduces to

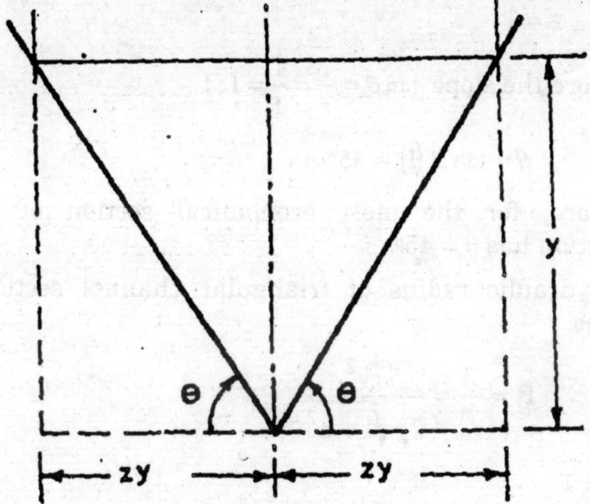

Fig. 4.10. Triangular channel section

$$A = zy^2 \tag{4.24}$$

$$P = 2y\sqrt{1+z^2} \tag{4.25}$$

From equation (4.24)

$$y = \sqrt{\frac{A}{z}}$$

Substituted in equation (4.25)

$$p = 2\sqrt{\frac{A}{z}}\sqrt{1+z^2}$$

or, $p = 2\sqrt{A}\sqrt{\dfrac{1+z^2}{z}} = 2\sqrt{A}\sqrt{z^{-1}+z}$ (4.26)

Differentiating the equation (4.26) with respect to z and equated to zero.

$$\frac{dp}{dz} = 0 = 2\sqrt{A}\frac{d}{dz}\sqrt{\left(z^{-1}+z\right)}$$

$$-\frac{1}{z^2} + 1 = 0$$

or, $z^2 = 1$

or, $z = 1$ (4.27)

Therefore the slope $\tan\theta = \dfrac{1}{z} = \dfrac{1}{1} = 1:1$

$$\theta = \tan^{-1}(1) = 45^\circ \tag{4.28}$$

Therefore, for the most economical section of triangular channel section has $\theta = 45^\circ$.

The hydraulic radius of triangular channel section can be expressed as

$$R = \frac{A}{P} = \frac{zy^2}{2y\sqrt{1+z^2}}$$

Put z = 1

$$R = \frac{y^2}{2y\sqrt{2}} = \frac{y}{2\sqrt{2}} \tag{4.29}$$

Thus it can be seen that the most economical triangular channel section will be half square described on a diagonal and having equal slopping sides.

4.6.4 Circular channel section

Circular channel section, only partly full are frequently encountered in civil engineering practice, particularly as drains and sewers. The same principles are used as if it were in a channel completely open at the top. The two important points in this analysis are that the area of flow cannot be maintained constants and the wetted perimeter varies with the depth of flow. Thus in the case of circular channel two separate conditions may be derived for the maximum discharge and the maximum average velocity of flow.

(a) Condition for maximum velocity for circular channel section

The circular pipe flowing partly full shown in Fig. 4.11 has a radius r and depth of water is y.

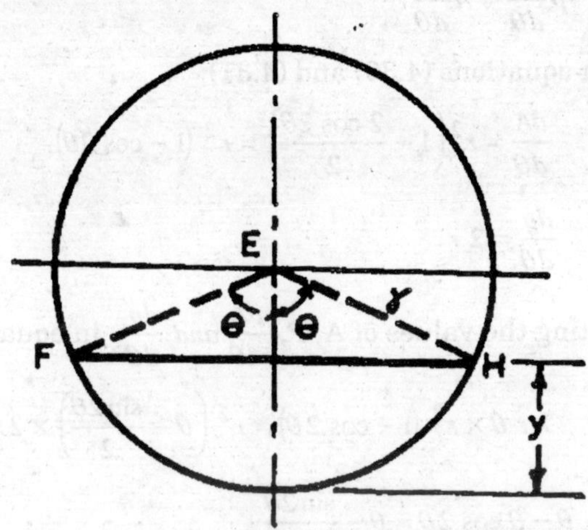

Fig. 4.11. Circular channel section

The velocity of flow in a given circular channel will depend upon the depth of the water. As the velocity is proportional to the hydraulic radius, its maximum value may be obtained by differentiating (A/P) and equated to zero.

For any depth y

Area of flow A = Sector EFGH – triangle EFH

$$A = \pi r^2 \left(\frac{2\theta}{2\pi}\right) - 2\left[\frac{1}{2} r \sin\theta \, r \cos\theta\right]$$

$$= r^2\,\theta - r^2 \sin\theta\cos\theta = r^2\left[\theta - \frac{\sin 2\theta}{2}\right] \qquad (4.30)$$

Wetted perimeter $p = 2\,\pi\, r\left(\frac{2\theta}{2\pi}\right) = 2\,\tau\,\theta$ (4.31)

Using the Chezy's formula

$$v = C\sqrt{RS}$$

and v will be a maximum for the value of θ which makes (A/P) a maximum

$$\frac{d\left(A/P\right)}{d\,\theta} = \frac{p\dfrac{dA}{d\theta} - A\dfrac{dp}{d\theta}}{p^2} = 0$$

or, $$p\frac{dA}{d\theta} = A\frac{dp}{d\theta} \qquad (4.32)$$

But from equations (4.30) and (4.31)

$$\frac{dA}{d\theta} = r^2\left(1 - \frac{2\cos 2\theta}{2}\right) = r^2\left(1 - \cos 2\theta\right) \qquad (4.33)$$

and $$\frac{dp}{d\theta} = 2\,r \qquad (4.34)$$

Substituting the values of A, P, $\frac{dA}{d\theta}$ *and* $\frac{dp}{d\theta}$ in equation (4.32)

$$2\,r\,\theta \times r^2\left(1 - \cos 2\theta\right) = r^2\left(\theta - \frac{\sin 2\theta}{2}\right) \times 2r$$

or, $$\theta - \theta\cos 2\theta = \theta - \frac{\sin 2\theta}{2}$$

or, $$\frac{\sin 2\theta}{\cos 2\theta} = 2\theta$$

or, $$\tan 2\theta = 2\,\theta \qquad (4.35)$$

This equation is satisfied when 2θ ≈ 257.5°.

The depth of flow in the channel for maximum velocity is given by

$$y = r - r\cos\theta = r\left[1 - \cos\theta\right] = r\left[1 - \cos\frac{257.5^\circ}{2}\right]$$

$$= r\left[1 - \cos 128.75^\circ\right] = \mathrm{r}\left[1 + 0.62\right] = 1.62\,\mathrm{r} \qquad (4.36)$$

That is, the depth of flow (y) for maximum mean velocity of flow in circular channel section is equal to 0.81 times the diameter of the channel.

y = 0.81 (diameter of the channel) (4.37)

(b) Condition of maximum discharge for circular channel section

According to chezy's equation

$$v = C\sqrt{RS} = C\sqrt{\frac{A}{p}S}$$

and $\quad Q = Av = C\sqrt{\frac{A^3}{p}S}$

Therefore, the discharge is a maximum when (A^3/P) is a maximum. Thus differentiating (A^3/P) with respect to θ and equating to zero for a maximum, we get

$$\frac{d\left(A^3/P\right)}{d\,\theta} = \frac{p\,3A^2\dfrac{dA}{d\theta} - A^3\dfrac{dp}{d\theta}}{p^2} = 0$$

$$P3A^2\frac{dA}{d\theta} = A^3\frac{dp}{d\theta}$$

$$3P\frac{d\,A}{d\theta} = A\frac{dp}{d\theta} \qquad (4.38)$$

Subtituing the values of A, P, $\dfrac{dA}{d\theta}$ *and* $\dfrac{dp}{d\theta}$ from the equations (4.30), (4.31), (4.33) and (4.34) we get

$$3\,(2r\theta)\left[r^2\,(1-\cos 2\theta)\right] = r^2\left(\theta - \frac{\sin 2\theta}{2}\right)2r$$

or, $\quad 3\,\theta\left[(1-\cos 2\theta)\right] = \theta - \dfrac{\sin 2\theta}{2}$

or, $\quad 6\theta - 6\theta\cos 2\theta = 2\,\theta - \sin 2\,\theta$

or, $\quad \sin 2\theta + 4\theta - 6\theta\cos 2\theta = 0$

or, $\quad \sin 2\theta - 6\theta\cos 2\theta + 4\,\theta = 0 \qquad (4.39)$

This equation is satisfied where $2\theta = 302^\circ\,24'$

The depth of flow in the channel for maximum discharge is given by

$$y = r - r\cos\theta = r\left(1 - \cos 151^\circ\,12'\right)$$

$$= r\,(1 + 0.8763) = 1.8763\ r \tag{4.40}$$

Thus the maximum discharge through a circular channel section the the depth of flow (y) is equal to the (0.938) diameter of the channel

$$d = 0.938 \text{ (diameter of the channel)} \tag{4.41}$$

The capacity of circular channel section is decreased as the top of the section is approached; both velocity and flow rate decreases with increased depth. Fig. 4.12 indicates that this result which is due to increasing perimeter as the pipe approaches the full state.

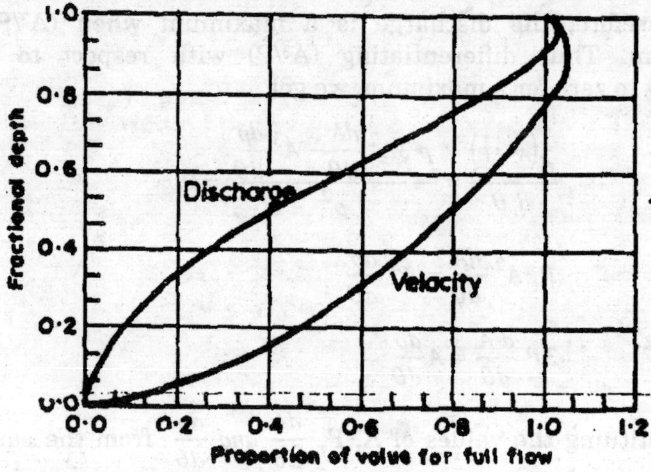

Fig. 4.12. Flow characteristics for a circular channel section flowing part full.

4.7 OPEN CHANNEL SECTION FOR CONSTANT VELOCITY AT ALL DEPTHS OF FLOW

For ordinary channel section the discharge fluctuates within wide range during the year; thereby velocity of flow also varies. While the depth of flow is large; it produces excessive velocity which may damage the channel lining. On the other hand, the velocity may become small if the depth of flow is low, causing deposition of suspended materials. Therefore, the section is to be so designed that it carries no silting, no scouring velocity and constant velocity is maintained which is independent of depth.

Assuming that the hydraulic radius R is constant for any depth of flow.

Let the section shown in Fig. 4.13 be a constant velocity section whose shape is to be designed.

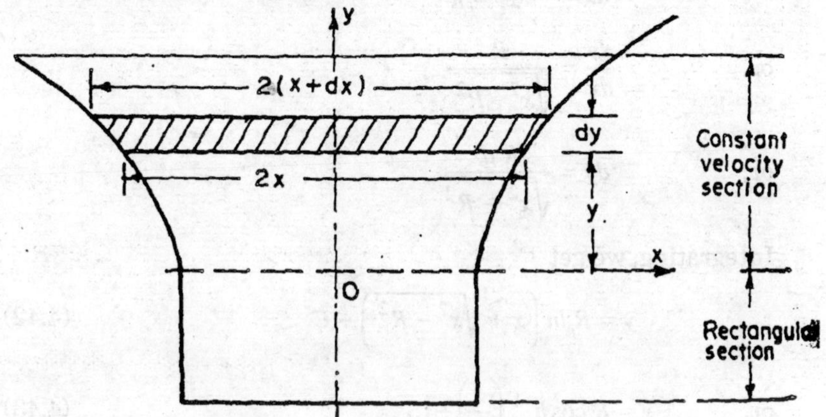

Fig. 4.13. Constant velocity section

Taking the origin at O, and considering a small portion of the cross section of depth dy.

Area of cross section $dA = 2x\,dy$

Wetted perimeter $dp = 2\sqrt{(dx)^2 + (dy)^2}$

Hydraulic radius $R = \dfrac{dA}{dp} = \dfrac{2\,x\,dy}{2\sqrt{(dx)^2 + (dy)^2}}$

or, $$R\sqrt{(dx)^2 + (dy)^2} = x\,dy$$

Squaring on both the sides,

$$R^2\left[(dx)^2 + (dy)^2\right] = x^2\,(dy)^2$$

or, $$R^2\left[1 + \left(\frac{dy}{dx}\right)^2\right] = x^2\,(dy)^2$$

or, $$R^2 + R^2\left(\frac{dy}{dx}\right)^2 = x^2\left(\frac{dy}{dx}\right)^2$$

or, $$\left(\frac{dy}{dx}\right)^2\left(x^2 - R^2\right) = R^2$$

or, $$\left(\frac{dy}{dx}\right)^2 = \frac{R^2}{x^2 - R^2}$$

or, $$\frac{dy}{dx} = \frac{R}{\sqrt{x^2 - R^2}}$$

or, $$dy = \frac{R\,dx}{\sqrt{x^2 - R^2}}$$

Integrating, we get

$$y = R\,ln\left(x + \sqrt{x^2 - R^2}\right) + C \tag{4.42}$$

or, $$y = R\cosh^{-1}\left(\frac{x}{R}\right) + C_1 \tag{4.43}$$

At y = 0, x = R, therefore,

$$y = 0 = R\,ln\left(R + \sqrt{R^2 - R^2}\right) + C$$

$$\therefore \quad C = -R\ \ln R$$

Substituting the value of R in equation (4.42)

$$y = R\,ln\left(x + \sqrt{x^2 - R^2}\right) - R\,ln\,R$$

$$= R\left[ln\left(x + \sqrt{x^2 - R^2}\right) - ln\,R\right]$$

The constant c_1 is calculated as

$$0 = R\cos h^{-1}(1) + C_1$$

or, $C_1 = 0$

Substituting in equation [4.42]

$$y = R\cos h^{-1}\left(\frac{x}{R}\right) \tag{4.44}$$

The constant velocity section is generally used in sewers where there is a large variation in the quantity of sewage. In such sewers, the bottom portion (trapezoidal or triangular) is designed for the minimum discharge which flows during the lean period. As soon as the discharge increases further, the curved section becomes effectivel and discharge the increased flow at the constant velocity.

4.8 SPECIFIC ENERGY AND CRITICAL DEPTH

The concepts of specific energy and critical depths are extremely useful in the solution of many problems in open channel flow. The depth at which the specific energy was minimum for a given discharge as the critical depth and the corresponding flow as the critical flow. In this article how they can be used for engineering applications.

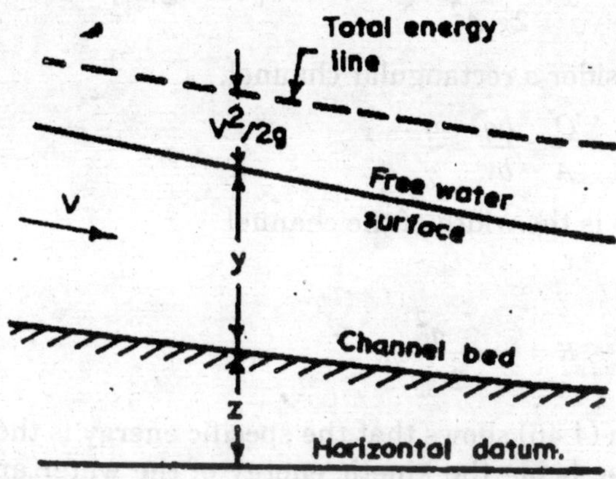

Fig. 4.14. The total energy of the flow in a channel

4.8.1 Total energy and specific energy

Consider the channel cross section shown in Fig. 4.14. The velocity v is assumed to be uniform over this section, the depth of flow being y and the elevation of the bed z above an arbitrarily chosen horizontal datum. The Bernoulli's equation gives the total energy of the streamline at the bed as

$$\text{Total energy H} = z + y + \frac{v^2}{2g} \tag{4.45}$$

If a length of channel is considered in profile a line may be drawn above the water surface to represent the total energy of the flow. This is known as the total energy line.

In gradually varied flow these three lines total energy, water surface and channel bed, may have different slopes with the condition that the total energy line must always slope downwards in the direction of flow unless energy is added from some outside source.

The concept of specific energy was introduced by Bakhmeteff in 1912. Specific energy is defined as the energy of the flow referred to the channel bed as datum.

$$E = y + \frac{v^2}{2g}$$

$$y + \frac{Q^2}{2g\,A^2} \qquad [\because v = Q/A]$$

Now consider a rectangular channel,

$$\frac{Q}{A} = \frac{bq}{by} = \frac{q}{y}$$

Where, b is the width of the channel.

Hence

$$E = y + \frac{q^2}{2\,gy^2} \tag{4.46}$$

Equation (4.46) shows that the specific energy is the sum of two parts : the one being the kinetic energy of the water and the other its depth.

Therefore,

$$E = E_P + E_k \tag{4.47}$$

4.8.2 Relationship between specific energy and depth

From equation (4.46) it can be seen that the for a given channel section and discharge the specific energy is a function only of a depth. The relationship between E and y may be shown graphically (Fig. 4.15) by plotting the specific energy against the depth of flow for a constant discharge q per unit width. This is called the specific energy curve.

From equation (4.46) taking q as constant

$$(E - y)\,y^2 = \frac{q^2}{2g} = \text{Constant}$$

or, $$(E-y)=\frac{Constant}{y^2}$$

Again, this is a cubic equation for the depth y for a given E. Considering only positive solutions, then the equation is a curve with two asymtotes :

$$\text{as } y \to 0,\ E \to \infty$$

$$\text{as } y \to \infty,\ E \to y$$

Fig. 4.15 shows that the potential energy line is a straight line and passes through the origin making an angle of 45°, with either axis, as long as the slope in the channel is not excessive. The kinetic energy curve represent the term $\frac{q^2}{2gy^2}=\frac{Constant}{y^3}$ and is asymptotic to the two axies. The specific energy curve has two limbs; the upper for large depths is asymtotic to the line E = y while the lower, for small depths, is asymtotic to the axis y = 0.

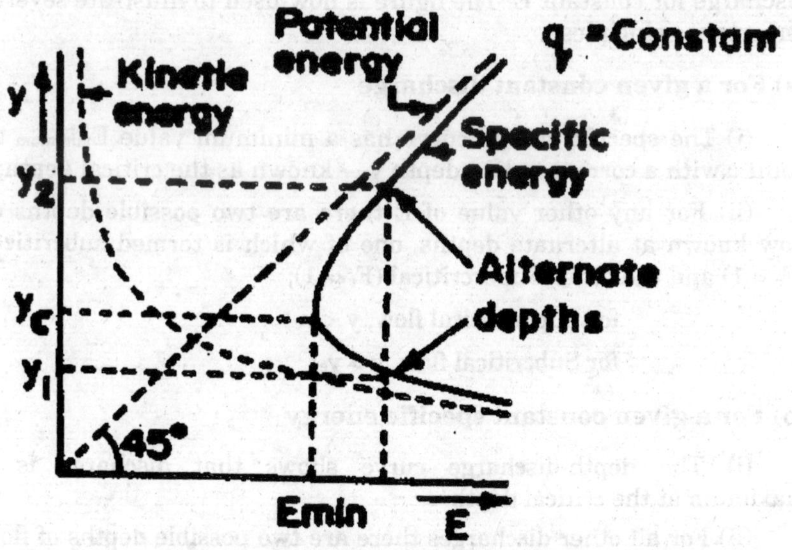

Fig. 4.15. Specific energy curve

For a particular value of specific energy E there will be two possible depths of flow y_1 and y_2.

These are known as alternate depths.

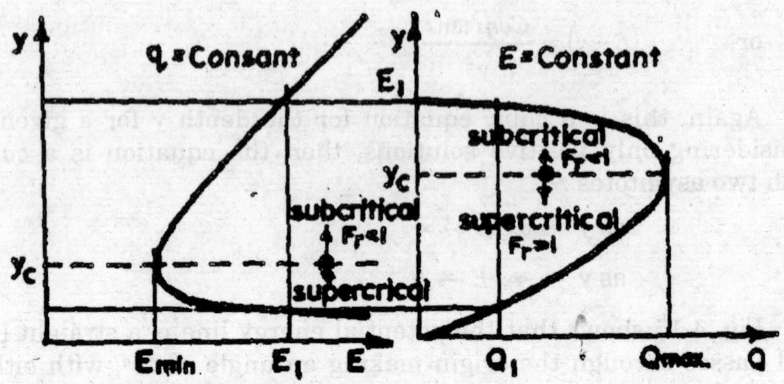

Fig. 4.16. Variation of specific energy and discharge with depth.

4.8.3 Critical depth (y_c)

In Fig. 4.16, the specific energy curve (for constant discharge) has been redrawn alongside a second curve of depth against discharge for constant E. The figure is now used to illustrate several important principles.

(a) For a given constant discharge

(i) The specific energy curve has a minimum value $E_{minimum}$ at point c with a corresponding depth y_c – known as the critical depth;

(ii) For any other value of E there are two possible depths of flow known at alternate depths, one of which is termed subcritical ($F_r < 1$) and the other Supercritical ($F_r > 1$);

for Supercritical flow, $y < y_c$

for Subcritical flow, $y > y_c$

(b) For a given constant specific energy

(i) The depth-discharge curve shows that discharge is a maximum at the critical depth.

(ii) For all other discharges there are two possible depths of flow (Sub and Supercritical) for any particular value of E.

In both cases it can be seen that there is a "critical depth" at which the two roots of the equation (4.46)

$$y^3 - Ey^2 + \frac{q^2}{2g} = 0$$

coincide and at which the energy required for a given discharge is a minimum and the discharge for a given energy is a maximum.

The critical velocity is the velocity in a channel when the fluid flows at critical depth.

4.8.4 The general equation of critical flow

The critical depth is defined as that at which the specific energy of a flow is minimum or discharge is maximum. Therefore, differentiation of the approximate form of the specific energy equation (4.46) will evaluate y_c.

Considering a general case of a channel of irregular cross section (Fig. 4.17) carrying a discharge Q at a maximum depth y and cross sectional area A. Equation (4.45) given the specific energy as

$$E = y + \frac{Q^2}{2\,gA^2}$$

and differentiating this expression with respect to the depth :

$$\frac{dE}{dy} = 1 + \frac{Q^2}{2g}\left(-\frac{2}{A^3}\right)\frac{dA}{dy}$$

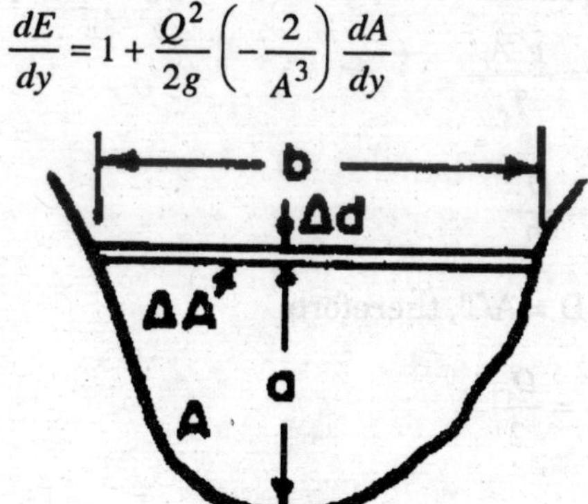

Fig. 4.17. Relationship between change in depth and change in cross sectional area.

when, $\frac{dE}{dy} = 0,\ y = y_c$ and other variables will have values corresponding to critical flow. Hence

$$\frac{Q^2}{g}\left(\frac{1}{A^3}\right)\frac{dA}{dy} = 1$$

(From Fig. 4.17)

$$\delta A = T\,\delta\,y$$

therefore, $\frac{\delta A}{\delta y}$ = T and at the limit $\frac{dA}{dy}$ = T. The general equation for critical flow then becomes

$$\frac{Q^2 T_c}{g A_c^3} = 1$$

or, $$\frac{Q^2}{g} = \frac{A_c^3}{T_c} \tag{4.48}$$

The critical velocity is obtained by putting $v_c = Q/A_c$ in the above equation (4.48) giving

$$v_C = \sqrt{\frac{g A_c}{T_c}} \tag{4.49}$$

The above conditions are used for the flow to occur at the critical depth.

Squarig the equation (4.49) on both the sides

$$v_C^2 = \frac{g A_c}{T_c}$$

or, $$\frac{v_C^2}{2g} = \frac{D_c}{2}$$

We know that D = A/T, therefore,

$$\frac{v_C^2}{2g} = \frac{D_c}{2}$$

or, $$\frac{v_C}{\sqrt{g D_c}} = 1 \tag{4.50}$$

or, $F_r = 1$

From the above analysis we concluded that the Froude number of flow is unity for critical flow. If it is less than one for subcritical and for supercritical flow more than one.

Rectangular Channel

Writing equation (4.46) for rectangular channel.

$$E = y + \frac{q^2}{2gy^2}$$

For a minimum value $\frac{dE}{dy} = 0$

$$\frac{dE}{dy} = 1 - \left(\frac{q^2}{2g}\right)\left(\frac{2}{y^3}\right) = 0$$

or, $$\frac{q^2}{g} = y^3$$

or, $$y = \left(\frac{q^2}{g}\right)^{1/3}$$

For the critical flow

$$y_c = \left(\frac{q^2}{g}\right)^{1/3} \tag{4.51}$$

Substituting in equation (4.46)

$$E_{min} = y_c + \frac{1}{2y_c^2}\left(y_c^3\right) \quad = y_c + \frac{y_c}{2} = \frac{3}{2}y_c \tag{4.52}$$

or, $$y_c = \frac{2}{3}E_{min} \tag{4.52 (a)}$$

The same expressions can be obtained by keeping E as constant and q varies.

Using equation (4.46)

$$E = y + \frac{q^2}{2gy^2}$$

or, $$2gEy^2 = 2gy^3 + q^2$$

or, $$q^2 = 2gy^2(E - y)$$

or, $$q = \sqrt{2g(E - y)y^2}$$

For discharge to be a maximum, differentiating the above equation with respect to y and equating to zero.

$$\frac{dq}{dy} = 2Ey - 3y^2 = 0$$

or, $$E = \frac{3}{2}y$$

or, $$E_{min} = \frac{3}{2}y_c \tag{4.52}$$

Substituting the value of E_{min} in the basic equation,

$$E_{min} = y_c + \frac{v_c^2}{2g}$$

or, $$\frac{3}{2}y_c = y_c + \frac{v_c^2}{2g}$$

or, $$\frac{v_C^2}{2g} = \frac{y_c}{2}$$

or, $$v_C^2 = g\ y_c$$

or, $$\frac{v_C}{\sqrt{g\ y_c}} = 1$$

or, $F_r = 1$.

Since Froude number is unity, the depth of flow which for a given discharge makes specific energy a minimum must be the critical depth.

Alternatively from the equation (4.51)

$$y_c = \left(\frac{q^2}{g}\right)^{1/3}$$

$q = v_c\, y_c$ for the rectangular channel, therefore,

$$y_c^3 = \left(\frac{v_c^2\ y_c^2}{g}\right)$$

or, $$y_c = \frac{v_c^2}{g}$$

or, $$\frac{v_c}{\sqrt{g\, y_c}} = 1$$

or, $F_r = 1$.

It will be noticed, that at low depths of flow, and gradually increasing the depth, the specific energy decreases to a minimum, at the critical depth where the Froude's number is unity, and then increases again as the depth of flow is still further increased. Thus except at the critical depth, there are two depths of flow for the same value of specific energy, one less than, and one greater than the critical depth.

The horizontal line through the critical depth divides the flow into two distinct types, that above the line being termed tranquil or subcritical or slow flow, whilst that below the line is termed shooting or supercritical or rapid flow.

4.8.5 An application of the critical depth line

The application of critical depth is illustrated in Fig. 4.18, which shows the local bed rise on the free surface elevation for various initial depths of flow. The critical depth is dependent only on the discharge. Hence, for a given discharge the critical depth line may be drawn on the longitudinal channel section as shown in Fig. 4.18. If the upstream flow is initially subcritical, then a depression of the water surface level will occur. Conversely, if the upstream water level is initially supercritical, then an increase in the water surface level will occur. As the upstream water level approaches the critical depth, then the change in elevation of the free surface becomes more marked. A practical example of this phenomenon often occurs in river under flood conditions. The normal depth of flow may be nearly critical, and any undulations in the bed result in large standing wave formations.

If the local bed rise is sufficiently large, then criticl flow will occur. The discharge is maximum for critical flow, and hence, under these circumstances, the local bed rise is acting as a `choke' on the flow, so called because it limits the discharge.

4.8.6 Channel Transitions

A transition is a small intermediate portion between two channels. The two channels may or may not have the same cross section. Transitions are usually required when there is a change in the cross section of the channel. The variation of the cross section may be accomplished either by changing the width or the depth or both. Transition are also provided when there is a change in the alignment of the channel.

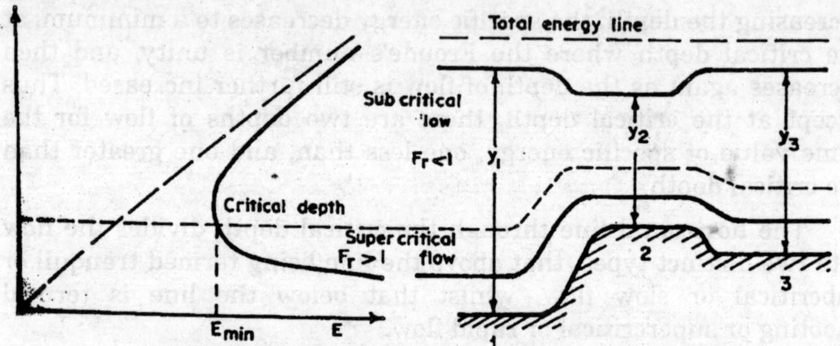

Fig. 4.18. Effect of bed rise on water surface elevation

The purpose of the transition is to change the conditions gradually so that a minimum of the energy loss occurs at that change.

Transitions in open channels may broadly be classified into those in which flow conditions are transformed from

1. Subcritical to Subcritical
2. Subcritical to Supercritical
3. Supercritical to Subcritical and
4. Supercritical to Supercritical

(1) Purely Subcritical transition

Consider a channel of rectangular cross section which has a reduced width as shown in Fig. 4.19. Since the flow is converging between section 1-1 and 2-2 it may appropriately be assumed that no loss of energy takes place between the two sections. Further if it is assumed that the bottom of the channel is horizontal the specific energy will be equal at both the sections.

Say we want to change slow flow conditions from A to B in Fig. 4.19. The path labelled A 1B, A 2B and AB are only three of the infinite number of theoretically possible ways of achieving this. In moving from A to 1 the channel width is maintained constant (q =

Constant) and bottom is raised by ΔZ. The loss in specific energy ΔE is compensated by a gain in potential head ΔZ thus keeping the total energy (H = E + Z = Constant). From 1 to B channel width is contracted until the new unit discharge q_2 is attained. The order of operation is reversed along A 2 B. A more practical approach especially if the length of transition should be restricted is to combine the channel width contraction with floor raising along path AB. The process in transforming the flow from B to A are reversible.

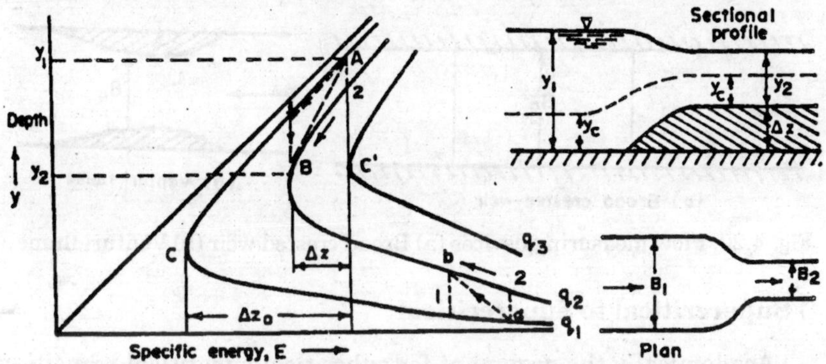

Fig. 4.19. Flow in a rectangular channel with reduction in width

(2) Subcritical to Supercritical

In changing from subcritical condition to supercritical conditions the flow must pass through a control section. A control section is that section at which critical flow conditions exist. Two of the possible ways by which flow conditions corresponding to A (Fig. 4.19) can be transformed to critical conditions at c or c′ are to raise the channel floor by an amount ΔZ_o or contract the channel width. The former method gives rise to broad crested weir and the latter to a venturi flume.

Broad crested weirs [Fig. 4.20 (a)] and venturi flumes (Fig. 4.20 (b)) or their combination are used extensively in measuring discharge through small rivers and canals.

(3) Supercritical to Subcritical

Like the change from subcritical to supercritical flow the transforming from supercritical flow to subcritical flow involves a critical phase. However, in the latter case there is an expansion of flow so that eddies are formed and energy is lost. This phenomenon gives rise to the hydraulic jump.

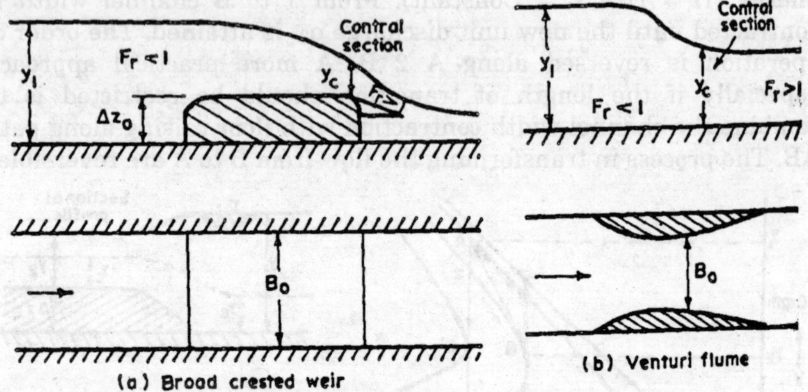

Fig. 4.20. Flow measuring devices (a) Broad crested weir (b) Venturi flume

(4) Supercritical to Supercritical

Academically the argument for subcritical-subcritical transition can be applied to supercritical-supercritical transitions. The fundamental difference is that whereas a rise in channel bed and contraction of channel width produce lowering of the liquid surface in the former case, the produce a rise in the latter case. This may be inferred from Fig 4.19, following the paths a1b and a2b. The phenomenon is however complicated by the fact that disturbances in supercritical flow produce shock waves whose continuous reflection from the side walls of the channel produce an irregular liquid surface of standing waves downstream of the source of disturbance.

Open channel shocks are analogous to shock waves produced in supersonic gas flow. If the shock is strong enough subcritical conditions are produced downstream of the shock front giving rise to an oblique hydraulic jump.

4.8.7 Critical depth meters

Since the control section provides a definite relationship between the depth and the discharge, it can be used for the measurement of flow in open channels. Various flow measurement devices have been developed using this principle. Some of devices commonly used for measurement of flow in open channels are described below.

(a) Broad crested weir

If the channel floor is raised by a height equal to or greater than ΔZ_c over a length sufficient enough for parallel flow to occur over the hump, as already discussed earlier, the flow over the hump will be critical. Such a structure is called a Broad-crested weir and provides an excellent means of measuring the discharge in open channels (Fig 4.20 (a)).

Let the width at the control section be B_o. The discharge per unit width referred to the critical depth is

$$q^2 = g\, y_c^3$$

$$\therefore \text{ Discharge } Q = B_o\, q = B\sqrt{g\; y_c^{3/2}}$$

$$\text{But the critical depth } y_c = \frac{2}{3} E$$

where E is the approach energy measured relative to the crest of the weir (bed level)

$$Q = B_o \left(\frac{2}{3}\right)^{3/2} \sqrt{g}\; (E)^{3/2} = 1.70\; B_o\; (E)^{3/2} \qquad (4.53\text{ a})$$

In order to account for losses of energy due to friction a factor c_d less than unity known as the coefficient of discharge is applied, yielding

$$Q = 1.70\; c_d\; B_o\; E^{3/2} \qquad (4.53\text{ b})$$

where $$E = (y_1 - \Delta Z_o) + \frac{v_1^2}{2g}$$

(b) Venturi flume

Venturi flume and Broad crested weir are very similar in operation in that the flow is contracted to a throat and then reexpanded. This may be compared with the nozzle used in a pipe system, in that it is the throat condition which controls the volume flow rate and in operation the critical depth should occur at the throat. Further since the velocity of flow at the throat is less than the critical velocity, standing wave or hydraulic jump will not be formed at any section in the venturiflume.

The discharge Q flowing through the channel can be calculated by measuring the depth of flow at the entrance and the throat of the flume and applying the following formula

$$Q = k\frac{Aa\sqrt{2g\left(y_1 - y_o\right)}}{\sqrt{A^2 - a^2}} \tag{4.54}$$

(c) Standing wave flume

A standing wave flume (Fig. 4.21) is a structure in a channel which has a narrowed throat having a hump (raised floor) at the bottom is acted as a broad crested weir. The critical flow is obtained on a hump on the channel bed in subcritical flow if the hump is of a certain minimum height. The broad crested weir, performs quite well even under submerged flow condition. The raised crest may, however, be a disadvantage in heavily silt-laden streams, since the slit may deposit in the quite waters upstream of the crest. The critical flow can also be obtained by contracting the channel width beyond a certain limit and this principle has been used in the development of a flow measuring device which is relatively free from the problem of silt deposition. By suitable choice of throat width and level difference between the throat and the downstream reach, critical flow is obtained in the throat section and supercritical flow followed by a hydraulic jump (some times called standing wave, downstream. Such a flume is know as a standing wave flume. This is also known as critical depth flume.

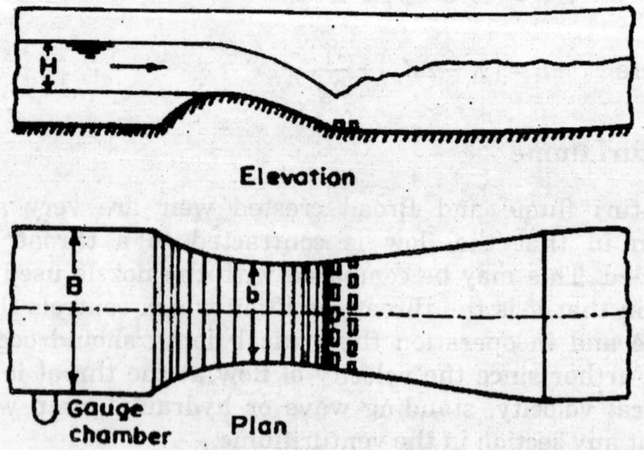

Fig. 4.21. Standing wave flume

The discharge Q is calculated by the formula

$$Q = Q = C\ B\ y_1^{3/2} \tag{4.55}$$

Where B = width at throat

C = discharge coefficient of flume = $\left(\frac{2}{3}\right)^{3/2} \sqrt{g} \approx 1.706$

(d) Parshall flume

The Parshall flume is one of the most extensively used standing wave flume in U.S.A., which was developed by R.L. Parshall in 1920.

Most extensive experiments on standing wave flumes have been carried out by Parshall and his studies have led to the evoluation of standardised flumes often called Parshall flumes.

The standard dimensions of the Parshall flume for throat widths ranging from 0.3 m to 2.5 m are given in Fig. 4.22. The flume operates with fixed discharge relationship.

$$Q = C\ B\ y_1^n \tag{4.56}$$

Where C and n are the constants of the flume which are obtained by caliberations.

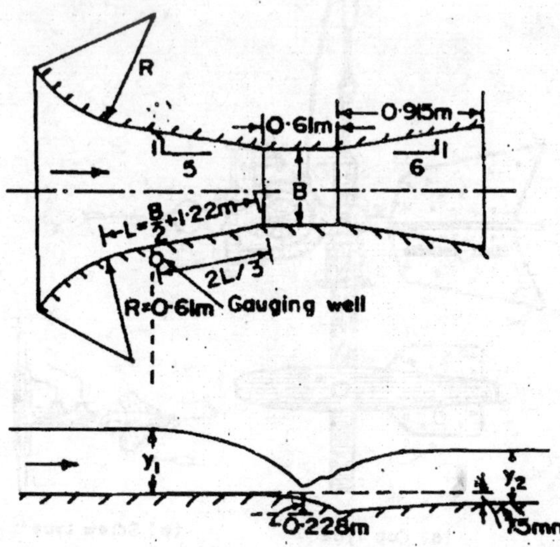

Fig. 4.22. Dimensions of Parshall flume (0.3 m < B < 2.5 m)

4.9 DETERMINATION OF AVERAGE-(MEAN) VELOCITY IN FLOW THROUGH CHANNELS

The discharge is equal to the area multiplied by the average velocity; therefore, the velocity of flow is required to be determined. The velocity of flow in channels may be measured by the following methods.

(1) Pitot tube (2) Current meter (3) Floasts

(1) Pitot tube

Pitot tube is already discussed in the earlier chapter.

(2) Current meter

According to the shape of the revolving element the current meters may be classified as cup type and screw type or propeller type.

In a cup type current meter the wheel or the revolving element has the forms of a series of conical cups, mounted on a spindle which is held vertical at right angle to the direction of flow as shown in Fig 4.23(a).

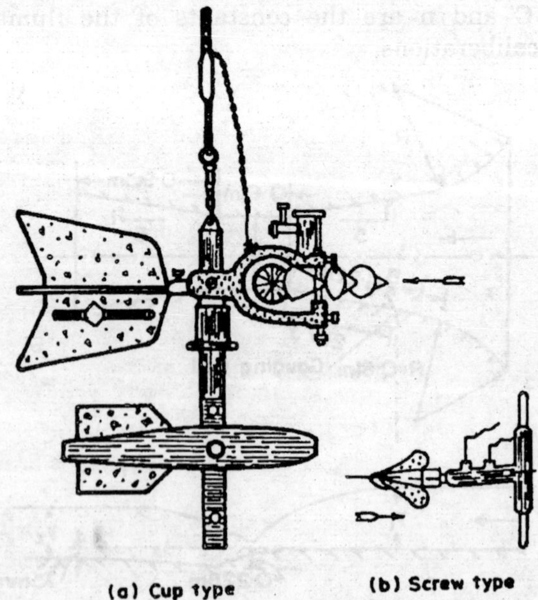

Fig. 4.23. Current meters.

In a screw or propeller type current meter the revolving element consists of a shaft, with its axis parallel to the direction of flow, which carries a number of curved vanes or propeller blades mounted around the periphery of the shaft as shown in Fig 4.23(b).

The current meter is suspended by means of a cable and is held vertically immersed in the flowing stream of water to the requried depth, such that the wheel or the revolving element is facing towards the upstream direction. The wheel is thus rotated by the dynamic thrust exerted on it by the flowing water. The number of revolutions of the wheel per unit time are proportional to the velocity of the flowing water. Thus after setting the current meter to the required point in the flowing stream of water the number of revolutions of the wheel per unit time are noted and from the calibration curve the velocity of flow of water may be determined.

(3) Floats

A float is a small object made of wood or other suitable material which is lighter than water and thus capable of floating on the water surface. It provides a simple way of measuring the velocity of flow of water in rivers and channels.

Two good methods for determining the velocity of flow are (1) Double float and (2) Rod float (Fig 4.24 (a) and (b)).

Double float method consists of a spherical surface float to which is attached a hollow metal sphere, heavier than water and supended from the float by a cord of known length. The depth of the lower float which is completely submerged may be regulated by the length of the cord. The velocity is then obtained by noting the time taken by the float to traverse a known distance.

A still better than the double float is the rod float. It consists of a vertical wooden rod which is weighted at the bottom to keep it vertical with its top end emerging out of free water surface when floating. The length of the rod float should be such that it reaches the bottom of the stream without touching the sand, weed or mud at the bottom of the stream. Sometimes the rod floats are made telescopic so that the length may be adjusted to suit any depth of flow. The rod floats will travel with a velocity equal to the average velocity of flow and hence a road float gives the average velocity of flow at the section.

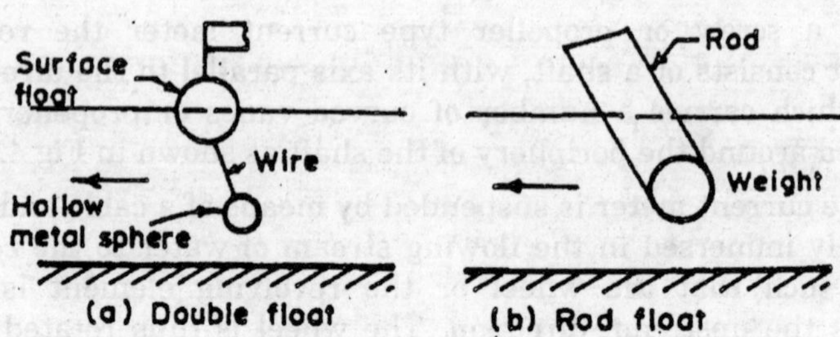

Fig. 4.24. Float methods

4.10 GRADUALLY VARIED FLOW

Gradually varied flow is non uniform flow in which the change of depth in the channel occur gradually, in the direction of flow. As a result the streamlines in any restricted locality can be considered straight and parallel and the resulting pressure distribution will therefore be hydrostatic. This restriction on the flow conditions also enables the Bernoulli's equation to be used to evaluate the fluid energy.

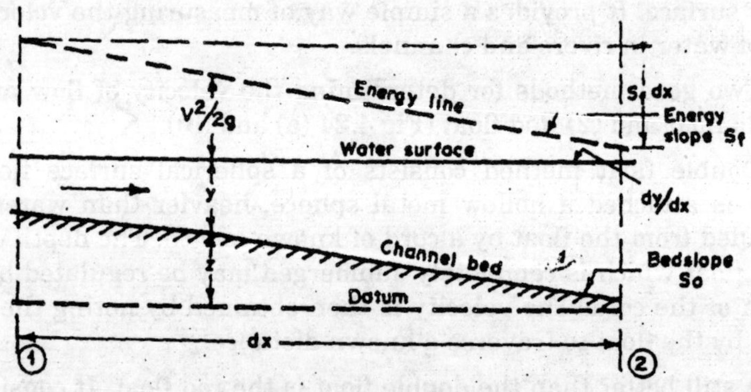

Fig. 4.25. Gradually varied flow

Because gradually varied flow involves small changes of depth it is concerned with long length of channel. Energy losses due to channel boundary friction which control uniform channel flow will therefore be the only significant means of energy dissipations. The results obtained for uniform flow are then valid if used to describe conditions locally in gradually varied flow [Fig. 4.25].

In natural river channels changes in cross section and profile are frequent or even continuous and, as a result, flow in such channels is nearly always of this type.

4.10.1 General equation of gradually varied flow or Dynamic equation gradually varied flow

The gradually varied flow equation can be derived fom the basic energy equation with the following assumptions :

(1) The channel is prismatic

(2) Uniform flow formulae (Chezy's or Mannings) may be used to for finding the slope of the gradually varied flow and the corresponding roughness.

(3) The pressure distribution is any vertical in hydrostatic.

(4) The bottom slope of the channel is very small.

(5) The energy correction factor is unity.

(6) The roughness coefficient is independent of the depth of flow and it's constant throughout the channel reach considered.

Consider an energy head at a point in a channel carrying gradually varied flow (Fig. 4.25)

$$H = z + y + \frac{v^2}{2g}$$

or, $$H = z + y + \frac{Q^2}{2g\,A^2} \qquad \left[v = \frac{Q}{A}\right]$$

Differentiating w.r.t.z., we get

$$\frac{dH}{dx} = \frac{dz}{dx} + \frac{dy}{dx} + \frac{d}{dx}\left(\frac{Q^2}{2gA^2}\right)$$

$$\frac{dH}{dx} = \frac{dz}{dx} + \frac{dy}{dx} + \frac{d}{dx}\left(\frac{Q^2}{2gA^2}\right)\frac{dy}{dx}$$

$$= \frac{dz}{dx} + \frac{dy}{dx}\left[1 - \frac{Q^2}{g\,A^3}\,\frac{dA}{dy}\right] \qquad (4.57)$$

$$\frac{dA}{dx} = \frac{dA}{dy}\cdot\frac{dy}{dx} = T\,\frac{dy}{dx}$$

or, $$\frac{dA}{dy} = \text{T}$$

But $\frac{dH}{dx} = -S_f$

and $\frac{dz}{dx} = -S_o$

The negative sign for S_f and S_o indicate that as x increases H and z decreases.

$\frac{dy}{dx}$ is the slope of the water surface with respect to the channel bottom.

Substituting in the above differential equation (4.57) we get

$$-S_f = -S_o + \frac{dy}{dx}\left(1 - \frac{Q^2 T}{gA^3}\right)$$

or, $$\frac{dy}{dx} = \frac{S_o - S_f}{1 - \frac{Q^2 T}{gA^3}} \tag{4.58}$$

This equation is known as the general equation of gradually varied flow.

But $\frac{A}{T} = D$

and $v = \frac{Q}{A}$

therefore, $\frac{Q^2 T}{gA^3} = F_r^2$

Substituting in equation (4.58), we get

$$\frac{dy}{dx} = \frac{S_o - S_f}{1 - F_r^2} \tag{4.59}$$

In a rectangular channel section, A = by and Q = q b. giving

$$\frac{dy}{dx} = \frac{S_o - S_f}{1 - \frac{q^2}{gy^3}} \tag{4.60}$$

The chezy's or manning's formula can be used suitably to evaluate S_f.

If the chezy formula is used.

$$S_f = \frac{v^2}{C^2 R}$$

and
$$\frac{dy}{dx} = \frac{S_o - \dfrac{v^2}{C^2 R}}{1 - \dfrac{q^2}{gy^3}} \tag{4.61}$$

in which R is the hydraulic radius of the cross section and C the chezy constant. If the Manning formula is used.

$$S_f = \frac{v^2 n^2}{R^{4/3}}$$

and
$$\frac{dy}{dx} = \frac{S_o - \dfrac{v^2 n^2}{R^{4/3}}}{1 - q^2 / gy^3} \tag{4.62}$$

Eliminating q from equations (4.61) and (4.62) gives

$$\frac{dy}{dx} = \frac{S_o - \dfrac{v^2}{C^2 R}}{1 - \dfrac{v^2}{gy}} \qquad [4.63\text{ (a)}]$$

and
$$\frac{dy}{dx} = \frac{S_o - v^2 n^2 / R^{4/3}}{1 - v^2 / gy} \qquad [4.63\text{ (b)}]$$

The above equations [4.63(a) and 4.63(b)] are useful in computational solutions.

4.10.2 Special values of varied flow equation

Without considering any particular problem it is possible to define certain types of surface profile and significant points by inspection of the varied flow equation (4.58).

(1) $\frac{dy}{dx} = S_o$ The limiting case when the velocity is zero. There will be no flow in the channel and the water surface will be horizontal. Whilst its occurence does not constitute open channel flow it forms a limit which the flow often approaches asymptotically. It can also occur as a transitory value at lower depths.

(2) $\frac{dy}{dx} = 0$ Constant depth, uniform flow, this condition is obtained when $S_f = S_o$ and this special case of the general equation is already discussed previously.

(3) $\frac{dy}{dx} = \pm\infty$ Water surface vertical. The equation yields this value when $\frac{Q^2T}{gA^3} = 1$ which occurs at critical depth. Hence, except when the channel slope is critical the water surface must be vertical at the critical depth. This implies that when the water surface has to pass from above the critical depth to below, or vice versa, it does so rapidly. This result is borne out in reality, thus invalidating the basic assumption of gradually varied flow. Although the detailed shapes of the surface profiles in this region close to critical depth are not therefore predicted accurately by the varied flow equation, their forms approximately fairly closely to the theoretical profiles in many cases. When the surface profile has to pass from $y < y_c$ to $y > y_c$ a hydraulic jump is formed. This represents the extreme case of breakdown of the varied flow equation. A decrease of depth through the critical depth is known as hydraulic drop. Such a hydraulic drop terminates the drawdown curve. The energy loss in the hydraulic drop is small and of the same order as that in uniform superciritcal flow.

(4) $\frac{dy}{dx} = \pm ve$ Depth increasing downstream. In channels of mild slope in which $y > y_o$ the resulting profile is called a back water curve (Fig 4.26). It is produced by obstructions, such as weirs and bridgespiers, changes of bed slope (increasingly mild) and channels discharging into reservoirs under certain condiitons.

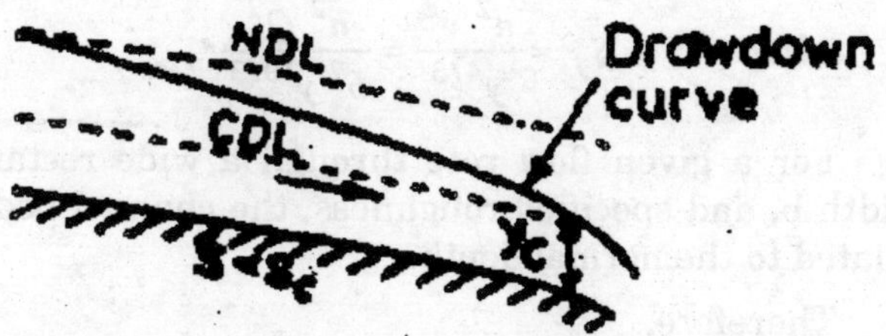

Fig. 4.26. Back water curve flow profile.

(5) $\frac{dy}{dx} = -ve$ Depth decreasing downstream. When the channel slope is mild the resulting profile is called a drawdown curve. It occurs at a free overall as in Fig 4.27 or where the channel is approaching a point at which the slope increases.

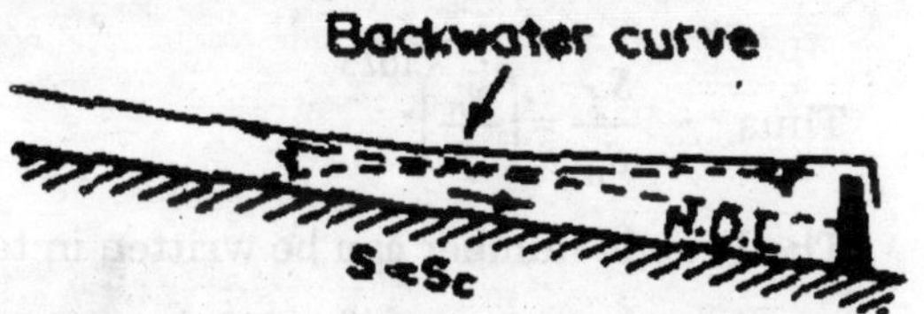

Fig. 4.27. Drawdown curve at a free overfall.

4.10.3 Classification of surface profiles

A good knowledge of surface profiles facilities computations and the student is well advised to familiarize himself with the following code of description and the principal details of all the surface profiles.

The Code letters refer to the slopes : Mild (M), Critical (C) and Steep (S), it is necessary to introduce : Horizontal (H) and Adverse (A).

The general behaviour of surface profiles can be studied most easily for wide rectangular channels : for such channels b > y and R ≈ y. For this case the surface profile is governed by

$$\frac{dy}{dx} = \frac{S_o - S_f}{1 - F_r^2} = \frac{S_o\left[1 - \left(\frac{S_f}{S_o}\right)\right]}{1 - F_r^2}$$

where $$S_f = \frac{n^2\, v^2}{y^{4/3}} = \frac{n^2\, Q^2}{b^2\, y^{10/3}} \tag{4.64}$$

For a given flow rate through a wide rectangular channel of width b, and specified roughness, the channel bed slope is uniquely related to the normal depth.

Therefore,

$$Q = \frac{R^{2/3}\, S_o^{1/2}}{n} A = \frac{y_n^{2/3}\, S_o^{1/2}}{n} b y_n$$

Solving for S_o, we get

$$S_o = \frac{n^2\, Q^2}{b^2\, y_n^{10/3}} \tag{4.65}$$

Thus $$\frac{S_f}{S_o} = \left(\frac{y_n}{y}\right)^{10/3} \tag{4.66}$$

The Froude number can be written in terms of critical depth.

$$F_r = \left(\frac{y_c}{y}\right)^{3/2} \tag{4.67}$$

The slope of the wide rectangular channel then can be written as

$$\frac{dy}{dx} = \frac{S_o\left[1-\left(\frac{y_n}{y}\right)^{10/3}\right]}{\left[1-\left(\frac{y_c}{y}\right)^{3/2}\right]} \tag{4.68}$$

Equation (4.68) shows that the sign of dy/dx depends only on the relation among the depths y, yn and yc. For a wide rectangular channel the critical slope is defined as

$$S_c = \frac{n^2\, g}{y_c^{1/3}} \tag{4.69}$$

From the above discussions we noted that (Fig. 4.28)

(M) Mild Slope (Fig. 4.28)

$$If\ y_n > y_c,\ then\ S_o < S_c$$

	Region 1	Region 2	Region 3
Mild slopes S<Sc	M1, Yn, Yc	M2, Yn, Yc	M3, Yn, Yc
Steep slope S>Sc	S1, Yc, Yn	S2, Yc, Yn	S3, Yc, Yn
Critical slope S=Sc	C1, Yn=Yc	C2, Yn=Yc	C3, Yn=Yc
Horizontal slope S=O	None, Yn, Yc	H2, Yn, Yc	H3, Yn, Yc
Adverse slope	None, Yc	A2, Yc	A3, Yc

Fig. 4.28. Classification of surface profiles.

(C) Critical Slope (Fig. 4.28)

$$If\ y_n = y_c,\ then\ S_o = S_c$$

(S) Steep Slope (Fig. 4.28)

If $y_n < y_c$, *then* $S_o > S_c$

(H) Horizontal Slope (Fig. 4.28)

In this case $S_o = 0$

(A) Adverse Slope (Fig. 4.28)

In this case $S_o < 0$

For a given bed slope the shape of the surface profile depends on the actual depth, y, relative to y_n and y_c. The possible range of depths of flow $0 < y < \infty$ are divided into three regions by the critical depth line (C.D.L.) and the normal depth line (N.D.L.) which of the two lines is the higher will depend upon the particular circumstances.

Region 1 Depth y greater than both y_n and y_c

$y > y_n$ and $y > y_c$

Region 2 Depth y lying between y_n and y_c

$y_n > y > y_c$

Region 3 Depth y less than both y_n and y_c

$y_n > y_c > y$

The possibility of three regions or zones in each of five slope categories suggests 15 different profile types. Several of these do not exist in reality for reasons that will appear in the following discussions with reference to Fig. 4.28.

This figure 4.28 shows the form of the surface profiles arranged in groups corre-sponding to the slope categories. They cannot of course exist in superposition.

Mild slopes–M curves

M_1 is called the backwater curve a term extended by some authors to include all properties in which the depth is increasing. This is perhaps the most commonly occuring profile being formed in channels at changes of section and at junctions with tributaries and lakes. It is asymptotic both to the normal depth line upstream and the horizontal downstream if it is not terminated earlier. The length of the M_1 profile can frequently be measured in kilometers and a practical limit to its extent is normally assumed when $(y - y_n) = 0.02$ m.

The M_2 profile will occur at the junction with a reservoir if the water level there is below the level of the normal depth in the channel exit. It may or may not pass through the critical depth at the junction, depending upon the reservoir level. The M_2 profile is known as a drawdown curve, and would occur at a free overfall.

The M_3 profile occurs below steep spillways and slucies and normally terminates in a jump.

Both M_2 and M_3 profiles are short by comparison with M_1.

The profiles may be shown in Fig. 4.29(a).

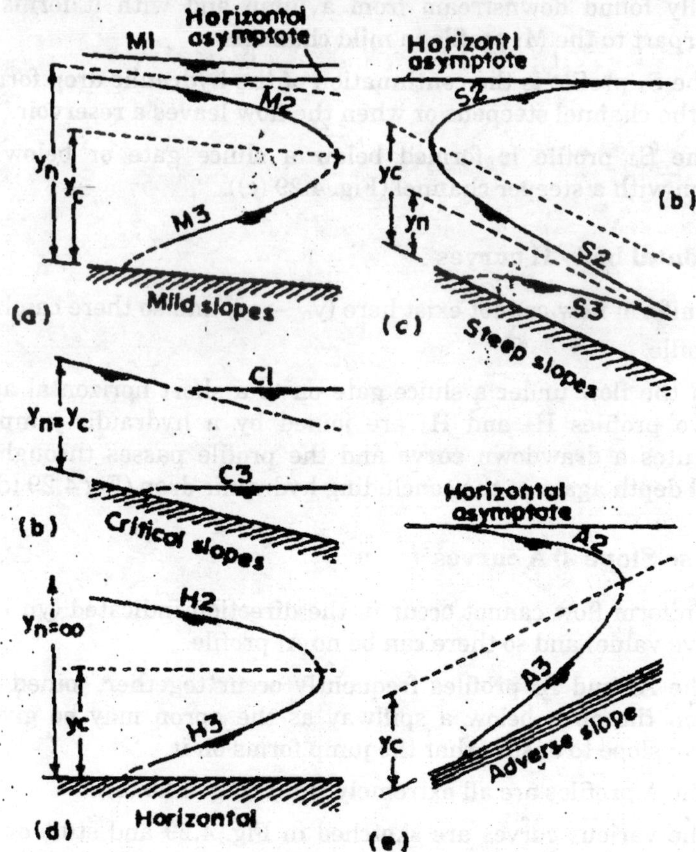

Fig. 4.29. Surface profiles

Critical Slope - C curves

C_1 and C_3 are close to being straight horizontal lines. Region 2 is reduced to a line $y = y_n = y_c$ and therefore theoretically C_2 corresponds to uniform flow at the normal depth. At the critical slope the water surface oscillates about yn for uniform flow, giving rise to a train of stationary surface waves ($F_r = 1$) (Fig. 4.29(b)).

Steep Slope - S curves

These flow conditions are found on spillways and chutes. The profiles are fairly short and all asymptotic downstream (velocity of flow exceeds gravity wave propogation velocity). The S_1 profile is normally found downstream from a jump and with it forms the counterpart to the M_1 profile in mild channels.

The S_2 profile is the continuation of the hydraulic drop formed. When the channel steepens or when the flow leaves a reservoir.

The S_3 profile is formed below a sluice gate or below the function with a steeper channel (Fig. 4.29 (c)).

Horizontal bed - H curves

Uniform flow cannot exist here ($y_n \rightarrow \infty$) and so there can be no H_1 priofile.

In the flow under a sluice gate on to a short horizontal apron the two profiles H_2 and H_3 are joined by a hydraulic jump. H_2 constitutes a drawdown curve and the profile passes through the critical depth again at the concluding hydraulic drop (Fig 4.29 (d)).

Adverse Slope -0 A curves

Uniform flow cannot occur in the direction indicated (yn has a negative value) and so there can be no A_1 profile.

The A_2 and A_3 profiles frequently occur together, joined by a jump on the open below a spillway as the apron may be given a negative slope to ensure that the jump forms on it.

The A profiles are all extremely short (Fig. 4.29 (e)).

The various curves are sketched in Fig. 4.29 and students are advised to learn to apply them to different physical situations of flow.

4.10.4 Methods of solution of the gradually varied flow equations

The three forms of the gradually varied flow equation are :

$$\frac{dy}{dx} = \frac{S_o - S_f}{1 - F_r^2} \tag{4.70}$$

$$\frac{dE}{dy} = S_o - S_f \tag{4.71}$$

$$\frac{dH}{dx} = - S_f \tag{4.72}$$

In order to enable the engineer to calculate the detailed water surface profile for gradually-varied flow the general equation (4.70) must be solved for a particular case.

There are three types of solution to the above equations (4.70), (4.71) and (4.72).

(a) direct integration method

(b) graphical integration method and

(c) numerical integration method.

These three methods of solution are discussed below :

(a) Direct integration method

Equation (4.70) may be solved directly only for regular channels. Integration methods have been developed by various workers, starting with Dupit in 1848, who found a solution for wide rectangular channels. In 1860, Bresse found a solution including changes a kinetic energy. Bakhmeteff, starting in 1912, extended the work to tropezoidal channels. Using Manning's resistance equation, various other workers have extended Bakhmeteff's method.

Full details of these methods may be found in chow (1959) or Henderson (1966). These are not discussed here as they have been superseded by numerical integration methods which may be used for both regular and irregular channels.

(b) Graphical integration method

The equation (4.70) may be written as

$$\frac{dx}{dy} = \frac{1 - F_r^2}{S_o - S_f}$$

Hence $\int_o^x dx = \int_{y_1}^{y_2} \frac{1 - F_r^2}{S_o - S_f} dy$

or, $x = \int_{y_1}^{y_2} f(y)\,dy$

where $f(y) = \frac{1 - F_r^2}{S_o - S_f}$

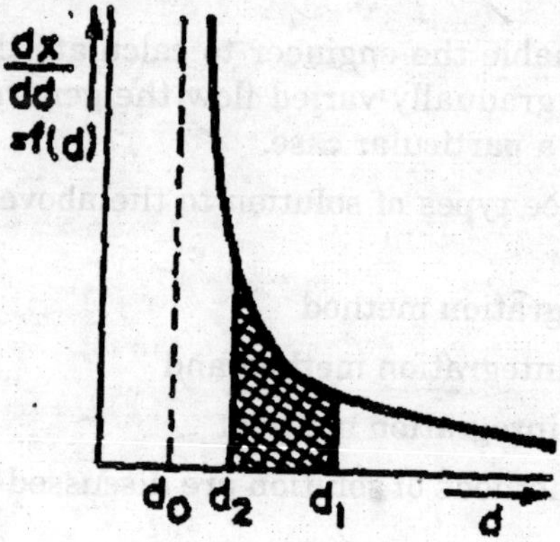

Fig. 4.30. Graphical solution gradually varied flow equation.

If a graph of y against f (y) is plotted, (Fig. 4.30), then the area under the curve is equivalent to x. The value of the function f(y) may be found by substitution of A, P, S_o and S_f for various y for a given Q. Hence, the distance X between given depths y_1 and y_2 may be found graphically.

Fig. 4.31 Shows the form of such a curve for the M_1 profile.

The area under the curve between the ordinates y_1 and y_2 can be measured by Planimetering, use of Simpsons rule, counting squares or any other equivalent method which gives the required accuracy. The value of the graphical integration method depends upon the relative ease with which the function $\frac{dx}{dy}$ can be computed. In general this method is more tedious and less satisfactory, but was more popular until the widespread use of computers, facilitated the use of the more versatile numerical methods.

(c) Numerical integration method

Using simple numerical techniques, all types of gradually varied flow problems may be quickly and easily solved using only a micro computer. A single program may be written which will solve most problems.

However, an aid to understanding, the numerical methods are discussed under three headings.

(i) the direct step method, for rectangular channels

(ii) the standard step method and

(iii) the standard step method for natural channels.

Of the great number of step methods available, the direct step method described below is perhaps the most simple, although applicable only to prismatic channels.

Equation (4.70) may be rewritten in finite difference from as

$$\Delta x = \Delta y \left(\frac{1 - F_r^2}{S_o - S_f} \right)_{mean} \tag{4.73}$$

Where `mean' refers to the mean value for the interval (Δ x). This form of the equation may be used to determine, directly, the distance between given difference of depth for any trapezonidal channel.

In the case of a prismatic channel the bed slope S_o will remain constant throughout but the total energy line will change its slope from one point to another so long as the flow is non uniform. The average value of S_f can be computed using the Manning formula based on average channel properties over the reach concerned. The method is best illustrated by an example.

ILLUSTRATIVE EXAMPLES

4.1 Water flows in a rectangular, concretc, open channel that is 12 m wide at a depth of 2.5 m. The channel slope is 0.0028. Find the water velocity and the flow rate. Take n = 0.013.

Solution

$$A = by = 12 \times 2.5 = 30\ m^2$$

$$P = b + 2y = 12 + 2 \times 2.5 = 12 + 5 = 7$$

$$R = \frac{A}{P} = \frac{30}{17} = 1.7647 \text{ m}$$

$$v = \frac{1}{n} R^{2/3} S^{1/2}$$

$$= \frac{1}{0.013} \times (1.7647)^{2/3} (0.0028)^{1/2} = 5.944 \text{ m/s} \qquad \text{Ans}$$

$$Q = A\ v = 30 \times 5.944 = 178.322 \text{ m}^3\text{/s} \qquad \text{Ans}$$

4.2 **What depth is required for 4 m³/s flow in a rectangular planed-wood (n = 0.012) channel 2 m wide with a bottom slope of 0.002 ?**

Solution

Let y be the depth of flow.

Therefore, A = by = 2y

$$P = b + 2y = 2 + 2y$$

$$R = \frac{A}{P} = \frac{2y}{2+2y} = \left(\frac{y}{y+1}\right)$$

$$Q = A \times v$$

or, $$4 = (2y)\left(\frac{1}{n} R^{2/3} S^{1/2}\right)$$

or, $$4 = (2y)\left[\frac{1}{0.012}\left(\frac{y}{y+1}\right)^{2/3} (0.002)^{1/2}\right]$$

$$\frac{1}{1.863} = \frac{y^{5/3}}{(y+1)^{2/3}}$$

$$0.5366 = \frac{y^{5/3}}{(y+1)^{2/3}}$$

Solving by trial and error method

$$y = 0.888 \text{ m} \qquad \text{Ans}$$

4.3 **There is a steady flow at 5 m³/s at a slope of 0.002. What is the proper width b of a rectangular channel if we wish to**

minimize the wetled perimeter for the sake of economy of construction. The channel is made of unfinished concrete (n = 0.0015)

Solution

For a rectangular section, the best hydraulic section has a width b equal to twice the depth. i.e. b = 2y.

$$A = by = b \times \frac{b}{2} = \frac{b^2}{2}$$

$$P = b + 2y = b + b = 2b$$

$$R = \frac{A}{P} = \frac{b^2/2}{2b} = \frac{b}{4}$$

$$Q = A \times v$$

or, $$5 = \left(\frac{b^2}{2}\right)\left(\frac{1}{n} R^{2/3} S^{1/2}\right)$$

or, $$5 = \left(\frac{b^2}{2}\right)\left[\frac{1}{0.0015} \times \left(\frac{b}{4}\right)^{2/3} (0.002)^{1/2}\right]$$

$$5.91589 \ b^{8/3} = 5$$

$$b^{8/3} = 0.84518$$

$b = (0.84518)^{3/8} = (0.8452)^{0.375} = 0.9388$ m Ans

Therefore minimum perimeter.

$P = 2b = 0.9388 = 1.8776$ m Ans

4.4 Water is to flow in a rectangular flume at a rate of 1.42 m^3/s and at a slope of 0.0028. Determine the dimensions of the channel cross section. (a) If b = 2y and (b) b = y

Compare the area. Take Manning's n = 0.017

Solution

(a) b = 2y (given)

$$A = by = \frac{b^2}{2}$$

$$P = b + 2y = 2b$$

$$R = A/P = \frac{b^2}{2 \times 2b} = \frac{b}{4}$$

$$Q = A\,v$$

$$Q = A\frac{1}{n}R^{2/3}S^{1/2}$$

$$1.42 = \frac{b^2}{2} \times \frac{1}{0.017} \times \left(\frac{b}{2}\right)^{2/3} (0.0028)^{1/2}$$

$$1.42 = 0.6176 \quad b^{8/3}$$

$$b^{8/3} = \frac{1.42}{0.6176} = 2.299$$

$$b = (2.299)^{3/8} = 1.366 \text{ m}$$ Ans

$$\therefore \; y = \frac{b}{2} = \frac{1.366}{2} = 0.683 \text{ m}$$ Ans

$$A = b^2/2 = \frac{(1.366)^2}{2} = 0.9329 \text{ m}^2$$

(b) $b = y$ (given)

$$A = by = b^2$$

$$P = b + 2y = b + 2b = 3b$$

$$R = \frac{A}{P} = \frac{b^2}{3b} = b/3$$

$$Q = A\,v$$

$$Q = A\frac{1}{n}R^{2/3}S^{1/2}$$

$$1.42 = b^2 \times \frac{1}{0.017} \times (b/3)^{2/3}(0.0028)^{1/2}$$

$$1.42 = 1.496 \quad b^{8/3}$$

$$b^{8/3} = \frac{1.42}{1.496} = 0.9489$$

$$b = (0.9489)^{3/8} = 0.9805 \text{ m}$$

$$b = y = 0.9805 \text{ m}$$

$$A = b^2 = (0.9805)^2 = 0.96145 \text{ m}^2$$

The cross sectional of part (a) has the small cross sectional area.

4.5 A rectangular channel, 8 m wide and 1.5 m deep has a slope of 1 in 1000 and is lined with smooth concrete plaster. It is desired to enhance (increase) the discharge to a maxium by changing the dimensions of the channel but keeping the same quantity of lining. Work out the new dimensions and the percentage increase in discharge. Assuming Manning's constant n = 0.015.

Solution

(a) $A = b\,y = 8 \times 1.5 = 12\ m^2$

$$P = b + 2y = 8 + 2 \times 1.5 = 11\ m$$

$$R = \frac{A}{P} = \frac{12}{11} = 1.09$$

$$v = \frac{1}{n} R^{2/3} S^{1/2} = \frac{1}{0.015}(1.09)^{2/3}\left(\frac{1}{1000}\right)^{1/2} = 2.23\ m/s$$

$$Q_1 = A\ v = 12 \times 2.23 = 26.79\ m^3/s$$

(b) $b = 2y$ for maximum discharge

$$P = b + 2y = 2y + 2y = 4y$$

Using the same quantity of lining

Therefore, $P = 11 = 4y$

$$y = \frac{11}{4} = 2.75\ m \qquad \text{Ans}$$

Therefore, $b = 2 \times 2.75 = 5.5\ m$ Ans

$$A = by = 5.5 \times 2.75 = 15.125\ m^2$$

$$R = \frac{A}{P} = \frac{y}{2} = \frac{2.75}{2} = 1.375\ m$$

$$v = \frac{1}{n} R^{2/3} S^{1/2} = \frac{1}{0.015}(1.375)^{2/3}\left(\frac{1}{1000}\right)^{1/2} = 2.6\ m/s$$

$$Q_2 = A\ v = 15.125 \times 2.6 = 39.42\ m^3/s$$

Therefore, percentage increase in discharge

$$= \frac{Q_2 - Q_1}{Q_1} \times 100 = \frac{39.42 - 26.79}{26.79} \times 100\% = 47.1\% \qquad \text{Ans}$$

4.6 In a channel of rectangular cross section with bed slope of 1 in 1000. The discharge is to be 1420 liters/s when the channel whose depth is half its width (breadth) is running full. The channel is roughly excavated, the value of Chezy's C is 28, if the bottom and side slopes are smoothly, Chezy's C is 84. Assuming that the cost of excavation per m^3 is 2.25 times the cost of smooth finishing per m^2, compare the cost of the two types of channels.

Solution: The discharge

$$Q = A\ C\sqrt{R\ S}$$

$$1.42 = by \times 28\sqrt{\frac{y}{2} \times \frac{1}{1000}}$$

$$1.42 = \frac{2.8}{\sqrt{5}}\ y^{5/2}$$

$$y^{5/2} = 1.134$$

$$y = (1.134)^{2/5} = 1.05 \text{ m}$$

Therefore, b = 2y = 2 × 1.05 = 2.1 m

After smooth lining, C = 84

$$Q = by \times 84\sqrt{\frac{y}{2} \times \frac{1}{1000}}$$

$$1.42 = \frac{8.4}{\sqrt{5}}\ y^{5/2}$$

$$y^{5/2} = 0.378$$

$$y = (0.378)^{2/5} = 0.6776 \text{ m} \approx 0.68 \text{ m}$$

$$b = 2y = 2 \times 0.6776 = 1.3552 \text{ m} \approx 1.36 \text{ m}$$

Cost of the excavation per running meter length of channel in first case is,

$$C_1 = 2.25\ k\ (1.05 \times 2.1) = 4.96\ k$$

Cost of the excavation and smooth finishing per meter length of channel in the second case is,

$$C_2 = 2.25\ k\ (0.68 \times 1.36) + k\ \{(1.36 + 2 \times 0.68)\}$$

$$= 2.07\ k + 2.72\ k = 4.79\ k$$

Therefore, $\frac{C_1}{C_2} = \frac{4.96k}{5.79k} = 1 : 0.966$ Ans

4.7 A rectangular open channel 5.5 m wide and 1.22 m deep has a slope of 1 to 1000 and is lined with rubble masonary (Manning's n = 0.017). We wish to increase the amount of water dishcarged as much as possible without changing the channel slope or rectangular form of the section. The dimensions of the section may be changed but the amount of excavation must not be changed. Determine (a) the discharge of the original channel; (b) new dimensions of the channel to give the maximum discharge; (c) the ratio of the new discharge to the original discharge. What is the new discharge?

Solution

(a) Original cross sectional area of flow

$$A = 5.5 \times 1.22 = 6.71 \text{ m}^2$$

$$R = \frac{A}{P} = \frac{6.71}{5.5 + 2 \times 1.22} = \frac{6.71}{7.94} = 0.845$$

$$v = \frac{1}{n} R^{3/2} S^{1/2}$$

$$= \frac{1}{0.017} (0.845)^{2/3} \left(\frac{1}{1000}\right)^{1/2} = 1.66259 \text{ m/s}$$

$$Q = A\ v = 6.71 \times 1.66259 = 11.156 \text{ m}^3/\text{s}$$ Ans

(b) The best hydraulic rectangular section has a b=2y i.e. y=b/2.

Since the cross sectional area must be unchanged

by = Constant = 6.71

or, $\frac{b^2}{2} = 6.71$

or, $b^2 = 13.42$

New width b = 3.663 m Ans

New depth $y = \frac{3.663}{2} = 1.8316$ m Ans

(c) The new hydraulic radius

$$R_{new} = \frac{A}{P} = \frac{6.71}{3.663 + 2 \times 1.8316} = \frac{6.71}{7.326} = 0.91587$$

Therefore new discharge over original discharge is written as

$$\left(\frac{R_{new}}{R_{original}}\right)^{2/3} = \left(\frac{0.91587}{0.845}\right)^{2/3} = (1.08387)^{2/3} = 1.055$$

Therefore, new dischage = 11.156 × 1.055 = 11.77139 m³/s Ans

4.8 A trapezoidal channel has a bottom width of 2.4 m, side slopes of 1.5 : 1 and the bed slope of 1 in 500. Find its discharge when the uniform depth of flow is 1.5 m. Take Chezy's Constant C = 61.

Solution

$$v = C\sqrt{RS} = 61\sqrt{\frac{A}{P} \times \frac{1}{500}} \quad (1)$$

$$A = (b + zy)\,y = [2.4 + 1.5\,(1.5)] \times 1.5 = 6.975 \text{ m}^2$$

$$P = b + 2y\sqrt{1 + z^2} = 2.4 + 2 \times 1.5\sqrt{1 + (1.5)^2}$$

$$= 2.4 + 3\sqrt{1 + 2.25} = 7.8 \text{ m}$$

$$R = \frac{A}{P} = \frac{6.975}{7.8} = 0.8932$$

Substituting in eqaution (1)

$$v = 61\sqrt{0.8932 \times \frac{1}{500}} = 61\sqrt{1.7865 \times 10^{-3}} = 2.578 \text{ m/s}$$

$$Q = A\ v = 6.975 \times 2.578 = 17.98\ m^3/s$$ Ans

4.9 A trapezoidal channel is to be designed to convey 280 m³ of water per minute. Determine the cross sectional dimensions of the channel if bed slope is 1 in 1600, the side slope at 45° and the cross section is to be minimum. Take Chezy's constant C = 50.

Solution: For minimum cross section

$$b + 2\ zy = 2y\sqrt{1 + z^2}$$

i.e. $b + 2y = 2y\sqrt{2}$

or, $b = 0.828\ y$

and $R = y/2$

$$A = (b + zy)\ y = (0.828\ y + y)\ y = 1.828\ y^2$$

$$P = b + 2\ zy\sqrt{1+z^2} = 0.828\ y + 2\sqrt{2}\ y = 3.656\ y$$

Now, $Q = A\ C\sqrt{RS}$

$$\frac{280}{60} = 1.828\ y^2 \times 50\sqrt{\frac{y}{2} \times \frac{1}{1600}}$$

$$0.0510 = y^{5/2}\sqrt{\frac{1}{3200}}$$

$$y^{5/2} = (2.888)$$

$\therefore$ $y = (2.888)^{2/5} = 1.528$ m **Ans**

$\therefore$ $b = 0.828\ y = 0.828 \times 1.528 = 1.265$ m **Ans**

4.10 **Determine the bed width of a trapezoidal channel of maximum discharge, having a depth of flow of 3 m and side slopes of 1 : 1. What will be the discharge if, the bed slope is 1 in 1600? Take Chezy's Constant C = 66.**

Solution: For maximum discharge

$$b + 2\ zy = 2y\sqrt{1+z^2}$$

or, $b + 2 \times 1 \times 3 = 2 \times 3\sqrt{1+1}$

or, $b + 6 = 6\sqrt{2}$

or, $b = 6\left(\sqrt{2} - 1\right) = 2.484$ m **Ans**

$$A = (b + ny)\ y = (2.484 + 1 \times 3)\ 3 = 16.452\ m^2$$

$$R = \frac{y}{2} = \frac{3}{2} = 1.5\ m$$

Therefore, $Q = A\ v$

$$= A\ C\sqrt{RS} = 16.452 \times 66\sqrt{1.5 \times \frac{1}{1600}} = 33.246\ m^3/s$$

4.11 Using Manning's formula, find the discharge of the trapezoidal concrete channel when n = 0.012, b = 0.6 m, y = 0.6 m, S = $\frac{1}{400}$ and side slopes are 1 : 1. What will be the discharge if the depth of flow is increased to 0.9 m?

Solution

$$A = (b + zy)\,y = (0.6 + 0.6) \times 0.6 = 0.72\ m^2$$

$$P = b + 2y\sqrt{1+z^2} = 0.6 + 2 \times 0.6\sqrt{1+1} = 0.6 + 1.2\sqrt{2}$$

$$= 0.6 + 1.697 = 2.3\ m$$

$$R = \frac{A}{P} = \frac{0.72}{2.3} = 0.313\ m$$

Now, $Q = A\,v = (0.72)\,\frac{1}{n}\,R^{2/3}\,S^{1/2}$

$$= \frac{0.72}{0.012}(0.313)^{2/3}\left(\frac{1}{400}\right)^{1/2} = 1.383\ m^3/s$$

In the second case, the depth of flow is 0.9 m, therefore,

$$A = (b + zy)\,y = (0.6 + 0.9)(0.9) = 1.35\ m^2$$

$$P = b + 2y\sqrt{1+z^2} = 0.6 + 2 \times 0.9\sqrt{1+1} = 0.6 + 1.8\sqrt{2}$$

$$= 3.15\ m$$

$$R = \frac{A}{P} = \frac{1.35}{3.15} = 0.428\ m$$

Hence, $Q = A\,v = (1.35) \times \frac{1}{n}(R)^{2/3}(S)^{1/2}$

$$= \frac{1.35}{0.012}(0.428)^{2/3}\left(\frac{1}{400}\right)^{1/2} = 3.1975\ m^3/s$$ **Ans**

4.12 The amount of water to be carried by a canal excavated in smooth earth (n = 0.03) is 10 m³/s. It has side slopes of 2 : 1 and the depth of water is to be 1.5 m or less. If the channel slope is 450 mm/km, what must be the width at the bottom?

Solution

$$A = (b + zy)\,y = (b + 2 \times 1.5) \times 1.5 = 1.5\,b + 4.5$$

$$P = b + 2\,y\sqrt{1 + z^2} = b + 2 \times 1.5\sqrt{1 + (2)^2}$$

$$= b + 3\sqrt{1 + 4} = b + 3\sqrt{5} = b + 6.7082$$

$$S = (450/1000)/1000 = 0.00045$$

Now, $Q = A\dfrac{1}{n} R^{2/3} S^{1/2} = \dfrac{1}{n}\dfrac{A^{5/3}}{P^{2/3}} S^{1/2}$

$$10 = \left(\frac{1}{0.03}\right)\frac{(1.5b + 4.5)^{5/3}}{(b + 6.7082)^{2/3}}(0.00045)^{1/2}$$

$$(14.142)\,(b + 6.7082)^{2/3} = (1.5b + 4.5)^{5/3}$$

Solving by trial and error, the value of b comes out to be 6.05 m Ans

4.13 Determine the proportions of the most efficient open channel of trapezoidal cross section with sides inclined at 60° to the bed. The flow velocity in such a channel is found from v = 50 $R^{0.57}$ $S^{0.47}$. Calculate the dimensions of the channel to discharge 27280 m³/ hour if the bed slope is 1 in 1760.

Solution: For most efficient channel,

$$b + 2\,zy = 2y\sqrt{1 + z^2}$$

where $z = \dfrac{1}{\tan\theta} = \dfrac{1}{\tan 60^\circ} = \dfrac{1}{\sqrt{3}}$

$$\therefore\; b + \frac{2y}{\sqrt{3}} = 2y\sqrt{1 + \frac{1}{3}}$$

or, $b + \dfrac{2y}{\sqrt{3}} = 2y\sqrt{4/3}$

or, $b = \dfrac{2y}{\sqrt{3}}$

and $R = y/2$

$$A = (b + zy)\,y = \left(\frac{2y}{\sqrt{3}} + \frac{y}{\sqrt{3}}\right)(y) = \sqrt{3}\,y^2\ m^2$$

$$P = b + 2y\sqrt{1+z^2} = \frac{2y}{\sqrt{3}} + 2y\sqrt{4/3} = 2\sqrt{3}\ y\ m$$

Now, Q = A v

$$\frac{27280}{3600} = \sqrt{3}\,y^2 \times 50\left(\frac{y}{2}\right)^{0.57}\left(\frac{1}{1760}\right)^{0.47}$$

$$7.577 = \sqrt{3} \times 50 \times y^{2.57} \times 0.6736 \times 0.0298$$

$7.7577 = 1.74\ y^{2.57}$

$y^{2.57} = 4.355$

$y = (4.355)^{1/2.57} = 1.772$ m Ans

Now, $b = \frac{2}{\sqrt{3}}(1.772) = 2.0469$ m Ans

4.14 What would be the cross section of greatest hydraulic efficiency for a trapezoidal channel with side slopes of 1.5 horizontal to 1 vertical if the design discharge is 10 m³/s and the channel slope is 0.0005? Take Manning's n = 0.025

Solution

Given, z = 1.5

For greatest hydraulic efficiency

$$b + 2\ zy = 2y\sqrt{1+z^2}$$

$$b + 2 \times 1.5\,y = 2y\sqrt{1+(1.5)^2}$$

$b + 3y = 3.6055\ y$

$b = 0.6055\ y$ (1)

$A = (b + zy)\ y = (0.6055\ y + 1.5\ y)\ y = 2.1055\ y^2$

$$R = \frac{y}{2}$$

Therefore, $Q = A\,\frac{1}{n}\,R^{2/3}\,S^{1/2}$

$$10 = 2.1055\,y^2 \times \frac{1}{0.025} \times \left(\frac{y}{2}\right)^{2/3}(0.0005)^{1/2}$$

$10 = 1.18638\ y^{8/3}$

$y^{8/3} = 8.42899$

$y = (8.42899)^{3/8} = 2.224$ m Ans

Form (1) $b = 0.6055 \times 2.224 = 1.3468$ m Ans

4.15 **A power canal of trapezoidal section has to be excavated through hard clay at the least cost. Determine the dimensions of the channel given, discharge equal to 14 m³/s, bed slope 1 : 2500 and Manning's rugovisty constant n = 0.02.**

Solution

For least cost

$$\tan\theta = \tan 60^\circ = \sqrt{3} \qquad z = \frac{1}{\tan\theta} = \frac{1}{\sqrt{3}}$$

$$b + 2\,zy = 2y\sqrt{1+z^2}$$

or, $$b + \frac{2y}{\sqrt{3}} = 2y\sqrt{4/3}$$

or, $$b = \frac{2y}{\sqrt{3}} \qquad (1)$$

$$A = (b + zy)\,y = \left(\frac{2y}{\sqrt{3}} + \frac{y}{\sqrt{3}}\right)(y) = \sqrt{3}\,y^2$$

$$R = \frac{y}{2}$$

Now, $Q = A\,v$

$$14 = A\frac{1}{n}R^{2/3}\,S^{1/2}$$

or, $$14 = \left(\sqrt{3}\,y^2\right)\left(\frac{1}{n}\right)\left(\frac{y}{2}\right)^{2/3}(S)^{1/2}$$

or, $$14 = \sqrt{3}\,y^2 \times \frac{1}{0.02}\left(\frac{y}{2}\right)^{2/3}\left(\frac{1}{2500}\right)^{1/2} = \frac{1.732}{2^{2/3}}\,y^{8/3}$$

$$y^{8/3} = \frac{14 \times 2^{2/3}}{1.732} = \frac{14 \times 1.5874}{1.732} = 12.831$$

$$y = (12.831)^{3/8} = 2.60375 \text{ m}$$ Ans

Therefore, from equation (1)

$$b = \frac{2y}{\sqrt{3}} = \frac{2 \times 2.60375}{1.732} = 3.0065 \text{ m}$$ Ans

4.16 An open channel of most economical section, having the form of a half hexagon with horizontal bottom is required to give a maximum discharge of 20.2 m³/s of water. The slope of the channel bottom is 1 in 2500 Taking Chezy's constant C = 60, determine the dimensions of the cross section.

Solution: For most economical section

$$b + 2\,zy = 2y\sqrt{1+z^2} \qquad (1)$$

For half hexagon, the angle θ = 60°

therefore, $\tan\theta = \tan 60° = \sqrt{3}$

and $\tan\theta = \frac{1}{z}$

or, $z = \frac{1}{\tan\theta} = \frac{1}{\sqrt{3}}$

Substituting in equation (1)

$$b + \frac{2y}{\sqrt{3}} = 2y\sqrt{4/3}$$

or, $$b = \frac{2y}{\sqrt{3}} \qquad (2)$$

and $$A = (b + zy)\,y = \left(\frac{2y}{\sqrt{3}} + \frac{y}{\sqrt{3}}\right)(y) = \sqrt{3}\,y^2$$

$$P = b + 2y\sqrt{1+z^2} = \frac{2y}{\sqrt{3}} + 2y\sqrt{4/3} = 2\sqrt{3}\,y$$

and $$R = \frac{y}{2}$$

Now, $Q = A v$

or, $20.2 = \sqrt{3}\, y^2\, C \sqrt{R\, S}$

or, $20.2 = \sqrt{3}\, y^2 \times 60 \sqrt{\frac{y}{2} \times \frac{1}{2500}}$

or, $1.4696\ y^{5/2} = 20.2$

$y^{5/2} = 13.744$

$y = (13.744)^{2/5} = 2.852$ m Ans

$b = \frac{2y}{\sqrt{3}} = \frac{2 \times 2.852}{\sqrt{3}} = 3.294\ m$ Ans

4.17 **A trapezoidal channel has side slopes 1 to 1. It is required to discharge 13.75 m³/s of water with a bed graident of 1 in 1000. If unlined the value of Chezy's C is 44. If lined with concrete, its value is 60. The cost per m³ of excavacation is four times the cost m² of lining. The channel is most efficient one. Find whether the lined channel or unlined channel will be cheaper. What will be the dimensions or that economical channel?**

Solution: For unlined channel.

$$Q = A\, C \sqrt{R\, S} = (b + zy)\, y\, C \sqrt{\frac{y}{2}\, S}$$

But $b = 2y\sqrt{1 + z^2} - 2\, zy = 2y\sqrt{2} - 2y$

$= 2\sqrt{2}\ y - 2y = 0.8284\ y = 0.83\ y$

Substituting the value of b in above equation.

$$Q = (0.83\, y + y)\, y \times 44 \sqrt{\frac{y}{2} \times \frac{1}{1000}}$$

$13.75 = 1.798\ y^{5/2}$

$y^{5/2} = 7.6434$

$y = (7.6434)^{0.4} = 2.255$

$\therefore$ $b = 0.83 \times 2.225 = 1.872$

For lined channel with concrete

$$Q = (b + zy)\, y\, C\sqrt{RS}$$

$$13.75 = (0.83\, y + y)\; y \times 60\sqrt{\frac{y}{2} \times \frac{1}{1000}} = 2.4552\, y^{5/2}$$

$$y^{5/2} = 5.4985$$

$$y = (5.4985)^{0.4} = 1.9774$$

Therefore, b = 0.83 y = 0.83 × 1.9774 = 1.64125 m

Cost of excavation per running meter length of channel for the first case

$$C_1 = 4\,k\,[(b + zy)\,y]$$

$$= 4\,k\,[(1.872 + 2.255) \times 2.255] = 37.225\,k$$

Where k is the cost of smooth finishing per m².

Cost of excavation and smooth finishing per meter length of channel in the second case is

$$C_2 = 4\,k\,[\,(b + zy)\,y] + k\;\;[(b + 2y\;\;\sqrt{1 + z^2}\;\,]$$

$$= 4\,k\,[(1.64 + 1.9774) \times 1.9774] + k\,[1.64 + 2 \times 1.9774\,\sqrt{2}\;]$$

$$= 4\,k \times 7.153 + 7.2329\,k = 28.612\,k + 7.2329\,k = 35.845\,k$$

$$\therefore \quad \frac{C_1}{C_2} = \frac{37.225\,k}{35.845\,k} = 1.03849 = 1 : 0.9629$$

The lined channel will be cheaper. The dimensions are

$$b = 1.64 \text{ m}$$

$$y = 1.9774 \text{ m}$$

4.18 (a) The normal depth of flow in a trapezoidal concrete lined channel is 2 m. The channel base width is 5 m and has side slopes of 1 : 2. Manning's n is 0.015 and the bed slope is 0.001. Determine the discharge, mean velocity and the Reynolds number. Take $\mu = 1.14 \times 10^{-3}$ kg/ms.

(b) If the discharge in the channel given in (a) part were 30 m³/s, find the normal depth of flow.

(c) During large foods, the water level in the channel given in part (a) exceeds the bankfull level of 2.5 m. The flood bands are 10 m wide and are grassed with side slopes of 1 : 3. The estimated Manning's n for these food banks is 0.035. Estimate the discharge for a maximum flood level of 4 m.

Solution : As per Fig. 4.31.

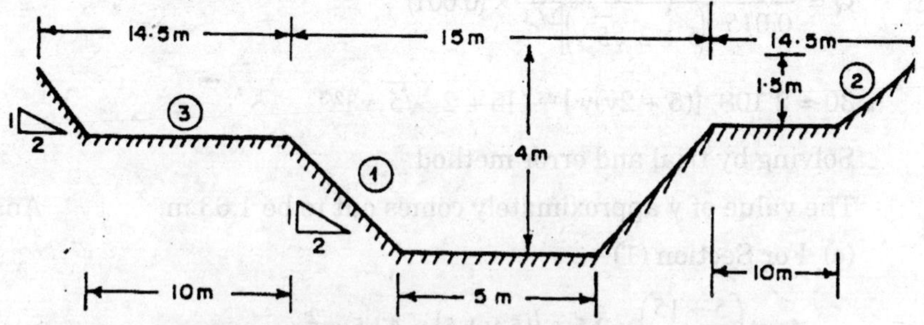

Fig. 4.31. Compound channel section P. 4.18.

$$A = (b + zy)\, y = (5 + 2 \times 2)\, 2 = 18 \text{ m}^2$$

$$P = b + 2y\sqrt{1 + z^2} = 5 + 2 \times 2\sqrt{1 + 4}$$

$$= 5 + 4\sqrt{5} = 13.944 \text{ m}$$

$$R = \frac{A}{P} = \frac{18}{13.944} = 1.29$$

$$Q = A\frac{1}{n} R^{2/3} S^{1/2}$$

$$= 18 \times \frac{1}{0.015} \times (1.29)^{2/3} (0.001)^{1/2} = 44.988 \ m^3/s \qquad \text{Ans}$$

$$Q = A \, v_{av}$$

$$v_{av} = \frac{Q}{A} = \frac{44.988}{18} = 2.499 \text{ m/s}$$

$$R_e = \frac{\rho R v_{av}}{\mu} = \frac{1000 \times 1.29 \times 2.499}{1.14 \times 10^{-3}} = 2.828 \times 10^6 \qquad \text{Ans}$$

(b) We know that

$$Q = A\frac{1}{n}\left(\frac{A}{P}\right) R^{2/3} S^{1/2} = \frac{1}{n}\left(\frac{A^{5/3}}{P^{2/3}}\right)\left(S^{1/2}\right)$$

Here $A = (5 + 2y)\, y$

$$P = 5 + 2\sqrt{5}\,y$$

Substituting in above equation

$$Q = \frac{1}{0.015}\frac{[(5+2y)y]^{5/3}}{\left[\left(5+2\sqrt{5}\,y\right)\right]^{2/3}} \times (0.001)^{1/2}$$

$30 = 2.108\ [(5 + 2y)y\,]^{5/3} / [5 + 2\ \sqrt{5}\ y\,]^{2/3}$

Solving by trial and error method.

The value of y approximately comes out to be 1.63 m. Ans

(c) For Section (1)

$$A_1 = \left(\frac{5+15}{2}\right) \times 2.5 + (15 \times 1.5) = 47.5\ m^2$$

$$P_1 = 5 + (2\sqrt{4} \times 2.5) = 16.18\ \text{m}$$

Section (2) and (3) have the same dimensions, hence

$$A_2 = A_3 = \left(\frac{10+14.5}{2}\right) \times 1.5 = 18.38\ \text{m}^2$$

$$P_2 = P_3 = 10 + \left(\sqrt{10} \times 1.5\right) = 14.74\ \text{m}$$

$$Q_1 = \frac{1}{n} A_1^{5/3} / P_1^{2/3}\,(S)^{1/2}$$

$$= \frac{1}{0.015} \times \frac{(47.5)^{5/3}}{(16.18)^{2/3}} \times (0.001)^{1/2} = 205.3\ m^3/s$$

$$Q_2 = Q_3 = \frac{1}{n}\frac{(A_2)^{5/3}}{(P_2)^{2/3}}(S)^{1/2}$$

$$= \frac{1}{0.035} \times \frac{(18.38)^{5/3}}{(14.74)^{2/3}} \times (0.001)^{1/2} = 19.2\ m^3/s$$

The total discharge

$$Q = Q_1 + Q_2 + Q_3 = 205.3 + 19.2 + 19.2 = 243.7\ \text{m}^3/\text{s}$$

4.19 A rough timber channel of Manning's n = 0.012 in the form of an equilateral triangle with apex down of 1.2 m sides is laid on a slope of 0.01. Calculate the uniform discharge occuring at a depth of 0.9 m.

Solution: As per Fig. 4.32

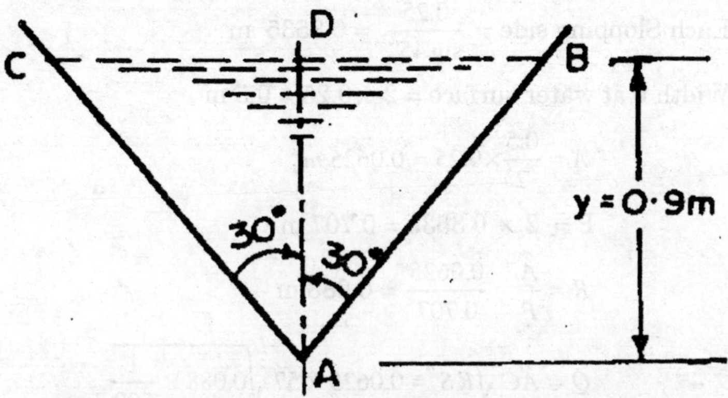

Fig. 4.32. Rough timber channel P. 4.19.

A = Area of triangular ABC = 2 × Area of triangle ADB

$$= \frac{2 \times AD \times BD}{2} = AD \times BD = AD \times AD \tan 30^{\circ}$$

$$= 0.9 \times 0.9 \tan 30^{\circ} = 0.46765\ m^2$$

P = AC + AB = AB + AB = 2 AB

$$= 2\sqrt{AD^2 + BD^2} = 2\sqrt{(0.9)^2 + (0.9 \tan 30^{\circ})^2}$$

$$= 2\sqrt{0.81 + 0.27} = 2\sqrt{1.08} = 2.07846$$

$$R = \frac{A}{P} = \frac{0.46765}{2.07846} = 0.22499$$

$$Q = A\frac{1}{n}R^{2/3}S^{1/2} = 0.46765 \times \frac{1}{0.012}(0.22499)^{2/3}(0.01)^{1/2}$$

∴ Q = 1.4416 m³/s Ans

4.20 **A water channel is V shaped with each side making an angle of 45° to the vertical. Calculate the discharge when the depth of water in the channel is 250 mm and the bed slope of channel is 1 in 500. Take Chezy's constant C = 57. What will be the depth of water in the channel to pass twice the above said discharge if the bed slope and value of C remain unaltered?**

Solution

Each Slopping side = $\frac{0.25}{\sin 45^\circ}$ = 0.3535 m

Width b at water surface = 2 × 0.25 = 0.5 m

$$A = \frac{0.5}{2} \times 0.25 = 0.0625\ m^2$$

$$P = 2 \times 0.3535 = 0.707 \text{ m}$$

$$R = \frac{A}{P} = \frac{0.0625}{0.707} = 0.088 \text{ m}$$

$$Q = AC\sqrt{RS} = 0.0625 \times 57\sqrt{0.088 \times \frac{1}{500}}$$

$$= 0.04726 \text{ m}^3\text{/s} = 47.26 \text{ liters/s}$$ Ans

Now $Q = 2 \times 0.04726 = 0.09452 \text{ m}^3\text{/s}$

If y be the new depth of flow, each slopping side = $\frac{y}{\sin 45^\circ} = \sqrt{2}\ y$

$$A = y^2$$

$$P = 2 \times \text{slopping side} = 2\sqrt{2}\ y$$

$$R = \frac{A}{P} = \frac{y^2}{2\sqrt{2}\ y} = \frac{y}{2\sqrt{2}}$$

But, $Q = AC\sqrt{R\ S}$

$$0.09452 = y^2 \times 57\sqrt{\frac{y}{2\sqrt{2}} \times \frac{1}{500}}$$

$$y^{5/2} = \frac{1.6583 \times 10^{-3}}{0.02659} = 0.062362$$

$$y = (0.062362)^{2/5} = 0.3295 \text{ m}$$

$\therefore$ y = 0.3295 m Ans

4.21 An open channel is to be designed to carry 1 m³/s at a slope of 0.0065. The channel material has an value of 0.011. Find the most efficient cross section for,

(a) rectangular section (b) triangular section

(c) trapezoidal section (d) semicircular section.

Solution

$$v = \frac{1}{n} R^{2/3} S^{1/2}$$

$$\frac{Q}{A} = \frac{1}{n}\left(\frac{A}{P}\right)^{2/3} (S)^{1/2}$$

or, $$\frac{A^{5/3}}{P^{2/3}} = \frac{Qn}{(S)^{1/2}}$$

$$= \frac{1 \times 0.011}{(0.0065)} = 0.136438 \qquad (1)$$

(a) rectangular section

A = $2y^2$

P = b + 2y = 4y $\qquad [\because b = 2y]$

Substituting in equation (1)

$$\frac{(2y^2)^{5/3}}{(4y)^{2/3}} = 0.136438$$

$$\frac{(2^{5/3})}{(4)^{2/3}} y^{8/3} = 0.136438$$

$$(1.2599)\, y^{8/3} = 0.136438$$

$y^{8/3} = 0.10829$

$y = (0.10829)^{3/8} = 0.4344$ m = 434.4 mm Ans

$\therefore$ b = 2y = 2 × 434.4 = 868.965 mm Ans

(b) triangular section

The most efficient triangular section has a 90° angle and 1 : 1 side slopes

A = y^2

and P = $2\sqrt{2}\, y$ = 2.828 y

Substituting in equation (1)

$$\frac{(y^2)^{5/3}}{(2.828\, y)^{2/3}} = 0.136438$$

$0.5\ y^{8/3} = 0.136438$

$y^{8/3} = 0.2728$

$y = (0.2728)^{3/8} = 0.6144 = 614.4$ mm Ans

Sides = $y\sqrt{2} = 0.6144 \times \sqrt{2} = 0.8689$ m = 868.9 mm Ans

(c) trapezoidal section

For most efficient section z, = $\sqrt{3}$

$$b + \frac{2}{\sqrt{3}} y = 2y\sqrt{1 + \frac{1}{3}}$$

$$b + \frac{2y}{\sqrt{3}} = 2y\sqrt{4/3}$$

$$b = \frac{2y}{\sqrt{3}} = 1.154\ y$$

$$A = (b + zy)\,y = \left(1.154\ y + \frac{1}{1.732}\,y\right) y = 1.732\ y^2$$

$$P = 2y\sqrt{1 + z^2} = 2y\sqrt{1 + 1/3} = 2y\sqrt{4/3} = 3.465\,y$$

Substituting in equation (1)

$$\frac{A^{5/3}}{P^{2/3}} = 0.136438$$

$$\frac{(1.732\ y^2)^{5/3}}{(3.465\,y)^{2/3}} = 0.136438$$

$1.09088\ y^{8/3} = 0.136438$

$y^{8/3} = 0.125$

$y = (0.125)^{3/8} = 0.4586$ m Ans

$b = 1.154\ y = 1.154 \times 0.4586 = 0.5292$ m Ans

(d) Semicircular section.

$$A = \frac{\pi}{8} y^2$$

$$P = \frac{\pi}{2} y$$

Substituting in equation (1)

$$\frac{\left(\frac{\pi}{8}y^2\right)^{5/3}}{\left(\frac{\pi}{2}y\right)^{2/3}} = 0.136438$$

$$(0.15584)\,y^{8/3} = 0.136438$$

$y^{8/3}$ = 0.87548

y = 0.951 m = 951 m **Ans**

4.22 **A 0.9 m diameter pipe is to have a maximum discharge of 0.7 m³/s. Calculate the bed slope. Taking Manning's n = 0.0149.**

Solution

$$Q = A\,v = A\frac{1}{n}R^{2/3}\,S^{1/2} = \frac{1}{n}\left(\frac{A^{5/2}}{P}\right)^{2/3} S^{1/2}$$

For maximum discharge $\left(\frac{A^{5/2}}{P}\right)$ must be a maximum and the corresponding value of θ is given by

θ = 151.5° = 2.64 radiam

Therefore, $A = r^2\left(\theta - \frac{\sin 2\theta}{2}\right) = (0.45)^2\left[2.64 + \frac{0.84}{2}\right] = 0.62$ **m²**

P = 2 r θ = 2 × 0.45 × 2.64 = 2.37 m

$$R = \frac{A}{P} = \frac{0.62}{2.32} = 0.262\ \text{m}$$

$$Q = \frac{1}{n}\frac{A^{5/3}}{P^{2/3}}S^{1/2} = \frac{1}{0.0149}\frac{(0.62)^{5/3}}{(2.37)^{2/3}}S^{1/2}$$

0.7 = 17.02 $S^{1/2}$

$S^{1/2}$ = 0.04112

S = 1.69 × 10^{-3} = 1 in 591.2 **Ans**

4.23 **Find the minimum diameter necesary if the channel has a slope of 1 in 5000 give a discharge of 1.4 m³/s. Take Chezy's Constant C = 66.**

Solution: For maximum discharge

$\theta = 154° = 2.68$ radian.

Depth for maximum discharge

$y = r(1 - \cos\theta) = r(1 - \cos 154°) = 0.95\ d$

$$A = r^2\left(\theta - \frac{\sin 2\theta}{2}\right) = \frac{d^2}{4}(2.68 + 0.394) = 0.77\ d^2$$

$P = 2r\theta = 2.68\ d$

$$Q = AC\sqrt{RS} = C\frac{A^{3/2}}{P^{1/2}}S^{1/2}$$

$$1.4 = 66\frac{(0.77)^{3/2}}{(2.68)^{1/2}}\frac{d^3}{d^{1/2}}\left(\frac{1}{5000}\right)^{1/2}$$

$(0.3852)\ d^{5/2} = 1.4$

$d^{5/2} = 3.634$

$d = (3.634)^{2/5} = 1.6756$ Ans

4.24 A Sewer a diameter 0.6 m, has a slope of 1 in 200. What will be the maximum velocity of flow that can occur, and what is the discharge at this velocity? Take Chezy's constant C = 55.

Solution: For maximum velocity

$\theta = 128.75° = 2.247$ radian

Depth of flow

$y = r - r\cos\theta = r(1 + 0.62) = 1.62\ r = 0.81\ d$

$$A = r^2\left(\theta - \frac{\sin 2\theta}{2}\right) = (0.3)^2\ [2.247 + 0.488] = 0.246\ m^2$$

$P = 2r\theta = 2 \times 0.3 \times 2.247 = 1.3482$ m

$$R = \frac{A}{P} = \frac{0.246}{1.3482} = 0.18246$$

$$v = C\sqrt{RS} = 55\sqrt{0.18246 \times \frac{1}{200}} = 1.66\ m/s$$

$Q = Av = 0.246 \times 1.66 = 0.4086\ m^3/s$

4.25 A Sewer pipe, is laid on a slope of 0.0018 and is to carry 2.76 m^3/s when the pipe flows at 80 percent of full depth.

Determine the required diameter of pipe. Take Manning's n = 0.014.

Solution : As per Fig. 4.33.

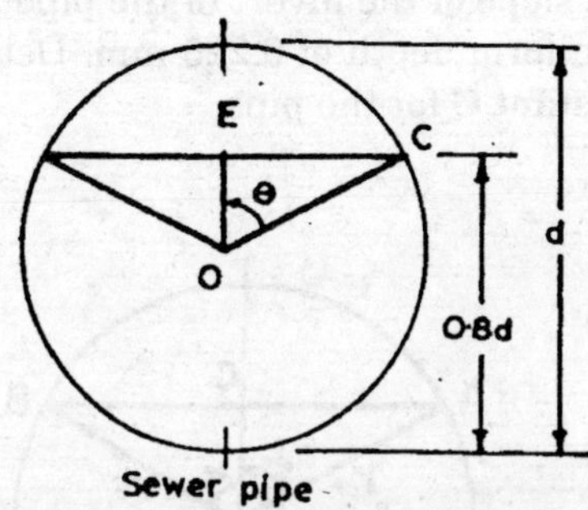

Fig. 4.33. Sewer pipe P. 4.25

$$OE = 0.8\ d - \frac{d}{2} = 0.3\ \text{d}$$

$$\cos\theta = \frac{OE}{OC} = \frac{0.3\ d}{0.5\ d} = 0.6$$

$$\theta = \cos^{-1}(0.6) = 53.13° = 53°7'\ 48''$$

$$\tan 53.13° = \frac{CE}{OE}$$

$$\therefore \quad \text{CE} = \text{OE} \tan 53.13° = (0.3\text{d})\ (1.33) = 0.4\text{d}$$

$$A = \{[360 - 2\,(53.13)]/360\}\left(\frac{\pi}{4} d^2\right) + 2\left[\frac{1}{2}\,(0.3\ d)\,(0.4\,d)\right]$$

$$= 0.6736\ \text{d}^2$$

$$P = \{[360 - (2)(53.13)]\ /\ 360\}\,(\pi\ d) = 2.214\ d$$

$$\text{Now,} \quad Q = A\,\frac{1}{n}\,R^{2/3}\,S^{1/2} = \frac{1}{n}\,\frac{A^{5/3}}{P^{2/3}}\,S^{1/2}$$

$$2.76 = \frac{1}{0.014}\left[\frac{\left(0.6736\,d^2\right)^{5/3}}{(2.214d)^{2/3}}\right]\left[(0.00018)^{1/2}\right]$$

$$2.88 = \text{d}^{8/3}\left[\frac{0.5176}{1.69876}\right]$$

$$2.88 = \text{d}^{8/3}\,(0.3047)$$

$d^{8/3} = (9.4518)$

$d = (9.4518)^{3/8} = 2.32$ m Ans

4.26 Water flows through a 300 mm diameter pipe at the rate of 35 liters/s. The slope of the invert of the pipe is 1 in 2000 and the flow is at uniform depth of 0.225 mm. Determine the value of Chezy's constant C for the pipe.

Solution

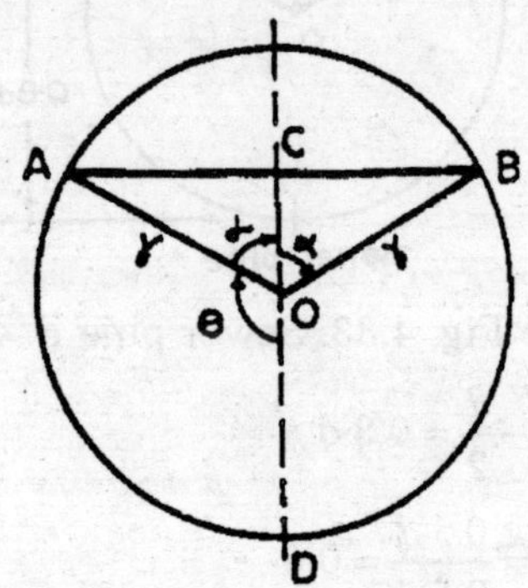

Fig. 4.34. Pipe flow P. 4.26

From Fig. 4.34, OC = CD – OD

$= y - r = 0.225 - 0.15 = 0.075$

From triangle OAC,

$$\cos\alpha = \frac{OC}{OA} = \frac{0.075}{0.15} = 0.5$$

$$\alpha = \cos^{-1}(0.5) = 60^{\circ}$$

$$\theta = 180 - \alpha = 180 - 60 = 120^{\circ}$$

$$= 120 \times \frac{\pi}{180} = 2.094 \textit{ radians}$$

$$A = r^2\left(\theta - \frac{\sin 2\theta}{2}\right)$$

$$= (0.15)^2\left[2.094 - \frac{\sin 2\ 40}{2}\right] = 0.0568\ m^2$$

$$P = 2\,r\,\theta = 0.3 \times 2.094 = 0.6282 \text{ m}$$

$$R = \frac{A}{P} = \frac{0.0568}{0.6282} = 0.0904 \text{ m}$$

$$Q = A v$$

$$35 \times 10^{-3} = 0.0568 \times v$$

$$v = 0.61619 \text{ m/s}$$

$$v = C\sqrt{R S}$$

$$0.61619 = C\sqrt{0.0904 \times 1/2000}$$

$$C = 91.65 \text{ m/s}$$

4.27 Calculate the quantity of water that will be discharged at a uniform depth of 0.9 m in a 1.2 m diamter sewer pipe which is laid at a slope of 1 in 1000. Assume Chezy's C = 58.

Solution: From Fig. 4.49

$$OC = CD - OD = 0.9 - 0.6 = 0.3$$

From triangle OAC,

$$\cos\alpha = \frac{OC}{OA} = \frac{0.3}{0.6} = \frac{1}{2}$$

$$\alpha = 60^{o}$$

$$\therefore \quad \theta = 120^{\circ} = 2.094 \text{ radians}$$

$$A = r^2\left(\theta - \frac{\sin 2\,\theta}{2}\right)$$

$$= (0.6)^2\left[2.094 - \frac{\sin 2\ 40^{o}}{2}\right] = 0.909\ m^2$$

$$P = 2\,r\,\theta = 1.2 \times 2.094 = 2.5128 \text{ m}$$

$$R = \frac{A}{P} = \frac{0.909}{2.5128} = 0.36174$$

$$v = C\sqrt{RS} = 58\sqrt{0.36174 \times \frac{1}{1000}} = 58 \times 0.019 = 1.1029$$

$$Q = Av = 0.909 \times 1.1029 \approx 1 \text{ m}^3/\text{s}$$ Ans

4.28 Water flows 1.2 m deep at an average velocity of 2.4 m/s in a rectangular channel.

(a) What is the Froude number? (b) What is the Specific energy? (c) What is the alternate depth? (d) What is the Specific thrust? (e) What is the conjugate depth?

Solution

(a) $F_r = \frac{v}{\sqrt{gy}} = \frac{2.4}{\sqrt{9.81 \times 1.2}} = 0.7$ Ans

(b) $E = y + \frac{v^2}{2g} = 1.2 + \frac{(2.4)^2}{2 \times 9.81} = 1.4935$ m Ans

(c) $E = y + \frac{q^2}{2g\ y^2} = y + \frac{(2.4 \times 1.2)^2}{2 \times 9.81\ y^2}$

$$1.4935 = y + \frac{8.2944}{19.62\,y^2}$$

or, $y^3 - 1.4935\ y^2 + 0.42275 = 0$

Solving we get $y = 0.758$ m Ans

(d) Specific thrust

$$= \frac{y^2}{2} + \frac{q^2}{gy} = \frac{(1.2)^2}{2} + \frac{(2.4 \times 1.2)^2}{9.81 \times 1.2}$$

$= 0.72 + 0.70459 = 1.42458$ m^2 Ans

(e) Specific thrust

$$= \frac{y^2}{2} + \frac{q^2}{gy} = \frac{y^2}{2} + (2.88)^2 / 9.81\ y$$

or, $y^3 - 2.848\ y^2 + 1.691 = 0$

Solving we get $y = 0.73$ m Ans

4.29 A rectangular channel 5.5 m wide conveys 16.5 m^3/s at a depth of 1.5 m.

(a) What is the critical depth?

(b) Is the given flow subcritical or superciritcal?

(c) What is the Froude number of the flow?

(d) What is the specific energy?

(e) If Manning's n = 0.012, what slope would be required to maintain uniform flow?

(f) Is this slope a mild or a steep slope?

Solution

(a) $$y_c = \left(\frac{q^2}{g}\right)^{1/3} = \left[\frac{(16.5/5.5)^2}{9.81}\right]^{1/3}$$

$= (9/9.81)^{1/3} = 0.9174$ m **Ans**

(b) Subcritical because $y > y_c$ **Ans**

(c) $$F_r = \frac{v}{\sqrt{gy}} = \frac{16.5/5.5 \times 1.5}{\sqrt{9.81 \times 1.5}}$$

$$= \frac{2}{\sqrt{14.715}} = \frac{2}{3.836} = 0.52137$$ **Ans**

(d) $$E = y + v^2/2g = 1.5 + \frac{(2)^2}{2 \times 9.81} = 1.70387$$

(e) $$Q = \frac{1}{n}\frac{A^{5/3}}{P^{2/3}} S^{1/2}$$

$$4.5 = \frac{1}{0.012} \times (8.25)^{5/3} / (8.5)^{2/3} S^{1/2}$$

$4.5 = 673.9526\ S^{1/2}$

$S^{1/2} = 0.02448 \therefore S = 6 \times 10^{-4}$

(f) Mild, since uniform flow is subcritical.

4.30 A rectangular channel 2.5 m wide has a bed slope of 1/400, n = 0.017, and conveys 9 m³/s. What is (a) The normal or uniform flow depth? (b) The critical depth for the given flow? (c) The flow regime?

Solution

$A = b\,y_n = 2.5\,y_n, \qquad P = b + 2\,y_n = 2.5 + 2\,y_n$

$$Q = A\,v = A\frac{1}{n} R^{2/3} S^{1/2} = \frac{1}{n}\frac{A^{5/3}}{P^{2/3}} S^{1/2}$$

$$9 = \left(\frac{1}{0.017}\right)\frac{(2.5\,y_n)^{5/3}}{(2.5 + 2\,y_n)^{2/3}}\left(\frac{1}{400}\right)^{1/2}$$

$$\frac{(2.5\, y_n)^{5/3}}{(2.5 + 2\, y_n)^{2/3}} = \frac{9}{2.941} = 3.06$$

Solving by trial and error method.

$y_n = 1.56$ m by trial Ans

(b) $y_c = \left(q^2 / g\right)^{1/3}$

$$= \left[\frac{(9/2.5)^2}{9.81}\right]^{1/3} = (1.321)^{1/3} = 1.0972 \text{ m}$$ Ans

(c) Q = by v

9 = 2.5 × 1.56 v

v = 2.3076 m/s

$$F_r = v/\sqrt{gy} = \frac{2.3076}{\sqrt{9.81 \times 1.56}} = 0.5898 \text{ Subcritical}$$

4.31 A channel of rectangular cross section is 1.2 m wide and the normal flow is 0.72 m³/s, with a depth of 0.6 m. If a stream lined hump Z metres high is installed on the bed, state if a standing wave will form when (a) Z = 100 mm and (b) Z = 300 mm. Give reasons.

Solution

$$q = Q/b = \frac{0.72}{1.2} = 0.6 \text{ m}^3\text{/s/m}$$

We know that

$$y_c = \left(q^2/g\right)^{1/3} = \left(\frac{(0.6)^2}{9.81}\right)^{1/3} = (0.03669)^{1/3} = 0.3323 \text{ m}$$

We know that

$$y_c = \frac{2}{3}E = \frac{2}{3}\left(y_o - z_{\min}\right)$$

$$0.33 = \frac{2}{3}\left(0.6 - z_{\min}\right)$$

$$z_{\min} = 0.105 \text{ m} = 105 \text{ mm}$$

(a) Since z given is less than z_{min} wave will not form.

(b) Since z given is more than z_{min} wave will form.

4.32 The bottom of a constnat width rectangular channel is raised gradually. The depth and velocity upstream of the raised bed are 3 m and 3 m/s, respectively. (a) For no losses, what is the maximum height the bottom may be raised without affecting the flow depth upstream? (b) For the condition of part (a), what is the horizontal component of force acting on the curved portion of the raised bed per unit width of channel? Neglect bed shear.

Solution

(a) $$E_1 = y_1 + \frac{v_1^2}{2g} = (3)^2/2g + 3 = 0.4587 + 3 = 3.4587 \text{ m}$$

$$y_c = \left(\frac{q^2}{g}\right)^{1/3} = \left[\frac{(3 \times 3)^2}{9.81}\right]^{1/3} = 2.021 \text{ m}$$

$$E_c = \frac{3}{2} y_c = \frac{3}{2} \times 2.021 = 3.0317 \text{ m}$$

$$\Delta z = E_1 - E_c = 3.4587 - 3.0317 = 0.4269 \text{ m}$$ Ans

(b) $$\frac{\rho g y_1^2}{2} - \frac{\rho g y_2^2}{2} - F = \rho q (v_2 - v_1)$$

$$\left[Here\ v_2 = v_c = \sqrt{g y_c} = \sqrt{9.81 \times 2.021} = 4.4526\ m/s\right]$$

$$\therefore \frac{9810 (3)^2}{2} - \frac{9810 (2.021)^2}{2} - F = 1000 \times 9 (4.4526 - 3)$$

$$44145 - 20034.18 - F = 13073.4$$

$$F = 11037.42 \text{ N} = 11.037 \text{ kN}$$ Ans

4.33 An open channel of constant width has its floor raised 45.7 mm at a given section. If the depth of the approximately flow is 457 mm calcuate the rate of flow indicated by :

(a) a 76.3 mm drop in the surface elevation over the raised bottom;

(b) a 76.3 mm increase in surface elevation over the raised bottom.

Neglect losses at the sudden change of section.

Calculate the critical depth in each case and plot it relatively to the water surface profile.

Solution

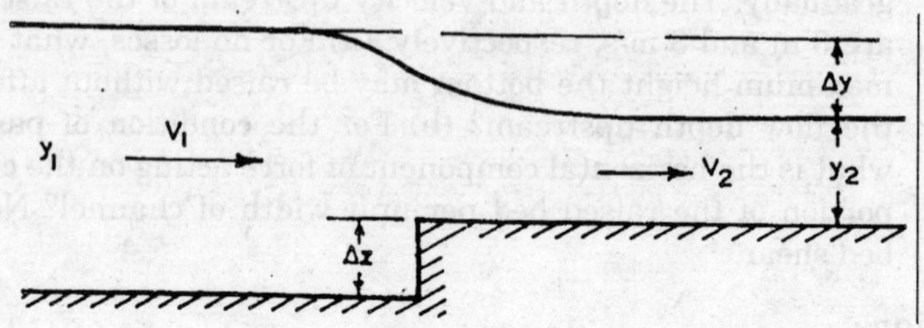

Fig. 4.35. Open channel of constant width floor raised P. 4.33.

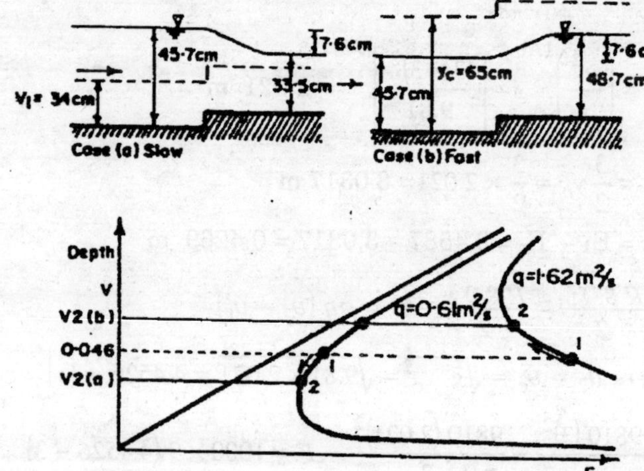

Fig. 4.36. Critical depth with reference to depth y² (Subcritical and supercritical flow). P. 4.33.

$$y_1 = y_2 + \Delta y + \Delta z \qquad (1)$$

Neglecting losses

$$E_1 = E_2 + \Delta z$$

or, $$y_1 + \frac{v_1^2}{2g} = y_2 + \frac{v_2^2}{2g} + \Delta z$$

$$y_1 + \frac{v_1^2}{2g} = y_1 - \Delta y + \frac{v_2^2}{2g} \qquad \text{(2) [From equation (1)]}$$

From the continuity equation

$$v_1 \; y_1 = v_2 \; y_2$$

$$v_2 = v_1 \left(\frac{y_1}{y_2}\right)$$

Substituting in equation (2)

$$\frac{v_1^2}{2g}\left[1-\left(\frac{y_1}{y_2}\right)^2\right] = -\Delta y$$

(a) $\Delta y = 76.3$ mm

$$\frac{y_1}{y_2} = \frac{0.457}{0.457-0.046-0.076} = \frac{0.457}{0.335} = 1.364$$

Therefore, $v_1^2 = \frac{2\times 9.81\times 0.076}{1-(1.364)^2} = -\frac{1.49112}{-0.86098} = 1.73286$

$\therefore \quad v_1 = 1.3163$

Discharge per unit width

$$q = y_1\, v_1 = 0.457\times 1.3163 = 0.6015\ m^2/s$$

$$y_c = \left(q^2/g\right)^{1/3} = \left[\frac{(0.6015)^2}{9.81}\right]^{1/3}$$

$= (0.0368)^{1/3} = 0.33286 = 332.8$ mm

(b) $\Delta y = -76.3$ mm

$$\frac{y_1}{y_2} = \frac{0.457}{0.457-0.046+0.076} = \frac{0.457}{0.487} = 0.93839$$

Therefore, $v_1^2 = \frac{-2g\,\Delta y}{1-\left(\frac{y_1}{y_2}\right)^2} = \frac{2\times 9.81\times(-0.0763)}{1-(0.93839)^2} = \frac{1.497}{0.1194} = 12.5368$

$\therefore$ $v_1 = 3.54$

$q = y_1\ v_1 = 0.457 \times 3.54 = 1.618\ m^2/s$

$$y_c = \left(\frac{q^2}{g}\right)^{1/3}$$

$$= \left[\frac{(1.618)^2}{9.81}\right]^{1/3} = (0.26686)^{1/3} = 0.6438\ m = 643.8\ \text{mm}$$

EXERCISES

A. THEORY

4.1 Working from first principles, derive the Chezy formula for uniform or normal flow in an open channel, and show the location of the hydraulic gradient line.

4.2 In the eqution v = $\sqrt{R\,S_o}$ as applied to flow through open channels, on what factors do the value of C depend?

4.3 Find the best ratio of width to depth for an open channel rectangular channel stating clearly the assumptions on which your theory is based.

4.4 State the Manning's formula for flow in channels giving typical values of roughness n.

4.5 Find the proportions of a rectangular channel of depth y and width b which will make the discharge a maximum for given cross sectional area A. Use the Chezy formula.

4.6 Using the Chezy formula, find the proportions of trapezoidal channel which will make the discharge a maximum for a given area.

4.7 Show that the sides and base of a trapezoidal section are tangential to a semi circle whose centre is at the water surface.

4.8 Prove that for a maximum discharge to occur through a channel of trapezoidal shape for a given cross section and fixed bed slope, can be expressed as :

(a) $\dfrac{b + 2\,zy}{2} = y\sqrt{1 + z^2}$

(b) $R = y/2$

Side slopes equals 1 vertical to n horizontal.

4.9 Derive the necessary relation of bed-width and depth for the most economical section of a trapezoidal channel with side slope z horizontal to one vertical.

4.10 Derive the condition for the best side slope of the most economical trapezoidal section.

4.11 A trapezoidal channel having a cross sectional area A_1, wetted perimeter P, Mannings coefficient n and laid to a slope S_0, carries a certain discharge slope S_0, carries a certain discharge Q_1 at a depth of flow equal to y. To increase the discharge the base width of the channel is widened by x, Keeping all other parameters same Prove that

$$\left(\frac{Q_2}{Q_1}\right)^3\left(1+\frac{x}{P_1}\right)^2=\left(1+\frac{xy}{A_1}\right)^5$$

Where Q_2 is the new discharge in the channel.

4.12 Show the relationship connecting Chezy's C and Mannings h.

4.13 Explain most economical (a) rectangular section (b) trapezoidal section (c) v-section (d) circular section.

4.14 Show that for a circular conduit running part full, R is maximum for a given A when y = 0.813 D, where D is the conduit diameter.

4.15 Show that for a circular culvert of diameter D the velocity of flow will be a maximum when the depth of flow at a centre is 0.81 D. Use Chezy's formula.

4.16 Find an expression for a theoretical depth for maximum velocity in a closed circular channel in terms of diameter d.

4.17 Find an expression for the theoretical depth for maximum discharge in a closed circular channel in terms of diameter d.

4.18 Prove that for a channel of circular section, the depth of flow,

(a) y = 0.81 D for maximum velocity and

(b) y = 0.95 D for maximum discharge.

Where y = depth of flow and D is the diameter of circular channel.

4.19 Define (a) Specific energy.

(b) Critical depth.

(c) Minimum Specific energy.

(d) Supercritical flow.

(e) Subcritical flow.

4.20 Define specific energy and show by plotting a specific energy diagram, the occurrence of a minimum specific energy for a given discharge rate.

4.21 What do you understand by critical depth of an open channel when the flow in it is not uniform.

4.22 Describe specific energy diagrams. How are they useful for the explanation of flow phenomena in an open channel?

4.23 Explain the meaning of the terms shooting flow and streaming or traquil flow in connection with open channel.

4.24 Define critical depth in as many ways as you can. Show that critical depth in rectangular channel is given by

$$y_c = 3\sqrt{q^2/h}$$

4.25 Show that in a rectangular channel (a) the critical depth is two thirds of specific energy and (b) the Froudes number at critical depth is unity.

4.26 Explain (a) specific energy curve (b) critical depth with reference to open channel flow.

4.27 Show that in an open channel minimum specific energy at a given discharge and maximum discharge at a given specific energy occurs when Froude number is unity.

4.28 Explain the terms specific energy and specific force. Prove that the critical depth at which minimum value of specific energy occurs in a wide rectangular channel is expressed as $\left(q^2/g\right)^{1/3}$ where q = discharge per unit width.

4.29 Define specific energy and critical depth as applied to flow in an open channel show that the critical depth for a given flow corresponds to the condition of minimum specific energy.

4.30 Draw a neat sketch of the specific energy curve for an open channel with constant discharge. Define critical flow in an open channel and derive a general criterion for such flow to occur in a channel.

5

Rapidly Varied Flow

5.1. INTRODUCTION

In the rapidly varied flow the depth changes considerably over a relatively short distance (increasing or decreasing). Flow from sleep slope to mild slope channels, flow downstream of over flow structures (spillways and weirs) and flow through under flow structures (Sluice gate) are some of the examples of rapidly varied flow.

The hydraulic jump phenomenon is an important type of rapidly varied flow. It is also an example of a stationary surge wave.

5.1.1 Hydraulic jump

When there is a rapid change from super critical flow (F_r>1) to sub critical flow (F_r < 1), then there is a sudden increase in the depth of flow. This is called a hydraulic jump. The resulting flow transition is rapid, and involves a large energy loss due to turbulence. Under these circumstances, a solution to the hydraulic jump problem cannot be found using a specific energy diagram. Instead, the momentum equation is used.

In any particular case the position of the hydraulic jump will depend upon a combination of the upstream and downstream flow condition. The overall form as well as the detailed performance of the jump depends upon the upstream and downstream depths and the velocity of the approaching flow. The conditions are best described quantitatively by the Froude number of the flow.

5.1.2 Forms of the hydraulic jump

There are essentially five different forms of the hydraulic jump which may occur on a horizontal apron and which may be encountered in the design of energy dissipation devices. It is important to note that the energy dissipation and interenal characteristics of the hydraulic jump vary considerably with each form. The following forms, illustrated by Fig. 5.1 a to e, were made as a result of extensive experiments conducted by the Bureau of Reclamation. Each form has been classified in relation to the kinetic flow factor of the entering stream.

1. Undular jump [Fig. 5.1 (a)]

The flow expands smoothly and then oscillates so that the jump has the form of a smooth initial wave followed by a train of waves of decreasing amplitude. Little turbulence is induced by such a jump and the energy loss is therefore small. An undular jump is formed when the F_{r_1} value lies between 1 and 1.7. When $F_{r_1} = 1$ (the limiting case) no jump occurs and the flow is critical ($y_1 = y_2 = y_c$).

2. Weak jump [Fig 5.1 (b)]

For F_{r_1} between 1.7 and 2.5 ($1.8 < y_2/y_1 < 3$) the hydraulic jump changes progressively. The increased rate of energy dissipation results in the first few wave crests "breaking" and isolated reverse rollers form on the upstream facing slope of these wave just below the crest.

3. Oscillating jump [Fig. 5.1(c)]

For F_{r_1} = 2.5 to 4.5, the jump formed is known as an oscillating jump. In this case of entering jet of water oscillates back and forth from the bottom to the surface and back again.

4. Steady jump [Fig. 5.1 (d)]

For F_{r_1} = 4.5 to 9 the jump formed is well stablized and is called a steady jump. For this jump the energy dissipation ranges for 45 to 70 per cent.

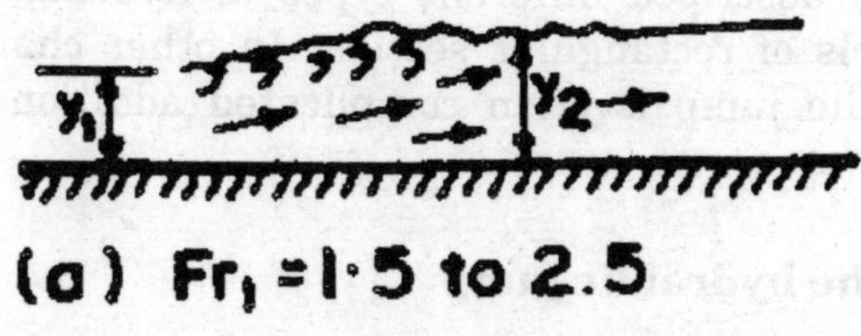

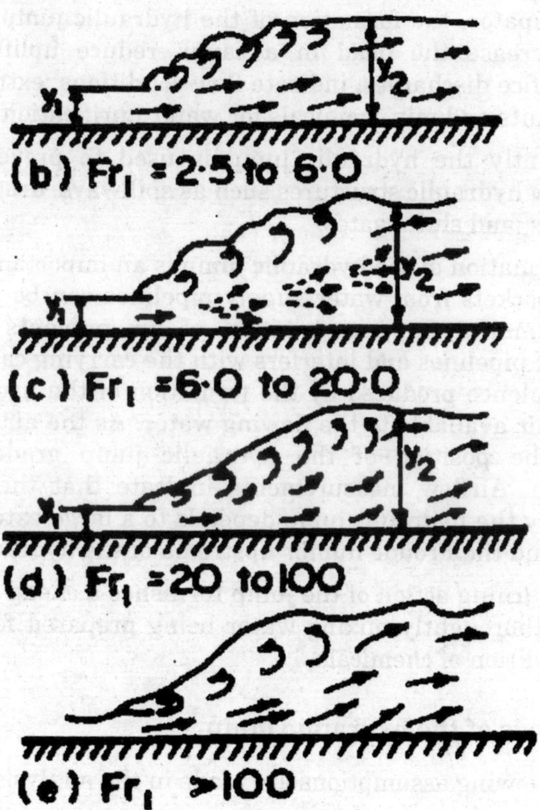

Fig. 5.1. Forms of hydraulic jump

5. Strong jump [Fig. 5.1 (e)]

For F_{r_1} >9 the jump formed is called a strong jump. In this case a rough surface prevails which continues downstream for a long distance. The jump action is quite rough but is effective since the energy dissipation may reach 85 per cent.

The above described different types of hydraulic jump refer only to channels of rectangular section. In other channel sections the shape of the jump is often complicated additionally by cross section.

5.1.3 Uses of the hydraulic jump

Although the most common use of the hydraulic jump is as an energy dissipator, the formation of the hydraulic jump can also be used to increase the head in a canal, reduce uplift pressures, increase orifice discharges, indicate flow conditions, extract air from closed conduits and mix chemicals for water purification.

Frequently the hydraulic jump is used to prevent scouring action below hydraulic structures such as spillways, drop structures, outlet works, and sluice gates.

The formation of the hydraulic jump is an important method by which air pockets from water supply pipelines can be removed. At pipeline summits the accumulation of air prevents the proper operation of pipelines and interfers with the carrying capacity of the lines. Turbulence produced by the formation of the hydraulic jump will make air available to the flowing water. As the air is gradually removed, the position of the hydraulic jump gradually moves downstream. Airflow measurements inidcate that the rate of air entrained by the hydraulic jump depends to a large extent upon the discharge and the Froude numbr upstream of the hydraulic jump.

The churning action of the jump furnishes a cheap and efficient means for thoroughtly mixing water being prepared for treatment with the addition of chemicals.

5.1.4 Analysis of the hydraulic jump

The following assumptions are made in the analysis :

(1) The channel is rectangular in shape, having parallel side constants.

(2) A horizonal floor exists.

(3) All friction losses are neglected.

(4) The hydraulic jump is assumed to take place instantaneously.

(5) Streamline flow exists immediately before and after the jump.

(6) The flow before the jump and after the jump is nearly uniform and one dimensional.

Fig. 5.2 (a) shows a section of a channel in which hydraulic jump occurs and at Fig. 5.2 (b), the static pressure distribution diagram for the section 1 and 2.

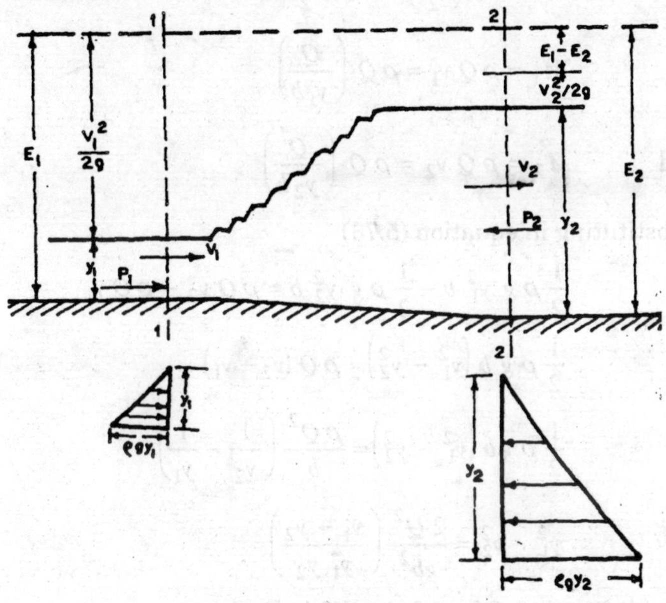

Fig. 5.2. Hydraulic jump

(a) Section of a rectangular channel at hydraulic jump occurs

(b) Static pressure distribution at sections (1) and (2)

Let y_1 be the depth before the jump, v_1 be the mean velocity at section (1), y_2 be the depth after the jump and v_2 be the mean velocity at the section (2).

When a jump occurs there is a change of momentum since the flow is slowed down. The force producing this change is due to the difference in hydrostatic pressure resulting from the change of depth.

The net force in the x direction = $P_1 - P_2$

and the momentum change = $M_2 - M_1$

hence

$$P_1 - P_2 = M_2 - M_1 \tag{5.1}$$

For a rectangular channel,

$$P_1 = \rho\, g \left(\frac{y_1}{2}\right) y_1\, b = \frac{1}{2} \rho\, g\, y_1^2\, b$$

and $$P_2 = \rho\, g \left(\frac{y_2}{2}\right) y_2\, b = \frac{1}{2} \rho\, g\, y_2^2\, b$$

$$M_1 = \rho\, Q\, v_1 = \rho\, Q \left(\frac{Q}{y_1 b}\right)$$

and $$M_2 = \rho\, Q\, v_2 = \rho\, Q \left(\frac{Q}{y_2 b}\right)$$

Substituting in equation (5.73)

$$\frac{1}{2} \rho\, g\, y_1^2\, b - \frac{1}{2} \rho\, g\, y_2^2\, b = \rho\, Q\, v_2 - \rho\, Q\, v_1$$

or, $$\frac{1}{2} \rho\, g\, b \left(y_1^2 - y_2^2\right) = \rho\, Q \left(v_2 - v_1\right)$$

or, $$\frac{1}{2} \rho\, g\, b \left(y_1^2 - y_2^2\right) = \frac{\rho\, Q^2}{b} \left(\frac{1}{y_2} - \frac{1}{y_1}\right)$$

or, $$y_1^2 - y_2^2 = \frac{2\, Q^2}{g b^2} \left(\frac{y_1 - y_2}{y_1\, y_2}\right)$$

Substituting q = Q/b and simplifying

or, $$\left(y_1 - y_2\right)\left(y_1 + y_2\right) = \frac{2\, q^2}{g} \left(\frac{y_1 - y_2}{y_1\, y_2}\right)$$

or, $$\left(y_1 + y_2\right)\left(y_1\, y_2\right) = \frac{2 q^2}{g} \tag{5.2}$$

Equation (5.2) is the momentum equation for hydraulic jump.

Equation (5.2) can be considered as quadratic equation in terms of y_1 and y_2 and its solution given the relations between the conjugate depths or sequent depths y_1 and y_2 as

$$y_2^2 + y_1\, y_2 - \frac{2\, q^2}{g\, y_1} = 0$$

$$\text{or,}\quad y_2 = \frac{-y_1 \sqrt{y_1^2 + \dfrac{8q^2}{g y_1}}}{2} = -\frac{y_1}{2} + \sqrt{\frac{y_1^2}{4} + \frac{2q^2}{g y_1}}$$

$$= -\frac{y_1}{2} + \sqrt{\frac{y_1^2}{4} + \frac{2 v_1^2 \, y_1^2}{g y_1}} = -\frac{y_1}{2} + \frac{y_1}{2}\sqrt{1 + \frac{8 v_1^2}{g y_1}}$$

$$= -\frac{y_1}{2} + \frac{y_1}{2}\sqrt{1 + 8 F_{r_1}^2} = \frac{y_1}{2}\left(\sqrt{1 + 8\, F_{r_1}^2} - 1\right) \qquad (5.3)$$

Using equation (5.3), y_2 may be evaluated in terms of y_1.

5.1.5 Energy loss across a hydraulic jump

$$\Delta\, E = E_1 - E_2 = \left(y_1 + \frac{v_1^2}{2g}\right) - \left(y_2 + \frac{v_2^2}{2g}\right)$$

$$= \left(y_1 + \frac{Q^2}{2g\,(b y_1)^2}\right) - \left(y_2 + \frac{Q^2}{2g\,(b y_2)^2}\right)$$

$$= \frac{Q^2}{b^2\,(2g)\,y_1^2} - \frac{Q^2}{b^2\,(2g)\,y_2^2} - (y_2 - y_1)$$

$$= \frac{q^2}{2g}\left[\frac{1}{y_1^2} - \frac{1}{y_2^2}\right] - (y_2 - y_1)$$

$$= \frac{q^2}{2g\; y_1^2\; y_2^2}\left(y_2^2 - y_1^2\right) - (y_2 - y_1)$$

Substituting the value of $\left(\dfrac{q^2}{g}\right)$ from equation (5.2)

$$\Delta\; E = \frac{(y_1 + y_2)(y_1\, y_2)}{4\; y_1^2\; y_2^2}\left(y_2^2 - y_1^2\right)(y_2 - y_1)$$

$$= \frac{(y_1 + y_2)^2}{4\; y_1\; y_2}(y_2 - y_1)(y_2 - y_1)$$

$$= \left(\frac{y_2 - y_1}{4\, y_1\; y_2}\right)\left[(y_1 + y_2)^2 - 4\, y_1\; y_2\right]$$

$$= \left(\frac{y_2 - y_1}{4\, y_1\, y_2}\right)\left[y_1^2 + 2\, y_1\, y_2 + y_2^2 - 4\, y_1\, y_2\right]$$

$$= \left(\frac{y_2 - y_1}{4\, y_1^2\, y_2^2}\right)\left[y_2^2 - 2\, y_1\, y_2 + y_1^2\right]$$

$$= \frac{(y_2 - y_1)(y_2 - y_1)^2}{4\, y_1\, y_2} = \frac{(y_2 - y_1)^3}{4\, y_1\, y_2} \tag{5.4}$$

Another form of the energy loss

$$\Delta E = E_1 - E_2 = y_1 + \frac{v_1^2}{2g} - \left(y_2 + \frac{v_2^2}{2g}\right) = \frac{v_1^2}{2g}\left(1 - \left(\frac{y_1}{y_2}\right)^2\right) + (y_1 - y_2)$$

$$\text{or,} \quad \frac{\Delta E}{y_1} = \frac{F_{r_1}^2}{2}\left[1 - \left(\frac{y_1}{y_2}\right)^2\right] + \left[1 - \frac{y_2}{y_1}\right] \tag{5.5}$$

Solving equation (5.5) for F_{r_1} in terms of y_2/y_1 and substituting into the basic equation, we obtain

$$\frac{\Delta E}{y_1} = \frac{\left[\frac{y_2}{y_1} - 1\right]^3}{4\, y_2 / y_1} \tag{5.6}$$

Since ΔE is positive, equation (5.6) shows that y_2/y_1 must be greater than one.

The specific energy E_1, can be written as

$$E_1 = \frac{v_1^2}{2g} + y_1 = y_1\left[\frac{v_1^2}{2g\, y_1} + 1\right] = \frac{y_1}{2}\left[F_{r_1}^2 + 2\right]$$

Nondimensionalizing ΔE on E_1, then

$$\frac{\Delta E}{E_1} = \frac{\left[y_2 / y_1 - 1\right]^3}{2\, \frac{y_2}{y_1}\left[F_{r_1}^2 + 2\right]} \tag{5.7}$$

The depth ratio is given in terms of F_{r_1}. Thus $\Delta E/E_1$ can be written as a function of upstream froude number.

The result is

$$\frac{\Delta E}{E_1} = \frac{\left[\sqrt{1 + 8\,F_{r_1}^2} - 3\right]^3}{8\left[\sqrt{1 + 8\,F_{r_1}^2} - 1\right]\left[F_{r_1}^2 + 2\right]} \tag{5.8}$$

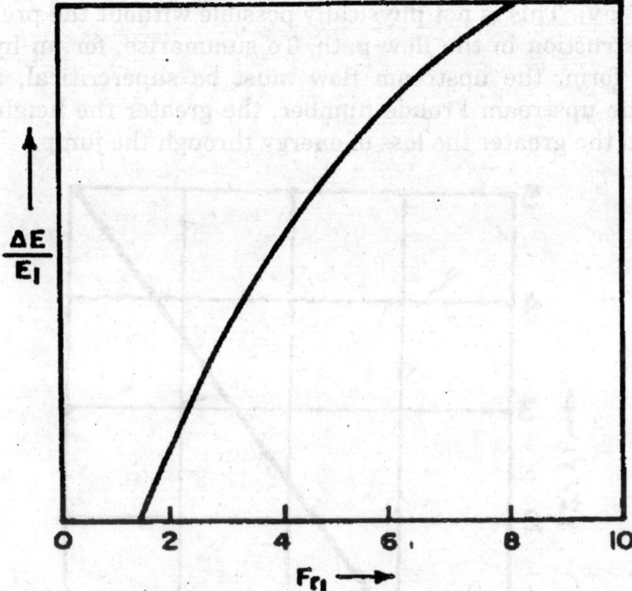

Fig. 5.3. Graph between $\frac{\Delta E}{E_1}$ v/s F_{r1}

The effect of upstream Froude number on energy loss is shown in dimensionless form in Fig. 5.3. Experimental data also correlated well with the prediction of equation (5.8). Fig. 5.3 shows that more than 70 percent of the energy of the upstream flow is dissipated in jumps with $F_{r_1} > 9$.

Inspection of equation (5.8) shows that if $F_{r_1} = 1$, the $\Delta E = 0$.

5.1.6 Significance of the hydraulic jump equations

In the preceding section, the hydraulic jump equation was deliberately formulated in terms of the upstream Froude number. The usefulness of this form of the equation is now demonstrated.

For critical flow F_{r_1} = 1, and substitution into equation (5.3) gives $y_2 = y_1$. This is an unstable condition resulting in an undular standing wave formation. For supercritical flow upstream, $F_{r_1} > 1$ and hence $y_2 > y_1$. This is the required conditions for the formation of a hydraulic jump. For subcritical flow upstream, $F_{r_1} < 1$ and hence $y_2 < y_1$. This is not physically possible without the presence of some obstruction in the flow path. To summarise, for an hydraulic jump to form, the upstream flow must be supercritical, and the higher the upstream Froude number, the greater the height of the jump and the greater the loss of energy through the jump.

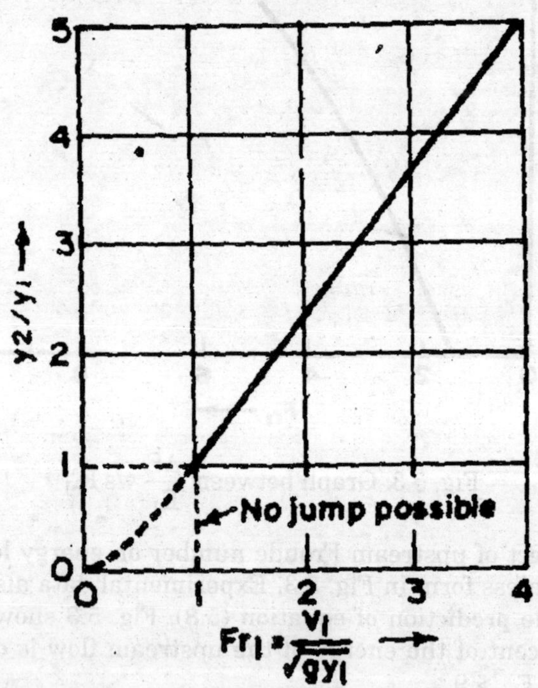

Fig. 5.4. Depth ratio across a hydraulic jump as a function of F_{r1}

The depth ratio y_2/y_1, across the hydraulic jump is shown as a function of the upstream Froude number in Fig. 5.4. The portion of the curve for $F_{r_1} < 1$ is dashed in recognition of the fact that to have a hydraulic jump the flow must be supercritical.

Fig. 5.5 shows a graph between the dimensionless head loss across a hydraulic jump as a function of upstream Froude number.

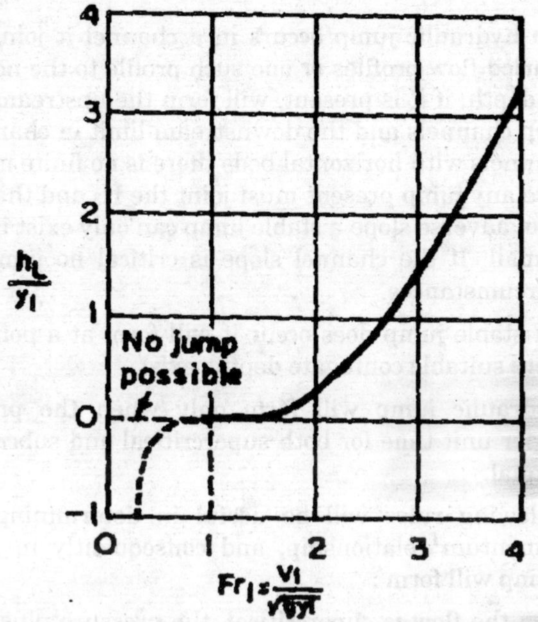

Fig. 5.5. Dimensionless energy loss across a hydraulic jump as a function of F_{r1}.

5.1.7 Height and length of the hydraulic jump

The height of the jump can be defined as the difference between the depth of water upstream and downstream of the jump. Hereafter, it will be designated by y_J. Thus

$$y_J = y_2 - y_1 \tag{5.9}$$

The considerable variation found in the results of different investigators of this aspect of the hydraulic jump is due to the difficulty in defining systematically the downstream termination. Variation in the location of this section does not significantly affect other properties of the jump but is clearly of primarily importance in the length measurement. The cummulative evidence of experimental results indicate that under most circumstances the length of the jump lies within the limits.

$5 < \frac{L}{y_2} < 6.9$ in which L is the length of the jump.

5.1.8 Location of hydraulic jump

When a hydraulic jump occurs in a channel it joins either two gradually varied flow profiles or one such profile to the normal depth. The normal depth, if it is present, will form the upstream limit of the jump in steep channels and the downstream limit in channels of mild slope. In channels with horizontal beds there is no finite normal depth and therefore any jump present must joint the H_3 and the H_2 profiles. In channels of adverse slope a stable jump can only exist if the slope is extremely small. If the channel slope is critical no jump can occur under any circumstances.

When a stable jump does occur it will form at a point along the channel where suitable confugate depths exist.

The hydraulic jump will form only when the pressure plus momentum per unit time for both supercritical and subcritical stages of flow are equal.

The following rules will be useful in determining the pressure-plus-momentum relationship, and consequently in determining where the jump will form :

(1) When the flow is supercritical, the pressure plus momentum increases as the depth decreases and decreases as the depth increases.

(2) When the flow is subcritical, the pressure plus momentum increases as the depth increases and decreases on the depth decreases.

(3) If the flow is accelerating when approaching the jump, the pressure plus momentum curve rises in the direction of flow.

(4) If the flow is decelerating when approaching the jump, the pressure plus momentum curve drops in the direction of the flow.

(5) If the flow is uniform when approaching the jump, the pressure plus momentum curve will be horizontal.

(6) If the flow is accelerating when leaving the jump, the pressure plus momentum curve drops in the direction of flow.

(7) If the flow is declerating when leaving the jump, the pressure plus momentum curve rises in the direction of flow.

(8) If the flow is uniform when leaving the jump the pressure plus momentum curve remains horizontal.

One of the problems that will be encountered by the designer is to find the location of the hydraulic jump at the junction of steep and mild slopes. This case is always present for a chute with a break in in grade that change from a steep to a mild slope. In the first case, shown in Fig. 5.6 (a), the jump may occur in either the steep or mild portion of the channel.

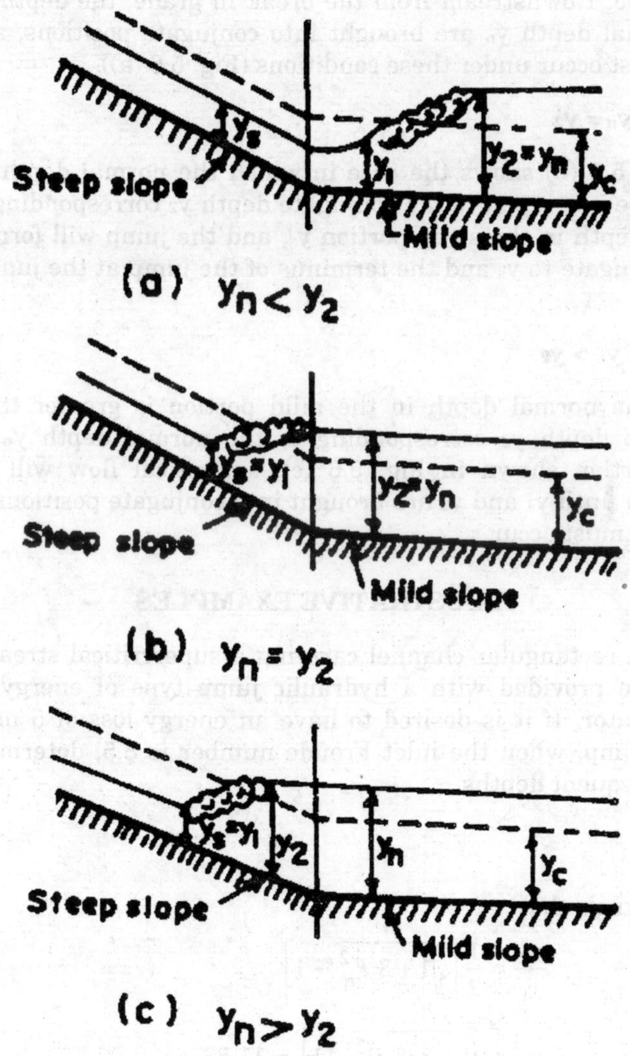

Fig. 5.6. Location of hydraulic jump is steep and mild channels.

Case I. $y_n < y_2$

If the normal depth in the mild portion is less than the upper conjugate depth y_2 corresponding to normal depth in the steep portion, supercritical flow will continue in the lower stage until extended to the mild portion, thereby increasing the depth on the mild slope. Downstream from the break in grade, the depth y_1 and the normal depth y_n are brought into conjugate positions, and the jump must occur under these conditions (Fig. 5.6 (a)).

Case II. $y_n = y_2$

Fig. 5.6 (b) shows the case in which the normal depth on the mild slope just equal to the conjugate depth y_2 corresponding to the normal depth in the steep portion y_s, and the jump will form when y_n is conjugate to y_s and the terminus of the jump at the junction of the slope.

Case III. $y_n > y_2$

If the normal depth in the mild portion is greater than the conjugate depth y_2 corresponding to the normal depth y_n in the steep portion shown in Fig. 5.6 (c), subcritical flow will extend upstream until y_1 and y_2 are brought into conjugate positions where the jump must occur.

ILLUSTRATIVE EXAMPLES

5.1 A rectangular channel carrying a supercritical stream is to be provided with a hydraulic jump type of energy disssipator. If it is desired to have an energy loss of 5 m in the jump, when the inlet Froude number is 8.5, determine the sequent depths.

Solution

We know that

$$\frac{y_2}{y_1} = \frac{1}{2}\left[\sqrt{1 + 8\,F_{r_1}^2} - 1\right]$$

$$= \frac{1}{2}\left[\sqrt{1 + 8\,(8.5)^2} - 1\right] = 11.53 \qquad (1)$$

and $\Delta E = \dfrac{(y_2 - y_1)^3}{4\ y_1\ y_2}$

$$5 = \frac{(y_2 - y_1)^3}{4\ y_1\ y_2}$$

$$20\ y_1\ (11.53\ y_1) = (11.53\ y_1 - y_1)^3 \qquad \text{[From (1)]}$$

$$230.6\, y_1^2 = y_1^3\ (11.53 - 1)^3 = 1167.978\ y_1^3$$

$$\therefore \quad y_1 = \frac{230.6}{1167.978} = 0.1974 \text{ m} \qquad \text{Ans}$$

and $y_2 = 11.53 \times 0.1974 = 2.276$ m **Ans**

5.2 The flow of liquid in an open channel is 200 liters/s. An adjustable sluice gate spans the entire width the rectangular channel which is 1 m. Find the critical depth of flow corresponding to this discharge in the channel.

The opening of sluice is adjusted to cause a standing wave downstream such that the depth just before the wave is 100 mm. Determine the loss of energy due to water.

Solution

(a) Given Q = 200 liters/s, = 200×10^{-3} m³/s.

b = 1 m, $y_1 = 0.1$ m

$$\therefore \quad q = Q/b = \frac{0.2}{1} = 0.2 \text{ m}^3\text{/s /m.}$$

$v_1 = q/y_1 = 0.2/0.1 = 2$ m/s

$$y_c = \left(\frac{q^2}{g}\right)^{1/3} = \left[\frac{(0.2)^2}{9.81}\right]^{1/3} = 0.15975 \text{ m}$$

$$F_{r_1}^2 = \frac{v_1^2}{g\ y_1} = \frac{(2)^2}{9.81 \times 0.1} = 4.077$$

$$y_2 = \frac{y_1}{2}\left[\sqrt{1 + 8\ F_{r_1}^2} - 1\right] = \frac{0.1}{2}\left[\sqrt{1 + 8 \times 4.077} - 1\right]$$

$$= 0.05\ [4.7979] = 0.23989 \text{ m} \qquad \text{Ans}$$

(b) $\Delta E = \dfrac{(y_2 - y_1)^3}{4\, y_1\, y_2} = \dfrac{(0.23989 - 0.1)^3}{4 \times 0.1 \times 0.2398} = 0.285\ mN/N$ Ans

5.3 A hydraulic jump occurs in a 500 mm wide rectangular channel at the point where depth of water flow is 150 mm and the Froude number is 2.5. Make calculations for the specific energy, critical and sequent depths, loss of head and the energy dissipated.

Solution : Sequent depth is calculated from the formula

$$y_2 = \frac{y_1}{2}\left[\sqrt{1 + 8\, F_{r_1}^2} - 1\right]$$

$$= \frac{0.15}{2}\left[\sqrt{1 + 8\,(2.5)^2} - 1\right] = 0.4604\ m \quad \text{Ans}$$

$$\text{Loss of energy} = \Delta E = \frac{(y_2 - y_1)^3}{4\, y_1\, y_2} = \frac{(0.4606 - 0.15)^2}{4 \times 0.15 \times 0.4606}$$

$$= \frac{0.02996}{0.27636} = 0.10843\ \text{m} \quad \text{Ans}$$

We know that

$$F_{r_1} = \frac{v_1}{\sqrt{g\, y_1}}$$

$$2.5 = \frac{v_1}{\sqrt{9.81 \times 0.15}}$$

$\therefore$ $v_1 = 3.0326$ m/s

$\therefore$ $Q = b_1\, y_1\, v_1 = 0.5 \times 0.15 \times 3.0326 = 0.2274\ m^3/s$

$q = y_1\, v_1 = 0.45489\ m^3/s/m$

$$E_1 = y_1 + \frac{v_1^2}{2g} = 0.15 + \frac{(3.03)^2}{2 \times 9.81}$$

$$= 0.15 + 0.4679 = 0.6179 \quad \text{Ans}$$

$$y_c = \left(\frac{q^2}{g}\right)^{1/3} = \left[\frac{(0.45489)^2}{9.81}\right]^{1/3}$$

$= (0.021093)^{1/3} = 0.2763$ m Ans

$$P = \rho\, g\, Q\, \Delta E / 1000 = \frac{9810 \times 0.2274 \times 0.10843}{1000}$$

$= 0.2418$ kW Ans

5.4 A hydraulic jump occurs in a channel of rectangular cross section 6 m wide in which the rate of flow is 24 m³/s. If the depth of water before the jump is 475 mm, determine (a) the depth after the jump (b) Specific energy of water before and after the jump (c) loss in power due to the jump.

Solution

Given b = 6 m, Q = 24 m³/s, $y_1 = 0.475$ m

$$\therefore \quad q = Q/b = \frac{24}{6} = 4 \text{ m}^3/\text{s/m}$$

$q = v_1\ y_1$

$4 = y_1 . (0.475)$

$\therefore \quad v_1 = 8.421$ m/s

$$F_{r_1}^2 = \frac{v_1^2}{g\ y_1} = \frac{(8.421)^2}{9.81 \times 0.475} = 15.2184$$

(a) We know that

$$y_2 = \frac{y_1}{2}\left[\sqrt{1 + 8\,F_{r_1}^2} - 1\right] = \frac{0.475}{2}\left[\sqrt{1 + 8 \times 15.2184} - 1\right]$$

$= 0.2375\ [10.0791] = 2.39379$ m ≈ 2.4 m Ans

$$\text{(b)} \quad E_1 = y_1 + \frac{v_1^2}{2g} = 0.475 + \frac{(8.421)^2}{2 \times 9.81}$$

$= 0.475 + 3.614 = 4.089$ Nm/N Ans

$$v_2 = q / y_2 = \frac{4}{2.39} = 1.673$$

$$E_2 = y_2 + \frac{v_2^2}{2g} = 2.39 + \frac{(1.673)^2}{2 \times 9.81}$$

$= 2.39 + 0.142 = 2.532$ Nm/N Ans

(c) Power lost = $\rho g Q (E_1 - E_2)$ = 9810 × 24 (4.089 – 2.532)

= 366399.7 Watt ≈ 366.4 kW **Ans**

5.5 The normal depth of flow in a rectangular concrete channel, 6 m wide, is given by the formula $v = 2y^{2/3}$. A sluice discharge 11.3 m^3/s into the channel, at a velocity of 6 m/s. Calculate the height of the hydraulic jump.

Solution: Q is the discharge at vena contracta at the downstream side of sluice.

$$Q = byv$$

or, $11.3 = 6 \times y \times 6$

or, $y = 0.31$ m

On the downstream side of wave

$$v_2 = 2y_2^{2/3}$$

or, $\dfrac{q}{y_2} = 2 y_{2/3}$

or, $q = 2y_2^{5/3}$

$$\frac{11.3}{6} = 2y_2^{5/3}$$

$$y_2^{5/3} = (0.94166)$$

$$y_2 = (0.94166)^{3/5} = 0.9645 \text{ m}$$

$$y_c = \left(q^2/g\right)^{1/3} = \left[\frac{(11.3/6)^2}{9.81}\right]^{1/3} = (0.36156)^{1/3} = 0.712 \text{ m}$$

$$y_1 = -\frac{y_2}{2} + \sqrt{\frac{y_2^2}{4} + \frac{2q^2}{g\,y_2}} = \frac{0.96}{2} + \sqrt{\frac{(0.96)^2}{4} + \frac{2(1.88)^2}{9.81 \times 0.96}}$$

$$= -0.48 + \sqrt{0.98} = -0.48 + 0.99 = 0.5\,1 \text{ m}$$

Here $y_1 > y < y_c$, therefore wave will form.

Height of jump $y_J = y_2 - y_1 = 0.96 - 0.51 = 0.45$ m **Ans**

5.6 A 3 m wide rectangular channel conveys 7.5 m^3/s of water with a velocity of 5 m/s. Is there a condition for hydraulic jump to occur? If so, calculate the height, length and strength of the jump.

Solution

$$Q = b_1\ y_1\ v_1$$

$$7.5 = 3 \times 5 \times v_1$$

$$\therefore \quad v_1 = 0.5 \text{ m}$$

$$q = Q/b_1 = \frac{7.5}{3} = 2.5 \text{ m}^3\text{/s/m}$$

$$y_c = \left(q^2 / g\right)^{1/3} = \left[\frac{(2.5)^2}{9.81}\right]^{1/3} = 0.861 \text{ m}$$

Since $y_c > y_1 >$ a hydraulic jump would occur.

$$F_{r_1} = \frac{v_1}{\sqrt{g\ y_1}} = \frac{5}{\sqrt{9.81 \times 0.5}} = 2.26$$

$$y_2 = \frac{y_1}{2}\left[\sqrt{1 + 8\ F_{r_1}^2} - 1\right]$$

$$= \frac{0.5}{2}\left[\sqrt{1 + 8\,(2.26)^2} - 1\right] = 1.367 \text{ m}$$

Height of the jump = $y_2 - y_1 = 1.367 - 0.5 = 0.867$ m **Ans**

Length of the jump = $L = 6\,(y_2 - y_1) = 6 \times 0.867 = 5.2$ m **Ans**

Strength of the jump $= \dfrac{y_2}{y_1} = \dfrac{1.367}{0.5} = 2.734$ **Ans**

5.7 **For a hydraulic jump in a rectangular channel, show that the Froude number upstream F_{r_1} and the Froude number down stream F_{r_2} are related by the equation**

$$F_{r_2}^2 = \frac{8\ F_{r_1}^2}{\left(\sqrt{1 + 8\ F_{r_1}^2} - 1\right)^3}$$

Solution

$$F_{r_1} = \frac{v_1}{\sqrt{g\ y_1}} \ \textit{and}\ F_{r_2} = \frac{v_2}{\sqrt{g\ y_2}}$$

Therefore, $\dfrac{F_{r_2}}{F_{r_1}} = \dfrac{v_2}{\sqrt{g\ y_2}} \times \dfrac{\sqrt{g\ y_1}}{v_1} = \dfrac{v_2\sqrt{y_1}}{v_1\sqrt{y_2}}$

$$= \frac{q/y_2\sqrt{y_1}}{q/y_1\sqrt{y_2}} = \sqrt{\frac{y_1}{y_2}}\left(\frac{y_1}{y_2}\right)$$

$$\frac{F_{r_2}^2}{F_{r_1}^2} = \frac{y_1}{y_2}\left(\frac{y_1}{y_2}\right)^2 = \left(\frac{y_1}{y_2}\right)^3 = \left[\frac{1}{\dfrac{y_2}{y_1}}\right]^3$$

We know that

$$\frac{y_1}{y_2} = \frac{1}{2}\left[\sqrt{1+8\ F_{r_1}^2}\ -1\right]$$

$$\frac{F_{r_2}^2}{F_{r_1}^2} = \frac{8}{\left[\sqrt{1+8\ F_{r_1}^2}\ -1\right]^3}$$ Proved

5.8 A rectangular channel 1.8 m wide carries a discharge of 0.57 m³/s. The depth of flow on the upstream side of jump is 0.1 m. Find :

(a) the height of hydraulic jump;

(b) the length of hydraulic jump;

(c) the energy dissipated at the hydraulic jump

(d) value of Froude's number on the upstream side of jump;

(e) the ratio of Froude's number on the upstream and downstream sides of the jump and hence, show that

$$\frac{F_{r_1}}{F_{r_2}} = \left(\frac{y_2}{y_1}\right)^{3/2}$$

Solution

(a) $q = \dfrac{0.57}{1.8} = 0.3167\ \text{m}^3/\text{s/m}$

$y_1 = 0.1\ \text{m},\ q = v_1\ y_1$

$0.3167 = v_1 \times 0.1$

v₁ = 3.167 m/s.

$$F_{r_1} = \frac{v_1}{\sqrt{g\,y_1}} = \frac{3.167}{\sqrt{9.81 \times 0.1}} = 3.197 \qquad \text{Ans}$$

$$y_2 = \frac{y_1}{2}\left[\sqrt{1 + 8\,F_{r_1}^2} - 1\right] = \frac{0.1}{2}\left[\sqrt{1 + 8\,(3.197)^2} - 1\right]$$

$$= \frac{0.1 \times 8.099}{2} = 0.40495 \text{ m}$$

$$y_J = y_2 - y_1 = 0.40495 - 0.1 = 0.3049 \text{ m} \qquad \text{Ans}$$

(b) $L_J \approx 6\,y_2 \approx 6 \times 0.40495 \approx 2.4297\,m$ Ans

(c) $$E_L = E_1 - E_2 = \frac{(y_2 - y_1)^2}{4\,y_1\,y_2} = \frac{(0.3049)^3}{4 \times 0.1 \times 0.40495}$$

$$= \frac{0.02834}{0.16198} = 0.174989\,m \approx 0.175\,m \qquad \text{Ans}$$

(e) $v_1\ y_1 = v_2\ y_2$

$3.167 \times 0.1 = v_2 \times 0.40495$

$v_2 = 0.782$ m/s.

$$F_{r_2} = \frac{v_2}{\sqrt{g\ y_2}} = \frac{0.782}{\sqrt{9.81 \times 0.40495}} = 0.392$$

$$\frac{F_{r_1}}{F_{r_2}} = \frac{3.197}{0.392} = 8.1556 \qquad \text{Ans}$$

$$\frac{F_{r_1}}{F_{r_2}} = \frac{v_1}{\sqrt{g\ y_1}} \times \frac{\sqrt{g\,y_2}}{v_2} = \frac{v_1}{v_2}\left(\frac{y_2}{y_1}\right)^{1/2}$$

$$= \frac{q/y_1}{q/y_2}\left(\frac{y_2}{y_1}\right)^{1/2} = \left(\frac{y_2}{y_1}\right)^{3/2} \qquad \text{Proved}$$

5.9 A 6.1 m wide horizontal stream flows over a short stretch of stones. The flow upstream of the rough stretch is 0.91 m deep and has a speed of 1.52 m/s. If the stones are estimated to experiecne a drag of 4448 N; determine the depth and speed of the stream downstream of the rough stretch. Sketch the water surface profile and indicate the critical depth line. Is the flow super critical or subcritical?

Solution

Discharge per unit width

$$q = 1.52 \times 0.91 = 1.3832 \text{ m}^2/\text{s}$$

and $$y_c = \left(\frac{q^2}{g}\right)^{1/3} = \left[\frac{(1.3832)^2}{9.81}\right]^{1/3} = (0.195)^{1/3} = 0.5799 \text{ m}$$

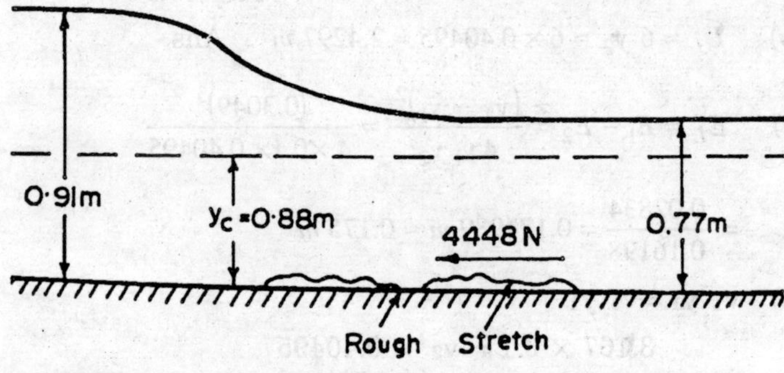

Fig. 5.7. Wide horizontal stream flow over a short stretch of stones. P. 5.9.

Applying momentum equation

$$6.1 \times 9.81 \times 10^3 \left(\frac{y_1^2}{2} - \frac{y_2^2}{2}\right) - 4.448 \times 10^3$$

$$= 10^3 \times 6.1 \times 1.38\,(v_2 - v_1)$$

From the continuity equation

$$v_2 = v_1 \left(\frac{y_1}{y_2}\right)$$

Therefore,

$$3.05 \times 10^3 \times 9.81\left(0.83 - y_2^2\right) - 4.448 \times 10^3$$

$$= 10^3 \times 8.42\, v_1 \left(\frac{0.91}{y_2} - 1\right)$$

Simplifying the above equation

$y_2^3 - 1.1\, y_2 + 0.39 = 0$

Solving by trial and error method

$y_2 = 0.77$ m

Therefore, $v_2 = \dfrac{q}{y_2} = \dfrac{1.38}{0.77} = 1.796$ m/s

$$F_{r_2} = \frac{v_2}{\sqrt{g\, y_2}} = \frac{1.796}{\sqrt{9.81 \times 0.77}} = 0.6536$$

The Froude number is less than 1 therefore the flow is subcritical and in other words the depth y_2 is greater than y_c therefore, the flow is subcritical.

5.10 A longitudinal section of a wide river shows a rough length followed by a long relatively smooth reach of the same slope S_o. The ratio of the corresponding Chezy coefficients is 1 : 2 $\sqrt{2}$. The slope joints a wide horizontal apron. The normal depth of flow on the rough section is 2.43 m for discharge per unit width is 7.7 m²/s. If the normal depth in the horizontal section is 3.05 m and the Froude number $\left(v^2 / g\ y_n\right)$ is 3/8, locate the toe of the hydraulic apron. Sketch the water surface profile showing the three sections of the river. Neglect side effects and take the Chezy coefficient for the horizontal apron as 45.7 m$^{1/2}$/s.

Solution: The critical depth

$$y_c = \left(\frac{q^2}{g}\right)^{1/3} = \left[\frac{(7.7)^2}{9.81}\right]^{1/3}$$

$$= \left(\frac{59.29}{9.81}\right)^{1/3} = (6.0438)^{1/3} = 1.8215 \text{ m}$$

The normal depth is given by Chezy's formula

$$q = y_n\, C \sqrt{y_n\, S_o}$$

Using the subscript r for the rough section and S for the smooth section

$$\frac{q_r}{q_s} = 1 = \left(\frac{y_{nr}}{y_{ns}}\right)^{3/2} \frac{C_r}{C_s}$$

or, $$1 = \left(\frac{y_{nr}}{y_{ns}}\right)^{3/2} \frac{1}{2\sqrt{2}}$$

or, $$\frac{y_{nr}}{y_{ns}} = \left(2\sqrt{2}\right)^{3/2} = 2$$

or, $$y_{ns} = \frac{y_{nr}}{2} = \frac{2.43}{2} = 1.215 \text{ m}$$

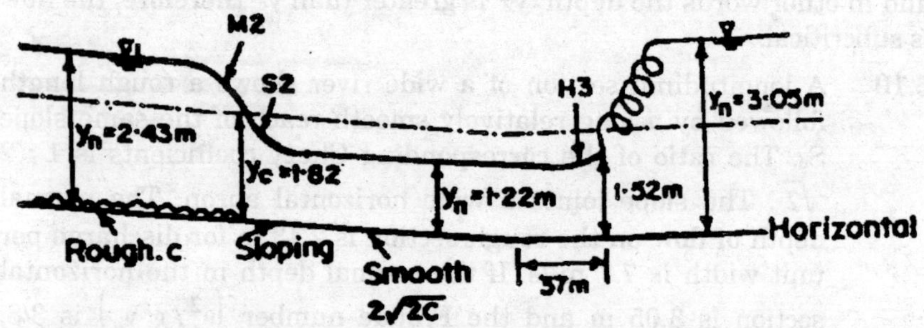

Fig. 5.8. Water surface showing three sections of river P. 5.10.

Flow in the rough section is subcritical but on the smooth section is supercritical. A jump forms on the horizontal apron. The conjugate depths are given by

$$\frac{y_1}{y_2} = \frac{1}{2}\left[\sqrt{8\ F_{r_1}^2 + 1} - 1\right] = \frac{1}{2}\left[\sqrt{8 \times \frac{3}{8} + 1} - 1\right]$$

$$= \frac{1}{2}[2 - 1] = \frac{1}{2}$$

$$y_1 = \frac{y_2}{2} = \frac{3.05}{2} = 1.525 \text{ m}$$

The depth on the horizontal apron will rise from 1.215 m to 1.525 m and a jump will form (Fig. 5.8).

$$\frac{dy}{dx} = \frac{S_o - S_f}{1 - \left(\frac{y_c}{y}\right)^3}$$

For Chezy's formula (wide rectangular channel)

$$S_f = \frac{q^2}{C^2 \; y^3}$$

In this problem $S_o = 0$

$$\text{Therefore,} \quad \frac{dy}{dx} = \frac{0 - \dfrac{q^2}{C^2 \; y^3}}{1 - \left(\dfrac{y_c}{y_o}\right)^3}$$

$$\text{or,} \quad -\frac{q^2}{C^2 \; y^3} \, dx = \left[1 - \left(\frac{y_c}{y_o}\right)^3\right] dy$$

Integrating

$$x_2 - x_1 = \frac{C^2}{q^2}\left[y_c^3 \; y - \frac{y^4}{y}\right]_{y=1.215}^{y=1.525}$$

$$= \frac{(45.7)^2}{(7.7)^2}\left[(1.82)^3 \; y - \frac{y^4}{y}\right]_{y=1.215}^{y=1.525}$$

$$= 35.225 \left[6.028 \; y - 0.25 \; y^4\right]_{y=1.215}^{y=1.525}$$

$$= 35.225\,[6.028\,(1.525 - 1.215)]$$

$$- 35.225 \;\; [0.25\,(1.525)^4 - (1.215)^4]$$

$$= 65.83 - 28.4379 = 37.392 \text{ m}$$

The hydraulic jump will form at 37.392 m from the end of the slope.

5.11 The spillway of a dam discharges 27.8 m²/s onto a horizontal stiling apron. The depth of high velocity flow at the entry point to the apron is 0.61 m. Both the apron and sill are sufficiently long to permit establishment of normal conditions. The sill is followed by a 1 to 10 sloping wide channel. (a) Calculate the height of the hydraulic jump and the height of sill necessary for its stabilization at the location shown. Neglect energy dissipation in flow over the sill. (b) Calculate the normal depth of flow in the downstream channel. (Manning's n = 0.015). (c) Sketch and label the surface profile at the various sections. (d) What practical difficulities would you expect in the operation of the structure?

Solution

(a) Upstream Froude number

$$F_{r_1}^2 = \frac{q^2}{g\ y_1^2} = \frac{(27.8)^2}{9.81 \times (0.61)^3} = 347$$

The conjugate depth ratio

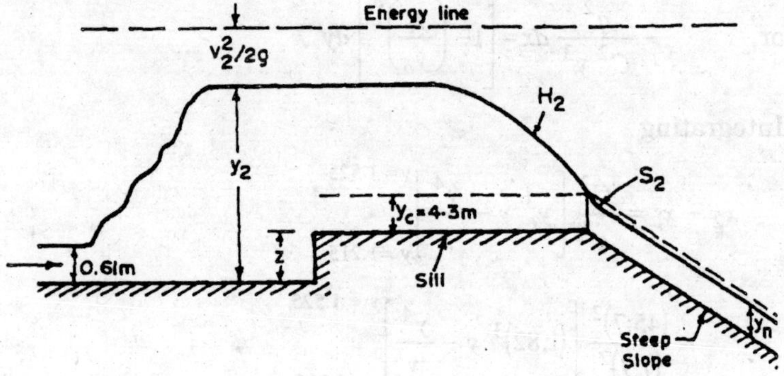

Fig. 5.9. Spillway discharge onto a horizontal apron to sloping channel P. 5.11.

$$\frac{y_2}{y_1} = \frac{1}{2}\left[\sqrt{1+8\ F_{r_1}^2} - 1\right] = \frac{1}{2}\left[\sqrt{1+8\ (347)} - 1\right]$$

$$= \frac{1}{2}[52.697 - 1] = \frac{1}{2}[51.697] = 25.8486$$

Therefore, $y_2 = 25.8486\ y_1 = 25.8486 \times 0.61 = 15.7676$ m

The specific energy after the jump

$$= \frac{1}{2}\left(\frac{q^2}{q\ y_2^2}\right) + y_2$$

$$= \frac{1}{2}\left[\frac{(27.8)^2}{9.81 \times (15.7676)^2}\right] + 15.7676$$

$$= \frac{1}{2} \times (0.3168) + 15.7676$$

$$= 0.1584 + 15.7676 = 15.926 \text{ m}$$

Since the end slope is steep, the downstream end of the sill will act as a control section. Flow condition there will be critical.

$$y_c = \left(\frac{q^2}{g}\right)^{1/3} = \left[\frac{(27.8)}{9.81}\right]^{1/3} = (78.78)^{1/3} = 4.2868 \text{ m}$$

Neglecting energy lost in flowing over the sill, the specific energy before the sill is equal to the specific energy on the sill plus z.

i.e. $\quad E = \frac{3}{2} y_c + z = \frac{3}{2} \times 4.2868 + z$

$$15.7676 = 6.43 + z$$

$$z = 9.3372 \text{ m}$$ **Ans**

(b) From Manning's formula

$$v = q / y_n = \frac{1}{n} y_n^{2/3} S_o^{1/2}$$

$$\frac{27.8}{y_n} = \frac{1}{0.015} (y_n)^{2/3} \left(\frac{1}{10}\right)^{1/2}$$

or, $\quad y_n^{5/3} = \frac{27.8 \times 0.015}{0.31622} = 1.3186$

$$y_n = (1.3186)^{3/5} = 1.18 \text{ m}$$ **Ans**

(c) As per Fig. 5.54

(d) Possible Practical difficulties

(i) The hydraulic jump may be submerged especially for relatively low flows thus reducing its energy dissipation. In such a case the fast incoming flow shoots under water in the stiling basin before dispersing (Jet dispersion).

(ii) Because of high velocities, cavitation problems may rise.

(iii) The hydraulic jump for high discharges may be pushed downstream and may in fact jump the sill unless the stiling basin is sufficiently long.

5.12 A rectangular channel is 1.5 m wide and has a discharge of 1700 liters/s. It has bed slope of 0.0006 and Manning's friction coefficient n = 0.012. At a certain section, the depth of flow is 900 mm, find the distance downstream to a section where the depth is 750 mm. Classify the surface profile.

Solution

At section 1 – 1

$y_1 = 0.9$ m

$A_1 = 1.5 \times 0.9 = 1.35$ m²

$P_1 = b_1 + 2\ y_1 = 1.5 + 2 \times 0.9 = 3.3$ m

$$R_1 = \frac{A_1}{P_1} = \frac{1.35}{3.3} = 0.409$$

$$v_1 = Q/A_1 = \frac{1700/1000}{1.35} = 1.259 \text{ m/s}$$

$$E_1 = y_1 + v_1^2 / 2g \quad = 0.9 + \frac{(1.259)^2}{2 \times 9.81} = 0.98$$

At section 2 – 2

$y_2 = 0.75$

$A_2 = 1.5 \times 0.75 = 1.125$ m²

$P_2 = b_2 + 2\ y_2 = 1.5 + 2 \times 0.75 = 3$ m

$$R_2 = \frac{A_2}{P_2} = \frac{1.125}{3} = 0.375$$

$$v_2 = Q / A_2 = \frac{1.7}{1.125} = 1.51 \text{ m/s}$$

$$E_2 = y_2 + v_2^2 / 2g = 0.75 + (1.51)^2 / 2 \times 9.81 = 0.866$$

For average section midway between section 1 – 1 and 2 – 2, we have,

$$v_{av} = \frac{v_1 + v_2}{2} = \frac{1.259 + 1.51}{2} = \frac{2.77}{2} = 1.385 \text{ m/s}$$

$$R_{av} = \frac{R_1 + R_2}{2} = \frac{0.409 + 0.375}{2} = \frac{0.784}{2} = 0.392$$

For average section

$$v_{av} = \frac{1}{n} R_{av}^{2/3} S_f^{1/2}$$

Where S_f is energy slope at mid section and is considered uniform over the length between section 1 – 1 and 2 – 2

$$\therefore\ 1.385 = \frac{1}{0.012}(0.392)^{2/3}\ S_f^{1/2}$$

$$S_f^{1/2} = \frac{1.385}{44.635} = 0.031$$

$S_f = 9.628 \times 10^{-4}$

We know that

$$L = \frac{E_2 - E_1}{S_o - S_f} = \frac{0.866 - 0.98}{0.0006 - 9.628 \times 10^{-4}}$$

$$= \frac{0.11378}{3.628 \times 10^{-4}} = 313.61\ \text{m}$$ Ans

$$q = Q/b = \frac{1.7}{1.5} = 1.133$$

$$y_c = \left(\frac{q^2}{g}\right)^{1/3} = 0.5076\ \text{m}$$

Also $q = y_c\ v_c$

$1.133 = 0.5076 \times v_c \quad \therefore\ v_c = 2.23$ m/s

$$R_c = \frac{A_c}{P_c} = \frac{1.5 \times 0.5076}{1.5 + 2(0.5076)} = \frac{0.7614}{1.5 + 1.0152}$$

$$= \frac{0.7614}{2.5152} = 0.3027$$

Therefore, $v_c = \frac{1}{n} R_c^{2/3}\ S_c^{1/2}$

$$2.23 = \frac{1}{0.012}(0.3027)^{2/3}\ S_c^{1/2}$$

$$2.23 = 37.57\ \ S_c^{1/2}$$

$$S_c^{1/2} = 0.0593$$

$$\therefore \quad S_c = 3.523 \times 10^{-3}$$

Since $S_c > S_o$ or $S_o < S_c$ the slope is mild.

Also, $Q = A_n\ v_n = 1.5\ y_n\ v_n$

$$1.7 = 1.5\ y_n \left(\frac{1}{n} R_n^{2/3}\ S_o^{1/2}\right)$$

where $R_n = \dfrac{A_n}{P_n} = \dfrac{1.5\ y_n}{1.5 + 2\ y_n}$

Therefore, $1.7 = 1.5\ y_n \left(\dfrac{1}{0.012} \dfrac{1.5\ y_n}{1.5 + 2\ y_n} . (0.0006)^{1/2}\right)$

By trial and error

$$y_n = 0.95$$

Now $y_1 > y_c < y_n$

Also $y_2 > y_c < y_n$

$\therefore$ the curve is M_2 curve.

5.13 In a rectangular channel, the discharge is 28 m³/s, width 15 m, depth 3 m, bed slope 1 in 6700 and Chezy's C = 61. Determine the actual slopes of the water surface in millimeters per kilometer.

Solution

$$A = by = 15 \times 3 = 45\ m^2$$

$$P = b + 2y = 15 + 2 \times 3 = 21\ m$$

$$R = \frac{A}{P} = \frac{45}{21} = 2.142\ m$$

$$v = Q/\ by = 28/\ 15 \times 3 = 0.623\ m/s.$$

$$v = c\ \sqrt{R\ S_f}$$

$$0.623 = 61\ \sqrt{2.142\ S_f}$$

$$S_f = 0.487 \times 10^{-4}$$

Bed slope $S_0 = 1/6700 = 1.4925 \times 10$ sup^{-4}

Therefore, $\dfrac{dy}{dl} = \dfrac{S_o - S_f}{1 - v^2/gy}$

$$= \frac{(1.4925 - 0.487) \times 10^{-4}}{1 - (0.623)^2/9.81 \times 3} = \frac{1 \times 10^{-4}}{1 - 0.013188}$$

$$= \frac{1 \times 10^{-4}}{0.9868} = 1.019 \times 10^{-4}\ m/m = 101.9 \text{ mm/ km Ans}$$

5.14 A rectangular channel 6.1 m wide has a bed slope of 1 in 1000 and it carries an intensity of discharge of 3.72 m³/s/m. The bed slope is suddenly made steeper downstream from a certain point in the bed. Find the depth of flow at this point where the two bed slopes meet and find the length of the slopes meet and find the length of the drop down curve on the upstream side of this point. Take Manning's n = 0.015.

Solution

At a point after which the slope becomes steeper, there will be critical depth of flow y_c. We know that

$$y_c = \left(\frac{q^2}{g}\right)^{1/3} = \left[\frac{(3.72)}{9.81}\right]^{1/3} = 1.1215 \text{ m}$$

$q = v_c\ y_c$

$3.72 = v_c \times 1.1215$

$\therefore \quad v_c = 3.316$ m/s

For the normal or uniform flow the depth of flow is y. therefore,

$$v = q/y = \frac{3.72}{y}$$

Using Manning's equation

$$v = \frac{1}{n} R^{2/3} S_o^{1/2}$$

$$\frac{3.72}{h} = \frac{1}{0.015}\left[\frac{6.1 \times y}{6.1 + 2y}\right]^{2/3}\left[\frac{1}{1000}\right]^{1/2}$$

$$\frac{1.7645}{y} = \left(\frac{6.1\,y}{6.1 + 2y}\right)^{2/3}$$

or, $$\frac{176.45}{(6.1)^{2/3}} = \frac{y\,(y)^{2/3}}{(6.1+2y)^{2/3}} = \frac{y^{5/3}}{(6.1+2y)^{2/3}}$$

or, $$\frac{1.7645}{3.3385} = \frac{y^{5/3}}{(6.1+2y)^{2/3}}$$

or, $$0.52854 = \frac{y^{5/3}}{(6.1+2y)^{2/3}}$$

Solving the above equation by trial and error method, the value of y comes out to be 1.64 m.

$$v_1 = \frac{3.72}{1.64} = 2.2683$$

$$E_1 = y_1 + \frac{v_1^2}{2g} = 1.64 + \frac{(2.2682)^2}{2 \times 9.81} = 1.64 + 0.262 = 1.9$$

$$E_2 = E_c = y_c + \frac{v_c^2}{2g} = \frac{3}{2}y_c = \frac{3}{2} \times 1.1215 = 1.68225$$

Here, $$v_{av} = \frac{2.2683 + 3.316}{2} = 2.79\ m/s$$

$$y_{av} = \frac{y_1 + y_2}{2} = \frac{1.1215 + 1.64}{2} = 1.38$$

$$R_{av} = \frac{6.1 \times 1.38}{6.1 + 2 \times 1.38} = 0.95\text{ m}$$

Using Manning's equation

$$v_{av} = \frac{1}{n} R_{av}^{2/3}\ S_f^{1/2}$$

$$2.79 = \frac{1}{0.015}(0.95)^{2/3}\ S_f^{1/2}$$

$$S_f^{1/2} = 0.0433$$

$$S_f = 1.875 \times 10^{-3} = \frac{1}{533.185}$$

Now, $$L = \frac{E_2 - E_1}{S_o - S_f}$$

$$= \frac{1.68225 - 1.9}{0.001 - 1.875 \times 10^{-3}} = \frac{0.21775}{8.75 \times 10^{-4}} = 248.857 \text{ m} \qquad \text{Ans}$$

5.15 **A long straight channel of rectangular section 3 m wide has a slope of 1 in 2000 and the Chezy's C is 77. The depth of water for normal flow is 1.8 m and the partial lowering of a sluice gate (the bottom of which is horizontal) raises the level upstream of the sluice by 0.6 m.**

Assuming that the flow remains unchanged, find how far back from the sluice the depth will be 2.3 m. Also find the maximum height in meters of the opening beneath the sluice if a standing wave is to be formed downstream of the sluice.

Solution

$$y_n = 1.8 \text{ m}, \; y_1 = 2.3 \text{ m}$$

$$y_2 = 1.8 + 0.6 = 2.4 \text{ m}$$

$$Q = A v = (b \; y_n) \; C \sqrt{R S_o}$$

$$= (3.1.8) \times 77 \sqrt{\frac{3 \times 1.8}{6.6} \times \frac{1}{2000}} = 8.41 \text{ m}^3/\text{s}$$

For non uniform flow

$$y_{av} = \frac{y_1 + y_2}{2} = \frac{2.3 + 2.4}{2} = 2.35 \text{ m}$$

$$A_{av} = b \; y_{av} = 3 \times 2.35 = 7.05 \text{ m}^2$$

$$Q = A_{av} \; v_{av}$$

$$8.41 = 7.05 \times v_{av}$$

$$\therefore \quad v_{av} = 1.192 \text{ m/s}$$

$$R_{av} = \frac{7.05}{7.7} = 0.9\,2 \text{ m}$$

Also, $$v_{av} = C \sqrt{R_{av} \, S_f}$$

$$1.1929 = 77 \sqrt{0.92 \, S_f}$$

or, $0.92\ S_f = 2.4 \times 10^{-4}$

or, $S_f = 2.6088 \times 10^{-4}$

$$\therefore \quad S_f = \frac{1}{3833.147}$$

$$\left(\frac{dy}{dl}\right)_{av} = \frac{S_o - S_f}{1 - \dfrac{v_{av}^2}{g\ y_{av}}} = \frac{0.0005 - 0.00026088}{1 - \dfrac{(1.1929)^2}{9.81 \times 2.35}}$$

$$= \frac{2.39 \times 10^{-4}}{1 - 0.06172} = \frac{2.39 \times 10^{-4}}{0.9382} = 2.548 \times 10^{-4}$$

$$L = \frac{y_2 - y_1}{\left(\dfrac{dh}{dl}\right)_{av}} = \frac{0.1}{2.548 \times 10^{-4}} = 392.389 \text{ m}$$

For standing wave on downstream side, $y_2 = 1.8$ m

Using hydraulic jump equation

$$y_1\ y_2\ (y_1 + y_2) = \frac{2q^2}{g}$$

$$1.8\ y_1\ (y_1 + 1.8) = \frac{2 \times (8.41/3)^2}{9.81}$$

$$y_1^2 + 1.8\ y_1 = 0.89$$

$$y_1^2 + 1.8\ y_1 - 0.89 = 0$$

$$y_1 = \frac{-1.8 \pm \sqrt{(1.8)^2 - 4 \times 1 \times (-0.89)}}{2}$$

$$= \frac{-1.8 \pm \sqrt{6.8}}{2} = 0.4038 \text{ m}$$

In order that the standing wave may form the height of opening should be less than 0.4038 m i.e the maximum value of the height of opening should be 0.4038 m.

5.16 **A channel of rectangular section 8 m wide is laid to a bed slope of 1 in 64 and carries a discharge of 40 m³/s. A barrier across the channel raises the water surface to 3 m just upstream of the barrier. Find the length of the water surface profile up to the hydraulic jump upstream. Taking Manning's**

n = 0.025. For calculation of the hydraulic radius the channel may be treated as wide channel. Use single step method.

Solution

$$q = 40/8 = 5\ \text{m}^3/\text{s/m}$$

Afflux = 3 m

The normal depth y_n is calculated as

$$Q = A\,v = (b\ y_n)\frac{1}{n}R^{2/3}\ S_o^{1/2}$$

$$40 = (8\ y_n)\frac{1}{0.025}\left(\frac{8\ y_n}{8+2\ y_n}\right)^{2/3}\left(\frac{1}{64}\right)^{1/2}$$

or, $y_n^5 = 1 + 0.0625\ y_n^2 + 5\,y_n$

Solving by trial and error method $y_n \approx 1.1$ m

$$\text{Critical depth } y_c = \left(q^2/g\right)^{1/3} = \left(\frac{(5)^2}{9.81}\right)^{1/3} = 1.3659$$

Since $y_c > y_n$; the given slope is "steep".

Depth after the jump $= y_2 = y_1 + Afflux = 1.1 + 3 = 4.1\ m$

$$R_1 = \frac{8\ y_1}{8+2\,y_1} = \frac{8\times 1.1}{8+2\times 1.1} = 0.8627$$

$$v_1 = q/y_1 = 5/1.1 = 4.545\ \text{m/s}$$

$$E_1 = y_1 + v_1^2/2g = 1.1 + \frac{(4.545)^2}{2\times 9.81} = 2.152$$

$$R_2 = \frac{8y_2}{8+2y_2} = \frac{8\times 4.1}{8+2\times 4.1} = 2.025$$

$$v_2 = q/y_2 = 5/4.1 = 1.2195$$

$$E_2 = y_2 + \frac{v_2^2}{2g} = 4.1 + \frac{(1.2195)^2}{2\times 9.81} = 4.175$$

$$R_{av} = \frac{R_1 + R_2}{2} = \frac{0.8627 + 2.025}{2} = 1.4435$$

$$v_{av} = \frac{v_1 + v_2}{2} = \frac{4.545 + 1.2195}{2} = 2.882$$

$$v_{av}^2 = \frac{1}{n} R_{av}^{2/3} \ S_f^{1/2}$$

$$2.882 = \frac{1}{0.025} (1.4435)^{2/3} \ S_f^{1/2}$$

$$S_f = 3.18 \times 10^{-3}$$

$$\therefore \ L = \frac{E_2 - E_1}{S_o - S_f} = \frac{4.175 - 2.152}{1/64 - 3.18 \times 10^{-3}} = \frac{2.023}{0.01244}$$

$$= 162.5825 \text{ m}$$ Ans

5.17 A weir is installed in a 30.5 m wide rectangular channel delivering 71 m³/s. The bed slopes at 1 : 1000 & the depth of water just upstream of the weir is 2 m. If Chezy C = 55 may be assumed constant, estimate the length of the back water curve.

Solution

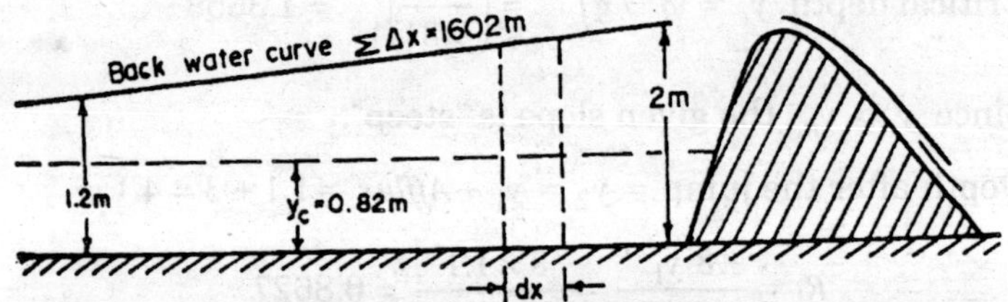

Fig. 5.10. Back water curve P. 5.17.

$$\text{Critical depth } y_c = \left(\frac{q^2}{g}\right)^{1/3} = \left[\frac{(71/30.5)^2}{9.81}\right]^{1/3} = \left[\frac{(2.3278)^2}{9.81}\right]^{1/3}$$

$$= (0.55239)^{1/3} = 0.82 \text{ m}$$

Normal depth

$$y_n = \left(\frac{q^2}{C^2 S_o}\right)^{1/3} = \left[\frac{(2.3278)^2}{(55)^2 \times \frac{1}{1000}}\right]^{1/3} = (1.79129)^{1/3} = 1.2144 \ m \approx 1.2 \ m$$

The length of the back water curve is the channel length over which the depth changes from 1.2 to 2 m.

Divide the change of depth into four increments of 0.2 m.

The finite difference form equation can be written as

$$\frac{dy}{dx} = \frac{S_o\left[1-\left(\frac{y_n}{y}\right)^3\right]}{1-(y_c/y)^3}$$

$$= \frac{0.001\left[1-\left(\frac{1.2}{y}\right)^3\right]}{1-\left(\frac{0.82}{y}\right)^3} = \frac{0.001\left[1-\frac{1.728}{y^3}\right]}{1-\frac{0.55}{y^3}}$$

$$\therefore dx = \left(\frac{1-0.55/y^3}{0.001-\frac{0.001728}{y^3}}\right)dy$$

Here y = average depth over the segement length dx

= depth at beginning of segment $y_1 + \frac{dy}{2}$

$y_2 = y_1 + dy_1$ = depth at the end of segment

The table below shows the details of the working procedure.

y_1 (m)	dy (m)	y (m)	y^3 (m^3)	$1-0.55/y^3$	$0.001-\frac{0.001728}{y^3}$	dx (m)	y_2 (m)
1.2	0.2	1.3	2.2	0.786	0.00021	744	1.4
1.4	0.2	1.5	3.37	0.832	0.00049	340	1.6
1.6	0.2	1.7	4.93	0.889	0.00065	273	1.8
1.8	0.2	1.9	6.85	0.92	0.00075	245	2.0
					Σ dx	1602	

The back water curve is about 1602 m long **Ans**

EXERCISES

A. THEORY

5.1 What is a hydraulic jump and under what conditions may it occur?

5.2 In a hydraulic jump show that

$$\frac{y_2}{y_1} = \frac{1}{2}\left[\sqrt{1 + 8\,F_{r_1}^2} - 1\right]$$

5.3 What is a hydraulic jump? Show that the relative change in depth through hydraulic jump depends upon the Froude's number of the approaching flow.

5.4 Define hydraulic jump and give three examples of its occurrence. Also derive the relation between the conjugate depths for a rectangular channel with a horizontal bed in the form.

$$\frac{2q^2}{g} = y_1\ y_2\ (y_1 + y_2)$$

5.5 Under what circumstances, hydraulic jump will be formed? Derive an expression for the loss of energy due to hydraulic jump in terms of depths before and after the jump.

5.6 If y_1 and y_2 are alternate depths and y_c = critical depth in a rectangular channel, show that

$$y_c^3 = \frac{2\ y_1^2\ y_2^2}{(y_1 + y_2)}$$

and hence specific energy

$$E = \frac{y_1^2 + y_1 y_2 + y_2^2}{(y_1 + y_2)}$$

5.7 Derive the expression for energy loss in hydraulic jump in terms of sequent depths.

5.8 Show that for hydraulic jump in a rectangular channel.

$$\Delta E = \frac{(y_2 - y_1)^3}{4\ y_1\ y_2}$$

5.9 For a hydraulic jump prove that formula

$$2\, y_c^3 = y_1\, y_2\, (y_1 + y_2)$$

5.10 For a hydraulic jump show that the ratio of Froude's number on the upstream and downstream sides of the jump and hence, show that

$$\frac{F_{r_1}}{F_{r_2}} = \left(\frac{y_2}{y_2}\right)^{3/2}$$

5.11 For a hydraulic jump show that the inlet Froude's number is given by

$$F_{r_1} = \sqrt{\frac{y_2}{2\, y_1}\left(\frac{y_2}{y_1} + 1\right)}$$

5.12 What do you understand by the terms (a) Venturi flume (b) Standing wave flume? Why is a hump sometimes used on the throat of a venturi flume.

5.13 Obtain the differential equation for water surface slope in the case of steady gradually varied flow in a prismatic channel. List all the assumptions made

(a) $$\frac{dy}{dL} = \frac{S_o - S_f}{1 - F_r^2} = \frac{S_o - S_f}{1 - \dfrac{Q^2 T}{gA^3}}$$

(b) If Manning's equation is used

$$\frac{dy}{dL} = \frac{1 - (y_o / y)^{10/3}}{1 - (y_c / y)^3}$$

(c) If Chezy's equation is used

$$\frac{dy}{dL} = \frac{1 - (y_o / y)^3}{1 - (y_c / y)^3}$$

5.14 Prove that for hydraulic jump, the nondimensional form of the energy.

$$\frac{\Delta E}{E_1} = \frac{\left[\dfrac{y_2}{y} - 1\right]}{2\, \dfrac{y_2}{y_1}\left[F_{r_1}^2 + 2\right]}$$

5.15 Prove that for hydraulic jump

$$\frac{\Delta E}{y_1} = \frac{\left[\frac{y_2}{y} - 1\right]}{4\left[\frac{y_2}{y_1}\right]}$$

5.16 Prove that for hydraulic jump the non dimensional ratio $\frac{\Delta E}{E_1}$ can be written as a function of upstream Froude number

$$\frac{\Delta E}{E_1} = \frac{\left[\sqrt{1 + 8 F_{r_1}^2} - 3\right]^3}{8\left[\sqrt{1 + 8 F_{r_1}^2} - 1\right]\left[F_{r_1} + 2\right]}$$

5.17 Draw graphs (a) Depth ratio v/s F_{r_1} (b) Dimensionless energy loss v/s F_{r_1} and explain.

5.18 Write down the rules for determing the location of the hydraulic jump.

5.19 Discuss three cases for location of hydraulic jump (a) $y_n < y_2$ (b) $y_n = y_2$ (c) $y_n > y_2$.

5.20 Name the various flow measure meet devices for open channel flow and explain the working broad crested weir.

5.21 Draw a neat sketch of venturi flume and explain its working.

5.22 Draw a neat sketch of standing wave flume and explain its working.

5.23 Draw a neat sketch of Parshall flume and explain its working.

5.24 Name the three methods for determination of average velocity in flow through channels and explain any one.

5.25 Draw a neat sketch of current meter and explain its working.

5.26 For gradually varied flow define surface profile and significant points.

5.27 Classify the surface profiles, discuss.

(a) Mild slope curves - M curves

(b) Steep slope curves - S curves

(c) Critical slope curves - C curves

(d) Horizontal slope curves - H curves

(e) Adverse slope curves - A curves

5.28 Name the various methods of solution of gradually varied flow equation and explain any one.

5.29 Derive the differential equation for steady gradually varied flow in open channels

$$\frac{dy}{dx} = \frac{S_o - S_f}{1 - F_{r_2}}$$

and list all the assumptions made in the derivation.

5.30 Prove that the length L, of the backwater curve for gradually varied flow in open channel is given by

$$L = \frac{E_2 - E_1}{S_o - S_f}$$

5.31 Derive the basic differential equation for the gradually varied flow.

5.32 What is backwater curve? Explain with the help of equation, gradually varied flow in an open channel.

B. UNSOLVED PROBLEMS

5.1 A horizontal bed channel 20 m wide has a hydraulic jump on it which has resulted a change of the depth of flow from 1 m to 3 m. Find the discharge (m³/s) flowing in the channel. [Q = 153.4 m³/s]

5.2 A broad channel carries a flow of 2.76 m³/s per metre width, the depth of flow being 1.524 m. Calculate the rise in floor level required to produce critical flow conditions. What is the corresponding fall in the surface level? [$y_1 - y_c$ = 1.524 – 0.919 = 0.605 m. $y_1 - y_2$ = 1.524 – 0.502 = 1.022 m]

5.3 At the foot of a 30 m wide spillway from a dam where the discharge velocity is 28.2 m/s and the depth is 0.96 m, a hydraulic jump is formed on a horizontal apron. Calculate the height of the jump and the total power dissipated in it. [y_2 = 12 m, $y_2 - y_1$ = 12 – 0.96 = 11.04 m]

5.4 A rectangular channel, 6.1 m wide carries 11.32 m³/sec and discharge onto a 6.1 m wide apron with no slope with a mean velocity of 6.1 m/s. What is the height of the hydraulic jump?

What energy is absorbed (lost) in the jump? [H = $y_2 - y_1$ = 1.065 m, 118.78 kW check 82.17 kW].

5.5 A rectnagular channel 4.88 m wide, carries a flow of 5.43 m³/s. The depth of water on the downstream side of the hydraulic jump is 1.28 m (a) What is the depth upstream? (b) What is the loss of head? [y_1 = 0.1387, E_1 = 3.417 E_2 = 1.319 m, $E_1 - E_2$ = 2.098 m]

5.6 The normal depth of flow in a rectangular concrete channel, 6 m wide, is given by the formula v = 2 $D^{2/3}$. A sluice discharges 11.3 m³/s into the channel, at a velocity of 6 m/s. Sketch the profile of the dowmstream of the sluice; calculate the height of the hydraulic jump and show how the position of the jump may be estimated. [y_2 = 0.96, y_1 = 0.51 m, y_c = 0.71 m]

5.7 If the velocity, when the water enters the channels, is 3 m/s and the depth of the flow is 0.15 m, obtain (a) the depth of flow immediately after the jump (b) the loss of specific energy due to the formation of the jump. [y_2 = 0.455 m, (b) E_L = 0.105 m]

5.8 In a rectangular channel 0.6 m wide a jump occurs where the Froude number is 3. The depth after the jump is 0.6 m. Estimate the loss of total head and power dissipated at the jump [0.225 m, 0.786 kW]

5.9 A 3 m wide rectangular channel conveys 7.5 m³/sec of the water with a velocity of 5 m/s. Is there a condition for hydraulic jump to occur? If so, calculate the height, length and strength of the jump. [$y_1 < y_c$ a jump will occur, H_f = 0.867 m, Strength of jump y_2 / y_1 = 2.734, L_f = 5.2 m = 6 × ($y_2 - y_1$)]

5.10 A hydraulic jump occurs in a 0.5 m wide rectangular channel at the point where depth of water flow is 0.15 m and the Froude number is 2.5. Make calculations for the specific energy, critical and sequent depths, loss of head and the energy dissipated [E = 0.618, y_c = 0.277 m, y_2 = 0.461, ΔE = 0.108 m, 0.24 kW]

5.11 A rectangular channel carrying a supercritical stream is to be provided with a hydraulic jump type of energy dissipater. If it is desired to have an energy loss of 5 m in the jump, when the inlet Froude number is 8.5 determine the sequent depths. [y_2 = 2.2769 m, y_1 = 0.19745 m]

5.12 Water is moving in a rectangular channel to have a Froude number $\sqrt{10}$. The channel is 5 m in width. The depth flow is 1 m. If the water undergoes a hydraulic jump. What is the Froude number after the jump and what percentage of the mechanical energy of the flow is dissipated from the jump? [F_{r_1} = 0.395, E_J/E_1 = 53.1%]

5.13 Water in a rectangular channel is seen to go through a hydraulic jump where the depth jumps from 3 m to 7 m. The width of the channel is 10 m. What is the volume flow rate? What are the initial and final Froude numbers? [F_{r_1} = 1.97, F_{r_2} = 0.553, Q = 321 m³/s]

5.14 Water flows over a spillway into a stiling basin 20 m wide. Before the jump the water has a depth of 1 m and a velocity of 18 m/s. Determine (a) the initial Froude number before the jump (b) the depth of flow after the jump and (c) the head loss in the jump. [F_{r_1} = 5.75, y_2 = 7.64 m, ΔE = 12.9 m]

5.15 Water flows in a wide channel at 10 m³/s and initial depth 1.25 m. If the flow undergoes a hydraulic jump, compte. (a) y_2 (b) v_2, (c) F_{r_2} (d) Δ E, (e) the percentage dissipation, (f) the power dissipated per unit width, and (g) the temperature rise due to dissipation. if C_p = 4200 J/kg K. [3.46 m, 2.89 m/s 0.496, 0.624 m, 13.8%, 61.1 kW/m, 0.0015 k]

5.16 A wide channel flow at depth 500 mm passes through a hydraulic jump and emerges at a depth of 2.8 m. Compute the velocities upstream and downstream of the jump. What is the critical depth for this flow? [9.52 m/s, 1.7 m/s, 1.32 m]

5.17 The flow downstream of a hydraulic jump in a rectangular channel is 9 m deep has a velocity of 3.5 m/s. What are the velocity and depth upstream of the jump? What is the critical depth for the flow? [2.04 m, 15.4 m/s, 4.65 m].

5.18 In a rectangular channel 3 m wide with a flow of 5.65 m³/s the depth is 0.3 m. If a hydraulic jump is produced, what will be the depth after it? What will be the loss of energy? [1.41 m, ΔE = 0.808 m]

5.19 A hydraulic jump occurs in a 5 m wide rectangular channel carrying 6 m³/s on a slope of 0.005. The depth after the jump is 1.4 m (a) What must be the depth before the jump? (b)

What are the losses of energy and power in the jump? [0.137 m, 2.626 m, 154 kW]

5.20 A hydraulic jump occurs downstream from a 15 m wide sluice gate. The depth is 1.5 m, and the velocity is 20 m/s. Determine (a) the Froude number and Froude number corresponding to the conjugate depth (b) the depth and velocity after the jump, and (c) the power dissipated by the jump. [F_{r_1} = 5.21, F_{r_2} = 0.288, y_2 = 10.34 m, v_2 = 2.9 m/s, 11.13 m of water, P = 49 × 10^3 kW]

5.21 A rectangular channel 2 m wide carries 2 m³/s of water in subcritical uniform flow at a depth of 0.8 m. A frictionless hump is to be installed across the bed. Find the critical hump height. [0.179 m]

5.22 A flow of 1.8 m³/s is carried in a rectangular channel 1.6 m wide at a depth of 0.9 m. Will critical depth occur at a section where (a) the frictionless hump 150 mm high is installed across the bed, (b) a friction less side wall constriction (with no hung) reduces the channel width to 1.3 m, and (c) the hump and the side wall constriction are installed together? [(a) Does not occur at the hump, (b) Does not occur at the constriction, (c) Does occur at the hump with constriction.]

5.23 Water flows at 8.5 m³/s in a 3 m wide open channel of rectangular cross section. The bottom slope is adverse i.e. rises 0.2 m per 100 m in the direction of flow. If the water depth decreases from 2.1 m to 1.65 m in a 150 m length of channel, determine Manning's n. [0.014].

5.24 A rectangular flume of planed timber (n = 0.012) is 1.5 m wide and carries 1.7 m³/s of water. The bed slope is 0.0006, and at a certain sec-tion the depth is 0.9 m. Find the distance (in one reach) to the section where the depth is 0.75 m. Is the distance upstream or downstream? Suppose that the slope of the flume is now increased to 0.01. With the same flow as before, find the depth 150 mm dowmstream from a section where the flow is 0.45 m deep. Use only one reach. Is the flow subcr-itical or supercritical? [315 m, 0.75 m depth is downstream, 0.383 m]

5.25 A rectangular channel is 3 m wide and ends in a free outfall. If the discharge is 10 m³/s, slope is 0.0025, and n = 0.016, find

y_n, y_c and the water surface profile for a distance of 150 m upstream from the outfall. [y_n = 1.34 m, y_c = 1.04 m, 224 m]

5.26 Water flowing at the normal depth in a rectangular concrete channel that is 12 m wide encounters an obstruction, causing the water level to rise above the normal depth at the obstruction and for some distance upstream. The water discharge is 126 m³/s and the channel bottom slope is 0.00086. The depth of water just upstream from the construction is 4.55 m. Find the distance upstream to the point where the water surface is at the normal depth. [2.95 m, 4568 m]

5.27 A weir is installed across a rectangular open channel there by raising the flow depth from 1.5 in a normal flow to 2.5 m at the weir. The width of the channel is 10 m and it is laid to a slope of 1 in 10000. Take the value of Manning's rugosity coefficient equal to 0.02. Find the approxi-mate length of the backwater curve cosidering the average velocity, average depth and average slope mid way between the two sections [L = 17378 m]

5.28 Water flows at a steady and uniform depth of 2 m in an open channel of rectangular cross section having base width equal to 5 m and laid at a slope of 1 in 1000. It is desired to obtain critical flow in the channel by providing a hump in the bed. Calculate the hump height and sketch the flow profile. Consider the value of Manning's rugosity coefficient n = 0.02 for the channel surface [E_1 = 2.146, E_{min} = 1.582, Z = E_1 – E_{min} = 0.564 m]

5.29 A channel of rectangular section 8 m wide is laid to bed slope of 1 in 64 and carries a discharge of 40 m³/s refer curve. A barrier across the channel raises the water surface to 3 m just upstream of the barrier. Find the length of the water surface profile upto the the hydraulic jump up stream. Taking Manning n = 0.025 for calculation of the hydraulic radius the channel may be treated as a wide channel. Use (single) step method [D ≈ 1.1 m, m_1 = 0.8627 v_1 = 3 m/s E_1 = 2.152, m_2 = 2.024, v_2 = 1.2195 m/s, E_2 = 4.175, J = 0.003183 – using v_{av} = $\frac{1}{n} m_{av}^{2/3} \, j^{1/2}$ L = 162.59 m]

5.30 What will be the actual slope of the water surface in millimeters per km at a section in rectangular channel at which the discharge is 28 m^3/sec, width 15 m, depth 3 m, bed slope 1 in 6700 and C = 61? [101.6 mm/km]

5.31 (a) 8.5 m^3/s of water flow in a channel of rectangular cross section 6 m wide with a slope of 1 in 2000 and a depth of 0.9 m. Calculate the value of C in Chezy's formula

(b) In order to raise the depth of water upstream a dam is built dowmstream. If the depth of water at the dam is 1.8 m, calculate how far upstream the depth will be 1.5 m. [C = 84.9,

$S_f = 1.165 \times 10^{-4}$, $\frac{dy}{dx} = 4.06 \times 10^{-4}$ m/m, L = 738 m]

5.32 A river 50 m wide and 4 m deep has natural bed and surface slope of 1 in 10000. Compute the approx, length of the back water curve produced by an afflux of 3 m. Take n = 0.03. [L = 52700 m]

6

Waves and Surges in Open Channel

6.1 INTRODUCTION

Unsteady channel flow is essentially non uniform flow and it can therefore be conveniently considered in two categories; gradually varied and rapidly varied unsteady flow.

The first catgory will have almost parallel streamlines, gradual changes in depth, vertical accelerations will play a minor part and channel friction forces a large one.

The second category will be characterised by very pronounced curvature of the stream-lines, rapid changes in depth and sometimes rapid changes in depth and sometimes discontinuous water surface profiles, large vertical acceleration and channel friction forces of secondary importance.

The addition of the time variable makes rigorous analysis very complex and only the most simple cases will be considered here.

River flood propagation, esturial flow, and surges resulting from gate operation or dam failures are practical examples of unsteady channel flows. Natural flood flows in rivers and the propagation of tides in estuaries are examples of gradually varied unsteady flow since the vertical component of acceleration is small. Surges are examples of rapidly varied unstead flow.

6.2. WAVES AND THEIR CLASSIFICATION

A wave may be defined as a temporal variation in the water surface which is propagated through a fluid medium.

The celerity of the wave is the speed of propagation of such a disturbance relative to the fluid.

The waves in open channel are also known as surface waves

Surface waves are classified as

(a) Capillary waves, as are found in small size models of channels.

(b) Gravity waves, as are found in big size channels. Gravity waves are further classified into two kinds :

Oscillatory wave and translatory waves.

(i) Oscillatory waves which oscillate but do not move, over a length of channel, with any velocity.

(ii) Translatory waves which move with certain velocity over appreciate length of the channel in which they are propagated.

Sea waves are oscillatory in nature, whereas the moving hydraulic jump and tidal bore are examples of translatory waves.

The translatory wave may be further classified as a solitary waves or a train of waves. The solitary wave has a rising limb, a single peak and a reces-sion limb and is followed and preceded by steady wave. A wave train is created by the sequence of several waves.

If the direction of wave movement is that of the channel slope, the wave direction is considered to be downstream. An upstream wave direction means the movement of the wave up the slope. When the wave causes the water surface to be higher than in the otherwise steady flow; it is called a positive wave; the wave is considered to be negative if the water surface is lowered.

Waves may also be classified as monoclinal (single-faced) or two faced depending on the shape of the wave. A two faced wave may be symmetrical or asymmetrical.

6.2.1 Celerity of small solitary wave

In Fig 6.1(a) is shown a solitary wave of small height h moving over still water of depth y in a rectangular channel. Let C be the celerity of the wave in the direction shown i.e. to the left. Movement of this wave is a case of unsteady flow in open channel. By applying C in opposite direction (i.e. to the right) to the wave and to the still water, we get steady flow in which the water is moving with velocity C and the wave is stationary at one point on the surface of the moving water below; this is shown in Fig 6.1 (b).

According to continuity equation of flow

$$v_1 y = v_2 (y + h)$$

or,
$$C y = v_2 (y + h)$$

or,
$$v_2 = C\left(\frac{y}{(y+h)}\right) \tag{6.1}$$

Applying specific energy equation to the two sections 1-1 and 2-2 of the equivalent steady flow of Fig. 6.1 (b), we have

$$E_1 = E_2$$

$$y + \frac{v_1^2}{2g} = (y + h) + \frac{v_2^2}{2g}$$

$$y + \frac{C^2}{2g} = y + h + \frac{C^2\left(y/(y+h)\right)^2}{2g} \quad \text{[Using equation (6.1)]}$$

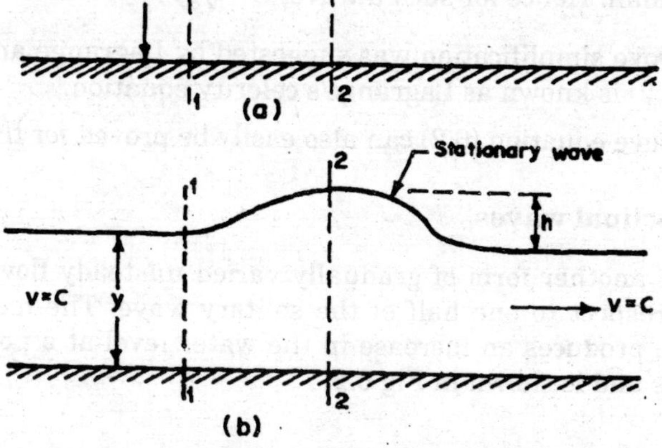

Fig. 6.1. Propagation of low wave in channel.

$$\text{or,} \quad \frac{C^2}{2g} = h + \frac{C^2}{2g}\left(\frac{y^2}{y^2 + 2\,yh + h^2}\right)$$

$$\text{or,} \quad \frac{C^2}{2g}\left[1 - \frac{y^2}{y^2 + 2\,yh + h^2}\right] = \text{h}$$

$$\text{or,} \quad \frac{C^2}{2g} = \left[\frac{y^2 + 2\,yh + h^2 - y^2}{y^2 + 2\,yh + h^2}\right] = \text{h}$$

$$\text{or,} \quad C^2 = \left[\frac{h\,(2y + h)}{y^2 + 2\,yh + h^2}\right] = 2\,\text{g}\,\text{h}$$

$$\text{or,}\, C^2 = \frac{2g\left(y^2 + 2\,yh + h^2\right)}{(2y + h)} = gy\left[\frac{y + 2h + h^2/y}{y + \frac{h}{2}}\right] \approx gh\left[1 + \frac{3}{2}\left(\frac{h}{y}\right)\right]$$

$$\therefore \quad C = \sqrt{gy}\left(1 + \frac{3}{2}\frac{h}{y}\right)^{1/2}$$

Using binomial expansion.

$$= \sqrt{gy}\left(1 + \frac{3}{4}\frac{h}{y}\right) \qquad (6.2)$$

It is to be noted that for a small wave, h is very small or, $\frac{h}{y}$ is very very small. Hence for such a wave, $C \approx \sqrt{gy}$.

The above simplification was suggested by Lagrange and hence equation (6.2) is known as Lagrange's celerity equation.

The above equation (6.2) can also easily be proved for the surge.

6.2.2 Monoclinal waves

This is another form of gradually varied unsteady flow similar in certain respect to one half of the solitary wave. The monoclinal rising wave produces an increase in the water level at a point after its passages and is shown in Fig 6.2.

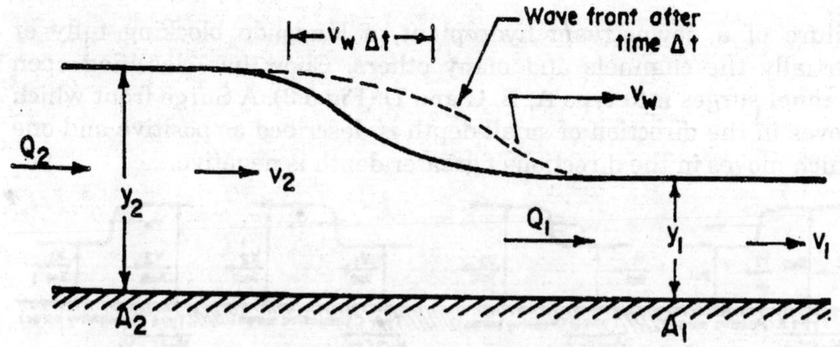

Fig. 6.2. Progression of a monoclinal rising wave along a channel

This type of wave is produced in a channel by a gradual increase in the discharge which is assumed in the case illustrated to reach a steady value of Q_2 in a finite time. If the initial steady discharge in the channel is Q_1 then the propagation velocity of the wave can be evaluated by contunuity considerations and is given by the expression

$$v_w = \frac{Q_2 - Q_1}{A_2 - A_1} \tag{6.3}$$

and in a rectangular or very wide channel

$$v_w = \frac{(y_2\, v_2 - y_1\, v_1)}{(y_2 - y_1)} = \frac{(q_2 - q_1)}{(y_2 - y_1)} \tag{6.4}$$

In a time interval Δt the wave front advances a distance $v_w\, \Delta t$ which must be greater than the corresponding distance travelled by any of the water particles, since v_w must be greater than both v_1 and v_2. If the channel is initially empty there will be an exception to this rule and v_w will be equal to v_2.

6.2.3 Surges

When a monoclinal wave has an abrupt change of surface curvature and slope at its forward end the conditions of gradually varied flow break down and the resulting rapidly varying wave profile is termed a surge.

A surge is a discontinuity in the liquid (depth of flow) which is propagated along the channel. Like all other types of wave a surge is produced by a disturbance at some point along the channel. Examples of such disturbances are the partial closing or opening of a gate, the failure of a downstream hydroplant, a landslide blocking fully or partially the channels and many others. Chow has classified open channel surges into type A, B, C and D (Fig 6.3). A Surge front which moves in the direction of small depth is described as positive and one which moves in the direction of greater depth is negative.

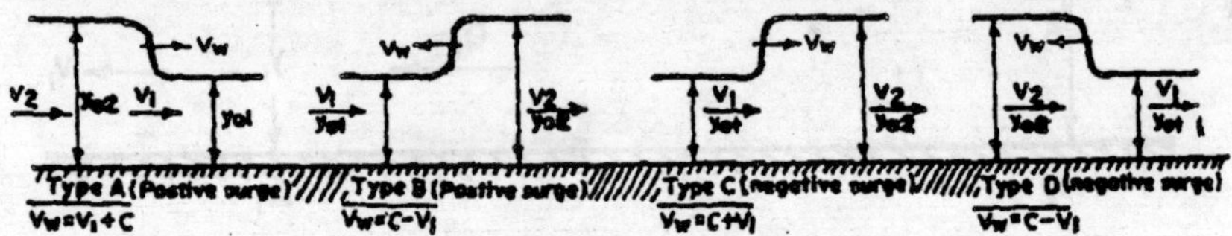

Fig. 6.3. Different type of surges

6.2.4 The upstream positive surge

Consider the propagation of a positive wave upstream in a frictionless channel resulting from a gate closure (Fig 6.4a).

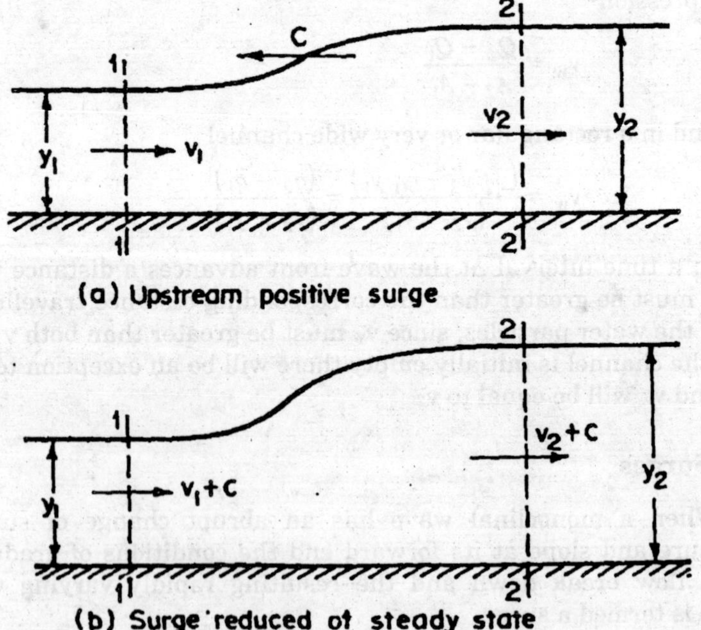

Fig. 6.4. Positive surge

The front of the surge wave is propagated upstream at a celerity, C, relative to the stationary observer. To the observer, the flow situation is unsteady as the wave front passes; to an observer travelling at a speed, C, with the wave the flow appears steady although non uniform. Fig 6.4b shows the surge reduced to steady state.

Applying continuity equation

$$A_1(v_1 + C) = A_2(v_2 + C)$$

or,
$$v_2 = \frac{A_1 v_1 - C(A_2 - A_1)}{A_2} \tag{6.5}$$

Applying momentum equation.

$$g\ A_1\ \bar{y}_1 - g\ A_2\ \bar{y}_2 + A_1(v_1 + C)(v_1 - v_2)$$

Where $\bar{y}_1$ are the respective depths of the centres of area. Substituting for v_2 from equation (6.5) in the above equation.

$$g\ (A_2\ \bar{y}_2 - A_1\ \bar{y}_1)\frac{A_2}{A_1\ (A_2 - A_1)} = (v_1 + C)^2$$

Solving for C, we get

$$C = \left[g\ A_2\ \frac{A_2\ \bar{y}_2 - A_1\ \bar{y}_1}{A_1\ (A_2 - A_1)}\right]^{1/2} - v_1 \tag{6.6}$$

In a special cae of a rectangular channel,

$$A = by\ ;\ \bar{y} = \frac{y}{2}$$

From equation (6.4);

$$C = \left[\frac{gy_2}{2}\ \frac{(y_2^2 - y_1^2)}{y_1(y_2 - y_1)}\right]^{1/2} - v_1$$

$$= \left[\frac{gy_2}{2y_1}(y_2 + y_1)\right]^{1/2} - v_1 \tag{6.7}$$

The hydraulic jump can be shown to be a stationary surge

Putting C = 0 in equation (6.7)

$$v_1^2 = \frac{gy_2}{2y_1}(y_2 + y_1)$$

or, $\frac{2v_1^2\, y_1}{g} = y_2^2 + y_1\, y_2$

or, $y_2^2 + y_1\, y_2 - 2\, F_{r_1}^2\, y_1^2 = 0 \qquad \left[F_{r_1}^2 = \frac{v_1}{gy_1}\right]$

Solving for y_2

$$y_2 = \frac{y_1}{2}\left(\sqrt{1 + 8\, F_{r_1}^2} - 1\right)$$

Which is identical with the equation.

In the case of a low wave where y_2 approaches y_1 equation (16.87) becomes

$$C = \sqrt{gy_1} - v_1 \qquad (16.8)$$

and in still water

$$C = \sqrt{gy_1} \qquad (16.9)$$

6.2.5 The downstream positive surge

This type of wave may occur in the channel downstream from a sluice gate the which the opening is rapidly increased (Fig 16.41).

Reducing the flow to steady state

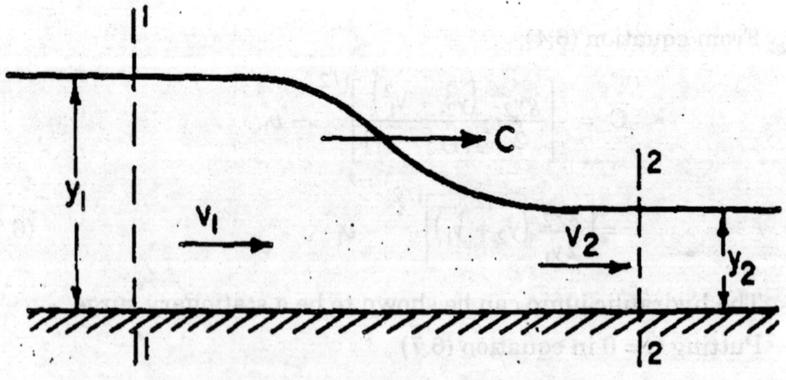

Fig. 6.5. Down stream positive surge.

Applying the continuity equation

$$(C - v_1)\, A_1 = (C - v_2)\, A_2$$

$$v_1 = \frac{C\,A_1 - C\,A_2 + v_2\,A_2}{A_1}$$

Applying momentum equation

$$g\,A_1\,\bar{y}_1 - g\,A_2\,\bar{y}_2 + (C - v_2)\,A_2\,(v_2 - v_1) = 0$$

Solving the above equation for c

$$C = \left[\frac{g\,(A_1\,\bar{y}_1 - A_2\,\bar{y}_2)}{(A_1 - A_2)}\left(\frac{A_1}{A_2}\right)\right]^{1/2} + v_2 \qquad (6.10)$$

In the case of rectangular channel

$$C = \left[\frac{g y_1}{2\,y_2}(y_1 + y_2)\right]^{1/2} + v_2 \qquad (6.11)$$

In terms of v_2

$$v_2 = \frac{(y_2 - y_1)}{C}\left(\frac{y_2 + y_1}{2\,y_1}\right)g$$

$$\text{or,} \quad v_2 = \frac{gh}{C}\left(\frac{y_2 + y_1}{2 y_1}\right) \qquad (6.12)$$

Thus the height of the surge is

$$h = \frac{C}{g}\left(\frac{2 y_1}{y_1 + y_2}\right) v_2 \qquad (6.13)$$

Again if the height of the surge is small relative to the depth of flow, then $y_1 = y_2$, and equations (6.12) and (6.13) may be approximately written as

$$v_2 = \frac{gh}{C} \qquad (6.14)$$

$$\text{and} \qquad h = \frac{v_2\,C}{g} \qquad (6.15)$$

6.2.6 Negative surge waves

The negative surge appears to a stationary observer as a lowering of the liquid surface. Such waves occur in the channel downstream from a control gate the opening of which is rapidly reduced or in the upstream stream channel as the gate is opened. The wave front can be considered to be composed of a series of small waves superimposed on each other. Since the upper most wave has the greatest depth it travels faster than those beneath, the retreating wave front therefore becomes flatter (Fig 6.6).

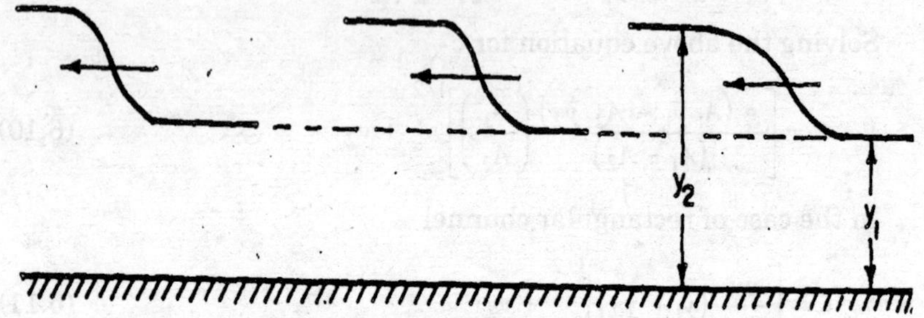

Fig. 6.6. Propagation of a negative surge.

Reducing the flow to steady state the contunity equation becomes

$$(v + C)\, y = (v - \delta v + C)(y - \delta y)$$

Fig 6.7 shows a small disturbance in a rectangular channel caused by a reduction in downstream discharge; the wave propogates upstream.

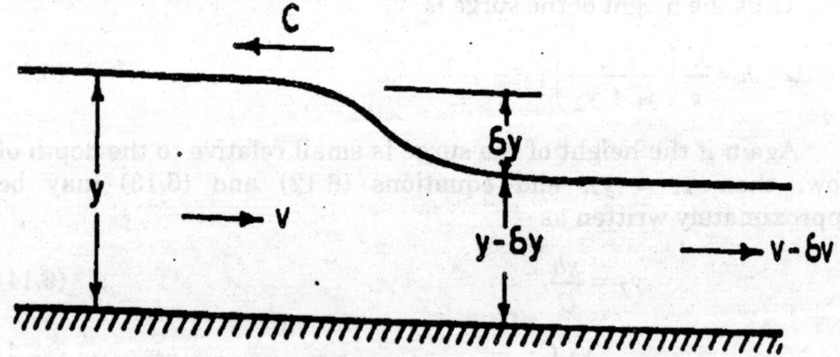

Fig. 6.7. Small distribution in a rectangular channel.

Solving the δy and neglecting the product of small quantities

$$\delta y = \frac{y\, \delta v}{v + C} \tag{6.16}$$

Applying the momentum equation

$$\frac{\rho g}{2}\left[y^2 - (y - \delta y)^2\right] + \rho\, y\,(v + c)\{v + C - (v - \delta\, v + C)\} = 0$$

Solving we get

$$\frac{\delta y}{\delta v} = \frac{(v + C)}{g}$$

or, $$\delta y = -\frac{\delta v\,(v + C)}{g} \tag{6.17}$$

Equating (6.16) and (6.17),

$$\frac{y\,\delta v}{(v + C)} = \frac{(v + C)\,\delta\, v}{g}$$

Solving we get

$$C = \sqrt{gy} - v \tag{6.18}$$

Substituting for (v + C) from equation (6.17) into (6.18) yields

$$\delta y = \frac{\delta v}{g}\sqrt{gy}$$

and in the limit as $\delta y \to 0$

$$\frac{dy}{\sqrt{y}} = -\frac{dv}{\sqrt{g}} \tag{6.19}$$

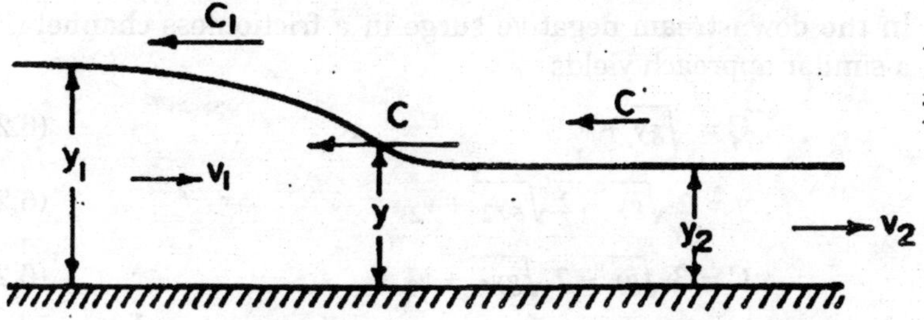

Fig. 6.8. Negative surge of finite height

For a wave of finite height (Fig. 6.8), integration of equation (6.19) yields

$$v = -2\sqrt{gy} + \text{constant}$$

when $y = y_1$, $v = v_1$, therefore

constant = $v_1 + 2\sqrt{gy_1}$

Therefore,

$$v = v_1 + 2\sqrt{gy_1} - 2\sqrt{gy} \quad (6.20)$$

From equation (16.100),

$C = \sqrt{gy} - v$ and substituting in equation (6.20) yields

$$C = 3\sqrt{gy} - 1\sqrt{gy_1} - v_1$$

The wave speed at the crest is therefore,

$$C_1 = \sqrt{gy_1} - v_1$$

and at trough

$$C_2 = 3\sqrt{gy_2} - 2\sqrt{gy_1} - v_1 \quad (6.21)$$

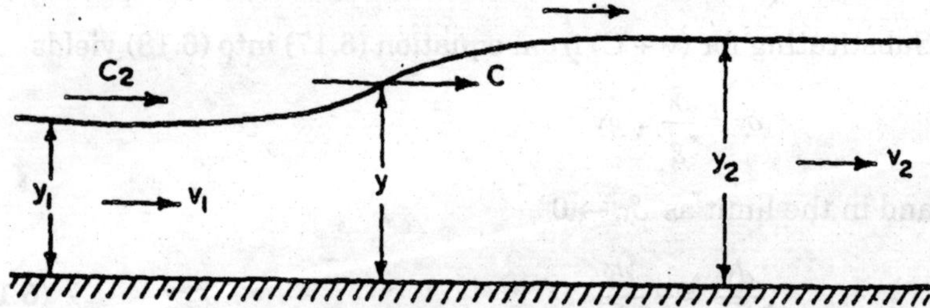

Fig. 6.9. Downstream negative surge.

In the downstream negative surge in a frictionless channel (Fig 6.9), a similar approach yields

$$C = \sqrt{gy} + v \quad (6.22)$$

$$v = 2\sqrt{gy} - 2\sqrt{gy_2} + v_2 \quad (6.23)$$

$$C = 3\sqrt{gy} - 2\sqrt{gy_2} + v_2 \quad (6.24)$$

$$C_1 = 3\sqrt{gy_1} - 2\sqrt{gy_2} + v_2 \quad (6.25)$$

$$C_2 = \sqrt{gy_2} + v_2 \quad (6.26)$$

ILLUSTRATIVE EXAMPLES

6.1 **Prove that the afflux caused at bridge is approximately loss of head at the bridge is given by**

$$y_1 - y_2 \approx x \approx \frac{Q^2}{2g}\left[\frac{1}{c^2\ B_2^2\ y_1^2} - \frac{1}{B_1^2\ y_1^2}\right]$$

A river 61 m broad, with practically vertical banks, has a bridge over it; there are 4 bridge openings, each 12.2 m wide. On the downstream side of bridge, the uniform depth of flow is 3.7 m, with a mean velocity of 1.8 m/s. Calculate the afflux caused by the bridge and the length of backwater curve if the bed slope of river is 1 in 2000. Take C in the afflux formula as 0.95 and Manning's Constant n = 0.0225.

Solution

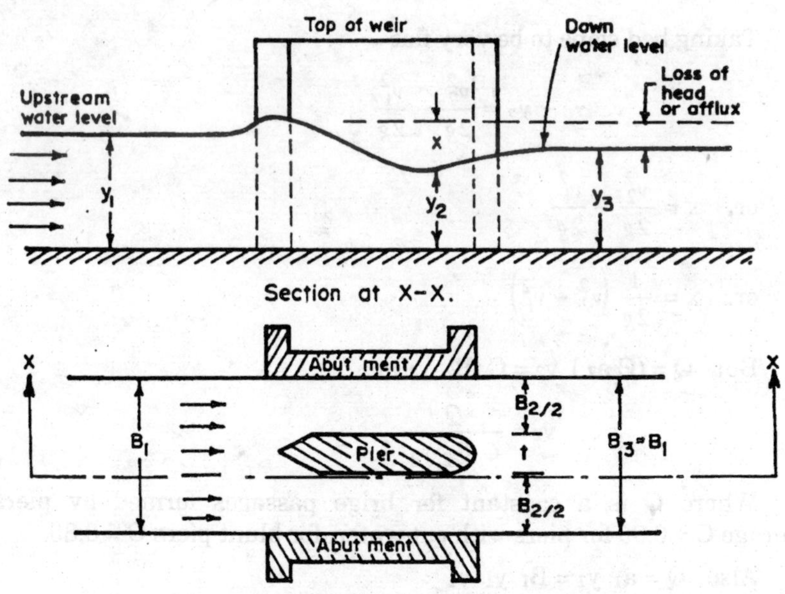

Fig. 6.10. Afflux at bridge P. 6.1.

At a bridge, the waterway is slightly decereased or constricted due to the obstruction caused by piers as shown in Fig 6.10. The bridge waterway may have one span and then the passage will be between the two abutments. When there are `n' number of spans, there will be (n – 1) number of piers equally spaced between the two abutments. Due to the constriction in waterway, there will be some heading up of water on the upstream side and there will be the usual back water curve. Afflux or heading up will be the difference between the water level on the upstream and downstream side of the bridge.

Afflux caused at bridge

$y_1 - y_3 \approx$ loss of head x at the bridge

For all practical purposes afflux h_l is taken equal to x which is equal to $(y_1 - y_2)$. This is on safer side because $x > h_l$.

Neglecting the friction in the converging channel or upstream side between section 1 – 1 and 2 – 2, we have

$$y_1 + \frac{v_1^2}{2g} = y_2 + \frac{v_2^2}{2g}$$

Taking bed slope to be very flat

$$\therefore \quad y_1 - y_2 = \frac{v_2^2}{2g} - \frac{v_1^2}{2g}$$

or, $$x = \frac{v_2^2}{2g} - \frac{v_1^2}{2g}$$

or, $$x = \frac{1}{2g}\left(v_2^2 - v_1^2\right)$$

But $Q = (C\, a_2)\, v_2 = C\,(B_2\, y_2)\, v_2$

$$\therefore \quad v_2 = \frac{Q}{C\, B_2\, y_2}$$

Where C is a constant for brige passages formed by piers; average C = 0.95 for piers with cut water; for blunt piers C = 0.85.

Also, $Q = a_1\, y_1 = B_1\, y_1\, v_1$

$$\therefore \quad v_1 = \frac{Q}{B_1\, y_1}$$

Hence $$x = \frac{1}{2g}\left[\frac{Q^2}{C^2\, B_2^2\, y_2^2} - \frac{Q^2}{B_1^2\, y_1^2}\right]$$

$$= \frac{Q^2}{2g}\left[\frac{1}{C^2\ B_2^2\ y_2^2} - \frac{1}{B_1^2\ y_1^2}\right] = \frac{Q^2}{2g}\left[\frac{1}{C^2\ B_2^2 (y_1 - x)^2} - \frac{1}{B_1^2\ y_1^2}\right]$$

As a first approximation

$$x \approx \frac{Q^2}{2g}\left[\frac{1}{C^2\ B_2^2\ y_1^2} - \frac{1}{B_1^2\ y_1^2}\right]$$

Taking $(y_1 - x) \approx y_1$ because y_1 is very large as compared to x.

(b) Given, $B_1 = 61$m, $B_2 = 4 \times 12.2 = 48.8$ m,

$y_3 = 3.7$ m, $v_3 = 1.8$ m/s, $B_3 = 61$ m,

So of the upstream and downstream beds of channel = $\frac{1}{2000}$, C = 0.95 and n = 0.0225.

Now, approximately x = $\frac{Q^2}{2g}\left[\frac{1}{C^2\ B_2^2\ y_1^2} - \frac{1}{B_1^2\ y_1^2}\right]$

$$x \approx \frac{v_1^2}{2g}\left[\frac{B_1^2}{B_2^2\ C^2} - 1\right]$$

$$\approx \frac{(1.8)^2}{2 \times 9.81}\left[\frac{(61)^2}{(48.8)^2\ (0.95)^2} - 1\right]$$

$$\approx 0.165\ [1.7313 - 1]$$

$$\approx 0.165 \times 0.7313$$

$$\approx 0.12 \text{ m}$$

For back water curve, we have

$v_1 = 1.8$ m/s, $y_1 = 3.7$ m, $y_2 = y_1 + x$

$= 3.7 + 0.12 = 3.82$ m

$$\therefore \quad v_2 = 1.8 \times \frac{3.7}{3.82} = 1.743 \text{ m/s}$$

$$E_1 = y_1 + \frac{v_1^2}{2g} = 3.7 + \frac{(1.8)^2}{2 \times 9.81} = 3.7 + 0.165 = 3.865$$

$$E_2 = y_2 + \frac{v_2^2}{2g} = 3.82 + \frac{(1.74)^2}{2 \times 9.81} = 3.82 + 0.154 = 3.9743$$

$$v_{av} = \frac{v_1 + v_2}{2} = \frac{1.8 + 1.74}{2} = 1.77\ m/s$$

$$R_{av} = \frac{61 \times 3.76}{61 + 2(3.76)} = 3.347$$

$$v_{av} = \frac{1}{n} R_{av}^{2/3} S_f^{1/2}$$

$$1.77 = \frac{1}{0.0225} \times (3.347)^{2/3} S_f^{1/2}$$

$$S_f^{1/2} = \frac{1.77}{99.45} = 0.017797$$

$$S_f = 3.1674 \times 10^{-4}$$

$$\text{Hence } L = \frac{E_2 - E_1}{S_o - S_f} = \frac{3.9743 - 3.865}{0.0005 - 3.1674 \times 10^{-4}}$$

$$= \frac{0.1093}{1.8325 \times 10^{-4}} = 596.44\ m$$ Ans

6.2 A river is flowing with a average velocity of 10 km/hr, at a average depth of 1.8 m. The incoming tide produces a change in depth to 2.4 m. Determine the velocity with which the resulting bore will travel upstream Prove the formula from first principles.

Solution

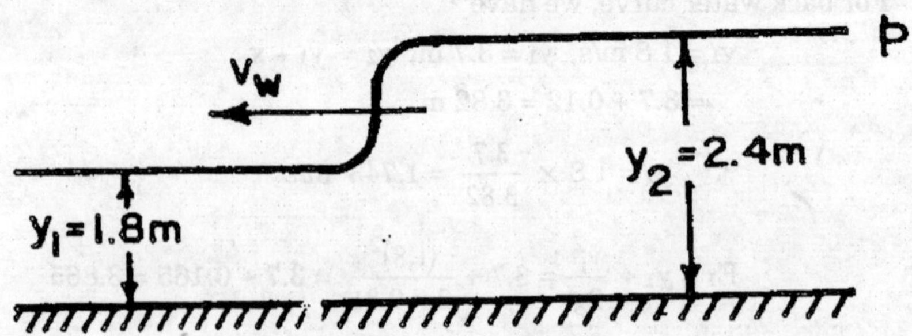

Fig. 6.11. River flow P. 6.2

We know that

$$C = \sqrt{g\ y_1}\left[\frac{y_2}{2y_1}\left(\frac{y_2}{y_1}+1\right)\right]^{1/2}$$

$$= \sqrt{9.81\times 1.8}\left[\frac{2.4}{2\times 1.8}\left(\frac{2.4}{1.8}+1\right)\right]^{1/2}$$

$$= \sqrt{9.81\times 1.8}\ \sqrt{1.55} = \sqrt{17.658}\ \sqrt{1.55}$$

$$= 4.2 \times 1.2472 = 5.241 \text{ m/s upstream}$$

Now, $v_w = v_1 \pm C = v_1 - C = 2.77 - 5.241 = -2.46$

[Negative sign means that the surface move upstream].

$\therefore \quad v_w = 2.46$ m/s upstream

6.3 A river is flowing with an average velocity of 6.4 km/hr and an average depth of 1.5 m. A sluice gate raised upstream, produced a sudden increase of depth to 2.1 m. Estimate the velocity with which the resulting surge travels downstream.

Solution

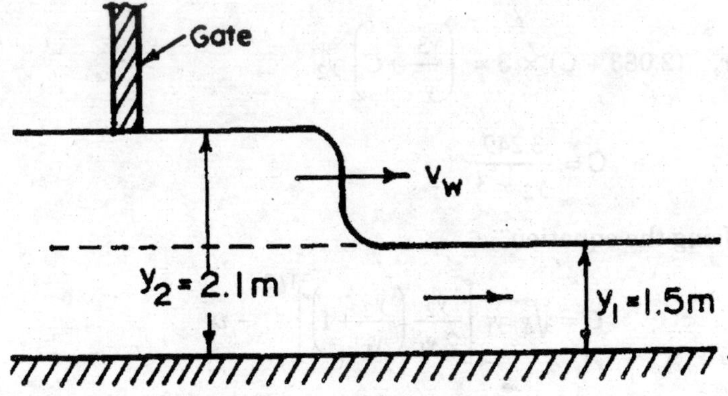

Fig. 6.12. Sluice gate P. 6.3.

$$C = \sqrt{g\ y_1}\left[\frac{y_2}{2\,y_1}\left(\frac{y_2}{y_1}+1\right)\right]^{1/2}$$

$$= \sqrt{9.81\times 1.5}\left[\frac{2.1}{2\times 1.5}\left(\frac{2.1}{1.5}+1\right)\right]^{1/2}$$

$$= \sqrt{14.715}\ \sqrt{1.68} \quad = 3.83 \times 1.296 = 4.972 \text{ m/s}$$

$$v_w = v_1 \pm C = v_1 + C = 1.78 + 4.97 = 6.75 \text{ m/s} \qquad \text{Ans}$$

6.4 **A rectangular channel 4 m wide conveys a discharge of 25 m³/s at a depth of 3 m. The downstream discharge is suddenly reduced to 12 m³/s by partial closure of a gate. Determine the initial depth and celerity of the positive surge wave.**

Solution: Using upstream surge wave

$$Q = A_1\ v_1 = 4 \times 3 \times v_1$$

$$v_1 = \frac{25}{12} = 2.083 \text{ m/s}$$

$$Q = A_2\ v_2$$

$$12 = 4 \times y_2 \times v_2$$

$$v_2 = \frac{3}{y_2}$$

Using the equation

$$(v_1 + C)y_1 = (v_2 + C)\ y_2$$

or, $$(2.083 + C) \times 3 = \left(\frac{3}{y^2} + C\right) y_2$$

$$C = \frac{3.249}{y_2 - 3}$$

Using the equation

$$C = \sqrt{g\ y_1}\left[\frac{y_2}{2\,y_1}\left(\frac{y_2}{y_1} + 1\right)\right]^{1/2} - v_1$$

$$\frac{3.249}{y_2 - 3} = \sqrt{9.81 \times 3}\left[\frac{y_2}{2 \times 3}\left(\frac{y_2}{3} + 1\right)\right]^{1/2} - 2.083$$

Solving by trial and error method

$y_2 = 3.75$ m, therefore, $C = 4.35$ m/s Ans

6.5 **A tidal channel, which may be assumed to be rectangular, 40 m wide, bed slope 0.0003, Manning's roughness coefficient, n = 0.022, conveys a steady freshwater discharge of 60 m³/s. A**

tidal bore observed to propagate upstream with a ceterity of 5 m/s. Determine the depth of the flow and the discharge immediately after the bore has passed, neglecting the density difference between the fresh water and saline water.

Solution: Using Manning's equation

$$v = \frac{1}{n} R^{2/3} S^{1/2}$$

$$\frac{60}{40 \times y} = \frac{1}{0.022} \left(\frac{40 \times y}{40 + 2y} \right)^{2/3} (0.0003)^{1/2}$$

$$\frac{1.90525}{y} = \left(\frac{40y}{40 + 2y} \right)^{2/3}$$

Solving by trial and error method y = 1.52 m

Therefore, y_1 = 1.52 m

and $v_1 = \frac{60}{40 \times 1.52} = 0.9868$ m/s

Using the equation

$$C = \sqrt{g\ y_1} \left[\frac{y_2}{2y_1} \left(\frac{y_2}{y_1} + 1 \right) \right]^{1/2} - v_1$$

$$5 = \sqrt{9.81 \times 1.52} \left[\frac{y_2}{2 \times 1.52} \left(\frac{y_2}{1.52} + 1 \right) \right]^{1/2} - 0.9868$$

Solving by trial and error method y_2 = 2.66 m.

Using the continuity equation

$$(v_1 + C)\, y_1 = (v_2 + C)\, y_2$$

$$v_2 = \frac{(v_1 + C)\, y_1}{y_2} - C = \frac{(0.9868 + 5) \times 1.52}{2.66} - 5 = -1.578 \text{ m/s}$$

Therefore, Q = $A_2\ v_2$

= b y_2 v_2 = (40 × 2.66 × (– 1.578)

= 167.8992 m³/s upstream

6.6 A rectangular tailrace channel, 15 m wide, bed slope 0.0002 and Manning's n = 0.017 conveys a steady discharge of 45 m³/s from a hydropower installation. A power increase results in a sudden increase in flow to the turbine to 100 m³/s. Determine the depth and celerity of the resulting surface wave in the channel.

Solution

$$v_2 = \frac{1}{n} R^{2/3} S^{1/2}$$

$$\frac{45}{15\,y_2} = \frac{1}{0.017}\left(\frac{15\,y_2}{15 + 2\,y_2}\right)^{2/3} (0.0002)^{1/2}$$

$$\frac{3.6062}{y_2} = \left(\frac{15\,y_2}{15 + 2\,y_2}\right)^{2/3}$$

Solving by trial and error method

$$y_2 = 2.42 \text{ m}$$

$$\text{Therefore, } v_2 = \frac{45}{15 \times 2.42} = 1.2396 \text{ m/s}$$

Using continuity equation

$$(C - v_1)\,y_1 = (C - v_2)\,y_2$$

Solving for C

$$C = \frac{\dfrac{Q_1\,y_1}{b y_1} - v_2\,y_2}{y_1 - y_2}$$

$$C = \frac{\dfrac{100 \times y_1}{15\,y_1} - 1.2396 \times 2.42}{y_1 - 2.42} = \frac{6.67 - 3}{y_1 - 2.42} = \frac{3.67}{y_1 - 2.42}$$

We know that

$$C = \sqrt{g\,y_1}\left[\frac{y_2}{2\,y_1}\left(\frac{y_2}{y_1} + 1\right)\right]^{1/2} - v_2$$

$$\frac{3.67}{y_1 - 2.42} = \sqrt{9.81\,y_1}\left[\frac{2.42}{2\,y_1}\left(\frac{2.42}{y_1} + 1\right)\right]^{1/2} + 1.2396$$

Solving by trial and error method $y_1 = 2.95$ m

Therefore, $C = \dfrac{3.67}{2.95 - 2.42} = \dfrac{3.67}{0.53} = 6.9245\ m/s$ Ans

6.7 A steady discharge of 25 m³/s enters a long rectangular channel 10 m wide, bed slope 0.0001, Manning's roughness coefficient 0.017, regula-ted by a gate. The gate is rapidly partially closed resulting in a reduction of the discharge to 12 m³/s. Determine the depth and mean velocity at a trough of the wave, the surface profile and the time taken for the wave front to reach a point 1 km downstream neglecting friction.

Solution: Using Manning's equation

$$v_2 = \frac{1}{n} R^{2/3}\ S^{1/2}$$

$$\frac{25}{10\,y_2} = \frac{1}{0.017}\left(\frac{10\,y_2}{10 + 2\,y_2}\right)^{2/3} (0.0001)^{1/2}$$

$$\frac{4.25}{y_2} = \left(\frac{10\,y_2}{10 + 2\,y_2}\right)^{2/3}$$

Solving by trial and error method $y_2 = 2.86$ m

Therefore, $v_2 = \dfrac{25}{10 \times 2.86} = 1.2396\ m/s$

and $v_1 = \dfrac{12}{10\,y_1} = \dfrac{1.2}{y_1}$

For a downstream negative surge in a frictionless channel

$$v_1 = v_2 - 2\sqrt{g}\left(\sqrt{y_2} - \sqrt{y_1}\right)$$

$$\frac{1.2}{y_1} = 0.874 - 2\sqrt{9.81}\left(\sqrt{2.86} - \sqrt{y_1}\right)$$

$$\frac{1.2}{y_1} = 0.874 - 6.264\left(1.69115 - \sqrt{y_1}\right)$$

Solving by trial and error method $y_1 = 2.637$ m Ans

Therefore, $v_1 = \dfrac{12}{10 \times 2.637} = 0.455$ m/s

and $C_2 = \sqrt{g\ y_2} + v_2 = \sqrt{9.81 \times 2.86} + 0.874$

$= 5.2968 + 0.874 = 6.17$ m/s

Therefore, time taken for 1 km distance

$$x = C_2 t$$

$$1 \times 1000 = 6.17\, t$$

$$t = 162.052 \text{ s} = 2.7 \text{ minutes} \qquad \text{Ans}$$

The surface profile equation is written as

$$x = C\,t = \left(3\sqrt{gy} - 2\sqrt{gy_2} + v_2\right)t$$

or, $$x = \left(9.4\sqrt{y} - 9.726\right)t \qquad \text{Ans}$$

6.8 Establish that the speed of a hydraulic bore relative to the upstream water surface in a gently sloping parabolic open channel whose width b is proportional to the square root of the depth of flow y is

$$C = \sqrt{0.4\,g\,\frac{\left(y_2^{5/2} - y_1^{5/2}\right)}{y_1^{3/2}\left[1 - \left(\frac{y_1}{y_2}\right)^{5/2}\right]}}$$

The subscripts 1 and 2 refer to the shallower and depper depths respectively. (The certroid of the parabolic lamina is 0.4 times the height measured from the base). Assume that the water surface is horizontal at both sections.

Water is released from under a sluice gate into a parabolic channel at the rate of 30.7 m³/s and with a normal depth of 1.22 m. The channel is described by

$$b = 5\sqrt{y}$$

Find the minimum right of way to avoid overtopping from the formation of a hydraulic jump. What is your criticism of the method of calculation?

Solution

The Speed of Propagation

$$A = \frac{2}{3}by = \frac{2}{3}\left(5\sqrt{y}\right)y = \frac{2\times 5}{3}y^{3/2} = \frac{10}{3}y^{3/2}$$

$$\bar{y} = 0.4\,y$$

$$C = \sqrt{\left(\frac{2}{3}\right) g \frac{\left(0.4\, y_2^{5/2} - 0.4\, y_1^{5/2}\right)}{\left(\frac{2}{3}\right) y_1^{3/2} \left[1 - \left(\frac{y_1}{y_2}\right)^{3/2}\right]}} = \sqrt{\frac{0.4\, g \left(y_2^{5/2} - y_1^{5/2}\right)}{y_1^{3/2} \left[1 - \left(\frac{y_1}{y_2}\right)^{3/2}\right]}}$$

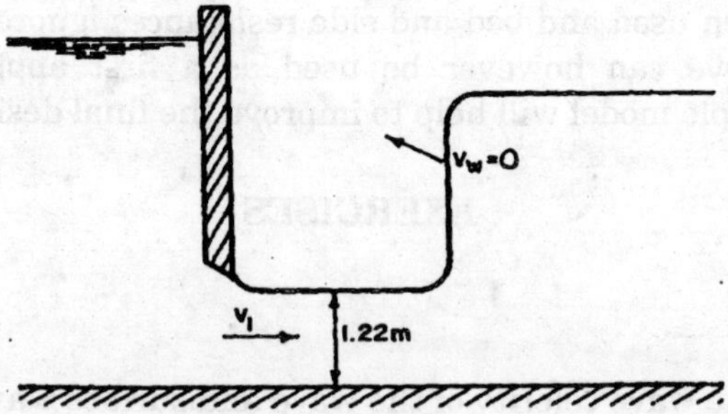

$V_w = O$

V_1 1.22m

Fig. 6.13. Gently slopping parabolic open channel P. 6.8.

For a real hydraulic jump $v_w = 0$, $v_1 = C$

$$A_1 = \frac{2}{3} \times 5\,(1.22)^{3/2} = 4.49\ m^2$$

$$v_1 = Q/A_1 = \frac{30.7}{4.49} = 6.834 \text{ m/s}$$

$$\frac{v_1^2}{g\ y_1} = 0.4 \left[\left(\frac{y_2}{y_1}\right)^{5/2} - 1\right] \Big/ \left[1 - \left(\frac{y_1}{y_2}\right)^{3/2}\right]$$

$$\frac{(6.834)^2}{9.81 \times 1.22} = 0.4 \left[\left(\frac{y_2}{y_1}\right)^{5/2} - 1\right] \Big/ 1 - \left(\frac{y_1}{y_2}\right)^{3/2}$$

or, $$\left(\frac{y_2}{y_1}\right)^{5/2} + 10 \left(\frac{y_1}{y_2}\right)^{3/2} = 11$$

Solving by trial and error

$$\frac{y_2}{y_1} = 2.3$$

$$y_2 = 2.3\ \ y_1 = 2.3 \times 1.22 = 2.806 \text{ m}$$

$$b = 5\sqrt{y} = 5\sqrt{2.806} = 8.375 \text{ m}$$

Maximum height of way is 8.375 m

By assuming that the water surface at each section is straight and horizontal, we have simplified the real situation considerably. The widening of the cross section involves transverse pressure gradients and velocities which thereby involve transverse wave action. The down-stream water surfaces at a number of sections will be irregular. This will affect the height and width of the stream. Average velocities have also been used and bed and side resistances ignored. The values obtained above can however be used as a first approximation in design; a simple model will help to improve the final design.

EXERCISES

A. THEORY

6.1 Define wave celerity of the wave and surface waves.

6.2 Classify the surface waves

6.3 Classify the gravity waves -

6.4 Derive the Lagrange's celerity equation.

6.5 Prove that

$$\text{Celerity C} = \sqrt{gy}\left[1+\frac{3}{4}\left(\frac{h}{y}\right)\right]$$

6.6 Prove that for monoclinal waves

$$v_\omega = \frac{Q_2 - Q_1}{A_2 - A_1}$$

6.7 Define surge and draw the sketches of different type of surges.

6.8 Describe the upstream positive surge and prove the formula

$$C = \left[\frac{gy_1}{2y_1}(y_2+y_1)\right]^{1/2} - v_2$$

6.9 Describe the downstream positive surge and prove the formula.

$$C = \left[\frac{gy_1}{2y_2}(y_1+y_2)\right]^{1/2} + v_2$$

6.10 Describe the negative surge waves and prove the formula

$$C = 3\sqrt{gy} - 2\sqrt{gy_1} - v_1$$

6.11 Describe the downstream negative surge.

B. UNSOLVED PROBLEMS

6.1 A rectangular channel, 12.2 m wide, carries 25.47 m³/s of the water. The slope of the channel is 0.00283. At section (1) the depth is 1.37 m and at section (2), 92 m downstream + depth is 1.52 m. What is the average value of roughness factor n?

6.2 A rectangular channel (n = 0.013) is 1.83 m wide and carries 1.87 m³/s of water. At a certain section F the depth is 975 mm. If the slope of the channel bed is constant at 0.000400, determine the distance where the depth is 823 mm. Use single step method [–547 m]

6.3 A uniform flow of 8.1 m³/s occurs at a depth of 1.5 m in a long rectangular channel 3 m wide. Calculate the minimum height of a flat top hump that can be built on the floor of the channel in order to produce a critical depth [0.492 m]

6.4 13.5 m³/s flow in a rectangular channel 7.5 m wide, lined with rubble masonary and laid on a slope of 1/10000. The channel ends in a free outfall and at a point in the channel, the depth is 1.86 m. How far upstream from this point the depth be 1.95 m? Take n = 0.017 [760 m].

6.5 A rectangular channel 7.2 m wide carries a discharge of 15.55 m³/s at the normal depth of 0.9 m, the constant bed slope being 1 in 540. A sluice gate built on the downstream side raises the depth of flow at an upstream point to 1.8 m, the discharge being same as above. Estimate the slope of the water surface at the point with respect to the bed. Also state, giving reasons whether a hydraulic jump may be expected upstream to the sluice. Chezy's formula may be used to determine the slope of the energy grade line [0.00028, dy/dx = 0.00818, y_c = 0.78 m, $y > y_n > y_c$. Mild slope MI, no hydraulic jump occurs].

6.6 A channel of rectangular section 7.5 m wide is laid along a bed sloope of 1 in 64 carries a discharge of 36.45 m³/s. A barrier across the channel raises the water surface to 3 m just upstream. Determine the water surface profile and draw it to a suitable scale. Use step method for the computation of surface curve. Assume Manning's n = 0.026 [77.7 m]

6.7 A long planned timber chute is 3 m wide and carries 4.05 m^3/s. The slope of the flume is 0.02 and the depth of flow at a point is 480 mm. Find normal depth and the point where the depth becomes or is within 1% of normal. How for is it from one point to another point and what type of backwater curve exists. Assuming Manning's n = 0.011 [44.3 m].

6.8 A wide rectangular channel of bed slope 0.0003 and Manning's roughness coefficient 0.02 carries a steady flow of 1.35 m^3/s per metre width of channel; determine whether a hydraulic jump will form downstream and, if so, find the distance from the gate to the jump. Also find the power absorbed in the jump per meter width of the channel. Use step method for calculations length of surface curve upto the jump [Length of surface curve upto jump = 1.8 m, P = 16.758 kW/m width].

6.9 A rectangular channel 30 m wide discharges 60 m^3/s at a uniform flow depth of 2.5 m into a reservori. The levels of water in the channel and reservoir at the reservori inlet are initially equal. Water in the reservoir is released rapidly so that the level falls at the rate of 1 m per hour. Neglecting friction and the channel slope determine the time taken for the level in the channel to fall 0.5 m at a section 1 km upstream from the reservoir. [36.45 min]

6.10 A rectangular channel 4 m wide conveys a discharge of 18 m^3/s at a depth of 2.25 m. Determine the depth and celerity of the positive surge wave resulting from (a) sudden, partial gate closure which reduces the downstream discharge to 10 m^3/s, and (b) sudden total gate closures. [y_2 = 2.815 m, C = 3.54 m/s, y_2 = 3.295 m, C = 4.311 m/s.]

6.11 A rectangular channel 10 m wide, bed slope 0.0001, Manning's roughness coefficient 0.015, receives inflow from a reservoir with a gated inlet. When a steady discharge of 30 m^3/s is being conveyed the gate is suddenly opened to release a discharge of 70 m^3/s. Calcuate the intial celerity and depth of the surge wave. [y_2 = 2.98 m, y_1 = 3.54 m, C = 7.14 m/s].

7

Impact of Free Jets

7.1. INTRODUCTION

A stream of fluid entering in a machine such as a hydraulic turbine, a pump or fan, has more or less defined direction. A force is always required to act upon the fluid to change its velocity either in direction or in magnitude. Newtons third law of motion states that to every action there is an equal and opposite reaction. According to this law an equal and opposite force is exerted by the fluid upon the body that causes the change. This force exerted by virtue of fluid motion is called a dynamic force and must be distinguished from hydrostatic pressure. Where as hydrostatic pressure implies no motion, dynamic force always involves a change in velocity and thus a change in momentum.

The major problem in turbomachinery is to find the power developed (or consumed) by (or in) a particular machine. A turbine produces power while a pump consumes power in order to run. The power is determined from the dynamic force or forces which are being exerted by the flowing fluid on the boundaries of flow passage and which are due to the change of momentum. These are determined by Newtons second law of motion.

When a steady jet of fluid issued from a nozzle, impinges on a solid smooth surface, a thin stream is formed, that glides along the surface upto the boundaries. Then it leaves the surface, approximately tangential to the surface. In doing so, the jet exerts a force on the surface, known as the impact of the jet. It may be evaluated by the impulse momentum principle.

In this chapter, different cases of forces exerted by free jets on stationary and moving plates or vanes of different shapes have been discussed.

7.2. FORCE EXERTED BY FLUID JET ON STATIONARY FLAT PLATE

Assuming no frictional resistance at the plate and no impact loss. The pressure every where is atmospheric and if small changes in elevation are neglected, the application of Bernoullis theorem will yield equal velocities in the two divided jets.

7.2.1. Flat plate normal to the jet

Let a jet of fluid of cross sectional area 'a' strike with a velocity V on a fixed plate as shown in Fig. 7.1. The jet after striking the plate will leave it tangentially i.e. the jet will get deflected through 90°. Hence the component of the velocity of jet in the direction of jet, after striking will be zero.

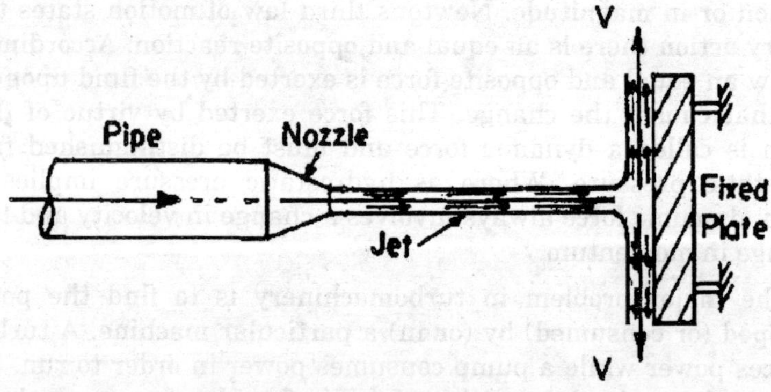

Fig. 7.1. Flat plate normal to the jet

The force exerted by the jet on the plate in the direction of jet.

$$F_x = \dot{m} \text{ (Initial velocity – Final velocity)}$$

$$= \rho a V (V - 0) \text{ [Final velocity after jet striking is zero]}$$

$$= \rho a V^2 \qquad (7.1)$$

If the force exerted on the jet by plate or force exerted by the plate on the jet is in the negative x direction.

7.2.2. Flat plate inclined at an angle θ to the jet

If the surface is inclined at an angle θ to the jet, the jet velocity can be resolved into two components, one normal to the surface and other parallel to it.

Let F_n is the force exerted by the jet on the plate in the direction normal to the plate (Fig. 7.2).

Jet velocity normal to plate v sin θ

$$\text{Normal force } F_n = \rho\ Q\,(V \sin\theta - 0)$$

$$= (\rho\ aV)\,(V \sin\theta)$$

$$= \rho\ aV^2 \sin\theta \qquad (7.2)$$

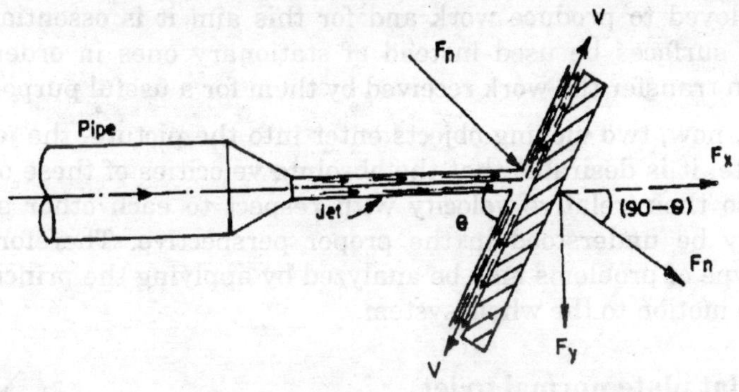

Fig. 7.2. Flat plate inclined at an angle θ

Component of this normal force F_n in the direction of the jet, $F_x = F_n \cos(90\text{-}\theta) = F_n \sin\theta$

$$\text{Therefore, } F_x = \rho\ aV^2 \sin^2\theta \qquad (7.3)$$

And,

$$F_y = \text{Component of } F_n\text{, perpendicular to flow}$$

$$= F_n \cos\theta$$

$$= \rho\ aV^2 \sin\theta \cos\theta \qquad (7.4)$$

The discharge flowing in the two directions can also be calculated by making use of the condition that change of momentum in the direction of plate must be zero. If the total discharge Q is divided into two portions Q_1 and Q_2, then

$$\rho\ Q_1\ V - \rho\ Q_2\ V = \rho\ Q\ V \cos\theta$$

$$Q_1 - Q_2 = Q \cos \theta \quad \text{(a)}$$

$$\text{Also } Q_1 + Q_2 = Q \quad \text{(b)}$$

Solving (a) and (b) the values of Q_1 and Q_2 are obtained as

$$Q_1 = \frac{Q}{2}(1 + \cos\theta) \quad (7.5)$$

$$Q_2 = \frac{Q}{2}(1 - \cos\theta) \quad (7.6)$$

7.3. FORCE EXERTED BY FLUID JET ON MOVING FLAT PLATE

For generation of power the force due to the impact of jet has to be employed to produce work and for this aim it is essential that moving surfaces be used instead of stationary ones in order that they can transfer the work received by them for a useful purpose.

As, now, two moving objects enter into the picture, the jet and the plate, it is desirable that the absolute velocities of these objects and also their relative velocity with respect to each other should correctly be understood in the proper perspective. Therefore, for these type of problems may be analyzed by ápplying the principle of relative motion to the whole system.

7.3.1. Flat plate normal to jet

Fig. 7.3 shows a flat plate held normal to the jet. Let V be the absolute velocity of the jet and u be the velocity of the flat plate.

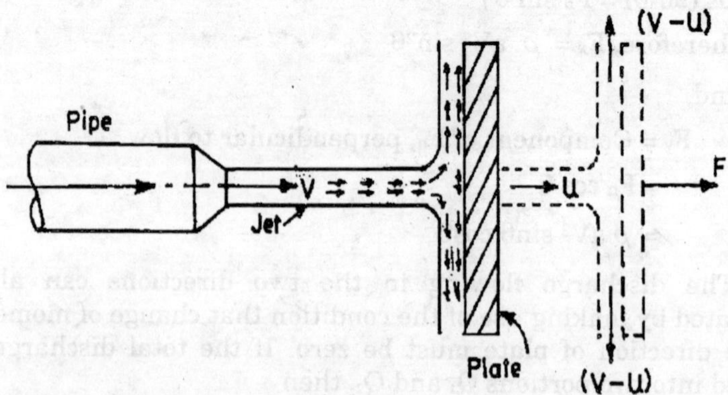

Fig. 7.3. Flat plate normal to jet.

In this case absolute velocity V will not be the effective velocity with which the jet strikes the plate, because as the jet is about to strike the plate, the plate has also moved away from the jet with a velocity u, and only that mass of fluid which really overtakes the plate will be striking it to cause the impinging action. This type of problem may therefore be analyzed by applying the principle of relative motion to the whole system. Therefore, the velocity of the jet relataive to the plate is (V – u).

The mass flow rate striking the plate = ρ Q

$$\dot{m} = \rho a\ (V - u)$$

The dynamic thrust exerted by the jet on the plate in the direction normal to the plate is given by

$$F = \dot{m}\ \text{x change in relative velocity}$$

$$= m\ [(V - u) - 0]$$

$$= \rho a\ (V - u)\ (V - u)$$

$$F = \rho a\ (V - u)^2 \quad (7.7)$$

The rate of doing work by the jet on the plate.

$$= F \times u$$

$$= \rho a\ (V - u)^2\ u \quad (7.8)$$

7.3.2. Flat plate inclined to jet

Fig. 7.4 shows a smooth flat plate inclined at an angle θ to the direction of jet.

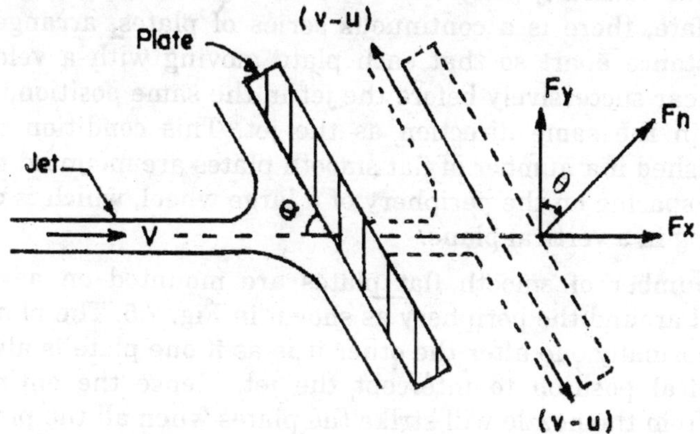

Fig. 7.4. Flat plate inclined to jet.

The force exerted by the jet on the plate in the direction normal to the plate is given as

$$F_n = \rho Q [(V - u) \sin \theta - 0]$$

$$= \rho a (V - u)(V - u) \sin \theta$$

$$= \rho a (V - u)^2 \sin \theta \qquad (7.9).$$

The force F_n can be resolved into two components

$$F_x = F_n \sin \theta$$

$$= \rho a (V - u)^2 \sin^2 \theta \qquad (7.10)$$

and

$$F_y = F_n \cos\theta$$

$$= \rho a (V - u)^2 \sin\theta \cos\theta \qquad (7.11)$$

Work done by the jet on the plate

$$= F_x u$$

$$= \rho a (V - u)^2 \sin^2\theta\, u$$

$$= \rho a u (V - u)^2 \sin^2 \theta \qquad (7.12)$$

7.3.3. Flat plate moving in series

The force exerted on a single flat plate is not practically feasible because the distance between the plate and nozzle will be constantly increasing at the rate of u per second, which in turn would require continuous lengthening of the jet at this rate in order to strike the plate. This difficulty may however, be overcome, if instead of a single plate, there is a continuous series of plates, arranged at a fixed distance apart so that each plate moving with a velocity u, may appear successively before the jet in the same position, and all moving in the same direction as the jet. This condition may be accomplished if a number of flat smooth plates are mounted radially at equal spacing on the periphery of a large wheel, which is capable of rotating in a vertical plane.

A number of smooth flat plates are mounted on a rotating wheel all around the periphery as shown in Fig. 7.5. The plates face the jet normally one after the other it is as if one plate is always in the vertical position to intercept the jet. Hence the entire fluid issuing from the nozzle will strike the plates when all the plates are considered. Therefore,

Mass flow rate of fluid striking the plates,

$$\dot{m} = \rho\, Q = \rho\, aV.$$

Force, $F = \dot{m}\,[(V - u) - 0]$

$$= \dot{m}\,(V - u)$$

$$= \rho\, aV\,(V - u) \qquad (7.13)$$

Work, $W = F \times u$

$$= \rho aV\,(V - u)\,u$$

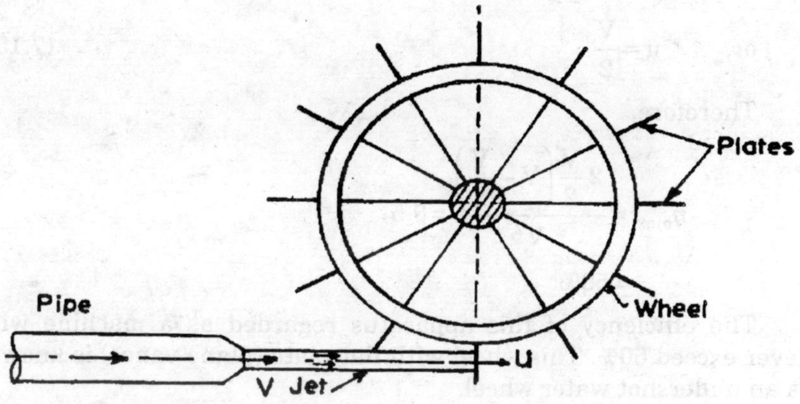

Fig. 7.5. Flat pates mounted on the periphery of a wheel.

The kinetic energy supplied by the jet = Input of the jet to the wheel.

Thus, input of jet $= \frac{1}{2}\dot{m}\,V^2$

$$= \frac{1}{2}\rho\, a\, V\, V^2$$

$$= \frac{1}{2}\rho\, a\, V^2$$

Therefore, efficiency of the wheel

$$\eta = \frac{\text{Output of jet}}{\text{Input of jet}}$$

$$= \frac{\rho\, a\, V\,(V - u)\, u}{\frac{1}{2}\rho\, a\, V^3}$$

$$= \frac{2u(V-u)}{V^2} \tag{7.14}$$

For maximum efficiency, differentiating it with respect to u and equating to zero.

$$\frac{d\eta}{du} = 0$$

Thus, $$\frac{d\eta}{du} = \frac{2(V-2u)}{V^2} = 0$$

or, $$u = \frac{V}{2} \tag{7.15}$$

Therefore,

$$\eta_{max} = \frac{2\frac{V}{2}\left(V - \frac{V}{2}\right)}{V^2} = 0.5$$

$$= 50\%$$

The efficiency of this apparatus regarded as a machine will never exceed 50%. This wheel with flat plate (plane vanes) is known as an undershot water wheel.

7.4. FORCE EXERTED BY A FLUID JET ON STATIONARY CURVED VANE

Modification is made in the construction of the undershot water wheel and the flat plates (straight vanes) are replaced by curved vanes which are bent only in one plane. The jet comes normal to the vane.

On the basis of the direction of the jet striking the curved vane or plane, two distinct cases as discussed below may be considered.

7.4.1. Jet striking a symmetrical stationary curved vane at the centre

Fig. 7.6 shows a jet of water strikes a fixed curved vane at its centre on the concave side with a velocity V. After striking the vane the jet comes out with a same velocity at its both the ends. The vane being smooth the velocity of the leaving jets will also be V. The jet after striking, will be deflected on each side through the angle (180

– θ). Accordingly the component of the velocity of the leaving jet in the direction of flow of incoming jet is V cos (180 – θ) = – V cosθ.

Applying the impulse momentum equation, then

F_x = Force on the vane in the direction of the jet

$= \dot{m}$ x change in velocity

$= \rho\, Q\, [V - (-V \cos\theta)]$

$= \rho\, aV\, (V + V\cos\theta)$

$$= \rho\, a\, V_2\, (1 + \cos\theta) \qquad (7.16)$$

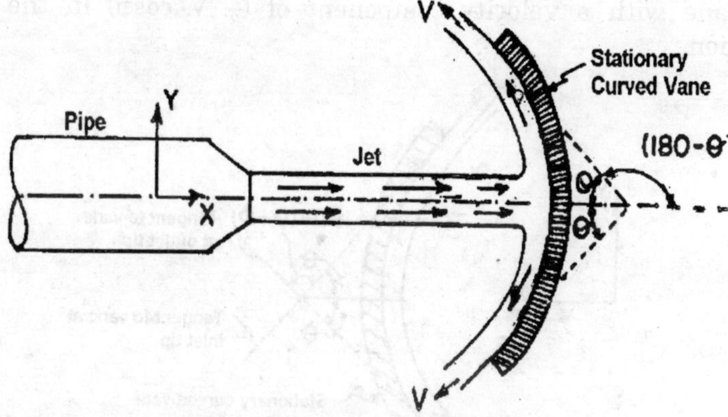

Fig. 7.6. Jet striking a stationary curved vane at its centre.

Due to symmetry of the vane, the net force in a direction perpendicular to the direction of the jet will be zero.

When θ = 90° the curved vane will be reduced to a flat plate, and the expression for F_x then becomes

$$F_x = \rho\, a\, V^2$$

Which is same as previously obtained in equation (7.1). By comparing equations (7.16) and (7.1) it will be observed that the force exerted by a jet in its direction of flow, on a curved vane is always greater than that exerted on a flat plate, because

$$\rho a\, V^2\, (1 + \cos\theta) > \rho\, a\, V^2$$

When θ = 0°, the vane becomes semicircular, the jet deflected back in the incoming direction, and

$$F = 2\rho\, a\, V^2 \qquad (7.17)$$

which indicates that the force exerted by a jet in its direction of flow, on a semicircular vane is twice that exerted on a flat plate, both plates being fixed in position.

7.4.2. Jet striking an unsymmetrical stationary curved vane tangentially at one of the tips

A jet of fluid strikes a fixed curved vane at one end, glides over the vane and then leaves it tangentially as shown in Fig. 7.7. The jet strikes the vane at an angle φ to the axis and leaves at an angle φ, the vane deftects the jet through an angle of 180 – (θ + φ). Velocity of the entering jet in the direction of the axis is V cosθ and it leaves the vane with a velocity component of (– V cosφ) in the same direction.

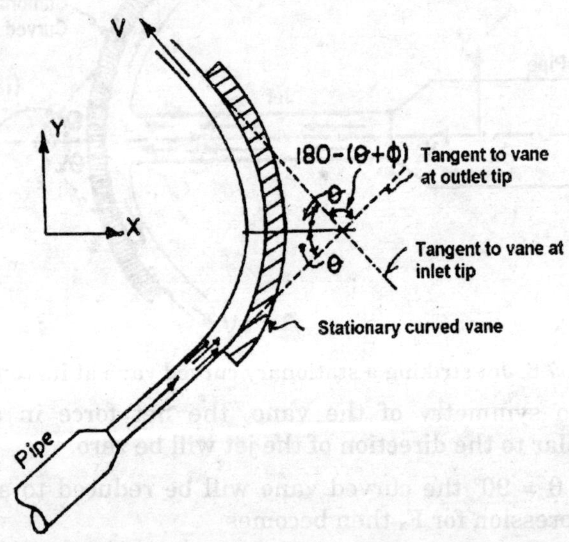

Fig. 7.7. Jet at one end of fixed unsymmetrical stationary curved vane.

Applying the impulse momentum theorem in the x and y direction we get,

$$F_x = \rho\, a\, V\left[V\cos\theta - (-V\cos\theta)\right]$$

$$= \rho\, a\, V^2 (\cos\theta + \cos\phi) \qquad (7.18)$$

and

$$F_y = \rho\, a\, V\,[\,V\sin\theta - (+V\sin\phi)]$$

$$= \rho\, a\, V^2 (\sin\theta - \sin\phi) \qquad (7.19)$$

The resultant force exerted by the jet on the vane will be $F = \sqrt{F_x^2 + F_y^2}$ and its inclination with the x axis is

$$\alpha = \tan^{-1}\left(\frac{F_y}{F_x}\right) \tag{7.20}$$

If the curvature of the vane is same at both inlet and outlet tips i.e., θ = φ, then the expression for F_x is reduced to

$$F_x = 2\,\rho\, a\, V^2 \cos\theta$$

and

$$F_y = 0 \tag{7.21}$$

If the friction and other losses on the vane surface are taken into account, the velocity of the jet at the outlet will be lesser than the velocity at the inlet. Therefore

$$F_x = \rho\, a\, V_1\,(V_1 \cos\theta + V\cos\phi) \tag{7.22}$$

$$F_y = \rho\, a\, V_1\,(V_1 \sin\theta - V_2 \sin\theta) \tag{7.23}$$

In both these cases since the vane is held stationary the wone done is equal to zero.

7.5. FORCE EXERTED BY A FLUID JET ON MOVING CURVED VANE

7.5.1. Curved vane, jet striking at the centre

A jet of fluid strikes at the centre of a curved vane is moving with a velocity u as shown in Fig. 7.8. The fluid glides over the vane and leaves the ends tangentially. The effective velocity with which the jet strikes the vane will be the velocity of the jet relative to the plane i.e. (V – u). The jet leaving the vane will have its velocity equal to the effective velocity of the incoming jet i.e. (V – u). Component of the outlet velocity in the x direction

$$= (V - u)\cos(180 - \theta)$$

$$= (V - u)\cos\theta$$

The force exerted by the jet on the plate in the direction of flow of incoming jet is given by

$$F_x = \rho\, a\,(V - u)\,[(V - u) - [-(V - u)\cos\theta]$$

$$= \rho\, a\, (V-u)^2\, [1+\cos\theta] \tag{7.24}$$

When $\theta = 90°$

$$F_x = \rho\, a\, (V-u)^2$$

Since the vane is symmetrical, net force in the y-direction is zero. Work done by the jet on the vane per second

$$= F \times u$$

$$= \rho\, a\, (V-u)^2\, (1+\cos\theta)\, u \tag{7.25}$$

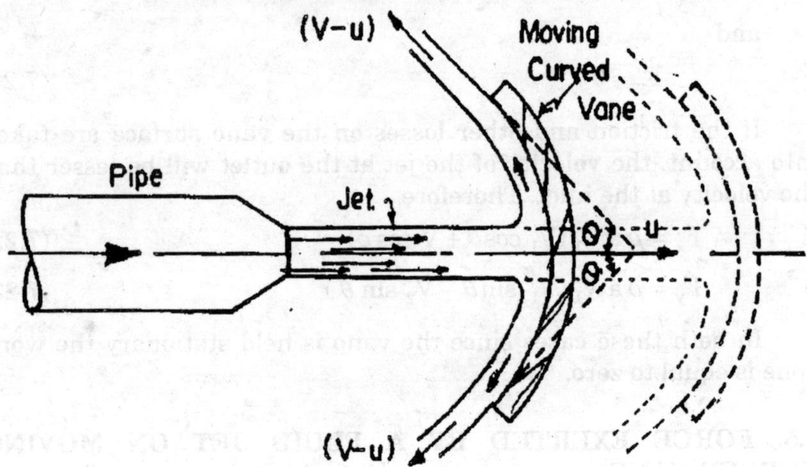

Fig. 7.8. Curved vane, jet strikes at centre.

The kinetic energy supplied by the jet per second.

$$\frac{1}{2}\rho\, a\, V^3 \tag{7.26}$$

The efficiency of the jet

$$\eta = \frac{\text{Output}}{\text{Input}} = \frac{\text{Work done by jet}}{\text{Kinetic energy supplied}}$$

$$= \frac{\rho\, a\, (V-u)^2\, (1+\cos\theta)\, u}{\dfrac{\rho\, a\, V^3}{2}}$$

$$= \frac{2\, u\, (V-u)^2\, (1+\cos\theta)}{V^3} \tag{7.27}$$

For maximum efficiency

$$\frac{d\eta}{du} = 0$$

or, $$\frac{d}{du}\left[\frac{2\,u\,(V-u)^2\,(1+\cos\theta)}{V^3}\right] = 0$$

or, $$\frac{2\,(1+\cos\theta)}{V^3} \cdot \frac{d}{du}\left[u\,(V-u)^2\right] = 0$$

Since, $$\frac{2\,(1+\cos\theta)}{V^3} \neq 0$$

Therefore, $V^2 - 4\,u\,V + 3\,u^2 = 0$

or, $(V-u)\;(V-3u) = 0$

i.e. $V = u$ or $V = 3u$

If, V = u, no force is exerted on the vane and no work is done. Thus to obtain the maximum work,

$$V = 3u \tag{7.28}$$

Substituting in equation (7.27) we get

$$\eta_{max} = \frac{2\,u\,(3\,u-u)^2\,(1+\cos\theta)}{(3\,u)^3}$$

$$= \frac{8}{27}(1+\cos\theta) \tag{7.29}$$

$$= \frac{8}{27}\left(2\cos^2\frac{\theta}{2}\right)$$

$$= \frac{16}{27}\left(\cos^2\frac{\theta}{2}\right) \tag{7.30}$$

If the vane is semi circular, i.e. θ = 0°

$$\eta_{max} = \frac{16}{27} = 0.592 = 59.259\% \approx 59.26\%$$

If instead of a single vane, there is a series of curved vanes are mounted on a wheel rotating about its axis then the mass of water striking the wheel per second is ρaV. Therefore, the force exerted by the jet on the vanes in the direction of flow of jet is given by

$$F = \rho\,a\,V\,[\,(V-u) - (-(V-u)\cos\theta]$$

$$= \rho \, a \, V (V - u)(1 + \cos\theta) \quad (7.31)$$

Work done per second

$$= F \times u$$

$$= \rho \, aV (V - u)(1 + \cos\theta) \, u$$

Kinetic energy of jet per second

$$= \frac{1}{2} \rho \, a \, V (V^2)$$

$$= \frac{1}{2} \rho \, a \, V^3$$

Therefore, the efficiency of the wheel is

$$\eta = \frac{\text{Work done per second}}{\text{Kinetic energy of jet per second}}$$

$$= \frac{2 u \rho \, a \, V (V - u)(1 + \cos\theta)}{\frac{1}{2} \rho \, a \, V^3}$$

$$= \frac{2 u (V - u)(1 + \cos\theta)}{V^2} = 0 \quad (7.32)$$

For, maximum efficiency

$$\frac{d\eta}{du} = 0$$

or $$\frac{d}{du}\left[\frac{2 u (V - u) \, (1 + \cos\theta)}{V^2}\right] = 0$$

From which we get

$$V = 2u \quad (7.33)$$

Substituting in equation (7.32)

$$\eta_{max} = \frac{1 + \cos\theta}{2}$$

For θ = 90°, the curved vanes are reduced to flat plates mounted on the periphery of a wheel. In this case

$$\eta_{max} = 0.5 = 50\%$$

Which is same as obtained previously.

For $\theta = 0°$, the curved vanes will become semicircular. Then the value of $\eta_{max} = 1$ or 100%. This is however a theoretical value of maximum efficiency for a wheel provided with semicircular curved vanes mounted on its periphery. Similar is the principle employed in the case of bucket of a Pelton wheel. The bucket of a Pelton wheel is double semi-ellipsoidal in shape. The jet of water impinges at the centre of the bucket and deflects through 160° to 170° depending upon the particular design.

7.5.2. Jet striking on unsymmetrical curved vane tangentially at one of the tips

When the jet impinges on the vane with velocity V_1, velocity of water over the vane will be equal to the relative velocity of the jet with respect to that of vane and may be found by the difference of vectors of V_1 and u_1 (Fig. 7.9). This velocity is represented by V_{r1}. Henceforth, the analysis is based on the construction of velocity diagrams (or velocity triangles) at inlet and outlet of the vane. Since the vane is now moving, a tangentially entering jet on a fixed vane will no longer remain tangential if admitted at that original angle. Again when the jet enters the vane, the vane keeps on moving and hence the jet inlet angle has to be so adopted that the jet enters the vane without causing any shock on the vane.

Absolute velocity of water is resolved into two components, one parallel to the direction of motion of the vane and known as the velocity of whirl, V_{w1}, since the drum or wheel rotates in that direction; the other normal to the direction of motion of the vane and known as the velocity of flow V_1, since the jet of water flows in and out of the vane in that direction i.e. it is the component of jet velocity which is responsible for flow of water along the vane (Fig. 7.9).

Water slides over the vane and leaves with relative velocity V_{r2}. The absolute velocity V_2 of exit water is found by constructing the velocity triangle at exit (Fig. 7.9).

The angle θ represents the angle between direction of relative velocity and direction of motion of the vane at inlet and φ the same angle at exit. Then, if water is to enter the vane without shock, the angle of the vane at inlet must be made equal to θ (Fig. 7.9).

Let suffices $_1$ and $_2$ indicate velocities at inlet and outlet from the vane respectively.

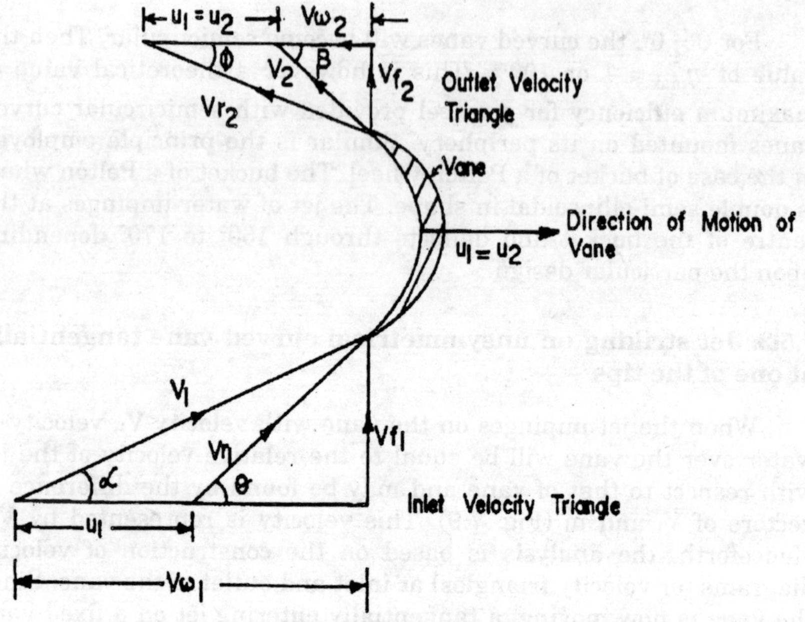

Fig. 7.9. Velocity triangles.

Let,

V_1 = Absolute velocity at inlet

u_1 = Velocity of vane at inlet

V_{r1} = Relative velocity of jet and vane at inlet (It is the vectorial difference between V_1 and u_1)

V_{w1} = The component of the velocity of the jet V_1 in the direction of motion, or velocity of whirl or tangential velocity at inlet.

V_{f1} = The component of the velocity of the jet V_1 perpendicular to the direction of motion or velocity of flow or axial velocity at inlet.

α = Angle with the direction of motion of the vane, at which the jet enters the vane.

β =Angle with the direction of motion of the vane, at which the jet leaves the vane.

θ =Vane anagle at inlet or angle V_{r1} makes with the direction of motion of the vane at inlet.

φ =Vane angle at outlet or angle V_{r2} makaes with the direction of motion of the vane at outlet.

The notation with suffix2 (i.e. V_2, u_2, V_{r2}, V_{f2}, V_{w2}) stand for the outlet triangle.

If the vane is smooth and is having velocity in the direction of motion at inlet and outlet are equal then we have,

$$u_1 = u_2 = u \text{ and } V_{r1} = V_{r2}.$$

It is more convenient to use the velocity triangle approach from which we can calculate velocity of whirl and velocity of flow very easily as comparison to the relative velocity approach. Force on the vane in the direction of the motion is given by

F_x = Mass of fluid flowing per second x change of whirl velocity.

$$F_x = \dot{m} \ (V_{w1} - V_{w2})$$

The direction of V_{w1} and V_{w2} are opposite i.e. β is an acute angle.

$$F_x = \dot{m}\,(V_{w1} + V_{w2}) \qquad (7.34)$$

when β is an obtuse angle

$$F_x = \dot{m} \ (V_{w1} - V_{w2}) \qquad (7.35)$$

when $\beta = 90°$

$$F_x = \dot{m} \ V_{w1} \qquad (7.36)$$

Therefore, general expression for F_x becomes.

$$F_x = \dot{m} \cdot (V_{w1} \pm V_{w2}) \qquad (7.37)$$

The work done per second by the jet on the vane

$$F_x \times u = \dot{m} \ (V_{w1} \pm V_{w2})\, u \qquad (7.38)$$

The case discussed above is not feasible. But the corresponding practical case may be that of a series of such vanes fixed radially to the rim of the wheel. Such types of wheels are usually provided for radial flow hydraulic turbines.

7.6. FLOW OVER RADIAL VANES

In radial flow hydraulic turbines, curved vanes are fixed radialy on a big wheel and the jet enters over the vanes either from the outer periphery towards the inner periphery of the wheel or vice-versa. Thus, while calculating the force or torque produced by the wheel, the variation of radial distance is also to be taken into account.

Consider a series of radial curved vanes mounted on a wheel as shown in Fig. 7.10. The jet of fluid strikes the vanes and the wheel starts rotating at a constant angular speed.

Let,

R_1 = Radius of wheel at entrance

R_2 = Radius of wheel at exit

ω = Angular velocity of wheel

u_1 = Tangential velocity of blade tip at inlet

u_2 = Tangential velocity of blade tip at outlet

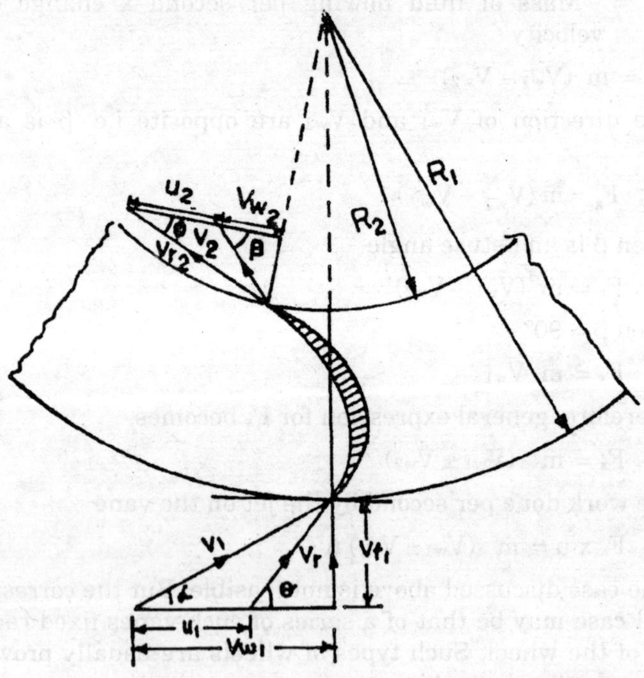

Fig. 7.10. Inlet and outlet velocity triangles on radial vane

Then, $u_1 = \omega R_1 = \dfrac{2\pi N}{60} R_1$

and $u_2 = \omega R_2 = \dfrac{2\pi N}{60} R_2$

Where, N represents the speed of the wheel in rpm. The velocity triangles at the inlet and the outlet tips of the vane may thus be obtained in the same manner as discussed in the article 7.5.

The mass of water striking per second for a series of vanes

$$\dot{m} = \rho Q = \rho\, a\, V_1 \cdot$$

Tangential momentum of fluid striking vanes at inlet

$= \dot{m}$ x component of V_1 in the tangential direction

$= \dot{m} \times V_1 \cos\alpha$

$= \dot{m} \times V_{w1}$

Tangential momentum of fluid striking the vanes at outlet

$= \dot{m}$ x component of V_2 in the tangential direction.

$= \dot{m} \cdot$ x $(-V_2 \cos\beta)$

$= \dot{m}$ x $(-V_{w2})$

$= -\dot{m}\ V_{w2}$

The negative sign is taken as the jet leaves the vane in the direction opposite to the motion of the wheel. Moment of momentum or angular momentum at inlet

$= \dot{m}\ V_{w1}\, R_1$

Moment of momentum or aangular momentum at outlet

$= -\dot{m}\, V_{w2}\, R_2$

Therefore, Torque on wheel = change of moment of momentum

$$T = [\,\dot{m}\, V_{w1}\, R_1 - (-\dot{m}\, V_{w2}\, R_2)]$$

$$= \dot{m}\ (V_{w1}\, R_1 + V_{w2}\, R_2)$$

Work done on the wheel per second or the power delivered to the wheel.

$$W = T \text{ x } \omega$$

$$= \dot{m}\, [V_{w1}\, R_1 + V_{w2}\, R_2]\ \omega$$

$$= \dot{m}\, [V_{w1}\, u_1 + V_{w2}\, u_2] \qquad (7.39)$$

When $\beta > 90°$, then V_{w2} is positive, therefore,

$$W = \dot{m}\, [V_{w1}\, u_1 - V_{w2}\, u_2] \qquad (7.40)$$

By combing the equations (7.39) and (7.40)

$$W = \dot{m}\left[V_{w1}\, u_1\right] \tag{7.41}$$

For radial discharge at the outlet tip of the vane $\beta = 90°$ and $V_{w2} = 0$. Then from equation (7.41)

$$W = \dot{m}\left[V_{w1}\, u_1\right] \tag{7.42}$$

The work done by the jet on the wheel will be the output of the system and the initial kinetic energy of the jet will be the input. Thus the efficiency will be expressed as

$$\eta = \frac{\text{work done per second}}{\text{kinetic energy per second}}$$

$$= \frac{\dot{m}\left(V_{w1}\, u_1 \pm V_{w2}\, u_2\right)}{\dot{m}\, V_1\, 2/2}$$

$$= \frac{2\left(V_{w1}\, u_1 \pm V_{w2}\, u_2\right)}{V_1^2} \tag{7.43}$$

7.7. EULER'S EQUATION OF FLUID MACHINES (Moment of Momentum or angular momentum equation)

If the equation (7.41) is taken for unit mass flow, then

$$E = \left(V_{w1}\, u_1 \pm V_{w2}\, u_2\right) \tag{7.44a}$$

Where, E is the rate of energy transfer per unit mass flow rate. If equation (7.41) is divided by the mass flow rate, then

$$H = \left(V_{w1}\, u_1 \pm V_{w2}\, u_2\right) \tag{7.44b}$$

Where, H is the energy transferred per unit mass, or the change of head of the fluid. Both forms of equation (7.44) have the same units but different symbols.

Equations (7.41) and (7.44a) and (7.44b) are forms of the Euler turbine equation or simply the Euler equation and are the basic relations for all forms of turbomachines, whether they be pumps, compressors or turbines.

The equation in its present form indicates the energy transfer to the wheel by the fluid, which gives motion to the wheel. This is the principle of motion of the turbines ($V_{w1}\, u_1 > V_{w2}\, u_2$, the energy transfer is positive from fluid to wheel).

Negative value of the above equation indicates the energy transfer by the wheel to the fluid, which can be used to raise the pressure energy of the fluid. This principle applies to pumps, compressors ($V_{w1}\, u_1 < V_{w2}\, u_2$, the energy transfer is from wheel to fluid and is negative).

Therefore, now energy transfer equation can be written separately for turbine and pump. For a turbine

$$E = \left(V_{w1}\, u_1 - V_{w2}\, u_2\right) \tag{7.45}$$

$$E = -\left(V_{w1}\, u_1 - V_{w2}\, u_2\right)$$

$$= \left(V_{w2}\, u_2 - V_{w1}\, u_1\right) \tag{7.46}$$

Which is Eulers pump equation.

7.7.1. The physical concept of Eulers equation

The Euler equation in terms of moment of momentum is fundamental, but it is useful to transform it into other forms which are not only convenient in certain aspects of design, but which are helpful in understanding the physical basis of the energy transfer.

From inlet velocity triangle (Fig. 7.10)

$$V_{f1}^2 + V_{w1}^2 = V_1^2$$

or, $$V_{f1}^2 = V_1^2 - V_{w1}^2 \tag{7.47}$$

and $$V_{f1}^2 + \left(V_{w1} - u_1\right)^2 = V_{r1}^2$$

or, $$V_{f1}^2 = V_{r1}^2 - \left(V_{w1} - u_1\right)^2 \tag{7.48}$$

Equating equations (7.47) and (7.48)

$$V_1^2 - V_{w1}^2 = V_{r1}^2 - (V_{w1} - u_1)^2$$

or, $$V_1^2 - V_{w1}^2 = V_{r1}^2 - V_{w1}^2 + 2\, V_{w1}\, u_1 - u_1^2$$

or, $$V_{w1}\, u_1 = \frac{1}{2}\left(V_1^2 + u_2^2 - u_2^2\right) \tag{7.49}$$

Similar procedure is followed for outlet velocity triangle (Fig. 7.10).

$$V_{w2}\, u_2 = \frac{1}{2}\left(V_{r2}^2 - V_1^2 - u_2^2\right) \tag{7.50}$$

Substituting the valves of V_{w1} u_1 and V_{w2} u_2 from equations (7.49) and (7.50) into equation (7.45), we get.

$$E = \frac{1}{2}\left[\left(V_1^2 - V_2^2\right) + \left(u_1^2 - u_2^2\right) + \left(V_{r2}^2 - V_{r1}^2\right)\right] \qquad (7.51)$$

Equation (7.51) is an alternative form of the Euler equation and will be referred to frequently hereafter. The pairs of terms in the round brackets throw light on the nature of the energy transfer and their relative value can abe interpreted to estimate the performance of a machine.

The first term is $\left(V_1^2 - V_2^2\right)/2$ and is readily seen to be the change of absolute kinetic energy or dynamic head occuring in the machine.

The second term $\left(u_1^2 - u_2^2\right)/2$ represents the change of fluid energy due to the movement of the rotation to another, i.e. the centrifugal energy. For an axial flow machine since $u_1 = u_2$ the term is equal to zero.

The third term in equation (7.51) is $\left(V_{r2}^2 - V_{r1}^2\right)/2$ represents a change of kinetic energy due to change of relative velocity. Velocity V_r is relative to the wheel and any change of this velocity results in a change of static head or pressure within the wheel itself, similar to that for the absolute velocity in the casing.

Of the three terms in equation (7.51), then, the first represents a change of dynamic head or pressure and the other two represents a change of static head or pressure. The total head or pressure is thus compounded of a dynamaic component and a static component and the relataive proportion of each may vary considerably in different designs.

ILLUSTRATIVE EXAMPLES

7.1. A nozzle forming at jet of 50 mm diameter is supplied with water under a constant head of 75 m. The jet is directed to strike perpendicularly on to a fixed plane plate. Calculate the force exerted by the jet on the plate. Take velocity coefficient of the nozzle as 0.92.

Solution : Velocity of jet of water

$$V = C_v \sqrt{2gH} = 0.92\sqrt{2 \times 9.81 \times 75} = 35.29 \text{ m/s}$$

See Fig. 7.1.

Change of velocity in the direction of jet

$$V - O = 35.29 - 0 = 35.29 \text{ m/s}$$

Force exerted by the jet on the plate in the direction of jet.

$$F_x = \rho a(V - O)$$

$$= \rho\, a V^2$$

$$= 1000 \times \frac{\pi}{4}\left(\frac{50}{1000}\right)^2 (35.29)^2$$

$$= 2445.3 \text{ N}$$ **Ans.**

7.2. **A 30 mm diameter jet exerts a force of 2000 N in the direction of the jet which is normal to the fixed plate. Find the velocity of jet of water and discharge.**

Solution : See Fig. 7.1.

$$F = \rho a V^2$$

$$2000 = 1000 \times \frac{\pi}{4}(0.03)^2 V^2$$

$$V^2 = \frac{2}{7.06858\ 10^{-4}} = 2829.42$$

$$V = 53.192 \text{ m/s}$$ **Ans.**

$$Q = a\,V$$

$$= \frac{\pi}{4}(0.03)^2 \times 53.192 = 0.0376 \text{ m}^3\text{/s}$$

$$= 37.6 \text{ liters/s}$$ **Ans.**

7.3. **A jet of water 80 mm diameter having a velocity of 18 m/s strikes normally a flat smooth plate. Determine the force on the plate (a) if the plate is at rest, (b) if the plate is moving in the same direction as the jet with a velocity of 4 m/s. Also find the work done per second on the plate in each case and the efficiency of the jet when the plate is moving.**

Soluton

(a) Area of jet $a = \frac{\pi}{4}(0.08)^2 = 5.0265 \times 10^{-3}\ m^2$

$$F = \rho a V^2 = 1000 \times 5.0265 \times 10^{-3} \times (18)^2$$

$$= 1628.6\ N$$ Ans.

$$W = F \times u = 1628.6 \times 0 = 0$$ Ans.

(b) $F = \rho a (V - u)^2$

$$= 1000 \times 5.0265 \times 10^{-3} (18 - 4)^2$$ Ans.

$$= 985.194\ N$$

$$W = F \times u = 985.194 \times 4 = 3940.776\ N\ m/s$$

$$= 3940.78\ Watt$$ Ans.

$$\eta = \frac{F u}{\frac{1}{2}\rho a V^3} = \frac{3940.78}{\frac{1}{2} \times 1000 \times 5.0265 \times (18)^3}$$

$$= 0.26886$$

$$= 26.886\%$$ Ans.

7.4. Jet of water 75 mm in diameter and having a velocity of 25 m/s strikes a fixed plate in such a way that the angle between the jet and the plate is 60°. Find

(a) Maximum force of the jet on the plate

(b) Force in the direction of the jet and

(c) Force normal to the jet direction.

Solution

$$a = \frac{\pi}{4} d^2 = \frac{\pi}{4}\left(\frac{75}{1000}\right)^2 = 4.4178 \times 10^{-3}\ m^2$$

$$\dot{m} = \rho a V = 1000 \times 4.4178 \times 10^3 \times 25$$

$$= 110.4466\ kg/s$$

(a) $F_n = m V \sin\theta$

$$= 110.4466 \times 25 \times \sin 60°$$

$$= 2391.24\ N$$ Ans.

(b) $F_x = F_n \sin\theta = 2391.24 \sin 60°$

$= 2070.874$ N Ans.

(c) $F_y = F_n \cos\theta = 2391.24 \cos 60°$

$= 1195.62$ N Ans.

7.5. **A 75 mm diameter jet having a velocity of 30 m/s, strikes a flat plate, the normal of which is inclined at 30° to the axis of the jet. Calculate the normal force exerted on the plate.**

(a) When the plate is stationary.

(b) When the plate is moving with a velocity of 12 m/s in the direction of the jet.

(c) Find also the work done and efficiency of the jet.

Solution

$$a = \frac{\pi}{4} d^2 = \frac{\pi}{4}\left(\frac{75}{1000}\right)^2 = 4.4178 \times 10^{-3} \text{ m}^2$$

$$\dot{m} = \rho a V = 1000 \times 4.4178 \times 10^{-3} \times 30$$

$$= 132.534 \text{ kg/s}$$

$$\theta = 90 - 30 = 60°$$

(a) $F = \rho a V^2 \sin\theta = \dot{m} V \sin\theta$

$= 132.534 \times 30 \sin 60°$

$= 3443.334$ N Ans.

(b) $F_{r1} = \rho a (V - u)^2 \sin\theta$

$= 1000 \times 4.4178 \times 10^{-3} (30 - 12)^2 \sin 60° = 1239.6$ N

$F_x = F_n \sin\theta = 1239.6 \sin 60° = 1073.5254$ N Ans.

(c) $W = F_x u = 1073.5254 \times 12 = 12882.3$ J/s

$$\text{Energy supplied} = \frac{1}{2}\dot{m}\,.\,V^2 = \frac{1}{2} \times 132.534 \times (30)^2$$

$$\therefore \eta = \frac{\text{Work done}}{\text{Energy supplied}} = \frac{12882.3}{59640.3} = 0.21599 = 21.6\% \text{ Ans.}$$

7.6. **A jet of 30 mm diameter strikes a symmetrical vane at the centre and gets deflected by 165°. If the velocity of the jet is 10 m/s and the vane is smooth, calculate the thrust on the vane. What will be the thrust if the jet strikes the same vane at one end and kaves at the other?**

Solution

$$a = \frac{\pi}{4}\left(\frac{30}{1000}\right)^2 = 7.06858 \times 10^{-4}\ m^2$$

$$\dot{m} = \rho\, a\, V = 1000 \times 7.06858\ \ 10^{-4} \times 10$$

F_x = 7.06858 kg/s

$$F_x = \dot{m}\ \ V\,(1 + \cos\theta)$$

= 7.06858 x 10 (1 + cos 15°)

= 138.963 N **Ans.**

When the jet strikes at one tip.

$$F_x = \dot{m}\ \ V\,(\cos\theta + \cos\theta)$$

$$= 2\,\dot{m}\ \ V\cos\theta$$

= 2 x 7.06858 x 10 x cos 15°

= 136.55 N **Ans.**

7.7. A jet of water 50 mm in diameter, moving with a velocity of 20 m/s, strikes a flat vane, which is normal to the axis of the stream. (a) Find the force exerted by the jet, if the vane moves with a velocity of 8 m/s. (b) Calculate the force exerted by the jet, if instead of one flat vane, there is a series of vanes, so arranged that each vane appears successively before the jet in the same position and always moving with a velocity of 8 m/s.

Solution

$$a = \frac{\pi}{4} d^2 = \frac{\pi}{4}\left(\frac{50}{1000}\right)^2 = 1.963 \times 10^{-3}\ m^2$$

$$\dot{m} = \rho\, a\,(V - u) = 1000 \times 1.963 \times 10^{-3} \times (20 - 8)$$

= 23.556 kg/s

(a) When a single vane is moving

$$F = \dot{m}\ \ (V - u)$$

= 23.556 x (20 – 8)

= 282.672 N **Ans.**

(b) When a series of vanes are moving

F = r a V (V – u)

= 1000 x 1.936 x 10–3 x 20 (20 – 8)

= 471.12 N Ans.

7.8. A jet of water 40 mm diameter enters a fixed curved vane with a velocity of 50 m/s at an angle 20° to the horizontal. Find the normal and tangentiala forces exertal by the jet, if it leaves the vane at an angle of 15° to the horizontal.

Solution

$$a = \frac{\pi}{4} d^2 = \frac{\pi}{4}\left(\frac{40}{1000}\right)^2 = 1.256 \times 10^{-3}\ m^2$$

$$\dot{m} = \rho\, a\, V = 1000 \times 1.256 \times 10^3 \times 50 = 62.83\ m/s$$

$$F_x = \dot{m}\,(V \cos\theta + V \cos\phi)$$

= 62.83 x 50 (cos 20° + cos 15°)

= 3141.5 x (0.93969 + 0.965925)

= 5986.5 N Ans.

$$F_y = \dot{m}\ (V \sin\theta - V \sin\phi)$$

= 62.83 x 50 (sin 20° – sin 15°)

= 62.83 x 50 (0.342 – 0.258)

= 261.376 N Ans.

7.9. A jet of water of diameter 70 mm strikes a curved vane at its centre with a velocity of 15 m/s. The curved plate is moving with a velocity of 6 m/s in the direction of the jet. The jet is deflected through an angle of 165°. Assuming the vane is smooth find.

(a) Force exerted on the vane in the direction of the jet.

(b) Power of the jet and

(c) Efficiency of the jet.

Solution

$$a = \frac{\pi}{4} d^2 = \frac{\pi}{4}\left(\frac{70}{1000}\right)^2 = 3.848 \times 10^{-3} \text{ m}^2$$

$$\dot{m} = \rho\, a\, (V - u) = 1000 \times 3.848 \;\; 10^{-3}\, (15 - 6)$$

$$= 34.692 \text{ kg/s}$$

(a) $F = \dot{m}\,(V - u)\,(1 + \cos\theta)$

$= 34.632 \times (15 - 6)(1 + \cos 15°)$

$= 311.688 \times 1.965925$

$= 612.755$ N Ans.

(b) Power $P = F \times u = 612.756 \times 6 = 3676.53$ Watt.

$= 3.676$ kW Ans.

(c) Efficiency

$$\eta = \frac{P}{\frac{1}{2}(\rho\, a\, V)\; V^2} = \frac{3676.53}{\frac{1}{2} \times 1000 \times 3.848 \times 10^{-3} \times (15)^3}$$

$$= \frac{3676.53}{6493.5} = 0.56618 = 56.618\%$$ Ans.

7.10. **Water flows inwards over a series of curved vanes which are fixed to the rim of a rotating wheel of outer diameter 1500 mm and inner diameter 750 mm. The angle between the jet and the wheel tangent at inlet is 30° and water leaves with a velocity of 10 m/s at an angle of 120° to the wheel tangent. Find the best angles of the vanes and the work done if the relatie velocity at outlet is 0.85 times that at inlet. The velocity of jet is 40 m/s and wheel makes 300 rpm.**

Solution

$$u_1 = \frac{\pi\, D_1\, N}{60} = \frac{\pi \times 1.5 \times 300}{60} = 23.56 \text{ m/s}$$

$$u_2 = \frac{\pi\, D_2\, N}{60} = \frac{\pi \times 0.75 \times 300}{60} = 11.78 \text{ m/s}$$

From the inlet velocity triangle (Fig. 7.10)

$$V_{w1} = V_1 \cos\alpha = 40 \cos 30° = 34.64 \text{ m/s}$$

$$V_{f1} = V_1 \sin\alpha = 40 \sin 30^\circ = 20 \text{ m/s}$$

$$\tan\theta = \frac{V_{f1}}{V_{w1} - u_1} = \frac{20}{34.64 - 23.56} = \frac{20}{11.08} = 1.805$$

$$\therefore \theta = \tan^{-1}(1.805) = 61.0135^\circ$$ Ans.

$$V_{r1} = \sqrt{V_{f1}^2 + (V_{w1} - u_1)^2} = \sqrt{(20)^2 + (11.08)^2}$$

$$= \sqrt{400 + 122.76} = \sqrt{522.7664}$$

$$= 22.864 \text{ m/s}$$

$$V_{r2} = 0.85\ V_{r1} = 0.85 \times 22.864 = 19.434 \text{ m/s}$$

From outlet velocity triangles (Fig. 7.10)

$$V_{w2} = V_2 \cos\beta = 10 \cos 60^\circ = 5 \text{ m/s}$$

$$\cos\phi = \frac{V_{w2} + u_2}{V_{r2}} = \frac{5 + 11.78}{19.434} = \frac{16.78}{19.434} = 0.8634$$

$$\phi = \cos^{-1}(0.8634) = 30.29^\circ$$

$$\text{Work done} = V_{w1}\, u_1 - V_{w2}\, u_2$$

$$= 34.642 \times 23.56 - 5 \times 11.78$$

$$= 816.165 - 58.9 = 757.265 \text{ J/kg}$$ Ans.

$$\text{Energy supplied} = V_r^2/2 = \frac{(40)^2}{2} = 800 \text{ J/kg}$$

$$\eta = \frac{\text{Work done}}{\text{Energy supplied}} = \frac{757.265}{800} = 0.94658 = 94.658\%$$ Ans.

7.11. A jet of water having a velocity of 40 m/s impinges on series of vanes moving with a velocity of 20 m/s. The jet makes an anagle of 30° to the direction of motion of the vanes when entering, and leaves at an angle of 120°. Draw velocity triangles at inlet and outlet and find

(a) Vane angles at inlet and outlet.

(b) Work done.

(c) Efficiency.

Solution

$$V_{f1} = V_1 \sin\alpha = 40 \sin 30^\circ = 20 \text{ m/s}$$

$$V_{w1} = V_1 \cos\alpha = 40 \cos 30^\circ = 34.64 \text{ m/s}$$

$$\tan\theta = \frac{V_{f1}}{V_{w1} - u_1} = \frac{20}{34.64 - 20} = \frac{20}{14.64} = 1.366$$

$$\theta = \tan^{-1}(1.366) = 53.79^\circ$$

$$V_{r1} = \sqrt{V_{f1}^2 + (V_{w1} - u_1)^2} = \sqrt{(20)^2 + (14.64)^2}$$

$$V_{r2} = V_{r1} = 24.78 \text{ m/s} \qquad \text{[No friction on the blades]}$$

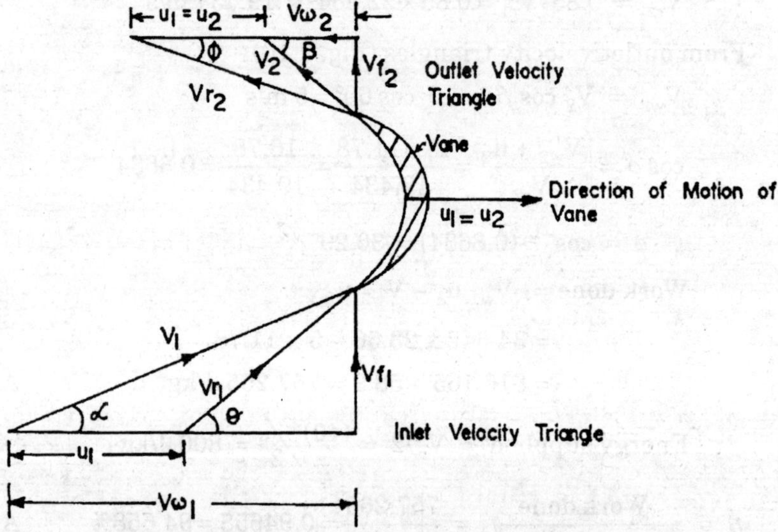

Fig. 7.11. Velocity triangles P. 7.11

Applying sine rule to the outlet velocity triangles.

$$\frac{V_{r2}}{\sin(180 - \beta)} = \frac{u_2}{\sin(\beta - \phi)}$$

$$\sin(\beta - \phi) = \frac{u_2}{V_{r2}} \sin(180 - \beta) = \frac{20}{24.78} \sin 60^\circ = 0.69897$$

$$\beta - \phi = \sin^{-1}(0.69897) = 44.34^\circ$$

$$\phi = 60 - 44.34 = 15.65^\circ$$

$$V_{w2} = V_{r2} \cos\phi - u_2 = 24.78 \cos 15.65^\circ - 20$$

$$= 23.86 - 20 = 3.86 \text{ m/s}$$

$$\text{Work done} = (V_{w1} + V_{w2})\ u_1 = (34.64 + 3.86) \times 20$$

$$= 770 \text{ J/kg.}$$ **Ans.**

$$\text{Efficiency } \eta = \frac{\text{Work done}}{\text{Energy input}}$$

$$= \frac{2\,(V_{w1} + V_{w2})\,u_1}{V_1^2}$$

$$= \frac{2 \times 770}{(40)^2} = 0.9625 = 96.25\%$$ **Ans.**

EXERCISES

A. THEORY

7.1. Derive expressions for the force exerted by fluid jet on stationary flat plate

(a) normal to the jet

(b) inclined to the jet.

7.2. Show that for a flat plate normal to the jet

$$F = \rho\, a\, V^2$$

7.3. Show that for a flat plate inclined at an angle θ to the jet

$$F = \rho\, a\, V^2 \sin^2\theta$$

7.4. Derive expressions for the force exerted by fluid jet on moving flat plate

(a) normal to the jet (b) inclined to the jet

7.5. Show that the rate of doing work by the fluid jet on the moving flat plate

$$W = \rho\, a\ (V - u)^2\, u$$

7.6. Show that the rate of doing work by the fluid jet on moving flat plate inclined to jet

$$W = \rho\, a\ (V - u)^2 \sin^2\theta$$

7.7. Show that the force exerted by the jet on the plate in the direction normal to the plate is

$$F_n = \rho a (V - u)^2 \sin\theta$$

7.8. Prove that the efficiency of the wheel

$$\eta = \frac{2u(V - u)}{V^2}$$

7.9. Show that for flat plate moving in series for maximum efficiency is given by

$$V = 2u$$

7.10. Derive expressions for jet striking an unsymmetrical stationary curved vane tangentially at one of the tips.

$$F_x = \rho a V^2 (\cos\theta + \cos\phi)$$

$$F_y = \rho a V^2 (\sin\theta - \sin\phi)$$

7.11. Derive an expression for force exerted by a fluid jet on moving curved vane

$$F_x = \rho a (V + u^2)(1 + \cos\theta)$$

7.12. Prove that the efficiency for curved vane, jet strikes at centre

$$\eta = \frac{2u(V - u)^2 (1 + \cos\theta)}{V^3}$$

7.13. Prove that for maximum efficiency condition for curved vane, jet strikes at centre is given by

$$V = 3u$$

7.14. Prove that the maximum efficiency for curved vane, jet strikes at centre is given by

$$\eta_{max} = \frac{8}{27}(1 + \cos\theta) = \frac{16}{27}\cos^2\frac{\theta}{2}$$

7.15. Prove that for maximum efficiency conditon

$$V = 3u$$

and $\eta_{max} = \dfrac{8}{27}(1 + \cos\theta)$

7.16. (a) Show that the rate of doing work by the fluid on a series of moving curved vane mounted on a wheel is given by

$$W = \rho a V (V - u)(1 + \cos\theta) u$$

(b) $$\eta = \frac{2u (V - u)(1 + \cos\theta)}{V^2}$$

7.17. Show that for the maximum efficiency condition for series of moving curved vane mounted on a wheel is given by

$$V = 2u$$

and the value of maximum efficiency is

$$\eta_{max} = \frac{1 + \cos\theta}{2}$$

7.18. Find the force of jet striking at the centre of a moving curved vane. What will be the work done and efficiency of the system? What will be the effect of mounting the vanes on a rotating wheel?

7.19. Determine the force exerted by a jet on a moving curved vane, when the jet strikes the vane at its one end tangentially. Also find an expression for efficiency of the system.

7.20. Prove that the work done per second on a series of moving curved vanes by a jet of water striking at one of the tips of the vaane is given by

$$W = \dot{m} u (V_{w1} \pm V_{w2})$$

7.21. Show that for a curved radial vane, the work done per second is given by

$$W = \dot{m} \left[V_{w1} u_1 \pm V_{w2} u_2\right]$$

7.22. Find an expression for the efficiency of a series of moving curved vanes when a jet of water strikes the vanes at one of the tips. Prove that maximum efficiency is when V = u and the value of maximum efficiency is 50%.

7.23. Show that the efficiency of a free jet striking normally on a series of flat plates mounted on the periphery of a wheel can never exceed 50 per cent.

7.24. Derive an equation of moment of momentum applicable to turbomachines for the calacultion of theoretical energy transfer. Transform the equation into the form which consists of centrifugal and other effects. Explain the physical significance of each term and discuss the term degree of reaction.

7.25. Prove that the conditions for best efficiency are: (i) The vane speed is half the jet speed and (ii) The angle of deviation is 180°. Explaian why these conditions are not fulfilled in practice.

7.26. Derive Euler's fundamental equation. Show also that for equation to be valid flow in front of runner and beyond the runner is that of free vortex.

7.27. What is the difference between an "Impulse force" and a "Reaction force"? Is the classification of hydraulic turbines into impulse type and reaction type anything to do with the above nomenclature? What is meant by degree of reaction?

7.28. A jet having a velocity V strikes a curved vane moving in the same direction as the jet with velocity u, so that the velocity of jet relative to vane is (V – u). The vane causes the jet to be reversed in direction. In flowing over the vane, the relative velocity head is reduced by 23%. Show that maximum velocity is obtained when V = 3u and this efficiency is 55.62%.

7.29. Prove that the maximum efficiency for semicircular vane is 59.26%.

7.30. Explain with reason that impact of a free jet gives greater efficiency on a curved vane than on a flat plate.

7.31. Derive the angular momentum equation for the flow of liquid in a radial turbine.

7.32. Show that the efficiency of a free jet striking normally a series of flat plates mounted on the periphery of a wheel never exceeds 50%.

7.33. For a series of curved vanes mounted radially on the periphery of a wheel show with usual notations.

$$\frac{V_1^2 - V_2^2}{2} + \frac{u_1^2 - u_2^2}{2} + \frac{V_{r2}^2 - V_{r1}^2}{2} = \left[V_{w1}\, u_1 - V_{w2}\, u_2\right]$$

Explain the nature and their relative value for the estimation the performance of a machine.

7.34. Show that, theoretically and under ideal conditions, the maximum achievable efficiency of a free jet striking a series of hemispherical vanes mounted on a rotating wheel periphery, can be 100%.

Enumerate the reasons due to which this cannot be achieved in practice.

B. UNSOLVED PROBLEMS

7.1. A 25 mm diameter jet exerts a force of 1.7658 kN in the direction of jet against a plane and smooth plate which is fixed such that it makes an angle of 30° with the axis of the jet. Find the discharge striking the plate [59 liters/s]

7.2. A jet of water 50 mm diamaeter, having a velocity of 9 m/s strikes normally a plane smooth plate. Find the thrust on the plate when the plate is stationary. What will be the thrust if the plate be moving in the direction of jet, with a velocity of 3 m/s? Find the work done on the plate in each case. [(a) 0 (b) 211.896 J/s]

7.3. Find the force exerted by a jet of water against a flat plate held normal to the axis of the jet if the diameter and the velocity of the jet are 40 mm and 25 m/s respectively [785 N].

7.4. A jet of water 60 mm in diameter strikes a flat plate at an angle of 60° to the plate. If the velocity of the jet is 20 m/s, find the force on the plate normal direction, if it is stationary. Also find the components of force in the direction of the jet and perpendicular to it. [979.45 N, 848.23 N, 489.73 N].

7.5. A jet of water 25 mm diameter strikes a flat plate normally at 30 m/s at a point 150 mm below the top of the plate. What force should be applied, 100 mm below the axis of the jet, in order to keep the plate vertical [265.14 N].

7.6. A rectangular plate weighing 40N is suspended vertically by a hinge on the top horizontal edge. The centre of gravity of the plate is 400 mm from the hinge. The horizontal jet of water 20 mm in diameter whose axis is 600 mm below the hinge impinges normally on the plate with a velocity of 6 m/s. Find the horizontal force, applied at the centre of gravity of the plate to maintain it in vertical position.

Further find the alteration in the velocity of the jet if the plate is deflected through 30° and the same horizontal force continu- es to act at the centre of gravity of the plate [16.96 N, 8.58m/s].

7.7. A square plate weighing 12 kg and of uniform thickness and 300 mm wide is hung, so that it can swing freely about its upper horizontal edge. A horizontal jet 20 mm diameter and having velocity of 15 m/s impinges on the plate. When the plate is vertical, the jet strikes the plate normally at its centre. Find the force, which must be applied at the lower edge of the plate, in order to keep the plate vertical. If the plate is allowed to swing freely, find the inclination to the vertical, which the plate will assume under the action of jet [35.316 N, 36°52'].

7.8. A 75 mm diameter jet, having a velocity of 12 m/s impinges on a plane smooth plate, the normal of which is inclined at 60° to the axis of the jet. Find the impact at right angles to the plate when it is stationary. What will be the impact if the plate moves with a velocity of 6 m/s in the direction of jet and away from jet? Also find the power and efficiency of the jet when the plate is moving. [317.84 N, 79.46 N, 0.238 kW, 6.25%].

7.9. A jet of water strikes a fixed symmetrical curved plate at the centre. The diameter of the jet is 80 mm and the velocity is 50 m/s. Calculate the force of the jet in the direction of the jet if it is deflected through an angle of 130° by the curved plate. [20.644 kN].

7.10. A jet of water delivers 110 liters/s on a curved vane, with a velocity of 12 m/s; the jet enters this curved vane (in the form of an angle) tangentially at the inlet tip; the vane is moving in the direction of jet with a velocity of 6 m/s. If the (inside) angle of the vane is 135°, find the impact on the vane in the direction of motion. Through what anagle will the jet be deflected (a) When the vane is stationary and (b) What it is in motion under its own velocity of 6 m/s, the jet coming out a the outlet tip? [193.25 N, 45°, 22.5°].

7.11. A jet of water 25 mm in diameter having a velocity of 6 m/s impinges on the concave side of a fixed surface of revolution which is symmetrical with regard to the axis of the jet. The jet is deflected through an angle of 120°. Find the dynamic thrust exerted by the jet on the concave surface. [26.48 N].

7.12. A horizontal jet of water 25 mm diameter, moving with a velocity of 6 m/s is intercepted by a curved vane which moves in the same direction as jet at 1.5 m/s. The vane, when stationary deflects the jet through 60°. Find the magnitude of horizontal and vertical component of the dynamic thrust on the vane. [4.97 N, 8.6328 N].

7.13. A 15 mm diameter nozzle having $C_v = 0.97$, is supplied with water under a head of 30 m. The jet impinges on a fixed curved vane, the water gliding on the vane tangentially and being deflected through 165°. Calculate the force on the vane in the direction of the jet, if

(a) there is no friction,

(b) the velocity of water leaving the vane is 0.8 of its impinging velocity. [192.379 N, 173.475 N]

7.14. A jet of water, discharging 23 liters/s with velocity of 21 m/s, impinges on a series of curved vane moving at 12 m/s in a direction making 25° to that of the jet. Determine the inlet angles of the vane so that the water enters the vanes without shock. If the water leaves the vanes at an angle of 150° to the direction of motion (when the vanes are stationary), calculate the force in the direction of motion assuming that all the water is effective and that its velocity relative to vane is reduced by 20% during its passage over the vane. [51°47', 340.4 N].

7.15. A jet of water having a velocity of 15 m/s impinges tangentially on a stationary vane whose section is in the form of a circular arc with a subtended angle (hence deflection angle) of 120°. Find the magnitude and direction of resultant thrust on the vane, when the discharge from the jet is 0.45 liters, assuming that the speed of water relative to the vane remains constant as water passes over the vane. If now the vane is allowed to move with a velocity of 6 m/s in the direction of jet, find the power developed for the same rate of discharge, issuing from the jet. [11.67 N, 30°7', 0.022 kW].

7.16. Water flows over a series of curved vanes of hydraulic turbine, the diameter of wheel between inlet tips of vanes is

1.8 m, that between outlet tips is 0.9 m, and the wheel rotates at 200 rpm. Jet enters, at an angle of 30° to the tangent to wheel at inlet, with a velocity of 40 m/s and leaves with a velocity of 1.5 m/s at an angle of 150° to the tangent to wheel at outlet tip. Find vane angles. [51°39', 4°].

8

Hydraulic Prime Movers

8.1. INTRODUCTION

Hydropower is extracted from the natural potential of usable water resources, and about one quarter of the worlds power requirement is at present derived in this way.

One of the most important factors affecting any hydropower development is the cost of the scheme. With the rising costs and shortage of resources, economic comparisons with various energy sources (e.g. thermal plants - oil, coal, nuclear etc.) have to be made. The fuel costs of hydroelectric plant are virtually nil. The fuel costs for power generation by a conventional thermal plant rise at least with inflation, makes hydroelectric plants economically advantageous, particularly in the long term. Cost of generation per kWh for hydro station and thermal station are in the ratio of 1 : 3.

Hydroelectric power projects involve a large amount of civil engineering construction work. One might imagine that hydroelectric power should be very cheap as water does not cost anything and there is no fuel cost. This, however, is not true. To store water at enough head, it is necessary to have a dam and civil engineering construction works. This costs money and increases investment costs and fixed costs which are not required for other types of power plants.

The latest technology advancement in hydroelectric power generation permit the selection of proper designs incorporating environmental and social requirements. Recent innovations in the development of special turbine-generator units, such as bulbs and slant axis units, also suggests considerable reductions in construction costs, thus augmenting the benefits-costing of a project. Therefore, the selection of the final design requires the comparison of many alternative proposal from several sites, which may incorporate expansion facility to meet future demands.

Recent estimates indicate that the total world hydropower production today is about 17% of the usable potential and will have risen to about 80% by the year 2020.

Installed capacities of total power supply and their hydropower components in various countries with the greatest percentage are : Norway (99.5%), Zambia (97.7%), Iceland (96%), the Netherlands (90%), Brazil (88%), Switzerland (87%), Morocco (85%), and Luxemburg (81%), whereas, countries such as Australia (28%), India (42%), Korea (15%), the U.K. (3.2%), the USA (13%) and the USSR (21%) have developed very limited hydropower schemes, despite having the greatest tidal potential along their coastlines.

8.2. HEADS

Head is the difference in elevation between two levels of water. Head can be characterised as: (a) Gross head and (b) Net or effective head.

8.2.1. Gross Head

The gross head (H_g) on a hydroelectric plant is the difference between head water elevation and tail water elevation. When the plant is not in operation. The head water elevation means the reservoir behind the dam.

$$\text{Gross head } (H_g) = \text{Head water elevation} - \text{Tail water elevation} \quad (8.1)$$

The gross head varies with conditions of flow in the stream and may be very different at time of minimum stream flow from what it is when the stream is in flood. Because of the variable in flow and operating conditions of the plant these levels vary.

Only a part of the gross head can be utilized in producing power. There are losses of head in the penstock and conduits or headrace supplying the turbines, losses in water passages of the plant, and losses in the tail race.

8.2.2. Net Head

The net head or effective head (H) on a hydroelectric plant is the gross head minus all the losses of head that may occur as water flows from the head race to the entrance of the turbine. The losses of head are mainly due to friction in penstocks, canal etc. Thus

$$H = H_g - h_f \qquad (8.2)$$

For an impulse turbine as shown in Fig. 8.1(a) the net head is given by

$$H = \left(\frac{P_1}{\rho g} + \frac{V_1^2}{2g} + Z_1\right) - Z_2 \qquad (8.3)$$

For a reaction turbine as shown in Fig. 8.1(b) the net head is given by

$$H = \left(\frac{P_1}{\rho g} + \frac{V_1^2}{2g} + Z_1\right) - \left(\frac{V_2^2}{2g} + Z_2\right) \qquad (8.4)$$

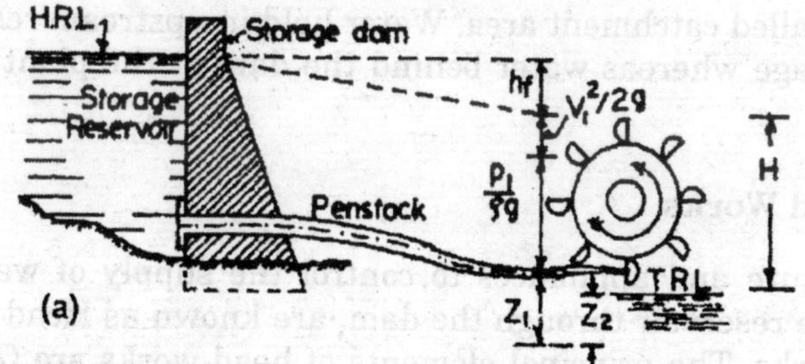

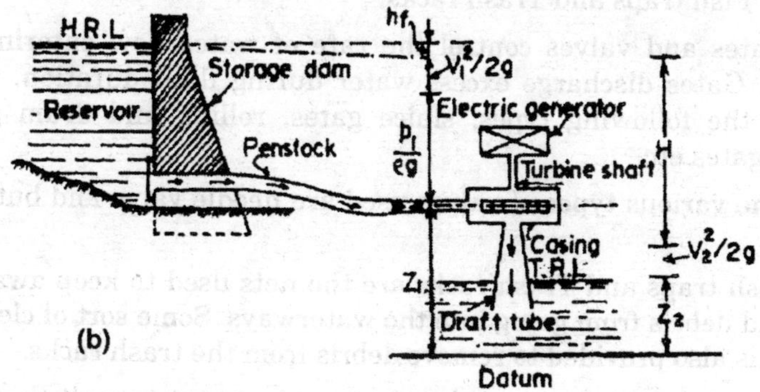

Fig. 8.1. General layout of hydro electric plant (a) Impulse turbine, (b) Reaction turbine.

8.3. COMPONENT OF HYDROPOWER PLANTS

The various appurtenances connecting the construction and operation of hydroelectric power plants are; dams (storage or diversion type control works); gates; valves; intakes; waterways (open channel or low pressure tunnel (head race); high pressure pipes (penstocks); fish passes; pressure relief valves; surge tanks or forebays; turbine governors; generators; the power house superstructure; a switch yard for transformers and switching equipment, cranes and finally transmission lines. The only accessories briefly described here are those that are of direct concern to the power plant engineer.

8.3.1. Water Reservoir

A water reservoir can be either natural or artificial. A natural reservoir is a lake in high mountains and an artificial reservoir is made by constructing a dam across the river.

The area behind the dam, which collects rain water, drain into a river is called catchment area. Water held in upstream reservoir is called storage whereas water behind the dam at the plant is called pondage.

8.3.2. Head Works

Structure and appliances to control the supply of water from the storage reservoir through the dam, are known as Head works or control works. The principal elements of head works are Gates and valves, Fish traps and Trash racks.

Gates and valves control the rate of water flow entering the intake. Gates discharge excess water during flood duration. Gates are of the following types; sluice gates, rolling and drum gates, radial gates etc.

The various types of valves used are needle valve and butterfly valves.

Fish traps and Trash racks are the nets used to keep away the fish and debris from going into the waterways. Some sort of cleaning device is also provided to remove debris from the trash racks.

In cold countries, some heating arrangement to melt the ice at the inlet is also provided.

8.3.3. Dams

Dam is a structure erected on suitable site to fulfills two fundamental functions namely; it (i) develop a reservoir which has a capacity to storewater, and (ii) builds up head and thus, potential for the river water. Dams may be of different types such as earth dam, masonary dam, timber dam and steel dam. Masonary dams are very common.

8.3.4. Water Ways

Waterway is a passage through which the water is carried from the storage reservoir to the power house. It may consist of tunnels, canals, flumes, forebays or surge tanks and penstocks.

(a) Tunnel – is made by cutting the mountains where canal or pipe line cannot be used due to topography. Tunneling provides a direct and short route for the water passages.

(b) Forebay – The open waterway usually terminates in a forebay which is an enlarged body of water from which the penstock convey the water to the turbines in the power house.

The purpose of a forebay is to store water rejected when the load on the plant is reduced and to supply water for initial increment of an increasing load while the water in the canal or pipe line is being accelerated. This may be developed by enlarging the channel just upstream of the intake for the penstocks.

(c) Penstocks – The penstocks are pipes of large diameter, usually of steel or concrete, used for conveying water from the source (reservoir or forebay) to the power house. They are usually high pressure pipe lines designed to withstand stresses developed due to static and water hammer pressures created by sudden changes in power demands (i.e. valve closures and openings according to power rejection and demand). The provision of such a high pressure line is very uneconomical if it is too long, in which case it can be divided into two parts, a long low pressure conveyance (tunnel) followed by short high pressure pipe line (penstock) close to the turbine unit, separated by a surge chamber which absorbs the waterhammer pressure rises and converts them into mass oscillations.

(d) Surge tanks – Surge tank is a structure introduced in the system in between the dam and the power house. Their primary purpose is the protection of the long pressure tunnel in medium and high head plants against high waterhammer pressure arising from sudden rejection or acceptance of load. The surge tank converts these fast (water hammer) pressure oscillations into much slower- and lower pressure fluctuations due to mass oscillation in the surge chamber.

The surge chamber (Fig. 8.2) dividing the pressure tunnel into a short high pressure penstock downstream and a long low pressure tunnel upstream thus functions as a reservoir for the absorption or delivery of water to meet the requirements of load changes. It quickly establishes the equilibrium of the flow conditions, which greatly assists the speed regulation of the turbine.

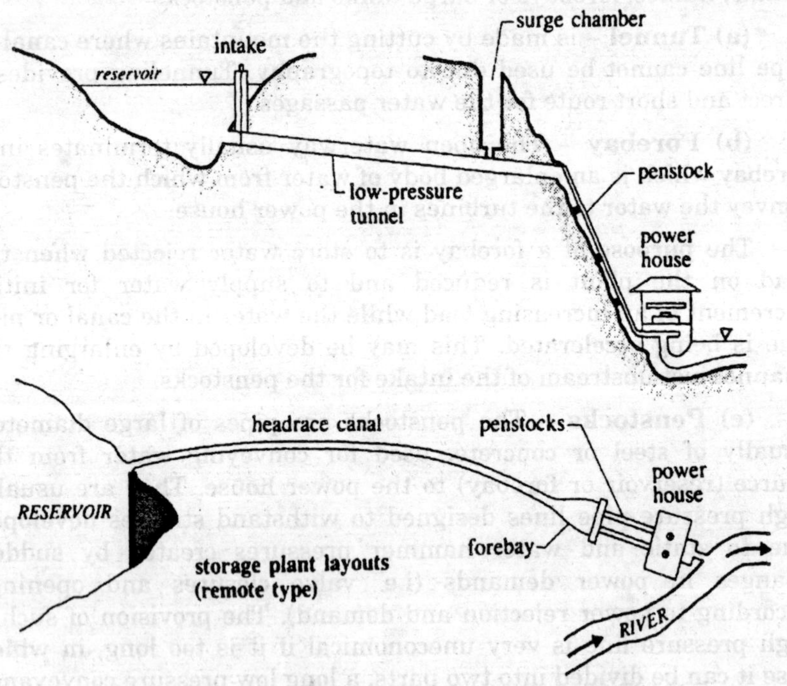

Fig. 8.2. Storage plant layouts (Ref. NOVAK)

If power house is located close to the head works, surge tanks are not required. In this case forebay naturally provided storage which is able to absorb the flow variations due to variations in system loads (Fig. 8.2).

(e) Power House – A power house consists of two sections, the substructure and the superstructure. The substructure supports the hydraulic and electric equipment and the superstructure housing the equipments. Superstructure mostly is a building, housing all operating equipments.

The power house is a building in which the turbines, alternators and auxiliary plant are housed. Here conversion of energy of water to electrical energy takes place.

8.3.5. Tail Race and Outlet Water Way

Tail race is a passage for discharging the water leaving the turbine into the river and in certain cases, the water from the tail race can be pumped back into the original reservoir. Water after doing work on turbine (reaction type) runner passes through the draft tube to tail race. The water held in the tail race is called as tail race water level.

8.3.6. Turbine Governors

The governor is a mechanism controlling the rotational speed of the turbogenerator unit. Constant speed must be maintained in order to obtain the a.c. supply with constant frequency. As the turbine and hence its interconnected generator tend to decrease or increase speed as the load varies, the maintenance of an almost constant speed requires regulation of the amount of water allowed to flow through the turbine by closing or opening the gates (or nozzles) of the turbine automatically, through the action of governor.

8.3.7. Generators

The generator is an electric machine coupled to the turbine shaft (either horizontally or vertically). The alternating current (a.c.) synchronous generator is widely used in hydro-electric power production practice. The turbine is governed to operate at constant speed and the generator is designed with the appropriate number of poles to produce the required frequency at the selected speed.

8.3.8. Transformer and transmission lines

The transformers connecting the power source (generators) and the receiving circuit (transmission lines) step up the voltage for transmission, thus reducing the power loss and permitting the use of smaller conductors (cables) in the transmission lines. The transformers are usually located in outside switchyards adjacent to the power house, as a necessary precaution to avoid high voltage and other hazards.

8.4. HYDRAULIC TURBINES AND THEIR SELECTION

Hydraulic turbines may be consider as prime movers of a water power development. Hydraulic turbine convert water energy (hydropower) into mechanical energy (shaft power). The shaft power developed is used in running electricity generators directly coupled to the shaft of the turbine, thus producing electric power. The power can also be used for direct mechanical purposes. Other source of weater power are waves and tides.

A hydraulic turbine is a rotary machine actuated by the impulse and/or reaction of a water current attacking the rotor (called the runner) which consists of a series of buckets or curved vanes or blades.

8.4.1. Impulse and Reaction turbines

Modern hydraulic turbines are basically divided into two main classes.

(a) Impulse turbine: In the impulse turbine all of the available potential energy (head) of the water is converted into kinetic energy with the help of contracting nozzle provided at the delivery end of the pipeline (penstock). After impinging on the curved buckets the water discharges freely (at atmospheric pressure) into the downstream channel (called tail race). The most commonly used impulse turbine is the Pelton wheel.

(b) Reaction turbine: In the reaction tubines only a part of the available energy of the water is converted into kinetic energy at the entrance to the runner, and a substantial part remains in the form of pressure energy. The runner casing (called the scroll case) has to be completely air tight and filled with water throughout the operation of the turbine. After leaving the runner, the water enters a draft tube which delivers the flow to the tail race. The most commonly used reaction turbines are Francis and Kaplan turbine.

8.4.2. Classification according to direction of flow

Turbines may also be classified according to the main direction of flow of water in the turbine as:

(a) Tangential flow turbines: When the flow is tangential to the wheel circle or pitch circle or mean diameter of wheel. The Pelton turbine is a tangential flow turbine (Fig. 8.3).

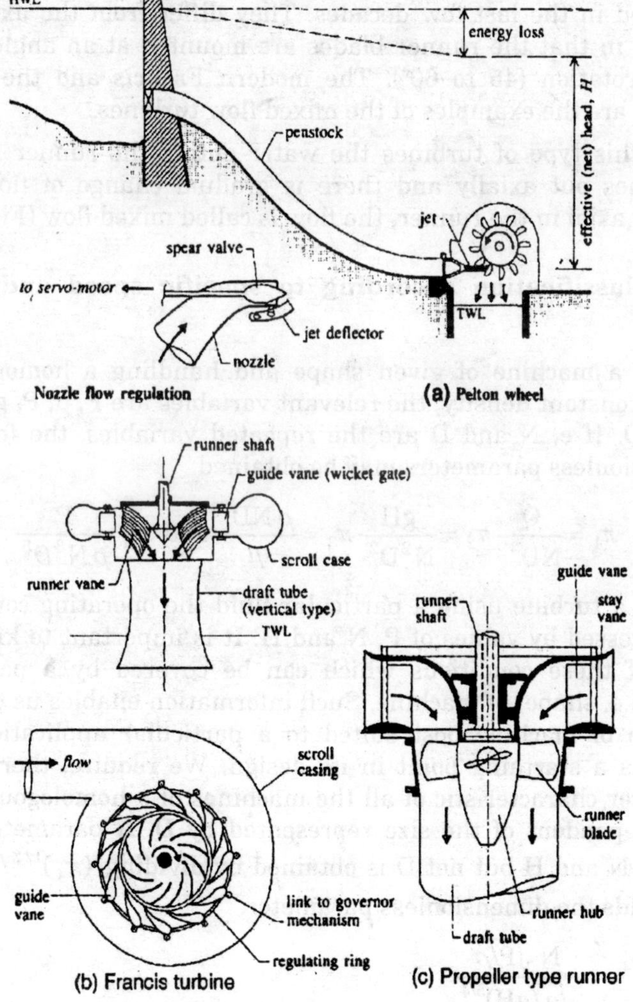

Fig. 8.3. Classification of turbines (Ref. Kay)

(b) Radial flow turbines: Radial flow is one in which the path of the flow of water in the radial direction i.e. perpendicular to the axis of shaft. The radial flow turbines are old Francis, Thompson and Girard. These turbines are not used now-a-days (Fig. 8.3).

(c) Axial flow turbines: In this turbine the path of the flow of water is in the axial direction i.e. parallel to the axis of the shaft. The propellor and Kaplan turbines are of the axial flow type.

(d) Mixed flow turbines: Mixed flow turbines have been developed in the last few decades. They differ from the axial flow turbines in that the runner blades are mounted at an angle to the axis of rotation (45 to 60°). The modern Francis and the Deriaz turbines are the examples of the mixed flow turbines.

In this type of turbines the water enters the runner radially and comes out axially and there is gradual change of flow from radial to axial in the runner, the flow is called mixed flow (Fig. 8.3).

8.4.3. Classification according to specific speed and speed factor

For a machine of given shape and handling a homogeneous fluid of constant density, the relevant variables are P, μ, P, g, Q, N., H and D. If e, N and D are the repeated variables, the following dimen- sionless parameters may be obtained.

$$\pi_1 = \frac{Q}{ND^{3'}}\ \pi_2 = \frac{gH}{N^2D^{2'}}\ \pi_3 = \frac{\rho ND^2}{\mu} \text{ and } \pi_4\ \frac{P}{\rho N^3 D^5}$$

For a turbine using a particular fluid the operating conditions are expressed by values of P, N and H. It is important to know the range of these conditions which can be covered by a particular design (i.e. shape) of machine. Such information enables us to select the type of machine best suited to a particular application, and serves as a startaing point in its design. We require, therefore, a parameter characteristic of all the machines of a homologous series and independent of the size represented by D. A parameter invo- lving P, N and H but not D is obtained by dividing $(\pi_4)^{1/2}/(\pi_2)^{5/4}$. This yields the dimensionless parameter.

$$\frac{N\sqrt{P/\rho}}{\sqrt{e}\,(gH)^{5/4}} \qquad (8.5)$$

For complete similarity of flow in machines of a homologous series, each of the dimensionless parameters must be unchanged throughout the series. Consequently the equation (8.5) must be unchanged. A particular value of this ratio therefore, relates all the combinations of P, N and H for which the flow conditions are similar in the machines of that homologous series.

For a given homologous series, therefore, we are concerned with a unique set of flow conditions, and thus a unique valve of equation (8.5) is obtained which may be designated K_n. Whatever the conditions of operation, provided only that they correspond to maximum efficiency, the machines of a particular homologus series, that is, of a particular shape, have a particular value of K_n. Machines of a different shape have, in general, a different value of K_n. Consequently, the value of K_n obtained from a set of values of P, N and H indicates the shape of machine that meets those conditions.

Hitherto the only incompressible fluid used in practical turbines has been cold water, and they have been operated only on the surface of the earth. The values of P and G have therefore, never been appreciably different from 1000 kg/m³ and 9.81 m/s² respectively. Consequently engineers have often omitted from Kn the constant P and G terms, and worked with values of $N\sqrt{P}/H^{5/4}$. The pression has been termed the specific speed Ns.

$$N_S = \frac{N\sqrt{P}}{H^{5/4}} \tag{8.6}$$

The rather, unfortunate, name arises from the concept of a hypothetical specific turbine-a member of the same homologous series as the actual turbine, but so reduced in size as to generate one unit of power (e.g. 1 kW) under one unit of head (e.g. 1 m) at maximum efficiency. The numeric, but not the units, obtained on substituting values in then corresponds simply to the speed at which the specific turbine would run.

The dimensionless parameter K_n is often known as the dimensionless specific speed to distinguish it from N_s. However, K_n is truly dimensionless only if N involves the radian measure of angle.

K_n and N_s are related by a conversion factor 1041.687636 = 1042.

$$N_s = 1042\,K_n \tag{8.7}$$

The speed factor, ku, is the ratio of the peripheral speed, u, of the buckets or vanes at the nominal diameter, D, to the theoretical velocity of water under the net head, H, acting on the turbine.

$$K_u = \frac{u}{\sqrt{2\,g\,H}} = \frac{DN}{84.6\,\sqrt{H}} \tag{8.8}$$

Table 8.1 suggests appropriate values of Ku which give the highest efficiencies for any turbine. Also shown in the table are the head and specific speed ranges and the efficiencies of the three main types of turbine.

Table 8.1. Ranges of Kn, Ns, H and Ku

Types of turbine	Kn (rev) (approx. values)	Ns (rpm, kW, m)	H (m)	Ku	η (%)
Pelton			Greater than		
(a) Single jet	0.0125 to 0.029	13 to 30			
(b) Multi jets	0.029 to 0.0575	30 to 52	300	0.43 to 0.48	85 to 90
Francis	0.0575 to 0.33	52 to 345	50 to 370	0.6 to 0.9	90 to 94
Kaplan	0.25 to 0.822	260 to 860	less than 60	1.4 to 2	85 to 95

8.4.4. Operating or normal speed N

All runners having the same geometrical shape, regardless of size, have the same specific speed. Charts or tables (Table 8.2) can be used for this selection. If the head and flow rate are known, the speed of the unit N can be calculated from equation (8.6).

Table 8.2. Q - H - Ns (after Novak etc. 1990) Q (m^3/s), H(m), P (kW) D (m). and N (rpm).

Types of turbine	Ns	H(m)	$Q_u = \frac{Q}{D^2\sqrt{H}}$	$N_u = \frac{ND}{\sqrt{H}}$
Pelton wheel				
1. Nozzle	7-11	1650-1800	0.011-0.007	39.4-39.8
2. Nozzle	11-17	700-1650	0.024-0.011	38.9-39.4

4. Nozzle	17-26	350-700	0.055-0.024	37.6-38.9
Francis				
Slow	51-107	410-700	0.35-0.1	63.6-60.8
Medium	107-150	240-41	0.59-0.35	67.5-63.6
	150-190	150-240	0.83-0.59	72.6-67.5
Fast	190-250	90-150	1.13-0.83	81-72.6
	250-300	64-90	1.28-1.13	92.2-81
Kaplan	240-450	50	1.22-0.93	145-85
	330-560	35	1.61-1.29	155-100
	390-690	20	2.0-1.6	170-110
	490-760	15	2.35-2.0	180-120
	570-920	6	2.45-2.35	200-135

Fig. 8.4 shows the type of the turbine as a function of both the specific speed N_s and the head H.

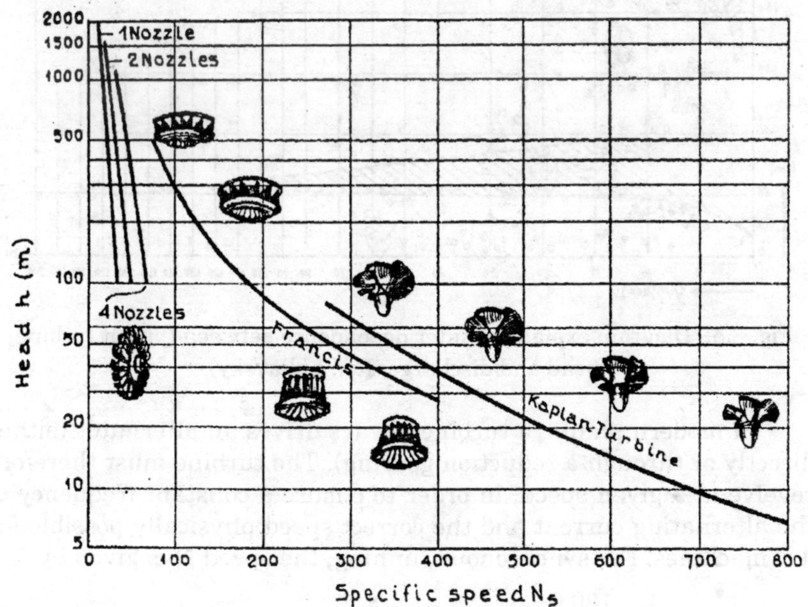

Fig. 8.4. Diagram showing interdependence between type of turbine, head and specific speed (Ref. Leliavsky).

On the other hand, Fig. 8.5 shows the type of turbine as a function of both the discharge and head.

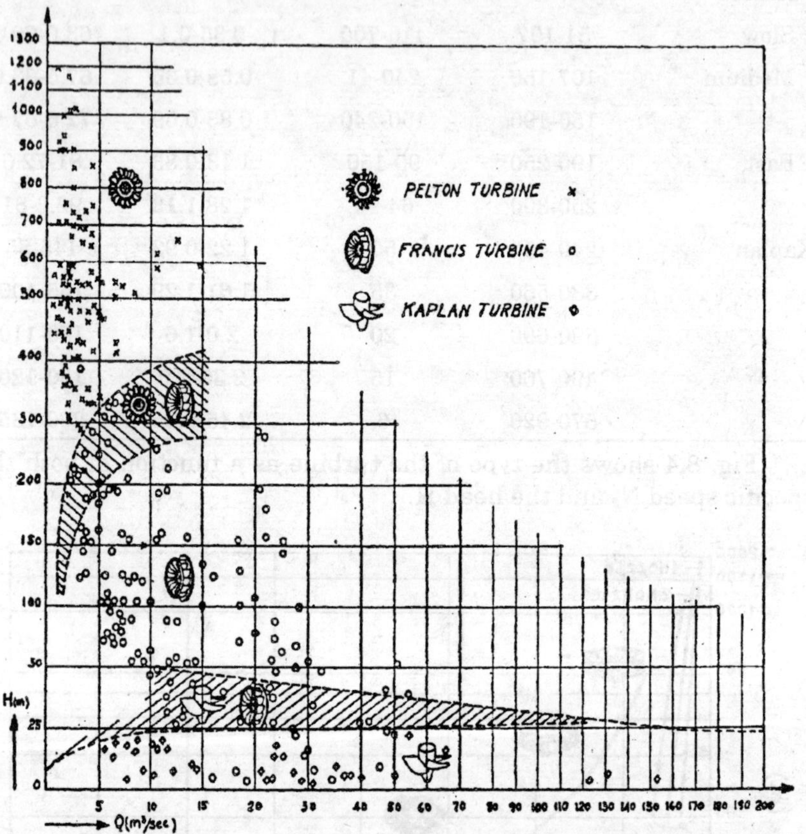

Fig. 8.5. Diagram explaining inter dependence between type of turbine, head and discharge (Ref. Leliavsky).

In modern plants, a turbine always drives an alternator (either directly or through a reduction gearing). The turbine must therefore revolve at a given speed, in order to ensure a constant frequency of the alternating current and the correct speed physically possible for d.c. machines. For synchronous running, the speed N is given by

$$N = \frac{120\,f}{P} \tag{8.9}$$

Where, f is the frequency of the a.c. supply in H_z (50 and 60 H_z), and p is the number of poles of the generator (divisible by 4 for heads upto 200m or by 2 for heads above 200m). The recommended (Mosonyi, 1959) normal speeds for 50 Hz a.c. supply machines are; 3000, 1500, 1000, 750, 600, 500, 375, 300, 250, 214, 188, 167, 150, 125, 107, 94, 83, 75 and 60 rpm.

The alternating current frequency for all power systems in the India is 50 Hz (corresponding to 50 cycles per second). To induce such a current, the simplest two pole rotor needs to make 50 x 60 = 3000 rpm. Water turbines rotate at a much lower speed, so the rotor must have several pairs of poles to maintain the standard frequency.

8.4.5. Turbine rating and performance

Hydraulic turbines are generally rated under maximum, minimum, normal and design head. The runner is designed for optimum speed and maximum efficiency at design head, which is usually selected as the head above and below which the average annual generation of power is approximately equal. But, in reality, head and load conditions change during operation.

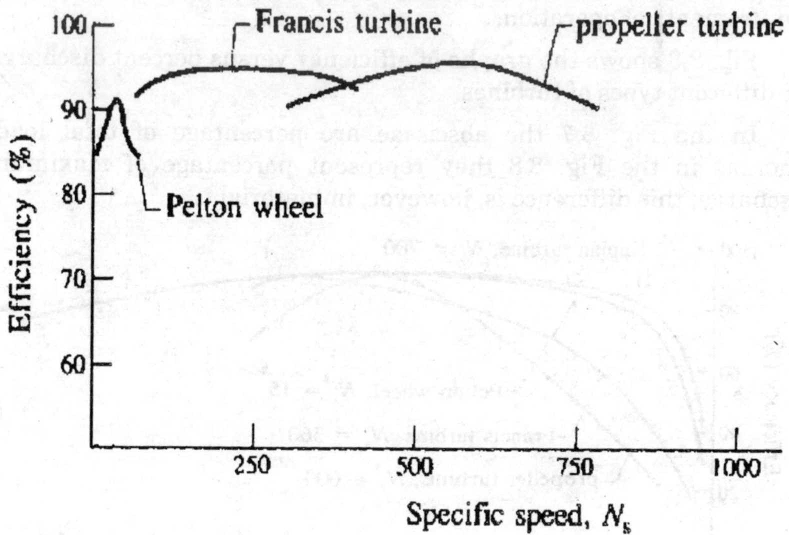

Fig. 8.6. Graph between N_s and η (Ref. Novak)

Maximum efficiency varies with the type of turbine, and in a particular type it varies with the specific speed, N_s, as shown in

Fig. 8.6. In general the higher the specific speed for a given head and power the smaller the size of the turbine and the generator and consequently lower the cost of the installation. However, too high a specific speed results in the fall of efficiency both for impulse as well as reaction turbines.

For Pelton wheel, very low specific speed increases the diameter and friction losses, while very high specific speed lead to uneconomic proportions and hence loss the efficiency. For a single jet Pelton $N_s = 12$ is about the best.

For reaction turbines, very low specific speed, increases disc friction and percent leakage area, while very high specific speed increases the exit losses and possible cavitation problems.

Fig. 8.7 shows the graphs of efficiency versus percent load for different types of turbines. Kaplan gives high efficiency over a considerable range of load. Pelton wheel has somewhat lower, but sustained good efficiency over a wide range of load. Francis turbines give lower part efficiencies as their specific speeds increases. The fixed blade propeller has the poorest part load efficiencies. Therefore, proper choice must be made according to the actual requirements of operation.

Fig. 8.8 shows the graphs of efficiency versus percent discharge for different types of turbines.

In the Fig. 8.7 the abscissae are percentage of total load, whereas in the Fig. 8.8 they represent percentage of maximum discharge; this difference is, however, immaterial.

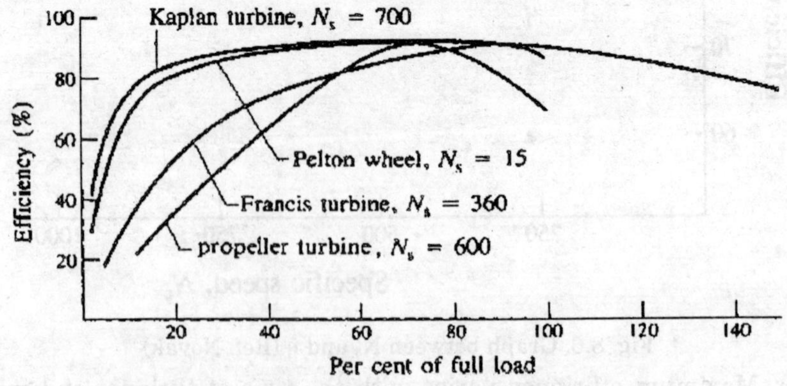

Fig. 8.7. Efficiency v/s percent load (Ref. Novak)

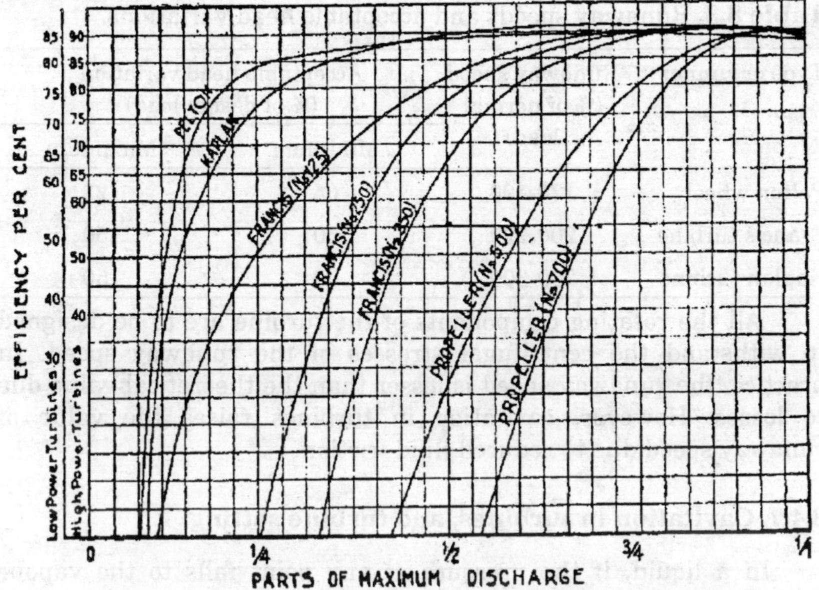

Fig. 8.8. Efficiency v/s percent discahrge (Ref. Leliavsky)

It is important to note here that the Kaplan curve is in all cases obviously the best, whilst the fixed blade propeller type is, as obviously, the worst. It follows that the Kaplan, though more expensive, is nevertheless suitable for developments working under a wide range of various load conditions, whereas the cheaper, propeller type is only admissible when such conditions remain constant throughout the seasons of the year (which, however, is rather exceptional).

8.4.6. Runaway speed

If the external load on the machine suddenly drops to zero (sudden rejection) and the governing mechanism fails at the same time, the turbine will tend to race up to the maximum possible speed, known as runaway speed. This limiting speed under no load conditions with maximum flow rate must be considered for the same design of the various rotating components of the turbo-generator unit. The suggested runaway speeds of the various runners for their appropriate design, considerations are given in Table 8.3.

Table 8.3. Runaway speeds and acceptable head variations.

Type or runner	Runaway speed (% of normal load)	Acceptable head variation (% of design load)	
		Minimum	Maximum
Pelton wheel	170-190	65	125
Francis turbine	200-220	50	150
Kaplan turbine	250-300	50	150

All the rotating components of the turbine are to be designed to withstand the centrifugal stresses of the runaway speed. In practice, the runaway speed is lesser than the theoretical value due to losses. However, cavitation in turbines raises the value of runaway speed due to reduced fluid friction.

8.4.7. Cavitation in turbines, and turbine setting

In a liquid, if the pressure at any point falls to the vapour pressure (at temperature concerned), the liquid there boils and small bubbles of vapour form in large numbers. These bubbles are carried along the flow, and on reaching a point where the pressure is higher they suddenly collapse as the vapour condenses to liquid again. A cavity results and the surrounding liquid rushes in to fill it. The liquid moving from all directions collides at the centre of the cavity, thus giving rise to very high local pressure (upto 1GPa). This alternate formation and collapse of vapour bubbles may be repeated with a frequency of many thousand time a second. The intense pressures, even though acting for only a very brief time over a tiny area can cause several damage to the surface. The material ultimately fails by fatigue, aided perhaps by corrosion, and so the surface becomes badly scored and pitted. Parts of the surface may even be torn away completely. Associated with cavitating flow there may be considerable vibration and noise; when cavitation occurs in a turbine or pump it may sound as though gravel were passing through the machine.

Cavitation results in pitting vibration, and reduction in efficiency and is certainly undesirable. Runners most seriously affected by cavitation are of the reaction type, in which the pressures at the discharge ends of the blades are negative and can approach the vapour pressure limits. Cavitation may be avoided by suitably designing, installing, and operating the turbine in such a way that the pressures within the unit are above the vapour pressure of the water. Turbine setting or draft head, $H_{s'}$ is the most critical factor in the installation of the reaction turbines. The recommended limits of safe specific speeds for various heads, based on experience with existing power plants, are shown in Fig. 8.9.

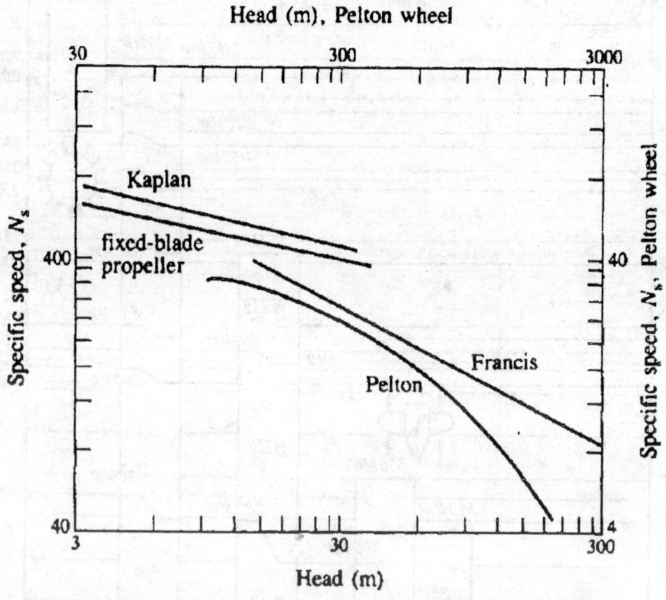

Fig. 8.9. Recommended limits of safe specific speeds (Ref. Novak)

8.4.8. Selecting the turbine type and number of units

For a given total plant capacity, total costs will generally increase with an increase in the number of units. Efficiencies of large units (turbine and generators) are higher than those of smaller ones of the same type. If the power demand is reasonably uniform it is practicable to install a small number of large units.

As far as the initial and maintenance costs are concerned, a major power plant should preferably have the least possible number of power generating units of maximum unit power. The maximum power for each turbine series at a given head is specified in the Chart of Fig. 8.10. However, other considerations (such as flexibility of operation, higher efficiency operation during low load demands, and minimum loss of capacity during shutdown for repair or maintenance) might dictate the use of multiple units where one unit would be feasible in terms of physical size. For example, appropriate regulations require that at least two Kaplan or Deriaz turbines or Four Francis or fixed blade propeller turbines be used for this reason.

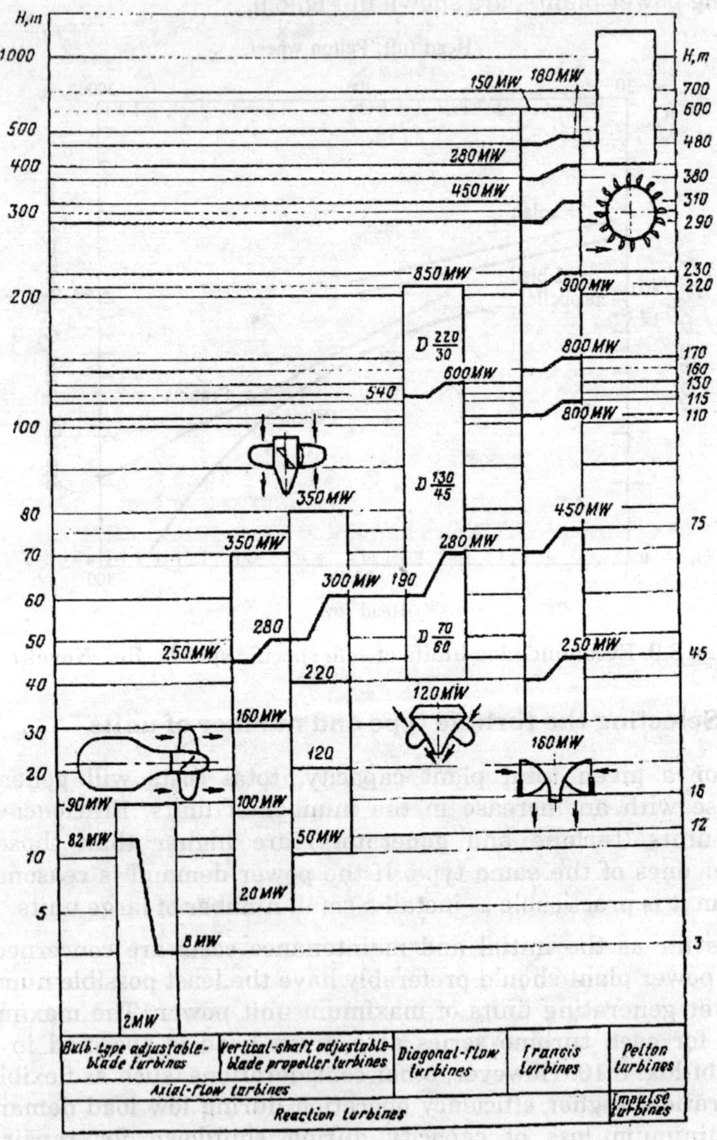

Fig. 8.10. Summary chart of water turbine application (Ref. Ilyinykh)

Kaplan and Deriaz turbines can be used for a very wide range of heads and loads. However, they are rather costly and, hence, more expensive power house substructure than their counterparts. When the head is expected to vary insignificantly, an appreciable efficiency can be obtained from Francis or fixed blade propeller turbines, especially at major hydroelectic plants using several power generating units at the same time. When choosing a turbine, the designer should remember that it must be able to operate at a reduced discharge during start-up periods.

8.5. TYPES OF ENERGY LOSS AND EFFICIENCIES

The highest possible efficiency of a hydraulic machine, as of any other machine, can only be realised if the energy losses are carefully analysed and reduced to their lowest limit. In a hydraulic machine the major losses can be included under the headings :

(1) Mechanical loss: This includes so called mechanical friction between fixed and moving parts, e.g. rubbing or sliding friction between a rotating shaft and its bearings.

(2) Disc friction: In rotodynamic machines, energy will be dissipated by hydraulic friction between liquid and the rotating faces of the wheel or rotor.

(3) Leakage loss : If liquid escapes or leaks from a high pressure zone of the machine to a low-pressure zone, it will carry with it energy and is subsequently wasted in eddying.

(4) Hydraulic losses : While the liquid is flowing through the fixed or moving passages of the machine, it will be subject to the type of wall friction loss or eddy loss. There is a possibility, too, that when the liquid leaves the machine, it may take with it considerable quantities of velocity energy which cannot afterwards be recovered this energy also is wasted.

These various types of energy dissipation will later be studied in connection with particular machine; at this stage, leakage losses will be discussed in general terms. Although it is true that the distinction between the various losses is only an arbitrary one-for types (2), (3) and (4) could all be classified as "hydraulic"-yet the distinction is permissible and useful.

8.5.1. Efficiencies

The losses can be related in terms of an energy balance through the turbine.

$$P = P_m + P_r + P_c + P_l + P_s \quad (8.10)$$

where, P_s = Shaft power output,

P_m = mechanical power loss,

P = runner power loss,

P_c = casing and draft tube loss,

P_l = leakage loss

and P = water power available.

Together, $P_r + P_c + P_l$ is the hydraulic power loss.

Runner power loss P_r is due to friction, shock at impeller entry and flow separation, and resulted in a head loss associated with a flow rate through the runner.

Leakage loss P_l is caused by a flow rate leaking past the runner and therefore not being handled by the runner.

Casing power loss P_c is due to friction, eddy and flow separation losses in the casing and draft tube.

The various energy losses are discussed above. The various efficiencies accordingly may be expressed as.

(1) Hydraulic efficiency (η_h) : The hydraulic efficiency of the turbine is the ratio of the power developed by the runner to the net power supplied by the water at the entrance to the turbine. These two powers differ by the amount of the hydraulic losses. That is

$$\eta_h = \frac{\text{Power developed by the runner}}{\text{Net power supplied at the turbine entrance}}$$

or $$\eta_h = \frac{\text{Power received by the runner}}{\text{Water power available at the inlet flange}}$$

Where,

Power received by runner = $\rho Q [V_{w1} u_1 \pm V_{w2} u_2]$

and water Power available = ρQgH

or Available water power

Therefore,

$$\eta_h = \frac{\rho\, Q\, [V_{w1}\, u_1 \pm V_{w2}\, u_2]}{\rho\, Q\, g\, H} \tag{8.11}$$

(2) Mechanical efficiency (η_m) : The mechanical efficiency of the turbine is the ratio of power available at the turbine shaft to the power developed by the runner. These two powers differ by the amount of the mechanical losses. e.g. bearing friction. That is

$$\eta_m = \frac{\text{Power available at the turbine shaft}}{\text{Power developed by the runner}}$$

or
$$\eta_m = \frac{\text{Shaft power or Brake Power}}{\text{Power received by the runner}}$$

$$= \frac{P}{[V_{w1}\, u_1 \pm V_{w2}\, u_2]} \tag{8.12}$$

(3) Volumetric efficiency (η_v) : The volumetric efficiency is the ratio of the quantity of water actually striking the runner and the quantity of water supplied to the turbine. These two quantities differe by the amount of water that slips directly to the tail race without striking the runner (q). That is

$$\eta_v = \frac{Q}{(Q + q)} \tag{8.13}$$

(4) Overall efficiency (η_o) : The overall efficiency of the turbine is the ratio of shat power (brake power) to the available water power (Water power available).

$$\eta_o = \frac{\text{Shaft power}}{\text{Water power available}}$$

$$\eta_o = \frac{P}{\rho\, Q\, g\, H} \tag{8.14}$$

It is evident from equation (8.14) that overall efficiency of the turbine is

$$\eta_o = \eta_h \times \eta_m \tag{8.15}$$

If the quantity of water leaving or not striking the runner blades or the volumetric efficiency is given then

$$\eta_o = \eta_h \times \eta_m \times \eta_v \quad (8.16)$$

EXERCISES

A. Theory

8.1. Describe briefly the development of turbine design.

8.2. Define gross head and net head for Impulse and reaction turbine.

8.3. Name the component of hydropower plants.

8.4. What are the various component of hydro power plants? Describe them briefly.

8.5. What are waterways? Describe their control works.

8.6. Define Head works. For what purpose are they installed.

8.7. Name the different types of gate and valves.

8.8. What are trashracks? State their uses. How is the debris removed from the trashracks? What method is used in cold countries?

8.9. What is a forebay? What is the difference between a forebay and a surge tank?

8.10. What is the purpose of using surge tank instead of forebay in a hydro electric power plant.

8.11. Why is a surge tank necessary in case of high and medium head plants?

8.12. What means are provided to save the penstock from water hammer, when the gates or valves of turbine are suddenly closed?

8.13. What is the difference between the impulse and reaction water turbines?

8.14. Classify the hydraulic turbines according to direction of flow.

8.15. Define specific speed (N_s) and non-dimensional specific speed and prove that

$$N_s = 1024\ K_n$$

8.16. Discuss the effect of the following factors in the selection of a turbine for hydroelectric plant.

(a) Head (b) Specific speed

(c) Speed (d) Cavitation and

(e) Part load operation.

8.17. Define speed factor φ. Give the approriate values of φ which gives the highest efficiencies for the turbines.

8.18. Draw a curve between N_s v/s H showing the interdependence between type of turbine, head and specific speed.

8.19. Draw a curve between Q v/s H and explain the interdependence between type of turbine, head and discharge.

8.20. Draw a curve between N_s v/s η and explain the inter depence between type of turbine, efficiency and specific speed.

8.21. Draw a curve between percent of load v/s η and explain the interdepence between the type of turbine, percent of load and efficiency.

8.22. Draw and explain a curve between percentage discharge v/s efficiency with reference to type of turbine.

8.23. Define run away speed. Give the acceptable head variation and run away speed for percentage of normal load.

8.24. What are the causes of cavitation. How the cavitation is avoided in the water turbines.

8.25. Draw a curve between H v/s N_s and explain the recommended limits of safe specific speed.

8.26. How the turbines and number of units are selected. Draw a summary chart of water turbine application.

8.27. Give an account of various types of losses occurring in a hydro power plants.

8.28. Define, hydraulic, mechanical, and volumetric of turbine. What is overall efficiency.

8.29. Prove that

$$\eta_o = \eta_h \times \eta_m \times \eta_v$$

9

Impulse Turbines

9.1. INTRODUCTION

In the impulse turbines, like the Pelton wheel, the water enters the turbine through a nozzle or nozzles, and the entire static head is converted into kinetic energy. In other words, there is no difference of pressure in the water at the inlet to the runner and at the discharge, because the ambient pressure is atmospheric; the casing of an impulse turbine does not perform any hydraulic function but merely prevents splashing and leads the water to the tail race.

Pelton turbines are regulated by decreasing or increasing the diameter of the jet, and this is effected by means of needle or spear which closes or opens the throat of the nozzle. The number of jets is not more than two for horizontal shaft turbines and is limited to six for vertical shaft turbines.

This is the only hydraulic turbine of impulse type having extensive use for high head installations. When it is necessary to transfer energy available at high head, it can transfer a large amount of energy at reduced capital cost.

9.2. THE PELTON WHEEL

Pelton wheels are high head, low specific speed, low discharge, tangential flow impulse turbine.

The Pelton wheel turbine consists of a wheel like runner (rotor) with spoon shaped buckets around the circumference of the wheel. Water from a high level is discharged through a stationary nozzle or nozzles and emerges as a high speed jet. The potential energy of the water is thus converted into kinetic energy of the jet. The water jet is directed at buckets, and its impulse causes the runner to rotate about its axis. After leaving the wheel, the water falls with a small residual velocity into the tail race.

9.2.1. Main Components and their functions

The essential component parts of a Pelton wheel are :

1. Nozzle and guide mechanism
2. Runner with buckets
3. Casing
4. Hydraulic brake

1. Nozzle and guide mechanism

The function of the nozzle of a Pelton wheel is to convert the available pressure energy into high velocity energy in the form of jet. The quantity of water required is proportional to the load on the turbine. Therefore, to control the flow through the nozzle, some sort of a regulating or governing mechanism is necessary. This is generally done by using a spear or needle inside the nozzle.

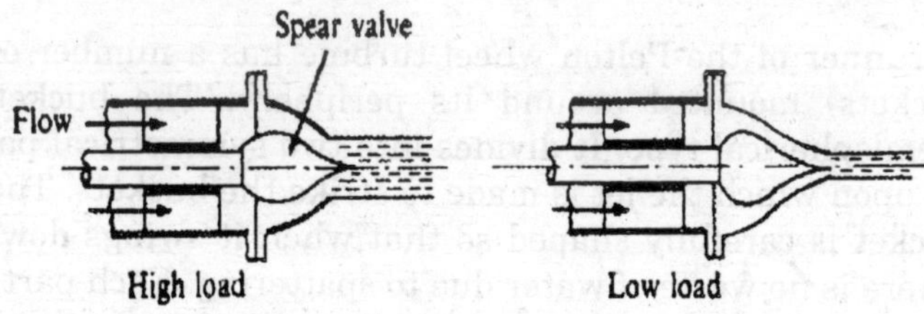

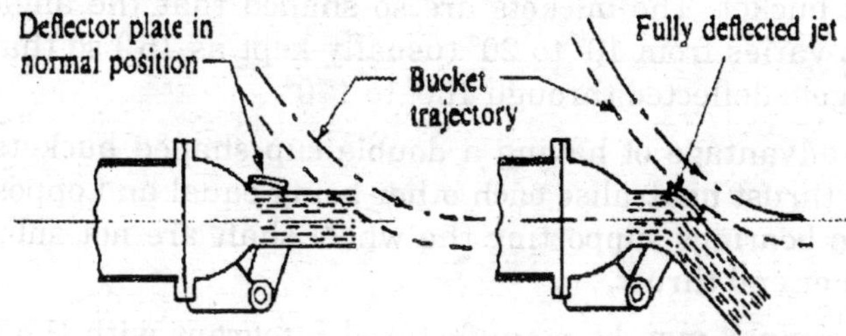

Fig. 9.1. Load control by a spear valve and deflector plate (Ref. Sayers)

Hydraulic turbines are usually coupled directly to an electric generator and, since the generator must run at a constant speed, the speed of the trubine must remain constant when the load changes. It is desirable to run at maximum efficiency so that the jet velocity must not change. The only way left to adjust to a change in flow rate, i.e. to change the nozzle area. The most effective way to change in flow rate by providing a spear or needle having a stream lined head. The movement of spear inside the nozzle changes the area of flow through it, thus varying the discharge (Fig. 9.1). The position of the spear is controlled by a servo mechanism that senses the load change. In effect, therefore, the device is simply one for varying the diameter of the jet.

The only draw back to the use of the spear is that it cannot be moved to close the nozzle rapidly except at the risk of generating serious dynamic or inertia pressures in the pipe line. It must consequently be supplemented by a pivoted deflector. For a sudden loss of load, a deflector plate rises to remove the jet totally from the buckets and to allow time for the spear to move slowly to its new load position (Fig. 9.1). This prevents excessive overspeeding.

2. Runner with buckets

The runner of the Pelton wheel turbine has a number of twin cups (buckets) mounted around its periphery. The buckets are almost hemispherical type. It divides into two symmetrical parts by a splitter upon which the jet is made to strike the buckets. The back of the bucket is carefully shaped so that when it swings down into the jet, there is no waste of water due to spattering. Each part of the jet glides over the buckets and comes out at the outer edge of the bucket (Fig. 9.2). The rear of the buckets is so shaped that the water coming out a bucket does not interfere with the jet striking the following bucket. The buckets are so shaped that the angle at the outlet tip varies from 10° to 20° (usually kept as 15°) so that the jet of water gets deflected through 160° to 170°.

The advantage of having a double cup shaped buckets is that the axial thrust neutralise each other being equal and opposite, and hence the bearings supporting the wheel shaft are not subjected to any axial or end thrust.

The runner may be manufactured integrant with the buckets, the material of construction being either cast iron for a peripheral velocity 30 m/s, or cast steel for 70 m/s. Most frequently, however, the buckets are cast integrant with the stems, and these are bolted to the disc of the wheel. This manufacturing method is more expensive but facilitates grinding and polishing of the buckets, their replacement in case of defect and the casting process itself. The buckets are in this case easily machinable than on a runner cast integral with them. Withregard to the alternating strees, the buckets are fastened to the wheel by means of fitted bolts, preferable with conical shanks. Difficulties are encountered in dimensioning the connections of the buckets to high speed runner (specific speed exceeding 26). In this case it is recommended to shrink two heated steel hoops on the buckets.

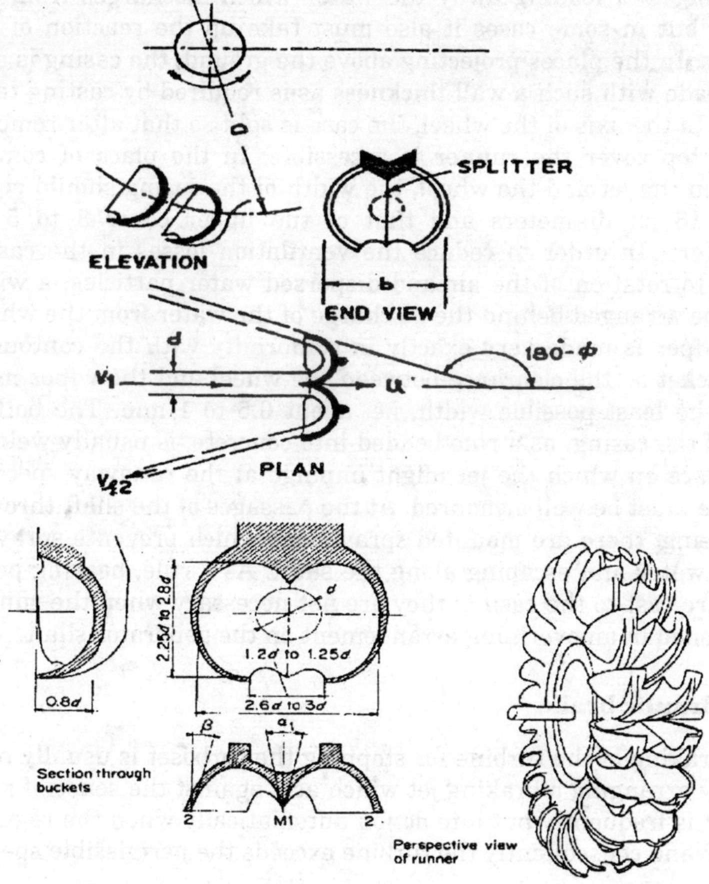

Fig. 9.2. (a) Three view of bucket, (b) Section through buckets (c) Perspective view of runner (Ref. Addison, Monition etc.)

The runner is fastened on the shaft by means of a key or feather, and in a overhung arrangement by a flange.

For high heads alloyed cast steel are employed. Great care must be devoted to the manufacture of the buckets. Their inner surface must exactly ground to shape and finally polished. The buckets must be free of cracks. The assembled runner must be carefully balanced statically and for higher speeds also dynamically.

3. Casing

The casing of the Pelton turbine has not only the function of collecting and leading away the water which discharges from the wheel, but in some cases it also must take up the reaction of the nozzles. In the places projecting above the ground, the casing is as a rule made with such a wall thickness as is required by casting technique. In the axis of the wheel, the case is split so that after removal of the top cover the runner is accessible. In the place of contact between the jet and the wheel, the width of the casing should equal 12 to 18 jet diameters and that of the upper cover 3 to 5 jet diameters. In order to reduce the ventilation losses in the casing owing to rotation of the air and dispersed water particles, a wiper must be arranged behind the discharge of the water from the wheel. This wiper is made very exactly in conformity with the contour of the bucket as the clearance between the wheel and the wiper must be of the least possible width, i.e. about 0.5 to 1 mm. The bottom part of the casing, as a rule bedded into concrete, is usually welded. The place on which the jet might impinge at the runaway speed of turbine must be well armoured. At the passages of the shaft through the casing, there are mounted spray rings which prevents spraying of the water and escaping along the shaft. As a rule, bearing pedestals are cast to the casing; they are not necessary when the runner is fastened in an overhung arrangement on the generator shaft.

4. Hydraulic brake

Braking of the turbine for stopping the turboset is usually realized by arranging a braking jet which acts against the sense of rotation. It is frequently put into action automatically when the regulator fails and consequently the turbine exceeds the permissible speed.

The brake consists of a small nozzle fitted in such a way that on being opened, it directs a jet on the back of the buckets to bring the revolving runner quickly to rest. The least diameter of the brake jet has been found to be equal to 0.6 times the least diameter of the main jet.

The typical Pelton wheel installation is shown in Fig. 9.3. Water comes from a storage reservoir through a penstock. At the Penstock end, there is a nozzle through which a jet issues and strikes on the buckets of a Pelton wheel. Pelton wheel is always fixed above highest possible water level in the tail race and there is no necessity of any draft tube.

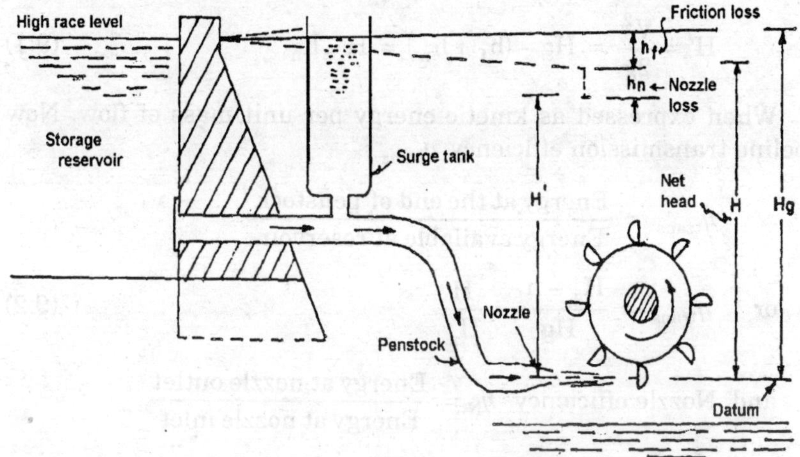

Fig. 9.3. Pelton wheel hydroelectric installation.

The water supply is from a constant head reservoir at elevation H_g above the centre line of the jet.

A pressure tunnel extends from the reservoir to a point almost vertically above the location of the turbine. A pipe of almost vertical slope called the penstock joins the end of the pressure tunnel to a nozzle, while a surge tank is installed at the upper end of the penstock to damp out flow control pressure and velocity transients. At the turbine end of the penstock is the nozzle, which converts the total head at inlet to the nozzle into a water jet with velocity V_1 at atmospheric pressure.

9.3. PELTON WHEEL LOSSES AND EFFICIENCIES

Head losses occur in the penstocks conveying the water to the nozzle and are composed of friction and bend losses. Losses also occur in the nozzle and these are expressed in terms of a velocity co-efficient c_v. Finally there are windage and bearing friction losses in the wheel itself. From Fig. 9.3, H_g is the gross head for turbine or total available head measured from the water level in the storage reservoir to the centre of nozzle or the jet.

As the fluid moves through the pressure tunnel and penstock up to entry to the nozzle, a frictional head loss h_f occurs. A further head loss h_n due to losses in the nozzle takes place so that the head available for power generation at exit from the nozzle is H′

$$H' = \frac{V_1^2}{2g} = Hg - (h_f + h_n) = H - h_n \quad (9.1)$$

When expressed as kinetic energy per unit mass of flow. Now Pipeline transmission efficiency η_{trans}

$$\eta_{trans} = \frac{\text{Energy at the end of penstock}}{\text{Energy available at reservoir}}$$

$$\text{or} \quad \eta_{trans} = \frac{H_g - h_f}{Hg} = \frac{H}{H_g} \quad (9.2)$$

$$\text{and, Nozzle efficiency} \quad \eta_N = \frac{\text{Energy at nozzle outlet}}{\text{Energy at nozzle inlet}}$$

$$\text{or,} \quad \eta_N = \frac{H'}{H_g - h_f} \frac{H'}{H} = \frac{V_1^2 / 2g}{H} = \frac{V_1^2}{2g\,H} \quad (9.3)$$

Therefore,

Nozzle and pipe transmission efficiency

$$= \eta_{trans} \times \eta_N$$

$$= \frac{H}{H_g} \times \frac{H'}{H} = \frac{H'}{Hg} = \frac{V_1^2}{2\,gH_g} \quad (9.4)$$

Also

$$\text{Nozzle velocity coeffcienct} = \frac{\text{Actual jet velocity}}{\text{Theoretical jet velocity}}$$

$$\text{or,} \quad c_v = \frac{V_1}{\sqrt{2g\,H}} \quad (9.5)$$

or, $$c_v^2 = \frac{V_1^2}{2g\,H}\,\eta_N$$

Therefore, $$\eta_N = c_v^2 \tag{9.6}$$

From equation (9.1)

Loss in the nozzle.

$$h_n = H - \frac{V_1^2}{2g}$$

$$= H - \frac{c_v^2\,(2\,g\,H)}{2g}$$

$$= H - c_v^2\,H$$

$$= H\,(1 - c_v^2) \tag{9.7a}$$

$$= \frac{V_1^2}{2g\,c_v^2}\,(1 - c_v^2) = \frac{V_1^2}{2g}\left(\frac{1}{c_v^2} - 1\right) \tag{9.7b}$$

The Pelton wheel has the following usual efficiencies

(1) Hydraulic efficiency

(2) Mechanical efficiency

(3) Overall efficiency

The hydraulic efficiency η_n of the Pelton turbine is defined as

$$\eta_h = \frac{\text{Hydraulic output}}{\text{Hydraulic input at the base of nozzle}} = \frac{\dfrac{V_1^2}{2} - \dfrac{V_2^2}{2}}{gH}$$

$$= \frac{\text{Power delivered by water runner}}{\text{Power available at the base of nozzle i.e. Available water power}}$$

$$\eta_h = \frac{\rho\,Q\,(V_{w1} \pm V_{w2})\,u_1}{\rho\,Q\,g\,H} \tag{9.8}$$

Equation (9.8) gives the hydraulic efficiency considering both runner and nozzle i.e. whole.

Mechanical efficiency η_m is defined as

$$\eta_m = \frac{\text{Work available from the shaft runner}}{\text{Work done on the turbine runner}}$$

$$= \frac{\text{Shaft power or Brake Power}}{\text{Power delivered by water runner}}$$

$$= \frac{P}{\rho Q (V_{w1} + V_{w2}) u_1} \tag{9.9}$$

The overall efficiency, η_o is defined as

$$\eta_o = \frac{\text{Shaft power}}{\text{Available water power}} = \frac{P}{\rho Q g H} \tag{9.10}$$

$$= \frac{P}{\rho Q g H} \times \frac{\rho Q (V_{w1} \pm V_{w2}) u_1}{\rho Q (V_{w1} \pm V_{w2}) u_1}$$

$$= \frac{\rho Q (V_{w1} \pm V_{w2}) u_1}{\rho Q g H} \times \frac{P}{\rho Q (V_{w1} \pm V_{w2}) u_1}$$

$$= \eta_o \times \eta_m \tag{9.11}$$

The mechanical efficiency of the pelton turbine can be assumed approximately equal to 95%, therefore, overall efficiency of the turbine can be calculated very easily by the equations (9.8) and (9.11).

9.4. THEORY OF PELTON WHEEL

The transfer of work from the jet of water to the buckets may be determined by applying Euler's turbine equation. Fig. 9.4 shows the velocity triangles at the tips of the bucket of a Pelton Wheel.

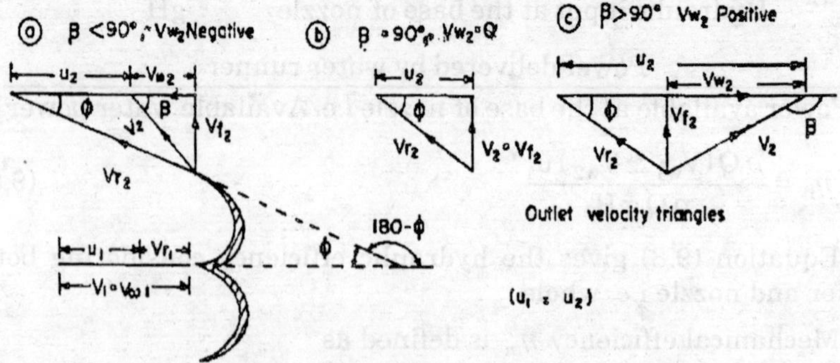

Fig. 9.4. Pelton wheel bucket with velocity triangles.

The work done and hydraulic efficiency can be obtained with the help of inlet and outlet velocity triangles.

Since the Pelton wheel is a tangential flow turbine the water enters the wheel tangentially i.e. $\alpha = 0^\circ, \theta = 0^\circ$ the inlet velocity triangle will be a straight line and thus

$$V_{r1} = V_1 - u_1 \text{ and } V_{w1} = V_1$$

At the outlet tip any one of the three velocity triangles as shown in Fig. 9.4 is possible depending upon the magnitude of u_1, corresponding to which it is slow, medium or fast runner. As the inlet and the outlet tips of the bucket are at the same radial distance, the tangential velocity of the bucket at both the tips is same i.e., $u_1 = u_2$.

The relative velocity V_{r2} with which the jet leaves the bucket will be somewhat less than the initial relative velocity V_{r1} at the inlet tip. This is because due to blade friction, although the inner faces of the buckets are polished so as to minimize frictional losses as water flows over them, such losses cannot be completely eliminated.

In addition to this some loss of energy will also take place as the jet strikes the splitter. These losses of energy reduce the relative velocity between the jet and the bucket, and hence

$$V_{r2} = K\,V_{r1} = K(V_1 - u_1) \tag{9.12}$$

Where K is a fraction slightly less than unity.

Now from the outlet velocity triangle (a) of Fig. 9.4. [$\beta < 90^\circ$ and V_{w2} is negative].

$$\begin{aligned} V_{w2} &= V_{r2}\cos\phi - u_2 \\ &= K\,V_{r1}\cos\phi - u_1 \\ &= K(V_1 - u_1)\cos\phi - u_1 \quad \text{[From equation (9.12)]} \end{aligned}$$

Now work done per second on the wheel is given as

$$\begin{aligned} \text{Work done} &= \dot{m}\left[V_{w1} \pm V_{w2}\right]u_1 \\ &= \dot{m}\left[V_1 + K(V_1 - u_1)\cos\phi - u_1\right]u_1 \\ &= \dot{m}\ \left[(V_1 - u_1)(1 + \cos\rho)\right]u_1 \end{aligned} \tag{9.13}$$

If from the outlet veloeity triangle Fig. 9.4(c) is considered, i.e. $\beta > 90°$ and V_{w2} is positive, we will get the same result as obtained from the equation (9.13).

The energy supplied to the wheel is in the form of kinetic energy of the jet is equal to $(\dot{m} V_1^2/2)$.

Therefore, the hydraulic efficiency of the Pelton wheel is given by

$$\eta_h = \frac{\dot{m}\left[(V_1 - u_1)(1 + K\cos\phi)\right]u_1}{\dot{m} V_1^2/2}$$

$$= \frac{2u_1(V_1 - u_1)(1 + K\cos\phi)}{V_1^2} \qquad (9.14)$$

The speed of the wheel for maximum hydraulic efficiency can be found by differentiating the equation (9.14) with respect to u_1 and equating to zero.

$$\frac{d\eta_h}{du_1} = \frac{d}{du_1}\left[\frac{2u_1(V_1 - u_1)(1 + K\cos\phi)}{V_1^2}\right]$$

or $$\frac{2(1 + K\cos\phi)}{V_1^2}\frac{d}{du_1}(V_1 u_1 - u_1^2) = 0$$

or $$\frac{2(1 + K\cos\phi)}{V_1^2}(V_1 - 2u_1) = 0$$

From which $V_1 = 2u_1$

$$\text{or, } u_1 = \frac{V_1}{2} \qquad (9.15)$$

Hydraulic efficiency of a Pelton wheel is maximum when the velocity of the runner is half the velocity of jet at the inlet.

Substituting the value of u_1 in equation (9.14) maximum hydraulic efficiency is obtained.

$$\eta_{h\,max} = \frac{2\left(\frac{V_1}{2}\right)\left(V_1 - \frac{V_1}{2}\right)(1 + K\cos\phi)}{V_1^2}$$

$$= \frac{2V_1^2(1 + K\cos\phi)}{4V_1^2}$$

$$= \frac{1 + K \cos \phi}{2} \tag{9.16}$$

Under ideal conditions, K=1 and the value of hydraulic efficiency and maximum hydraulic efficiency from the equations (9.14 and 9.16) gives

$$\eta_h = \frac{2\, u_1 (V_1 - u_1)(1 + \cos \phi)}{V_1^2} \tag{9.17}$$

and $$\eta_{h\,max} = \frac{1 + \cos \phi}{2} \tag{9.18}$$

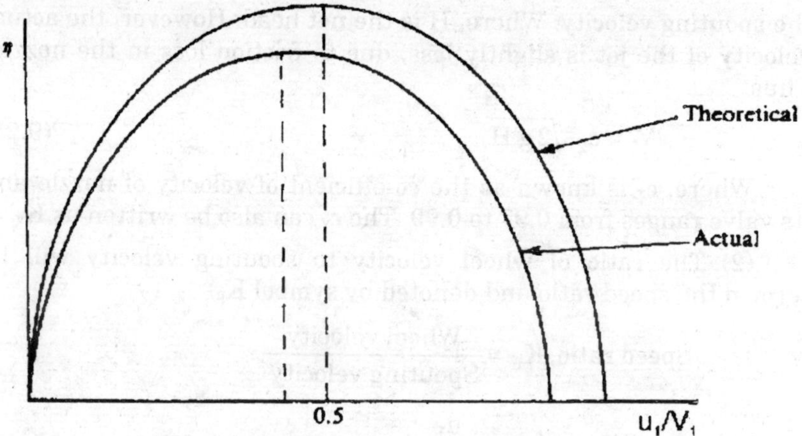

Fig. 9.5. Efficiencies and jet speed ratio of a Pelton wheel (Ref. Sayers)

If ϕ = 0°, the maximum hydraulic efficiency is 100%. In practice, the side clearance angle φ is usually kept at 10 to 20° (15° is the most common value) to avoid interference with the incoming jet and η_h is accordingly reduced (90 to 94%). Fig. 9.5 shows the theoretical efficiency as a function of speed ratio. The overall efficiency is lower than the theoretical as well as having a reduced speed ratio at maximum efficiency. This is due to penstock and nozzle losses.

For Pelton turbine (including nozzle), the hydraulic efficiency and maximum hydraulic efficiency are given

$$\eta_h = \frac{2\, u_1 (V_1 - u_1)\ (1 + K \cos \phi)}{V_1^2}\, c_v^2 \tag{9.19a}$$

$$= \frac{2\,u_1\,(V_1 - u_1)(1 + K\cos\phi)}{V_1^2}\,\eta_N \qquad (9.19b)$$

and $$\eta_{h\,max} = \frac{(1 + K\cos\phi)}{2}\,c_v^2 \qquad (9.20a)$$

$$= \frac{1 + K\cos\phi}{2}\,\eta_N \qquad (9.20b)$$

9.5. WORKING PROPORTIONS OF PELTON WHEEL

(1) The ideal velocity of the jet $\sqrt{2\,g\,H}$ is sometimes known as the spouting velocity. Where, H is the net head. However, the actual velocity of the jet is slightly less, due to friction loss in the nozzle. Thus

$$V_1 = c_2\sqrt{2\,g\,H} \qquad (9.21)$$

Where, c_v is known as the co-efficient of velocity of nozzle and its valve ranges from 0.97 to 0.99. The c_v can also be written as K_v.

(2) The ratio of wheel velocity to spouting velocity will be termed the speed ratio and denoted by symbol K_u.

$$\text{Speed ratio, } K_u = \frac{\text{Wheel velocity}}{\text{Spouting velocity}}$$

$$K_u = \frac{u_1}{\sqrt{2\,g\,H}}$$

or $$u_1 = K_u\sqrt{2\,g\,H} \qquad (9.22)$$

The value of K varies from 0.43 to 0.47. The maximum or theoretical value is 0.5.

It is a usual practice to give the ratio of u_1/V_1. Thus,

$$u_1 = K_u\sqrt{2\,g\,H}$$

and $$V_1 = c_v\sqrt{2\,g\,H}$$

Therefore, $$\frac{u_1}{V_1} = \frac{K_u}{c_v} \qquad (9.23)$$

The ratio of u_1/V_1 varies according to the values of K_u and c_v.

(3) Mean diameter or pitch circle diameter D of the Pelton wheel is obtained as follows. If the wheel rotates at a speed of N, then,

$$u_1 = \frac{\pi D N}{60} \tag{9.24}$$

(4) Discharge through the nozzle is given by

$$Q = \frac{\pi}{4} d^2 V_1 \tag{9.25}$$

The least diameter of the jet is obtained from the equation (9.25).

(5) The shaft power or brake power is calculated as follows.

$$P = \rho g Q H \eta_o \tag{9.26}$$

Where, $\eta_o = \eta_h \times \eta_m$

It is easy to calculate the hydraulic efficiency without a confusion of whirl velocity at outlet either positive or negative by the formula.

$$\eta_h = \frac{2 u_1 (V_1 - u_1)(1 + K \cos \phi)}{V_1^2}$$

It is a general practize to give mechanical efficiency but, if it is not given, it can be assumed approximately 95%.

(6) The ratio of wheel diameter to jet diameter is known as jet ratio and it is denoted by a symbol **m**.

$$m = \frac{D}{d} \tag{9.27}$$

For maximum efficiency the jet ratio varies from 11 to 14 but average value of 12 is used.

(7) For the design of the complete bucket we can employ reference values obtained on the base of detailed layouts and verified by the experiments and experience.

The routine design the dimension of the bucket are determined according to the diameter of the cross section of the jet. The shape of the bucket is approximately defined by the following dimensions.

Length of the bucket = 2.5 to 2.8 times jet diameter.

Width of the bucket = 2.8 to 4 times jet diameter

Depth of the bucket = 0.95 of the jet diameter.

Buckets are designed according to the ratio d/D or m=D/d

$$\frac{d}{D} = \frac{1}{16} \text{ to } \frac{1}{14} = \text{to} \quad \text{for current cases of normal design}$$

$$\frac{d}{D} = \frac{1}{10} \text{ to } \frac{1}{6} = \text{to} \quad \text{for extreme cases of very carefullyand}$$

specially design buckets.

Sometimes, the ratio d/D is also taken according to the relation

$$d = \frac{75\,D}{1000} + 5\text{ mm}$$

If for the given case the ratio d/D assumes a higher magnitude than indicated, two or more inlet nozzles are employed for one wheel. The best energy characteristics are obtained at

$$\frac{d}{D} = \frac{1}{10} \text{ to } \frac{1}{18}$$

however, greater values (up 1/7 to 1/8) are also tolerable and sometimes when head is very high and power is low, d/D may be considerably smaller (say 1/30 to 1/40) when leads to increasing diameter and reducing the speed.

At full opening η_o may be 87 to 88%. Now

$$N_s = (245 \text{ to } 253)\ \sqrt{n}\ (d/D)$$

Ranges of N_s for most favourable ratio (d/D = 1/10 to 1/18) are given in Table 9.1.

Table 9.1. Relation between number of nozzle (jets) n and specific speed Ns

Pelton turbine	Number of nozzle (jets) 'n'	Specific speed Ns
One nozzle	1	14 to 23
Two nozzle	2	20 to 34
Three nozzle	3	26 to 40
Four nozzle	4	32 to 47
Six nozzle	6	38 to 58

To determine D and N for different values of number of jets (nozzles) n are given in Table 9.2.

Table 9.2. Relation between wheel diameter, speed of rotation and number of jets

Number of jets (nozzles) n	1	2	3	4	5
Wheel diameter D (n)	6.3	4.45	3.63	3.15	2.57
Speed of rotation N (rpm)	161.8	229.2	280.9	323.7	396.8

The number of buckets for a Pelton wheel should be such that the jet is always completely intercepted by the buckets so that the volumetic efficiency of the turbine is very close to unity. The number of buckets is usually more than 15. Certain empirical formulae have been developed for determining the number of buckets. One such formula which is widely used has been given by **Taygun** according to which the number of buckets Z is approximately given by

$$Z = \left(\frac{D}{2d} + 15\right)$$

The above equation holds good for all values of m = D/d ranging from 6 to 35.

If the actual number differs from the value given by the above relation, the efficiency will not be maximum.

9.6. NUMBER OF JETS AND WHEELS PER UNIT

The specific speed of a Pelton wheel is calculated for a single jet is given by

$$N_s = \frac{N\sqrt{P}}{H^{5/4}}$$

where, P = Brake power for one wheel one jet.

The simplest case is that with one nozzle and one wheel. In some cases, however, to produce more power with the same wheel, number of nozzles are used. Though theoretically six nozzles can be used with one Pelton wheel, it is a good practice not to use more than two nozzles per runner when runner is horizontal and not more than four nozzles per runner when runner is vertical.

Maximum number of nozzles so far used in a turbine is six in a vertical installation and two in a horizontal installation. To increase the specific speed and more power, multi jets are essential but for such an arrangement makes the governing of turbine complicated and more expensive.

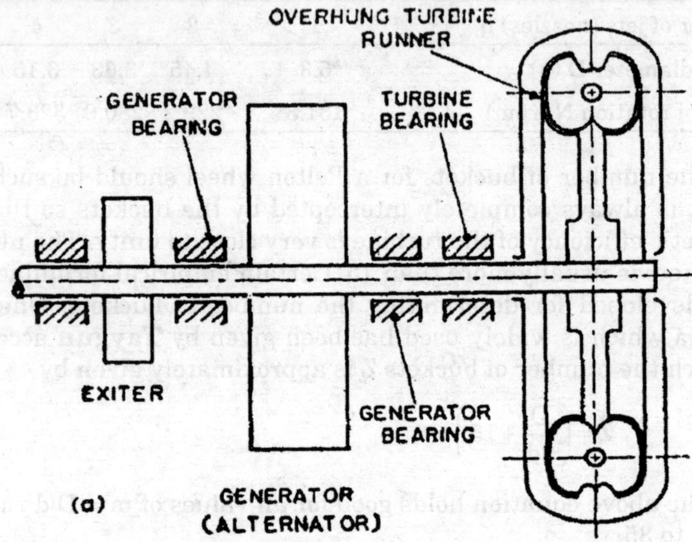

Fig. 9.6. Single overhung unit

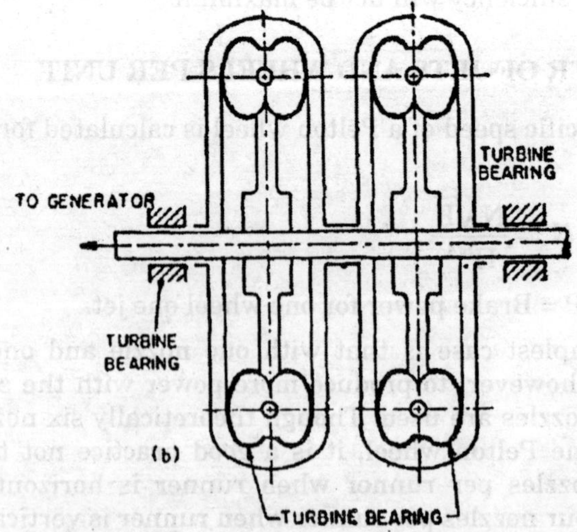

Fig. 9.7. Double Runner arrangement

For one wheel and one nozzle, the turbine shaft is seated in one bearing and other bearing is for the generator, or the runner may be fastened in an overhung position on the generator shaft. When two wheels are used, they are mounted outboard at either end of a horizontal shaft carried by two heavy bearings, with the electric generator rotor mounted in between. Each of the two wheels may have two jets, making four in all (Figures 9.6 to 9.10).

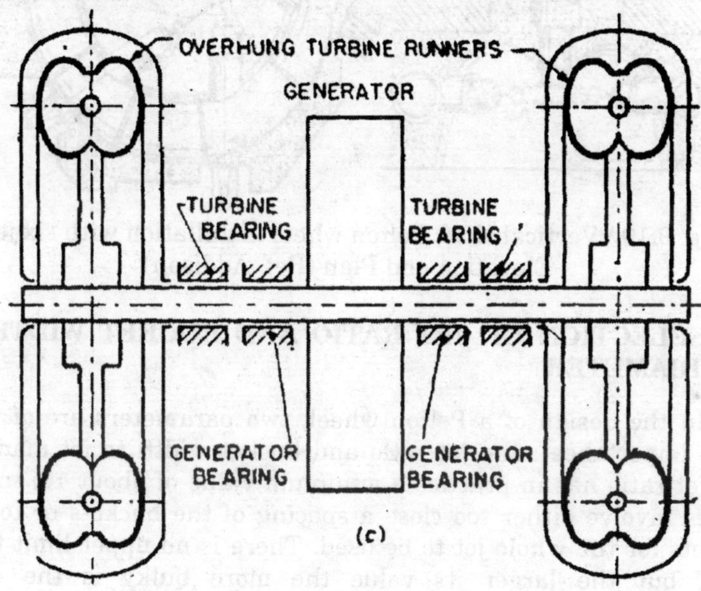

Fig. 9.8 Double overhung arrangement

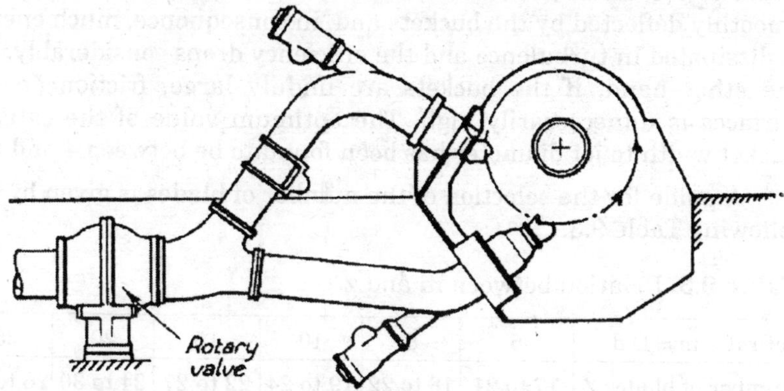

Fig. 9.9. Pelton wheel with two jets (Ref. Addison)

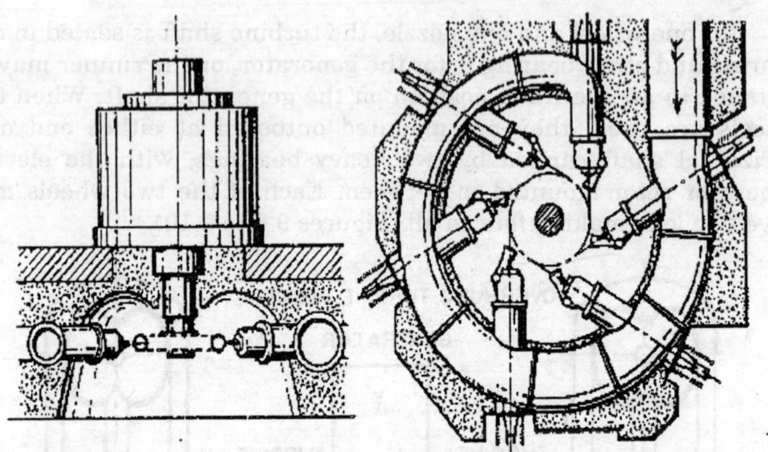

Fig. 9.10. Vertical shaft Pelton wheel installation with six jets Elevation and Plan (Ref. Addison)

9.7. SELECTION OF JET RATIO AND BUCKET WIDTH TO JET DIAMETER

In the design of a Pelton wheel, two parameters are of particular importance : the jet ratio and bucket width to jet diameter. The jet ratio has in practice a minimum value of about 10; smaller values involve either too close a spacing of the buckets or too few buckets for the whole jet to be used. There is no upper limit to the ratio, but the larger its value the more bulky is the entire installation; the smaller ratios are thus desirable. If the bucket width is too small in relation to the jet diameter, the fluid is not smoothly deflected by the buckets and, in consequence, much energy is dissipated in turbulence and the efficiency drops considerably. On the other hand, if the buckets are unduly large, friction on the surfaces is unnecessarily high. The optimum value of the ratio of bucket width to jet diameter has been found to be between 4 and 5.

A guide for the selection of the number of blades is given by the following Table 9.3.

Table 9.3. Relation between **m** and **z**

Jet ratio m = D d	6	8	10	15	20	25
Number of blades Z	17 to 21	18 to 22	19 to 24	22 to 27	24 to 30	26 to 33

9.8. SELECTION OF SPEED

Pelton turbines operate over a wide range of speed. However, there is a particular speed at which, when operated, maximum efficiency can be achieved.

By increasing the speed of a turbine its specific speed is also increased. High specific speed will result in smaller size of turbine. Due to increase in specific speed, the diameter of jet will decrease and this will increase the value of jet ratio for the same turbine. An increase in jet ratio will increase the efficiency of the turbine (Table 9.4). However, there is one disadvantage due to increase of specific speed and that is the increase in number of nozzle which will complicate the governing problem.

Table 9.4. Relation between m, Ns and η_o

$m = \frac{D}{d}$	6.5	7.5	10	20
$N_s = \frac{N\sqrt{P}}{H^{5/4}}$	30	27.5	20.6	5.6
η_o (%)	82	86	89	90

If the speed of the Pelton turbine is made higher, the speed of the electric machine which is directly coupled to the turbine will also increase and thus a less number of pair of poles are required reducing the electric generators, cost. But the materials of construction for the parts of high speed turbine and electric generator will have to withstand high stresses resulting from high speed which means more costly material will have to be used.

9.9. SELECTION OF HIGH HEAD

Pelton turbine is best suited for high heads. It is efficient and reliable when employed under these conditions. Pelton turbine is less suited for smaller heads. To develop a given output power under a smaller head the rate of flow would need to be greater with a consequence increase in jet diameter. The increase of jet diameter in turn requires an increase of wheel diameter. On the other hand the jet velocity and consequently peripheral velocily of runner will be reduced because of the low head. These two factors i.e., increase of runner diameter and decrease of runner velocity will make the turbine bulky and slow running for low heads. In fact, for low heads, turbines of the reaction type are more suitable.

9.10. RELATION BETWEEN SPECIFIC SPEED AND JET RATIO

An expression for specific speed N_s and jet ratio m = D/d for Pelton turbine may be obtained as follows :

$$u_1 = \frac{\pi D N}{60} = K_u \sqrt{2gH}$$

$$\therefore N = \frac{60 K_u \sqrt{2gH}}{\pi D}$$

$$= \frac{84.6 K_u \sqrt{H}}{D}$$

and $$P = \rho g Q H \eta_o$$

$$= 1000 \times 9.81 \, Q H \eta_o \text{ Watts}$$

$$= 9.81 \, Q H \eta_o \text{ kW}$$

$$= 9.81 \times \frac{\pi}{4} d^2 V_1 H \eta_o$$

$$= 9.81 \times \frac{\pi}{4} \times c_v \sqrt{2gH} \, d^2 H \eta_o$$

$$= 34.13 (c_v d^2 \eta_o H^{1/2})$$

Specific speed, $$N_s = \frac{N\sqrt{P}}{H^{5/4}}$$

$$= \frac{84.6 K_u \sqrt{H}}{D H^{5/4}} \left[34.13 c_v d^2 \eta_o H^{1/2}\right]^{1/2}$$

$$= \frac{494.24 K_u \sqrt{H} \left(c_v d^2 \eta_o H^{3/2}\right)^{1/2}}{DH^{5/4}}$$

$$= \frac{494.24 K_u (c_v \eta_o)^{1/2}}{(D/d)}$$

$$= \frac{494.24 K_u (c_v \eta_o)^{1/2}}{m} \quad (9.28)$$

Assuming $\eta_o = 87\%$, $K_u = 0.46$ and $c_v = 0.98$

Substituting in equation (9.28), we get

$$N_s = \frac{210}{m} = 210\left(\frac{d}{D}\right) \tag{9.29}$$

or,
$$\frac{d}{D} = \frac{N_s}{210} = 0.00476\, N_s \tag{9.30}$$

Thus the ratio, jet diameter to wheel diameter, varies directly as the specific speed of the wheel.

The above relationship is for single jet Pelton wheel. Since the value of d/D is more or less constant and in most of the cases it is equal to 1/12' almost all the Pelton turbines have the same specific speed.

For a multiple jet Pelton turbine having 'n' number of jets, the expression for specific speed may be obtained in the same manner. In this case the discharge will be (nQ) and hence the power will be (nP). Therefore, in this case the specific speed will be $\sqrt{n}$ times the specific speed for a single jet Pelton turbine.

9.11. LIMITATIONS OF THE PELTON TURBINE

There is some limitations of running a Pelton turbine for a very high head. When the head is very high, it becomes necessary to reduce the r.p.m. to limit the centrifugal stresses on the buckets. Thus the normal specific speed range diminishes at a very high head, and it may not be possible to run a turbine at a specific speed corresponding to the point of maximum efficiency.

A turbine should run at some constant synchronous speed if it drive an alternating current generator. For 50 H_z cycle current, the rotational speed of the shaft expressed in rpm is given by N = 3000/p, where ρ is the number of poles in a generator. A generator will not be economical if the shaft speed is lower than 75 rpm.

If the available head for a Pelton turbine is rather low and has to run within its normal range of specific speed, either the power or the rpm has to be reduced. To generate large power at low head, the rpm may fall below the economic limit.

ILLUSTRATIVE EXAMPLES

9.1 A Pelton turbine required to develop 6615 kW when working under a head of 300 m. The wheel may rotate at 500 rpm. Assuming the jet ratio as 10 and the overall efficiency of 85% calculate the following

(a) quantity of water required

(b) diameter of wheel

(c) number of jets.

Solution

Assuming $c_v = 0.97$ and $K_u = 0.45$

$$P = \rho g Q H \eta_0$$

$$6615 \times 10^3 = 1000 \times 9.81 \times Q \times 300 \times 0.85$$

$$Q = 2.644 \text{ m}^3/\text{s}$$ Ans.

$$V_1 = c_v \sqrt{2gH} = 0.97\sqrt{2 \times 9.81 \times 300} = 74.41\ m/s$$

$$u_1 = K_u \sqrt{2gH} = 0.45\sqrt{2 \times 9.81 \times 300} = 34.52\ m/s$$

$$u_1 = \frac{\pi D N}{60}$$

$$34.52 = \frac{\pi D \times 500}{60}$$

$$D = 1.318\ m$$ Ans.

$$\therefore d = 0.1318\ m$$

$$Q = n\frac{\pi}{4}d^2 V_1$$

$$2.644 = n \times \frac{\pi}{4}(.1318)^2 \times 74.41$$

$$n = 2.6044 = 3$$ Ans.

9.2 A Pelton turbine develops 10 MW under a head of 500 m. Taking the overall efficiency of the turbine as 90%, jet ratio as 12, speed ratio as 0.45, find the nozzle diameter, wheel diameter and the speed.

Solution

$$P = \rho\, g\, Q\, H\, \eta_o$$

$$10 \times 10^6 = 1000 \times 9.81 \times Q \times 500 \times 0.9$$

$$Q = 2.265\, m^3/s$$

$$V_1 = \sqrt{2\, g\, H} \quad [Assuming\ C_v = 1]$$

$$= \sqrt{2 \times 9.81 \times 500}$$

$$= 99.045\, m/s$$

$$Q = \frac{\pi}{4} d^2\, V_1$$

$$2.265 = \frac{\pi}{4} d^2 \times 99.045$$

$$d = 0.1706\ m$$ Ans.

$$D = 12\, d = 12 \times 0.1706 = 2.047$$ Ans.

$$u_1 = K_u \sqrt{2\, g\, H} = \frac{\pi\, D\, N}{60}$$

$$0.45 \sqrt{2 \times 9.81 \times 500} = \frac{\pi \times 2.047 \times N}{60}$$

$$44.57 = 0.107\, N$$

$$N = 415.8\, rpm$$ Ans.

9.3 A Pelton wheel driven by two similar jets transmits 3750 kW to the shaft when running at 375 rpm. The head from the reservoir level to nozzle is 200 m and the efficiency of power transmission through the pipe line and nozzle is 90%. The centrelines of the jets are tangentially to a 1.45 m diameter circle. The relative velocity decrease by 10 per cent as the water traverses the bucket surfaces which are so shaped that they would, if stationary, deflects the jet through an angle of 165°. Neglecting the windage losses. Find:

(a) the efficiency of the runner, and

(b) the diameter of each jet.

Solution

$$H_g = 200\ m$$

$$\eta_t = \frac{H_g - h_f}{H_g}$$

$$0.9 = \frac{200 - h_f}{200}$$

$$\therefore h_f = 20\ m$$

$$H = H_g - h_f = 200 - 20 = 180\ m$$

$$V_1 = c_v \sqrt{2\,g\,H} \qquad [Assuming\ c_v = 0.98]$$

$$= 0.98\sqrt{2 \times 9.81 \times 180} = 58.23\ m/s$$

$$u_1 = \frac{\pi\ D\ N}{60} = \frac{\pi \times 1.45 \times 375}{60} = 28.47\ m/s$$

$$\therefore \frac{u_1}{V_1} = \frac{28.47}{58.23} = 0.4889$$

The hydraulic efficiency is given by

$$\eta_h = 2\left(\frac{u_1}{V_1}\right)\left(1 - \frac{u_1}{V_1}\right)(1 + K\cos\varphi)$$

$$= 2\,(0.4889)\,(1 - 0.4889)\,(1 + 0.9\cos 15^\circ)$$

$$= 0.9342 = 93.42\%$$

Assuming mechanical efficiency as 95%

$$\therefore \eta_0 = \eta_h \times \eta_m$$

$$= 0.9342 \times 0.95$$

$$= 0.8875 = 88.75\%$$

$$P = \rho\,g\,Q\,H\,\eta_0$$

$$3750 \times 10^3 = 1000 \times 9.81 \times Q \times 180 \times 0.8875$$

$$Q = 2.3928\ m^3/s$$

Therefore,

$$Q = n\frac{\pi}{4}d^2 V_1 = 2\times\frac{\pi}{4}d^2\times 58.23$$

$$d^2 = 0.02616$$

$$\therefore d = 0.1617\ m = 161.7\ mm$$ Ans.

Aliter

The hydraulic efficiency can also be obtained by assuming $\frac{u_1}{v_1} = (0.46\ to\ 0.49)$ taking $\frac{u_1}{v_1} = 0.48$.

$$\eta_h = 2\left(\frac{u_1}{v_1}\right)\left(1-\frac{u_1}{v_1}\right)(1+K\cos\varphi)$$

$$= 2(0.48)(1-0.48)(1+0.9\cos 15°)$$

$$= 0.93317 = 93.317\%$$ Ans.

Remaining procedure is same as above.

9.4 The gross available head for a Pelton wheel is 600 m out of which one third is lost due to friction in the penstock which takes water to the nozzle of the Pelton wheel. The rate of flow of water through the nozzles fitted at the end of the penstock is 2 m³/s. The angle of deflection of the jet is 165°. The reduction in relative velocity while passing through buckets as 15 percent.

Take speed ration K_u = 0.45 and coefficient of velocity c_v = 0.978 Determine.

(a) Power developed by the turbine,

(b) Hydraulic efficiency,

(c) The unit power, and (d) The dimensionless specific speed.

Solution

$$H_g = 600\ m, h_f = \frac{1}{3}H_g = \frac{600}{3} = 200\ m$$

$$\therefore H = H_g - h_f = 600 - 200 = 400\ m$$

$$V_1 = c_v\sqrt{2gH} = 0.978\sqrt{2\times 9.81\times 400}$$

$$= 86.64\ m/s$$

$$u_1 = k_u\sqrt{2\,g\,H} = 0.45\sqrt{2\times 9.81\times 400}$$

$$= 39.86\ m/s$$

Therefore, $\dfrac{u_1}{v_1} = \dfrac{39.86}{86.64} = 0.46$

(b) $\eta_h = 2\left(\dfrac{u_1}{V_1}\right)\left(1-\dfrac{u_1}{V_1}\right)(1+K\cos\varphi)$

$$= 2\,(0.46)(1-0.46)(1+0.85\cos 15^\circ)$$

$$= 0.9041$$

$$= 90.41\%$$ Ans.

Assuming mechanical efficiency as 95%. Therefore,

$$\eta_0 = \eta_h \times \eta_m = 0.9041\times 0.95 = 0.8588$$

$$= 85.88\%$$

(a) $P = \rho\,g\,Q\,H\,\eta_0$

$$= 1000\times 9.81\times 2\times 400\times 0.8588$$

$$= 6740.6\ kW$$ Ans.

(c) $P_u = \dfrac{P}{H^3/2} = \dfrac{6740.6}{(400)^3/2} = 0.8425$ Ans.

(d) Assuming $\dfrac{d}{D} = \dfrac{1}{10}$

$$Q = \frac{\pi}{4}d^2\,V_1$$

$$2 = \frac{\pi}{4}d^2\times 86.64$$

$$d^2 = 0.02939$$

$$\therefore d = 0.1714\,m$$

$$\therefore D = 10d = 1.714\ m$$

$$u_1 = \frac{\pi D N}{60}$$

$$39.86 = \frac{\pi \times 1.714 \times 4}{60}$$

$N = 444\ rpm = 7.4\ rps.$

K_n = Dimensionless specific speed

$$= \frac{N\sqrt{P/\rho}}{(g H)^{5/4}} = \frac{7.4\sqrt{6740.6}}{(9.81 \times 400)^{5/4}}$$

$= 0.01956$ Ans.

9.5 A double jet Pelton wheel required to develop 5400 kW has a specific speed of 20 and is supplied through a pipe line 800 m long from a reservoir, the level of which is 350 m above the nozzles.

Allowing for 5 per cent friction loss in the pipe line, calculate:

(a) the speed,

(b) diameter of the jets,

(c) mean diameter of bucket circle;

(d) diameter of the supply pipe.

Assuming c_v for the jets is 0.98, bucket speed is equal to 0.46 of jet speed, overall efficiency of wheel 85 percent and $\lambda = 4f = 0.024$.

Solution

Given data

$$N_s = 20, l_p = 800 m,\ H_g = 350\ m$$

$$P = 5400\ kW, u_1 = 0.46\ V_1, c_v = 0.98$$

(a) $H = 0.95 \times H_g = 0.95\ kt\ 350 = 332.5\ m$

$$N_s = \frac{N\sqrt{P}}{H^{5/4}}$$

$$20 = \frac{N\sqrt{5400}}{(332.5)^{5/4}} = 0.0517\,N$$

$$\therefore N = 386.43\ rpm$$

$$V_1 = c_v\sqrt{2\,g\,H} = 0.98\sqrt{2\times 9.81\times 332.5} = 79.15\ m/s$$

$$u_1 = 0.46\,V_1 = 0.46\times 79.15 = 36.41\ m/s$$

(c) $u_1 = \dfrac{\pi\,D\,N}{60}$

$$36.41 = \frac{\pi\,D\times 386.43}{60}$$ Ans.

(b) $P = \rho\,g\,Q\,H\,\eta_0$

$$5400\times 10^3 = 1000\times 9.81\times Q\times 332.5\times 0.85$$

$$Q = 1.947\ m^3/s$$

$$Q = 2\times\frac{\pi}{4}d^2\,V_1$$

$$1.947 = 2\times\frac{\pi}{4}d^2\times 79.15$$

$$d^2 = 0.015565$$ Ans.

(d) $h_f = 5\%\ of\ 350 = 17.5\ m$

$$h_f = \frac{\lambda\,l_p\,Q^2}{12.1\,d_p^5}$$

$$17.5 = \frac{0.025\times 800\times (1.947)^2}{12.1 d_p^5}$$

$$d_p^5 = 0.3437$$

$$\therefore d_p = 0.80768\ m = 807.68\ mm$$ Ans.

9.6 **Water under a head of 270 m is available for a hydroelectric power station, and is to be delivered to the power house through three pipes 2.4 km long in which the friction loss is to be 24 m.**

It is decided to install a number of single jet Pelton wheels with a specific speed not exceeding 38, to produce a total shaft output of 18 MW. The wheel speed is to be 650 rpm and the ratio of bucket to jet speed 0.46. Assume that the overall efficiency of the wheel will be 87%, and that the nozzle will have a c_d of 0.94 and a c_v of 0.97.

Determine : (a) Number of Pelton wheels to be used.

(b) The wheel diameter

(c) The jet diameter

(d) The diameter of the supply pipes

Solution

(a) $H_g = 270\ m, h_f = 24\ m, \therefore H = H_g - h_f = 270 - 24 = 246\ m$

$$N_s = \frac{N\sqrt{P}}{H^{5/4}}$$

$$38 = \frac{650\sqrt{P}}{(246)^{5/4}}$$

$$\therefore P = 3243.98\ kW.$$

$$\text{Number of wheels} = \frac{18000}{3243.98}$$

$$= 5.5487 = 6$$ Ans.

(b) $V_1 = c_v\sqrt{2\,g\,H} = 0.97\sqrt{2 \times 9.81 \times 246} = 67.38\ m/s$

$$u_1 = 0.46\,V_1 = 0.46 \times 67.38 = 30.998\ m/s$$

$$\therefore u_1 = \frac{\pi\,D\,N}{60}$$

$$30.998 = \frac{60D \times 650}{60}$$

$$\therefore D = 0.9108\ m$$

$$\text{Power for one wheel} = \frac{18000}{6} = 3000\,kW$$

$$P = \rho\,g\,Q\,H\,\eta_0$$

$$3000 \times 10^3 = 1000 \times 9.81 \times Q \times 246 \times 0.87$$

$$Q = 1.4288\ m^3/s$$

$$\therefore Q = c_d \times \frac{\pi}{4} d^2 \times V_1$$

$$1.4288 = 0.94 \times \frac{\pi}{4} d^2 \times 67.38$$

$$\therefore d = 0.169\ m$$ Ans.

(d) Total discharge for 6 wheels $= 6\,Q = 6 \times 1.4288$

$$= 8.5728\ m^3/s$$

Number of pipes = 3

Discharge per pipe $= \dfrac{8.5728}{3} = 2.8576$

$$\therefore h_f = \frac{\lambda\, l_p\, Q^2}{12.1\, d_p^5}$$

$$24 = \frac{0.024 \times 2.4 \times 1000 \times (2.8576)^2}{12.1\, d_p^5}$$

$$\therefore d_p = 1.1012\ m$$ Ans.

9.7 **A Pelton wheel operates under a head of 274 m. The jet diameter is 100 mm, the coefficient of velocity for the nozzle is 0.97, and the ratio of bucket speed to jet speed is 0.47. The angle of deflection of the water in passing over the buckets is 160° and the relative velocity is reduced by 10% in the process. Calculate the diameter of the runner, the hydraulic efficiency, the power developed at the buckets and the shaft power if the speed is 600 rpm and the mechanical efficiency 90%.**

Solution

$$H = 274\ m$$

$$V_1 = c_v \sqrt{2\,g\,H} = 0.97\sqrt{2 \times 9.81 \times 274}$$

$= 71.12\ m/s$

$u_1 = 0.47\ V_1 = 0.47 \times 71.12 = 33.426\ m/s$

$\therefore u_1 = \dfrac{\pi D N}{60}$

$33.426 = \dfrac{\pi D \times 600}{60}$

$D = 1.064\ m$ Ans.

$\eta_h = 2\left(\dfrac{u_1}{V_1}\right)\left(1 - \dfrac{u_1}{V_1}\right)(1 - k \cos \gamma)$

$= 2\,(0.47)\,(1 - 0.47)\,(1 - 0.9 \cos 160°)$

$= 0.9195 = 91.95\%$ Ans.

$\eta_0 = \eta_h \times \eta_m = 0.9195 \times 0.9 = 0.8275$

$= 82.75\%$

$Q = \dfrac{\pi}{4} d^2\ V_1 = \dfrac{\pi}{4}\,(0.1)^2 \times 71.12$

$= 0.558\ m^3/s$

Shaft power $P = \rho\, g\, Q\, H\, \eta_0$

$= 1000 \times 9.81 \times 0.558 \times 274 \times 0.8275$

$= 1242.42\ kW$ Ans.

The transfer of power from the jet of water to the buckets can be calculated by applying Eulers turbine equation

$= \dot{m}\,[V_{w1} + V_{w2}]\,u_1$

$= \dot{m}\,[(V_1 - u_1)(1 + k \cos \varphi)]\,u_1$

$= 1000 \times 0.558\,[(71.12 - 33.42)\,(1 + 0.9 \cos 20°) \times 33.52]$

$= 18648.36\,(37.7 \times 1.8457)$

$= 1297623.2\ W$

$= 1297.6\ kW$ Ans.

9.8 In a hydroelectric scheme a number of Pelton wheels are to be used under the following conditions: total output required 30 MW; gross head 245 m, speed 6.25 rps; 2 jets per wheel; C_v of nozzle 0.97; maximum overall efficiency (based on conditions immediately before the nozzle) 81.5%; dimensionless specific speed not to exceed 0.022 revolution per jet; head lost to friction in pipe line not to exceed 12 m.

Calculate :

(a) the number of wheel's required

(b) the diameter of the jets and wheels

(c) the hydraulic efficiency, if the blades deflected the water through 165° and reduce its relative velocity by 15%.

(d) the percentage of the input power which remain as kinetic energy of the water at discharge.

Solution

$$H_g = 245\ m, h_f = 12\ m \ \therefore H = H_g - h_f = 245 - 12$$

$$= 233\ m.$$

$$K_n = \frac{N\sqrt{P + \rho}}{(g\,H)^{5/4}}$$

Where P is in watts

N is in rps

H is in meter

$$\rho = 1000\ kg/m^3$$

and $g = 9.81\ m/s^2$

Alternately use $N_s = 1042\ K_n = \dfrac{N\sqrt{P}}{H^{5/4}}$

$$0.022 = \frac{6.25\sqrt{P/1000}}{(9.81 \times 233)^{5/4}}$$

$$\therefore P = 3094907.5\ W = 3094.907\ kW \text{ (for one jet)}$$

$$= 2 \times 3094.907 \text{ (for one wheel)}$$

$= 6189.814 kW$

Number of wheels required $= \dfrac{30000}{6189.814}$

$= 4.8466 \approx 5$ Ans.

Therefore, power per wheel

$$P = \frac{30000}{5} = 6000\ kW$$

$$P = \rho\, g\, Q\, H\, \eta_0$$

$$6000 \times 10^3 = 1000 \times 9.81 \times Q \times 233 \times 0.815$$

$$Q = 3.22\ m^3/s$$

Discharge per jet $= \dfrac{3.22}{2} = 1.61\ m^3/s$

$$V_1 = c_v\, \sqrt{2\, g\, H} = 0.97\, \sqrt{2 \times 9.81 \times 233}$$

$$= 65.584\ m/s$$

$$Q = \frac{\pi}{4} d^2 \times 65.584$$

$\therefore d = 0.1768\ m$ Ans.

$$u_1 = 0.46\ V_1 = 0.46 \times 65.584 = 30.168\ m/s$$

$$\therefore u_1 = \frac{\pi\, D\, N}{60} \text{ or, } 30.168 = \pi\, D \times 6.25$$

$D = 1.536\ m$ Ans.

$$\eta_h = 2\left(\frac{u_1}{V_1}\right)\left(1 - \frac{u_1}{V_1}\right)(1 + k \cos \varphi)$$

$$= 2 \times 0.46\,(1 - 0.46)\,(1 + 0.85 \cos 15°)$$

$$= 0.9041 = 90.41\%$$

Hydraulic efficiency (before the nozzle condition)

$$\eta_h = 2\left(\frac{u_1}{V_1}\right)\left(1 - \frac{u_1}{V_1}\right)(1 + K \cos \varphi)\ \eta_n \qquad \left[\eta_n = c_v^2\right]$$

$$= 2\,(0.46)\,(1 - 0.46)\,(1 + 0.85\cos 15°)\,(0.97)^2$$

$$= 0.8504 = 85.04\%$$ **Ans.**

(d) Using the outlet velocity triangle (Fig.)

$$V_{r1} = V_1 - u_1 = 65.584 - 30.168 = 35.416\ m/s$$

$$V_{r2} = 0.85 \times 35.416 = 30.1\ m/s$$

Therefore,

$$V_2^2 = u_2^2 + V_{r2}^2 - 2\,u_2\,V_{r2}\,\cos\varphi$$

$$= (30.168)^2 + (30.1)^2 - 2 \times 30.168 \times 30.1 \cos 15°$$

$$= 909.62 + 906.01 - 1750.2548$$

$$= 65.3752$$

$$\therefore V_2 = 8.085\ m/s$$

$$\text{K.E. ratio} = \frac{V_2^2}{V_1^2} = \frac{(8.085)^2}{(65.584)^2} = \frac{65.3752}{4301.26} = 0.015199$$

$$= 1.52\%$$ **Ans.**

9.9 The following data refer to a Pelton wheel : 4 nozzles each 50 mm in diameter with $c_v = 0.97$; reservoir head 300 m; head lost in friction 30m on 360 m of pipe line with $\lambda = 4f = 0.024$; bucket pitch circle diameter 0.83 m; bucket speed is equal to 0.46 of jet speed; bucket friction reduces the relative velocity by 15 percent; angle through which the buckets deflect the jet 165°; mechanical efficiency 94 per cent.

Determine:

(a) The diameter of penstock supplying the machine

(b) The speed of rotation

(c) The hydraulic and overall efficiencies of the machine

(d) The output power

(e) The unit power, unit speed and specific speed.

Solution

Given data

$H_g = 300\ m,\ h_f = 30\ m,\ l_p = 360\ m, d = 50\ mm,$

$n = 4, c_v = 0.97\ , \lambda = 4,\ f = 0.024, D = 0.83\ m,$

$u_1 = 0.46\ V_1 \quad k = 0.85$

$\eta_m = 94\% \quad \varphi = 15°$

$H = H_g - h_f = 300 - 30 = 270\ m$

$V_1 = C_v \sqrt{2\,g\,H} = 0.97\sqrt{2 \times 9.81 \times 270} = 70.6\ m/s$

$u_1 = 0.46\ V_1 = 0.46 \times 70.6 = 32.47\ m/s$

$$Q = n\,\frac{\pi}{4}\,d^2\,V_1$$

$$= 4 \times \frac{\pi}{4}(0.05)^2 \times 70.6$$

$$= 0.55\ m^3/s$$

(a) $$h_f = \frac{\lambda\,l_p\,Q^2}{12.1\,d_p^5}$$

$$30 = \frac{0.024 \times 360 \times (0.55)^2}{12.1\,d_p^5}$$

$$d_p^5 = 7.2 \times 10^{-3}$$

$\therefore d_p = 0.3728\ m = 372.8\ mm$ **Ans.**

(b) $$u_1 = \frac{\pi\ D\,N}{60}$$

$$32.47 = \frac{\pi \times 0.83 \times N}{60}$$

$N = 747.15\ rpm$ **Ans.**

$$\eta_h = \frac{2\,u_1\,(V_1 - u_1)(1 + k\cos\varphi)}{V_1^2}$$

$$= 2\left(\frac{u_1}{V_1}\right)\left(1 - \frac{u_1}{V_1}\right)(1 + K\cos\varphi)$$

$$= 2\,(0.46)\,(1 - 0.46)\,(1 + 0.85 \cos 15°)$$

$$= 0.9047$$

$$= 90.47\%$$ Ans.

$$\eta_0 = \eta_h \times \eta_m = 0.9047 \times 0.94 = 0.85$$

$$= 85\%$$ Ans.

(d) $P = \rho\, Q\, g\, H\, \eta_0$

$$= 1000 \times 0.55 \times 9.81 \times 270 \times 0.85$$

$$= 1238864 \; Watt$$

$$= 1238.864 \; kW$$ Ans.

(e) (i) $P_u = \dfrac{P}{H^{3/2}} = \dfrac{1238.864}{(270)^{3/2}} = 0.279$ Ans.

(ii) $N_u = \dfrac{N}{\sqrt{H}} = \dfrac{747.15}{\sqrt{270}} = 45.47$ Ans.

(iii) $N_s = N_u \sqrt{P_u}$

$$= 45.47\,\sqrt{0.279}$$

$$= 24$$ Ans.

Aliter

$$N_s = \frac{N\sqrt{P}}{H^{5/4}}$$

$$= \frac{747.15\,\sqrt{1238.864}}{(270)^{5/4}}$$

$$= 24$$ Ans.

9.10 A Pelton wheel is required to develop 4500 kW at 400 rpm, operating under an available head of 360 m. There are two equal jets and the bucket angle is 165°. Calculate the bucket pitch circle diameter, the cross sectional area of each jet and hydraulic efficiency of the turbine making the following assumptions :

(i) The loss in the nozzle is 3 percent of the available head.

(ii) The overall efficiency is 85% when the water is discharged from the wheel in a direction parallel to the axis of rotation.

(iii) The bucket friction reduces the relative velocity by 15 percent.

(iv) The bucket speed is equal to 0.46 of jet speed.

Solution

Head loss due to nozzle h_n

$$\therefore h_n = \frac{V_1^2}{2g}\left(\frac{1}{c_v^2} - 1\right)$$

$$0.03\,\frac{V_1^2}{2g} = \frac{V_1^2}{2g}\left(\frac{1}{c_v^2} - 1\right)$$

$$\frac{1}{c_v^2} = 1 + 0.03$$

$$c_v^2 = \frac{1}{1.03} = 0.97087$$

$$\therefore c_v = 0.9853$$

$$\eta_h = 2\left(\frac{u_1}{V_1}\right)\left(1 - \frac{u_1}{V_1}\right)(1 + K\cos\varphi)$$

$$2 \times 0.46\,(1 - 0.46) \times (1 + 0.85\,\cos 15°)$$

$$\therefore\ \eta_h = 0.9047 = 90.47\%$$ Ans.

$$P = \rho\,g\,Q\,H\,\eta_0$$

$$4500 \times 10^3 = 1000 \times 9.81 \times Q \times 360 \times 0.85$$

$$Q = 1.499\ m^3/s$$

$$V_1 = c_p\,\sqrt{2\,g\,H}$$

$$= 0.985\,\sqrt{2 \times 9.81 \times 360}$$

$$= 82.78\ m/s$$

$$Q = n\, a_n\, V_1$$

$$1.499 = 2 \times a_n \times 82.78$$

$$a_n = 9.0538 \times 10^{-3}\ m^2 = 9053.878\ mm^2$$ **Ans.**

$$u_1 = 0.46\, V_1 = 0.46 \times 82.78 = 38.078\ m/s$$

$$\therefore \quad u_1 = \frac{\pi\, D\, N}{60}$$

$$38.078 = \frac{\pi \times D \times 400}{60}$$

$$D = 1.818\ m$$ **Ans.**

9.11. **A Pelton wheel is to be designed for the following data:**

Shaft power = 5000 kW

Gross head = 300 m

Transmission efficiency = 90 percent

Mechanical efficiency = 95 percent

Speed of the wheel = 600 rpm

Ratio of wheel diameter to Jet diameter = 10

Ratio of Bucket velocity to Jet velocity = 0.46

The blades deflect the water through 160° and reduces its relative velocity by 10%.

The loss in the nozzle = 3% of the available head.

Find the number of jets, diameter of jet, diameter of the wheel and quantity of water required.

Solution:

$$\text{Nozzle loss} = h_n = \frac{V_1^2}{2g}\left[\frac{1}{c_v^2} - 1\right]$$

$$0.03\,\frac{V_1^2}{2\,g} = \frac{V_1^2}{2\,g}\left[\frac{1}{c_v^2} - 1\right]$$

$$\frac{1}{c_v^2} - 1 = 0.03$$

$$\frac{1}{c_v^2} = 1 + 0.03$$

$$c_p^2 = \frac{1}{1.03} = 0.9708$$

$$\therefore \quad c_v = 0.985$$

$$\text{Transmission efficiency} = \eta_t = \frac{H_g - h_f}{H_g}$$

$$0.9 = \frac{300 - h_f}{300}$$

$$\therefore h_f = 30\ m$$

$$\therefore H = H_g - h_f = 300 - 30 = 270\ m$$

$$V_1 = c_v \sqrt{2\,g\,H} = 0.985\sqrt{2 \times 9.81 \times 270}$$

$$= 71.69\ m/s$$

$$u_1 = 0.46\,V_1 = 0.46 \times 71.69 = 32.978\ m/s$$

$$\text{Hydraulic efficiency} \quad \eta_h = 2\left(\frac{u_1}{V_1}\right)\left(1 - \frac{u_1}{V_1}\right)(1 + K\cos\varphi)$$

$$\eta_h = 2\,(0.46)\,(1 - 0.46)\,(1 + 0.9\cos 20°)$$

$$= 0.9169 = 91.69\%$$

$$\therefore \eta_0 = \eta_h \times \eta_m = 0.9169 \times 0.95 = 0.871$$

$$= 87.1\%$$

$$P = \rho\,g\,Q\,H\,\eta_0$$

$$5000 \times 10^3 = 1000 \times 9.81 \times Q \times 270 \times 0.871$$

$$\therefore Q = 2.167\ m^3/s$$ **Ans.**

$$u_1 = \frac{\pi\,D\,N}{60}$$

$$32.987 = \frac{\pi\,D \times 600}{60}$$

$\therefore D = 1.04972\, m$ Ans.

$\therefore d = \frac{D}{10} = \frac{1.04972}{10} = 0.104972\, m$ Ans.

$$Q = n\,\frac{\pi}{4}\,d^2\,V_1$$

$$2.167 = n \times \frac{\pi}{4}(.104972)^2 \times 71.69$$

$$2.167 = n \times 0.6204$$

$$n = 3.49339 \approx 4$$

4 Jets will be used. Ans.

9.12. A 8 km long penstock (diameter 1m) carries water from a reservoir to an impulse turbine. Calculate the power that can be produced by the turbine if the machine is 80% efficient and the reservoir is at an elevation of 500m over the turbine. Assuming that λ = 4f = 0.015 for the penstock, neglect losses in the nozzle and take the jet diameter to be 0.1 m. Calculate the diameter of the turbine wheel to achieve a speed of 360 rpm. Assume ideal conditions for the bucket design.

Solution : Given

Given :

$$I_p = 8000\, m,\ d_p = 1 m,\ \eta_0 = 80\%, H_g = 500\, m$$

$$\lambda = 4\,f = 0.015, d = 0.1\, m,\ n = 360\ rpm$$

$c_v = 1$ (Neglect losses in the nozzle)

Ideal condition for bucket design i.e.

$$\frac{u_1}{V_1} = 0.5$$

$$Q = A_p\,V_p = a\,V_1$$

$$\frac{\pi}{4} d_p^2\,V_P = \frac{\pi}{4}\,d^2\,V_1$$

$$V_P = \left(\frac{0.1}{1}\right)^2 V_1 = 0.01\,V_1$$

$$h_f = \frac{\lambda\, l_p\, V_p^2}{2\, g\, d_p} = \frac{0.015 \times 8000 \times (0.01\,V_1)^2}{2 \times 9.81 \times 1}$$

$$= 6.1162 \times 10^{-4}\, V_1^2$$

$$= 6.1162 \times 10^{-4}\, c_v^2 \times 2\, g\, (H_g - h_f)$$

$$h_f = 0.012\, (H_g - h_f)$$

$$h_f = 0.012\, (500 - h_f)$$

$$h_f = 6 - 0.012\, h_f$$

$$\therefore\; h_f = \frac{6}{1.012}\, 5.9288\, m = 5.93\, m$$

Net heat $H = H_g - h_f = 500 - 5.93$

$$= 494.07\, m$$

$$V_1 = c_v\, \sqrt{2\, g\, H} = \sqrt{2 \times 9.81 \times 494.07}$$

$$= 98.456\, m/s$$

$$u_1 = 0.5\;\, V_1 = 0.5 \times 98.456 = 49.228\, m/s$$

$$u_1 = \frac{\pi\, D N}{60}$$

$$49.228 = \frac{\pi \times D \times 360}{60}$$

$\therefore D = 2.6116\, m$ **Ans.**

$$Q = \frac{\pi}{4} d^2\;\, V_1 = \frac{\pi}{4}\, (0.1)^2 \times 98.456 = 0.77327\, m^3/s$$

$$P = \rho\, Q\, g\, H\, \eta_0 = 1000 \times 0.77327 \times 9.81 \times 494.07 \times 0.8$$

$= 2998.33\, kW$ **Ans.**

9.13. Determine the power that can be achieved by a Pelton turbine with an efficiency of 80% if the elevation of the reservoir is 1800 m and the elevation at the nozzle is 800 m.

The diameter of the pipe is 0.6 m and it has a length of 10 km. The Pelton turbine has a diameter of 4 m and the penstock resistance coefficient (λ= 4f) is 0.02. The diameter of the jet is 0.1 m. Neglect losses in the nozzle and assume that the speed of the results from the conditions $V_1 = 2\ u_1$. Determine the angular speed of the wheel and the torque on the turbine shaft.

Solution :

$$Hg = HRL - TRL$$

$$= 1800 - 800 = 1000\ m$$

$$Q = A_p\ V_P = a\ V_1$$

$$V_P = \left(\frac{d}{dp}\right)^2 V_1 = \left(\frac{0.1}{0.6}\right)^2 V_1 = 0.0277\ V_1$$

$$h_f = \frac{\lambda\, l_p\, V_P^2}{2\,g\,d_p} = \frac{0.02 \times 10000 \times \left(0.0277\ V_1^2\right)}{2 \times 9.81 \times 0.6}$$

$$= 0.0131\ V_1^2$$

$$= 0.0131 \times c_v^2\ 2\,g\,\left(H_g - h_f\right) = 0.2572\,\left(1000 - h_f\right)$$

$$= 257.2 - 0.2572\ h_f$$

$$\therefore h_f = \frac{257.2}{1.2572} = 204.5829\ m$$

$$\therefore H = H_g - h_f = 1000 - 204.5829 = 795.41\ m$$

$$V_1 = c_v\ \sqrt{2\,g\,H} = \sqrt{2 \times 9.81 \times 795.41} = 124.92\ m/s$$

$$u_1 = 0.5\ V_1 = 0.5 \times 124.92 = 62.46\ m/s$$

$$u_1 = \frac{\pi\, D\, N}{60} = \frac{\pi \times 0.6 \times N}{60}$$

$$N = \frac{60\ \times 62.46}{\pi \times 4} = 298.22\ rpm$$

$$\omega = \frac{2\,\pi\, N}{60} = \frac{2\,\pi \times 298.22}{60} = 31.23\ rad/s$$ Ans.

$$Q = \frac{\pi}{4} d^2 V_1 = \frac{\pi}{4} (0.1)^2 \times 124.92$$

$$= 0.98115\ m^3/s$$

$$P = \rho Q g H \eta_0 = 1000 \times 0.98115 \times 9.81 \times 795.41 \times 0.8$$

$$= 61245.17\ kW$$

$$Power = T \times \omega$$

$$6124.517 = T \times 31.23$$

$$T = 196.11 \text{ kNm}$$

Ans.

9.14. Three identical, double jet Pelton wheels operate under a gross head of 400 m. The nozzle are 80 mm diameter with coefficient of velocity is 0.98. The pitch circle of the bucket is 1.3 m diameter and the bucket speed is 0.46 times the jet speed. The bucket deflect the jet by 165° and due to friction the relative velocity is reduced by 17%. The mechanical efficiency is 95%. The water from the reservoir is supplied to the turbines by means of two parallel pipes, each 600 mm diameter 480 m long, with λ = 4f = 0.032. If the quantity of water supplied to each turbine is 0.65 m³/s, calculate its specific speed and dimensionless specific speed (shape number).

Solution :

Total discharge $= 3 \times 0.65 = 1.95\ m^3/s$

$$h_f = \frac{\lambda\, l_p\, Q^2}{12.1\, d_p^5} = \frac{0.032 \times 480 \times (1.95)^2}{12.1 \times (0.6)^2} = 62.075\ m$$

$$H = H_g - h_f = 400 - 62.075 = 337.924\ m$$

$$V_1 = c_v \sqrt{2 g H} = 0.98 \times \sqrt{2 \times 9.81 \times 337.924}$$

$$= 79.796\ m/s$$

$$u_1 = 0.46\, V_1 = 0.46 \times 79.796 = 36.7\ m/s$$

$$\eta_h = 2\left(\frac{u_1}{V_1}\right)\left(1 - \frac{u_1}{V_1}\right)(1 + K \cos \phi)$$

$$= 2 \times 0.46\left(1 - 0.46\right)\left(1 + 0.83 \cos 15°\right) = 0.89509 = 89.51\%$$

$$\eta_0 = \eta_h \times \eta_m = 0.8951 \times 0.95 = 0.8503 = 85\%$$

$$P = \rho Q\ g\ H\ \eta_0$$

$$= 1000 \times 1.95 \times 9.81 \times 337.924 \times 0.85$$

$$= 5496.87\ kW$$

$$u_1 = \frac{\pi D N}{60}$$

$$36.7 = \frac{\pi \times 1.3 \times N}{60}$$

$$\therefore N = 539.17\ rpm$$

$$N_s = \frac{N\sqrt{P}}{H^{5/4}} = \frac{539.17\sqrt{5496.87}}{(337.924)^{1.25}}$$

$$= 27.59$$ **Ans.**

$$K_n = \frac{N_s}{1042} = 0.0264$$ **Ans.**

9.15. A Pelton wheel is supplied with water from a reservoir 274 m above the nozzle, through a long pipe. The combined efficiency of the pipe and nozzle is 85% and the friction losses in the pipe amount 19.8 m. If the bucket speed is 0.46 of the jet speed and if the water is deflected through 160° by the buckets and has its velocity reduced by 10% in the process, estimate the hydraulic efficiency of (a) the runner alone, (b) the turbine. If the mechanical efficiency of the turbine is 95%, estimate (c) the overall efficiency of the turbine alone, based on head at inlet to the nozzle, and (d) the overall efficiency of the pipe and turbine.

Solution:

Transmission or pipe efficiency

$$\eta_t = \frac{H_g - h_f}{Hg} = \frac{274 - 19.8}{274} = \frac{254.2}{274} = 0.9277 = 92.77\%$$

Nozzle efficiency $\eta_n = \dfrac{V_1^2}{2\,g\,H}$

Nozzle and pipe transmission efficiency i.e. combined efficiency $= \eta_n \times \eta_t$

$$\eta_c = \eta_t \times \eta_n$$

$$0.85 = 0.9277 \times \eta_n$$

$$\therefore \eta_n = 0.9162 = 91.62\%$$

(a) $$\eta_{h\ runner} = 2\left(\frac{u_1}{V_1}\right)\left(1-\frac{u_1}{V_1}\right)(1 + K\cos\phi)$$

$$= 2 \times 0.46 \times (1 - 0.46)\,(1 + 0.9\cos 20°)$$

$$= 0.9169 = 91.69\%$$ Ans.

(b) $$\eta_{h\ turbine} = \eta_{hr} \times \eta_n$$

$$= 0.9169 = \eta_{hr} \times \eta_n$$

$$= 0.84 = 84\%$$ Ans.

(c) Overall efficiency based on head at inlet to nozzle

$$\eta_0 = \eta_{ht} \times \eta_m$$

$$= 0.84 \times 0.95$$

$$= 0.798 = 79.8\%$$ Ans.

(d) Overall efficiency for pipe and turbine

$$\eta_0 = \eta_0 \times \eta_t$$

$$= 0.798 \times 0.9277$$

$$= 0.74 = 74\%$$ Ans.

9.16. A Pelton wheel of 1.5 m diameter is operated by two jets each of 95 mm diameter. The total head at the nozzles is 275 m. The nozzle velocity coefficient is 0.98, the bucket angle of deflection is 165° and the turbine develops 250 kW when the wheel is running at the theoretically best speed. Show how this speed is determined. Calculating the efficiency of the system and the velocity coefficient for the buckets.

Solution:

$$V_1 = c_v \sqrt{2\,g\,H}$$

$$= 0.98\sqrt{2 \times 9.81 \times 275}$$

$$= 71.985 \approx 72\ m/s$$

$$Q = n\frac{\pi}{4}d^2\,V_1$$

$$= 2 \times \frac{\pi}{4}\left(\frac{95}{1000}\right)^2 \times 72 = 1.02\ m^3/s$$

$$P = \rho\,Q\,g\,H\,\eta_0$$

$$2250 \times 10^3 = 1000 \times 1.02 \times 9.81 \times 275 \times \eta_0$$

$$\eta_0 = \frac{2250}{2753.6} = 0.817 = 81.7\%$$

Assuming mechanical efficiency as 96%

$$\eta_h = \eta_0 / \eta_m = \frac{0.817}{0.96} = 0.85 = 85\%$$

For best theoretical speed

$$\eta_{h\ \max} = \frac{1 + K\cos\phi}{2}$$

$$0.85 = \frac{1 + K\cos 15°}{2} = \frac{1 + 0.9659\,K}{2}$$

or, $1.7 = 1 + 0.9659\,K$

or, $0.9659\,K = 0.7$

$$\therefore K = \frac{0.7}{0.9659} = 0.72469$$ Ans.

9.17. In a hydroelectric power station, water is brought to the turbine in six penstocks each having a diameter of 1.5 m and each connected to a nozzle of 250 mm diameter at the end. There are in all six single jet, single wheel Pelton turbines working under a gross head of 600 m. The pitch circle diameter of each wheel is 3 m. Assuming that the jet is turned through an angle of 165° by the bucket. The length of each penstock may be taken as 5000 m long. Take

$$c_v = 0.98\,,\; k_u = 0.45\,, \lambda = 4f = 0.032, K = V_{r2} / V_{r1} = 0.85$$

Determine :

(a) Speed of the wheel.

(b) Specific speed of the Pelton wheel

(c) Power of each jet

(d) Hydraulic power of each turbine assuming volumetric efficiency as 98%.

(e) Brake power of each turbine assuming the mechanical efficiency as 95%.

Solution:

$$Q = A\,V_p = a\,V_1$$

$$\therefore V_p = \left(\frac{d}{dp}\right)^2 V_1 = \left(\frac{0.25}{1.5}\right)^2 = V_1 = 0.0277\,V_1$$

$$h_f = \frac{\lambda\, l_p\, V_p^2}{2\,g\,dp}$$

$$= \frac{0.032 \times 5000 \times (0.0277\,V_1)^2}{2 \times 9.81 \times 1.5}$$

$$= 4.1949 \times 10^{-3}\, V_1^2$$

$$= 4.1949 \times 10^{-3} \times c_v^2\;\; 2\,g\,(Hg - h_f)$$

$$= 0.079\,(Hg - h_f) = 0.079\,(600 - h_f)$$

$$= 47.427 - 0.079\,h_f$$

$$h_f = \frac{47.427}{1.079} = 43.95\,m$$

$$\therefore\; H = Hg - h_f = 600 - 43.95 = 556.05\,m$$

$$V_1 = c_v\,\sqrt{2\,g\,H} = 0.98\,\sqrt{2 \times 9.81 \times 556.05}$$

$$= 102.36\,m/s$$

$$u_1 = K_u \sqrt{2gH} = 0.45\sqrt{2 \times 9.81 \times 556.05}$$

$$= 47 \ m/s$$

$$u_1 = \frac{\pi D N}{60}$$

or, $$47 = \frac{\pi \times 3 \times N}{60}$$

$$\therefore N = 299.2 \ rpm$$ Ans.

$$Q = \frac{\pi}{4} d^2 V_1$$

$$= \frac{\pi}{4}(0.25)^2 \times 102.36$$

$$= 5.0245 \ m^3/s$$

Power at each jet

$$= \rho Q g H = 1000 \times 5.0245 \times 9.81 \times 556.05$$

$$= 27407.896 \ kW$$ Ans.

Hydraulic power per wheel

$$= \rho Q (V_1 - u_1)(1 + K \cos\phi) u_1$$

$$= 1000 \times 5.0245 (102.36 - 47)(1 + 0.85 \cos 15°) \times 47$$

$$= 23807.048 \ kW$$ Ans.

$$\eta_h = \frac{2 u_1 (V_1 - u_1)(1 + K \cos\phi)}{V_1^2}$$

$$= 2\left(\frac{u_1}{V_1}\right)\left(1 - \frac{u_1}{V_1}\right)(H \ K \cos\phi)$$

$$= 2(0.459)(1 - 0.459)(H \ 0.85 \cos 15°)$$

$$= 0.9047 = 90.47\%$$

$$\eta_0 = \eta_h \times \eta_m \times \eta_v = 0.9047 \times 0.95 \times 0.98 = 0.84227$$

$$= 84.227\%$$

Brake power per wheel $= P = \rho Q g H \eta_0$

$$= 1000 \times 5.0245 \times 9.81 \times 556.05 \times 0.84277$$

$$= 23085\ kW \qquad \text{Ans.}$$

$$N_s = \frac{N\sqrt{P}}{H^{5/4}} = \frac{299.2\sqrt{23554.346}}{(556.05)^{1.25}} \qquad \text{Ans.}$$

9.18. A single jet Pelton wheel is supplied with water from a reservoir 300 m above the centre of the nozzle through a pipe of 640 mm diameter and 5 km long. The friction coefficient for the pipe is $\lambda = 4f = 0.032$. The jet has a diameter of 80 mm and its velocity coefficient is 0.985. The wheel buckets have a speed of 0.47 of the jet speed and deflect the water through 165°. The relative velocity of the water is reduced 15 per cent in passing over the buckets. The mechanical efficiency of the wheel is 86%. Determine the hydraulic and overall efficiency of the system.

Solution

$$H = \frac{V_1^2}{2g} = Hg - (h_f + h_n)$$

or,

$$\frac{V_1^2}{2g} = Hg - h_f - h_n$$

Therefore,

$$Hg = \frac{V_1^2}{2g} + h_f + h_n \qquad (1)$$

$$Q = A_p V_p = \frac{\pi}{4} d^2 V_1$$

$$V_p = \left(\frac{d}{dp}\right)^2 V_1 = \left(\frac{80}{640}\right)^2 V_1$$

$$= 0.015626\ V_1$$

$$h_n = \frac{\lambda\, l_p\, V_p^2}{2g\ dp} = \frac{0.032 \times 5000 \times (0.015625\ V_1)^2}{2 \times 9.81 \times 0.64}$$

$$= 0.00311\ V_1^2$$

$$h_n = \frac{V_1^2}{2g}\left(\frac{1}{c_v^2} - 1\right)$$

$$= \frac{V_1^2}{2\times 9.81}\left[\frac{1}{(0.985)^2} - 1\right]$$

$$= \frac{0.0306}{19.62} V_1^2 = 0.001564\ V_1^2$$

Substituting in equation (1)

$$300 = \frac{V_1^2}{2g} + 0.00311\ V_1^2 + 0.001564\ V_1^2$$

$$= V_1^2\ [0.050968 + 0.00311 + 0.001564]$$

$$300 = 0.0556\ V_1^2$$

$$V_1^2 = 5391.572$$

$$\therefore V_1 = 73.427\ m/s$$

$$\therefore h_f = 0.00311 \times 5391.572 = 16.7677\ m$$

$$\therefore H = Hg - h_f = 300 - 16.7677 = 283.23\ m$$

$$u_1 = 0.47\ V_1 = 0.47 \times 73.427 = 34.51\ m/s$$

$$\eta_h = 2\left(\frac{u_1}{V_1}\right)\left(1 - \frac{u_1}{V_1}\right)(1 + K\cos\phi)$$

$$= 2 \times 0.47 \times (1 - 0.47)(1 + 0.85\cos 15°)$$

$$= 0.90724 = 90.724\%$$ **Ans.**

$$\eta_0 = \eta_h \times \eta_m = 0.90724 \times 0.86 = 0.7802$$

$$= 78.02\%$$ **Ans.**

EXERCISES

A. THEORY

9.1. What are the advantages and disadvantages of high speed Pelton turbine?

9.2. What are the limitations of Pelton wheel turbine?

9.3. Discuss the shape of bucket of a Pelton wheel turbine.

9.4. What are the essential component parts and their functions of a Pelton wheel turbine?

9.5. How will you control the load with the help of spear rod and deflector plate explain with the help of sketches?

9.6. Draw a sketch of a Pelton wheel turbine and discuss the main components.

9.7. Under what circumstances more than one nozzle should be used for a Pelton wheel turbine? What are the disadvantages of such an arrangement?

9.8. Why does a Pelton wheel turbine not posses any draft tube?

9.9. Draw inlet and outlet velocity trinagles for a Pelton wheel turbine and indicates the direction of various velocities.

9.10. Describe with the neat sketches the important features of the Pelton wheel bucket.

9.11. Draw three views of a Pelton wheel bucket. Why is the notch cut off from the bottom portion of the bucket?

9.12. What are the various losses in the Pelton wheel installations?

9.13. Prove that Nozzle efficiency $\eta_N = c_v^2$ for a Pelton wheel turbine.

9.14. Define pipe line transmission efficiency.

9.15. Prove that loss in the Nozzle $h_n = H(1 - c_v^2)$

9.16. Prove that loss in the nozzle is given by

$$h_n = \frac{V_1^2}{2g}\left(\frac{1}{c_v^2} - 1\right)$$

9.17. How will you select jet ratio for a Pelton wheel turbine?

9.18. How will you select bucket width to jet diameter for a Pelton wheel turbine?

9.19. How will you select the speed for a Pelton wheel turbine?

9.20. How will you select head for a Pelton wheel turbine?

9.21. Prove the formula

(a) $\eta_o = \dfrac{2\,u_1\,(V_1 - u_1)(1 + K\cos\phi)}{V^2}$

(b) $\eta_{h\,max} = \dfrac{1 + K\cos\phi}{2}$

(c) $V_1 = 2\,u_1$

(d) $\eta_h = \dfrac{2\,u_1\,(V_1 - u_1)(1 - k\cos\gamma)}{V_1^2}\eta_N$

(e) $\eta_{h\,max} = \left(\dfrac{1 - K\cos\gamma}{2}\right)c_v^2$

(f) $\eta_N = c_v^2$

(g) $\eta_n = H\left(1 - c_v^2\right) = \dfrac{V_1^2}{2g}\left(\dfrac{1}{c_v^2} - 1\right)$

B. Unsolved Problems

9.1. The following data were obtained from tests on a Pelton wheel turbine

Diameter of jet = 100 mm

Brake power = 415 kW

Head at nozzle = 120 m

Power 1055 in friction and windage = 21 kW

Discharge = 390 liters/s

$c_v = 0.98$

Assuming that the velocity of water at discharge from the buckets is 7.9 m/s, draw up a balance sheet showing the distribution of energy of the supply water.

[η_o = 89%, Nozzle loss = 41.5%, Discharge loss = 2.62%, Friction and windage loss = 4.41%]

9.2. Tests carried on a Pelton wheel gave the following results

Heat at the base of nozzle = 27 m

Discharge of nozzle = 280 liters/s

Diameter of jet = 125 mm

Power developed = 63.2 kW

Power absorbed in mechanical resistances like bearings, windage, etc. = 3.675 kW

Find (a) the power lost in nozzle, (b) the power lost due to hydraulic resistance (i.e. friction) in the runner. [1.15395 kW, 6.05 kW]

9.3. A Pelton wheel is working under a gross hend of 400 m. The water is supplied through penstock of diameter 1m and length 4 km from reservoir to the Pelton wheel. The friction factor for the penstock is given as $\lambda = 4f = 0.032$. The jet of water diameter 150 mm strikes the buckets of wheel and jet deflects through an angle 165°. The relative velocity of water at outlet is reduced by 15% due to friction between inside surface of the bucket and water. If the velocity of buckeis is 0.45 times the jet velocity at inlet and the mechanical efficiency is 85%, determine

(a) Power given to the runner (b) Shaft power

(c) Hydraulic efficiency and overall efficiency.

[P_{runner}=4891.217 kW, P_{shaft}=4156.2968, η_h =90.147%, η_0 =76.6%].

9.4. A double overhung Pelton wheel turbine is directly coupled to a 25 MW generator which has an efficiency of 95%. Find the power generated by each turbine. If each of the above turbine runner operates under an effective head a 300 m at the base of the nozzle, workout: size of the jet, mean diameter of the runner, synchronous and specific speed of each wheel. The following parametes constitute the design criteria; Pelton wheel efficiency = 85%, C_v = 0.97, k_u = 0.45, $d/D = 1/12$, f = 50 Hz and pair of poles = 16, [13158 kW, d = 300 mm, D = 3.6 m, N = 187.5 rpm, N_s = 17.23. $D_{revised}$ = 3.518 m].

9.5. The buckets of a Pelton wheel deflect the jet through an angle of 170°, and the relative velocity of the water is reduced by 12% by bucket friction, calculate from first principles the value of the hydraulic efficiency for a speed ratio of 0.47. The bucket circle diameter of the wheel is 900 mm, and there are two jets, the coefficient of velocity C_v being 0.98.

The actual efficiency of the wheel is 0.9 times the theoretical hydraulic efficiency. Under a gross head of 600 m the wheel develops 1250 kW, when the loss of head by pipe friction is 48 m. Calculate (a) the speed of rotation in rev/min (b) the diameter of the nozzles.

[η_h = 89.31%, N=1017.1 rpm, d = 42.3mm, $\eta_{h\ actual}$ = 80.4%]

9.6. The maximum overall efficiency of a Pelton wheel turbine is 82% when the speed ratio K_u = 0.45 and the coefficient of velocity C_v = 0.97 for the nozzle. Assuming that the loss due to windage and bearing friction is proportional to the square of rotational speed, estimate the friction factor for buckets, hydraulic efficiency and mechanical efficiency, if the jet turns through 165°. [K = 0.91, η_h = 88%, η_{mech} = 93%].

10

Reaction Turbines

10.1. INTRODUCTION

In a reaction turbine, a part of the available head is converted to velocity head before the runner is reached. Since there is a continuous pressure-drop when a fluid flows through the passage in the runner, the fluid must be admitted around the entire circumference of the runner.

Flow of the water in reaction turbines may be radial, axial, or mixed. In a turbine which has pure radial flow, the runner absorbs the energy from the water as it flows toward the shaft. This design is rarely used in modern practice.

The propeller type of turbine has axial flow. The water is flowing parallel to the turbine axis, as it imparts its energy to the wheel. The mixed flow (Francis) turbine has its blade curved in such a manner that the flow is first radial and then axial.

Reaction turbines may be installed with either horizontal or vertical shafts, but the vertical shaft mounting is the more commonly used. With a horizontal shaft installation, both generator and turbine must be located on the powerhouse floor. With a vertical-shaft installation, the generator is located on the powerhouse floor, but the turbine runner may be set as low as its operating characteristics may require, through the use of a long connecting shaft.

10.2. THE FRANCIS TURBINE

Francis turbines are medium head (40-60 to 500-700 m), medium specific speed (52 to 345), medium discharge, mixed flow (radial plus axial) reaction turbine (See Fig. 10.1).

The water from the penstock enters a scroll casing (spiral casing) which completely surrounds the runner. From the spiral casing the water passes through a speed ring or stay ring. The speed ring consists of an upper and a lower ring held together by series of fixed vanes called stay vanes. The number of stay vanes is usually taken as half the number of guide vanes. From the speed ring water passes through guide vanes or gates located around the circumference, and flows through the runner, from which it dishcarges into a draft tube sealed below the tailwater level. All the water passages are completely filled with water, which acts upon the whole circumference of the runner. Only a portion of the power is derived from the dynamic action due to the velocity of the water, a large part of the power being obtained from the difference in pressure acting on the front and back of the runner blades.

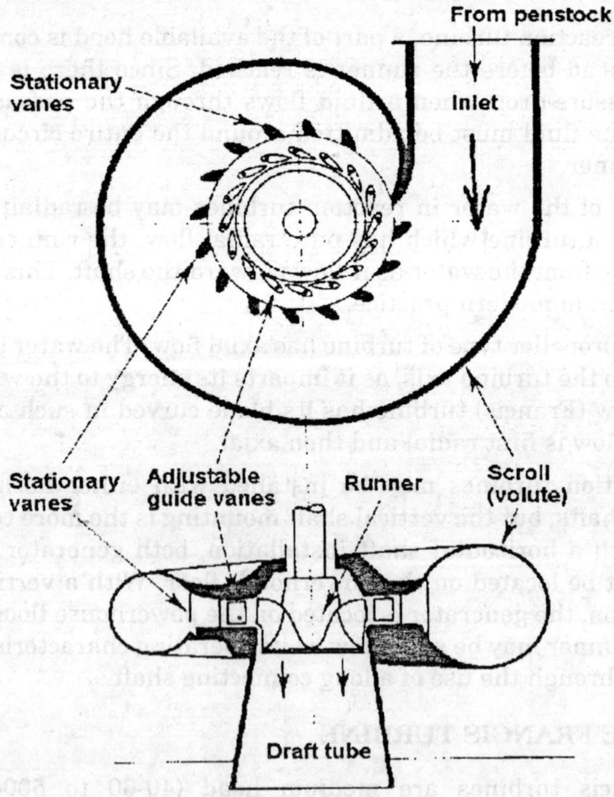

Fig. 10.1. Francis turbine sectional arrangement (Ref. Douglas)

The essential component parts of a Francis turbine are

(1) Scroll casing or spiral casing

(2) Speed rings or stay rings

(3) Guide vanes or wicket gates

(4) Runner

(5) Draft tube

10.2.1. Scroll casing or spiral casing

A scroll case is the conduit directing the water from the penstock to the runner in reaction type turbine installations.

The purpose of scroll casing is to provide an even distributioin of water around the circumference of the turbine runner, maintaining an approximately constant velocity for the water so distributed. In order to keep the velocity of water constant throughout its path around the runner, the cross sectional area of the casing is gradually decreased.

A full spiral case entirely enclosing the turbine with a nose angle, of 360° ensures most perfect slow conditions. However, in practice spiral case with a range between $320° < \varphi < 340°$ are also called full spiral cases.

This kind of spiral case will generally be used in medium and high head installations where discharge requirements are smaller.

For low head plants the entrance area should be large so as to allow large flows. This is achieved by choosing nose angles that are less than 320°. The spacing of the units is therefore governed by these large entry widths of the partial spiral scroll case.

For high head plants, a circular scroll case cross section is normally adopted; a metal casing is more suited to this type. For low head plants where the water quantities are large a rectangular section with rounded corners may be constructed in concrete. The approximate dimensions of a Francis turbine layout with steel scroll case, and a propeller turbine layout with concrete scroll, respectively, are shown in Fig. 10.2 (a) and (b).

The design of the shape of the sprial case is governed by the flow requirements. Initial investigations should be based on the following assumptions:

(1) a spiral case of constant height,

(2) an evenly distributed flow into the turbine, and

(3) no friction losses.

The spiral casing dimension are such that the water velocities are approximately :

For low specific speed $0.14\sqrt{2gH}$

For high specific speed $0.2\sqrt{2gH}$

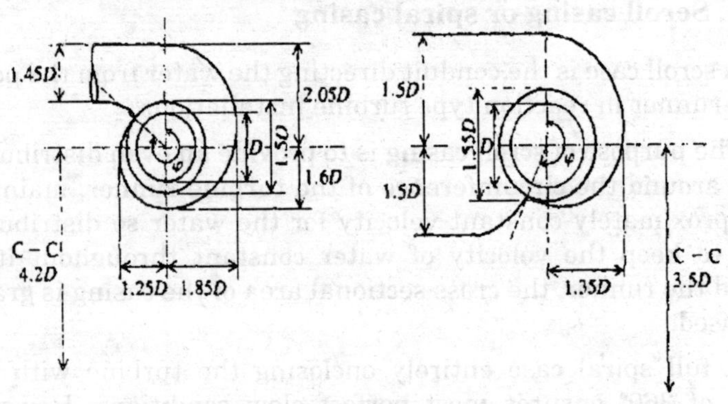

Fig. 10.2. Recommended dimensions of scroll casings (Ref. Novak)

It is desirable to provide streamlined dividing piers in entrance flumes of large widths to ensure as even a flow distribution as possible. The final design of the scroll case should preferable be checked by model tests, especially in the case of unconventioinal arrangement of large turbine units. An example of inner shaping of a concrete spiral casing is shown in Fig. 10.3.

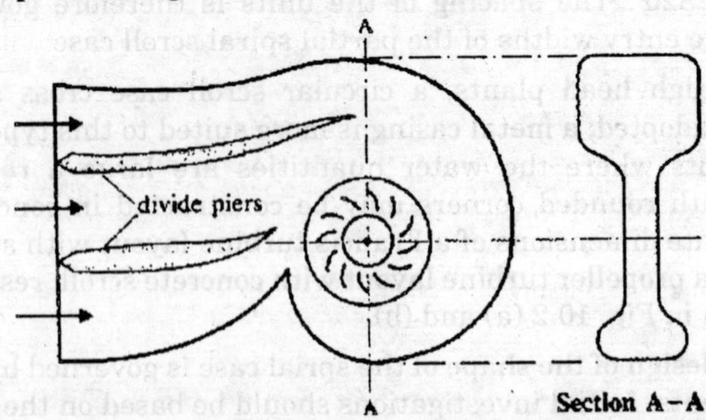

Fig. 10.3. Inner shape of a concrete scroll case with divided piers (Ref. Novak)

The field of application of different types of steel cases in mainly determined by the available head. Fig. 10.4 shows that large turbine use concrete (reinforced concrete) cases at heads within 4 to 80 m, and metal cases at 40 to 700 m.

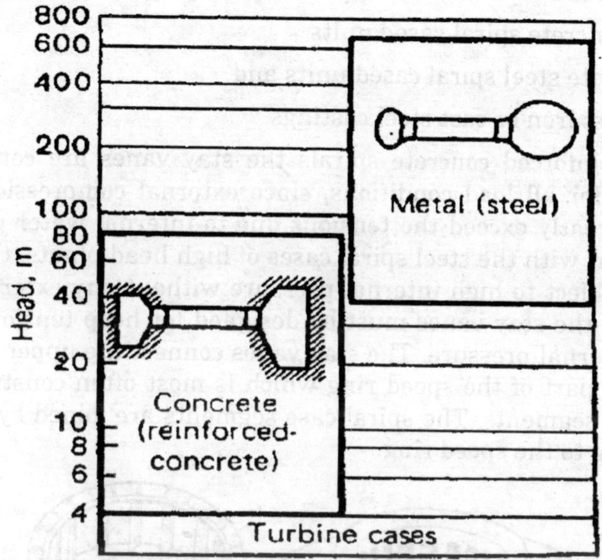

Fig. 10.4. Application of scrolls of different heads (Ref. Krivchenko)

The application limit for spiral (scroll) cases whose shells made from upto 40 mm thick high strength alloyed steel take up a maximum internal pressure p_{max}, including the water hammer, is determined from the condition

$$P_{max}\ D_{sp} < 1100 \text{ to } 1300 \tag{10.1}$$

where, p_{max} is expressed in MPa, and D_{sp} is the inlet diameter (section) of the spiral case in cm. If this condition (Equ. 10.1) is not satisfied, other solutions regarding the spiral case must be taken.

10.2.2. Speed rings

Speed rings constitute part of the spiral case. They are located at the inside, that is, near the runner inlet, and are applied mainly for structural reasons and sometimes also for modifying the intensity of the free vortex (with low heads they serve to reduce the moment of momentum, whereas with higher heads they increase it).

In its simplest construction it consists of upper and lower rings held together by vertical ribs, sometimes called statioinary guide vanes or stay vanes. These ribs carry the load from above and take the force of the internal water pressure.

There are three distinct types of speed rings.

(a) concrete spiral cased units

(b) plate steel spiral cased units and

(c) cast iron or cast steel castings.

At reinforced concrete spirals the stay vanes are compressed elements for all load conditions, since external compression forces usually greatly exceed the tensions due to internal water pressure. Howewver, with the steel spiral cases of high head plants, the spiral case is subject to high internal pressure without any external load. Therefore the stay vanes must be designed for hoop tension caused by the internal pressure. The stay vanes connect the upper part and the lower part of the speed ring which is most often constructed of cast steel segments. The spiral case segments are joined by welding or riveting to the speed ring.

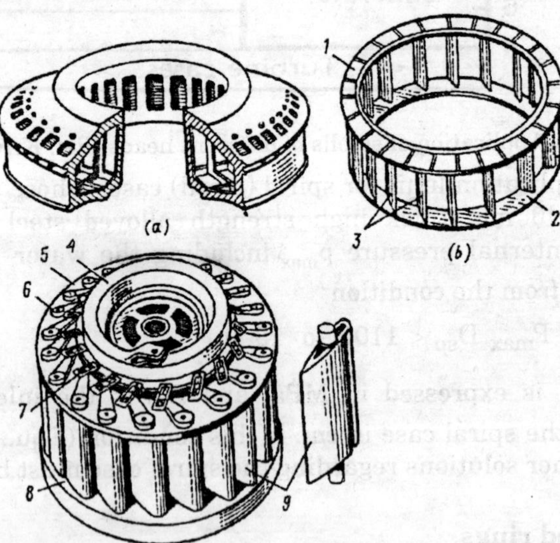

Fig. 10.5. Stationary components of a reaction turbine (a) cover, (b) stay ring, (c) wicket gate; 1- upper circular plate of stay ring; 2- lower circular plate of stay ring; 3- stay venes; 4 - operating ring; 5- upper ring of wicket gates; 6- levers; 7 - slotted guides; 8- lower ring of wicket gates; 9 - gates (Ref. Ilynikh)

In effect, the stay ring (Fig. 10.5b) forms the outlet of a turbine case. It resists heavy loads imposed by the equipment and the power house structure. It consists essentially of an upper and a lower circular plate joined by 10 to 16 stay vanes of streamlin shape. The stay rings of some turbines may have no lower plate.

10.2.3. Guide vanes or wicket gates

The movable guide vanes, sometimes called wicket gates. The guide vane assembly, together with the speed ring, is sometimes called the distributor.

Inside the stay ring are guide vanes intended to regulate the discharge and to direct the water flow. Also, the guide vanes serve to start up and shut donw the turbine and to control its power output and its rotational speed. As can be seen from Fig. 10.5 (c), the guide vanes have an upper and a lower ring resting in recesses provided in the circular plates of the stay ring. The rings are joined together by 20 to 32 blades (or gates) of streamlined shape. The gates are connected to an operating ring by slotted guides and levers rigidly fastened to their top pins.

The function of the operating (or shifting) ring is to open or close the gates which can rotate about their axes. The operating ring is installed in a recess provided in a turbine cover which, in turn, rests in a recess made in the upper ring of the guide vanes.

The guide vane mechanism for both the Francis and the propeller runners is of two general types :

(a) the open flume or inside type

(b) increased or outside type.

In the open flume the connecting links are exposed to the flow of the water, and the outside type the upper guide vane stems extend up through the cover plate outside the water. The open flume guide vanes are usually made of cast iron, although some manufacturers make them of plate steel, the plates being formed and welded. The guide vanes for outside gate mechanism are generally constructed of cast steel. These have the one pivot extended through the cover plate, and this pivot must transmit the full torque to operate the guide vanes, cast steel is desirable for strength.

Fig. 10.6 shows the guide blade mechanis for the Francis (mixed flow) turbine. Fig. 10.6 (a) shows the guide vane operating gear in the full closing position. If the regulating ring turns counterclockwise, all leaves will turn through the same angle, as well as the guide vanes -and the turbine will be opened. Fig. 4.9 (b) shows the operating gear in the full opening position. It follows that to change the turbine power it is necessary to turn the regulating ring.

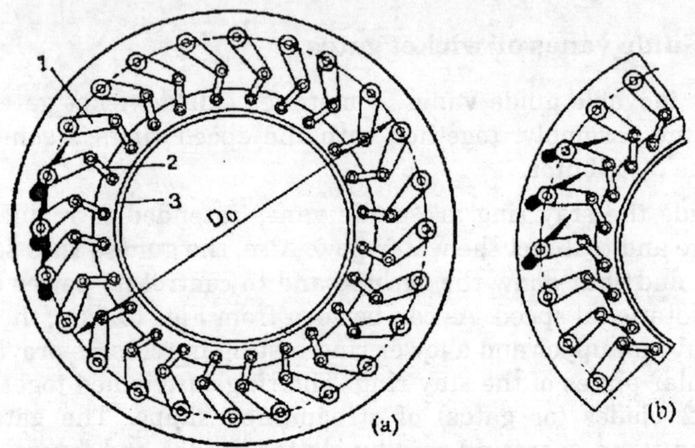

Fig. 10.6. Guide blade mechanism Schematic diagram. 1 - Levers, 2 - Shack les and pull rod; 3 - regulating ring. (Ref. Krivchenko).

10.2.4. Runner

The runner of a Francis turbine (Fig. 10.7) has the upper ends of the moving vanes attached to a conical crown and the lower ends attached to a band, thus completely enclosing the water passageway through the runner. The shrouded vanes made an angle with the radial planes and bend at the bottom. The number of blades usually between 14 and 19.

The clearance between wicket gate and the runner should be as small as possible in order to reduce leakage and to obtain thereby good volumetric efficiency clearance should not exceed 1 to 2 mm. Nevertheless, considerable water losses will occur under high pressure and, since the clearance cannot be reduced further, the application of labyrinth seals will become necessary.

In order to reduce the heavy axial thrust acting on the runner, connection must be established between the draft tube and the space enclosed by the crown plate and the cover plate. This is necessary because water leaks into the space beneath the cover through the top of the clearance.

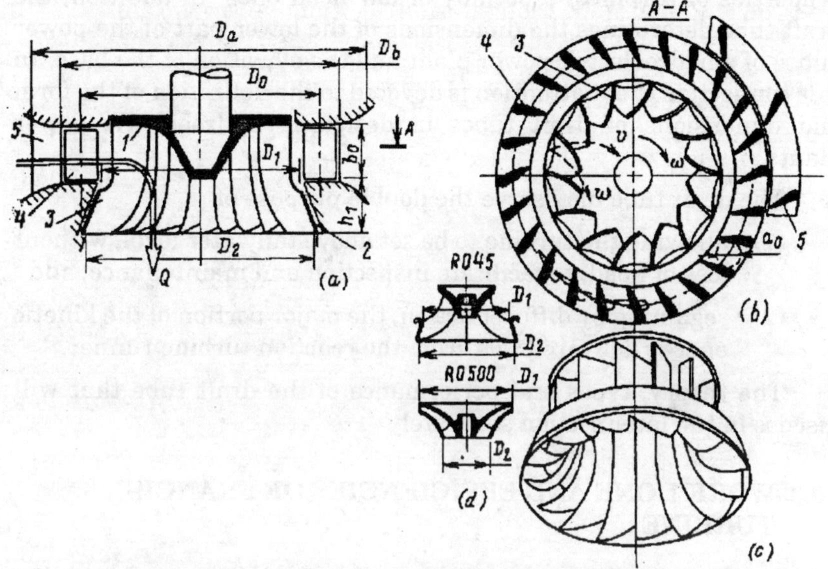

Fig. 10.7. Francis turbine (a) cross sectional evaluation (b) horizontal sectional through runner (c) general appearance (d) runners of various heads. 1 - crown; 2 - band; 3- buckets; 4 - wicket gates; 5 - stay vanes (Ref. Ilyinykh).

The turbine cover is provided with ribs extending downwards in order to prevent the runner from setting the water contained beneath the cover into rotation which could superfluously increase mechanical resistance. (The volume of water rotating inside the enclosed space is forced towards the periphery by the centrifugal force and exerts considerable pressure on the runner. The increase in axial load caused thereby may be expressed by a component proportional to the square of the peripheral velocity. The ribs are applied to prevent these phenomena).

Small runners are always cast integrally; their material is cast iron or bronze. Large runners are made of cast steel or of cast stainless steel.

10.2.5 Draft tubes

In reaction turbines water is discharged from the runner into a draft tube through which the water is discharged into the tail water pool. The draft tube affects materially the power generating properties of turbines, especially of low head ones. In addition, the draft tube determines the dimensions of the lower part of the power house of a hydroelectric power plant and the elevation of the base. In this connection great attention is devoted to the definition of the form and dimensions of draft tubes in designing hydroelectric power plants.

The draft tube may serve the double purpose of

(1) allowing the turbine to be set above tailwater level, without loss of head, to facilitate inspection and maintenance, and

(2) regaining by diffuser action, the major portion of the kinetic energy delivered to it from the reaction turbine runner.

The theory, types and performance of the draft tube that will discuss in the later section separately.

10.3. WORK DONE AND EFFICIENCIES OF FRANCIS TURBINE

In the reaction turbine, the net head (H) is equal to the difference of the total energy of water at inlet and the draft tube exit.

$$H = \left(\frac{P_s}{\rho g} + \frac{V_1^2}{2g} + Z_s\right) - \left(\frac{P_3}{\rho g} + \frac{V_3^2}{2g} + Z_3\right) \qquad (10.2)$$

where subscript 's' correspond to the spiral case.

If the draft tube exit is at tail race level i.e. open to atmospher (zero gauge pressure) and datum is also taken at that level, then

$$H = \left(\frac{P_s}{\rho g} + \frac{V_3^2}{2g} + Z\right) - \frac{V_3^2}{2g}$$

where, $Z = Z_s - Z_3$ and Ps is in gauge pressure. Neglecting the velocity V_s at the exit of the draft tube then

$$H = \frac{P_s}{\rho g} + \frac{V_s^2}{2g} + Z \qquad (10.3)$$

The general expression for the work done is given by according to Euler turbine equation

$$\text{Work done/kg} = \frac{[V_{w1}\, u_1 \pm V_{w2}\, u_2]}{g} \qquad (10.4a)$$

$$\text{Work done/s} = \rho Q [V_{w1} u_1 \pm V_{w2} u_2] \tag{10.4b}$$

Applying Bernoullis theorem (Fig. 10.8)

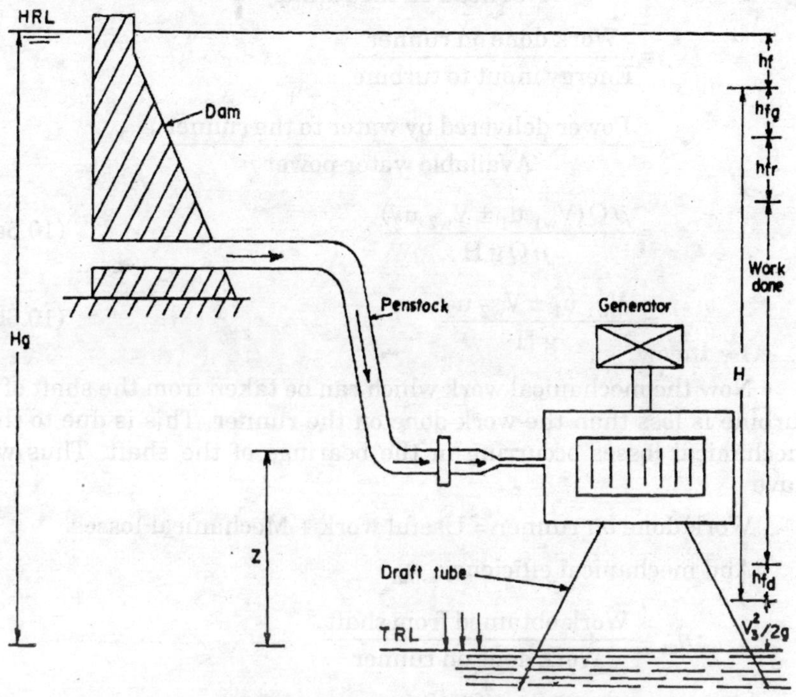

Fig. 10.8. Layout of Francis turbine

Energy supplied = Work done + kinetic energy rejected + sum of losses

$$Hg = \text{Work done} + \frac{V_3^2}{2g} + \Sigma h_f \tag{10.5}$$

The efficiency of turbine may be hydraulic, mechanical and overall.

Ordinarily, by efficiency of turbine we mean the overall efficiency unless mentioned otherwise or unless it is understood otherwise from the problem. We have,

$$\eta_0 = \eta_h \times \eta_m$$

Also, Hydraulic efficiency $\eta_h = \dfrac{\text{Hydraulic output}}{\text{Hydraulic input}}$

Hydraulic input = Hydraulic output + Hydraulic losses

$$\eta_h = \frac{\text{Head utilised to do work or runner}}{\text{Net head on the runner}}$$

$$= \frac{\text{Work done on runner}}{\text{Energy input to turbine}}$$

$$= \frac{\text{Power delivered by water to the runner}}{\text{Available water power}}$$

$$= \frac{\rho\, Q\, (V_{w1}\, u_1 \pm V_{w2}\, u_2)}{\rho\, Q\, g\, H} \quad (10.6a)$$

$$= \frac{V_{w1}\, u_1 \pm V_{w2}\, u_2}{g\, H} \quad (10.6b)$$

Now the mechanical work which can be taken from the shaft of a turbine is less than the work done on the runner. This is due to the mechanical losses occurring in the bearings of the shaft. Thus we have

Work done on runner = Useful work + Mechanical losses.

And mechanical efficiency, η_m.

$$\eta_m = \frac{\text{Work obtained from shaft}}{\text{Work done on runner}}$$

$$= \frac{\text{Shaft power}}{\text{Power delivered by water to turbine}}$$

$$= \frac{P}{\rho\, Q\, (V_{w1}\, u_1 \pm V_{w2}\, u_2)} \quad (10.7)$$

Now,

$$\eta_0 = \eta_h \times \eta_m$$

$$= \frac{\rho\, Q\, (V_{w1}\, u_1 \pm V_{w2}\, u_2)}{\rho\, Q\, gH} \times \frac{P}{\rho\, Q\, (V_{w1}\, u_1 \pm V_{w2}\, u_2)}$$

$$= \frac{P}{\rho\, Q\, gH} \quad (10.8)$$

or $= P = \rho\, QgH\, \eta_o$ (10.9)

10.4. VELOCITY DIAGRAMS FOR FRANCIS TURBINE

A section of the runner guide vane ring, showing the blades, vanes and velocity triangles, is shown in Fig. 10.9.

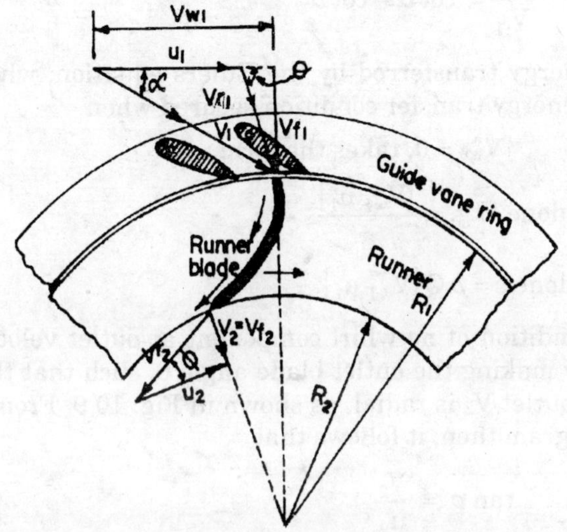

Fig. 10.9. Blades vanes and velocity triangles of Francis turbine (Ref. Douglas)

The direction of V_1 is governed by the guide vane angle α. It is chosen in such a way that the relative velocity meets the runner blade tangentially, i.e. it makes an angle θ with the tangent at blade inlet.

$$\tan\alpha = \frac{V_{f1}}{V_{w1}} \tag{10.10}$$

and $$\tan\theta = \frac{V_{f1}}{V_{w1} - u_1} \tag{10.11}$$

Eliminating V_{w1} from the two equations.

$$\tan\theta = \frac{V_{f1}}{\dfrac{V_{f1}}{\tan\alpha} - u_1}$$

$$\cot\theta = \frac{\dfrac{V_{f1}}{\tan\alpha} - u_1}{V_{f1}}$$

$$= \cot \alpha - \frac{u_1}{V_{f1}}$$

or, $$\frac{u_1}{V_{f1}} = \cot \alpha - \cot \theta \qquad (10.12)$$

The energy transferred by the Eulers equation, which, for the maximum energy transfer condition secured when

$V_{w2} = 0$, takes the form

$$\text{Work done / kg.} = \frac{[V_{w1}\, u_1]}{g} \qquad (10.13a)$$

$$\text{Work done/s} = \rho Q [V_{w1}\, u_1] \qquad (10.13b)$$

The condition of no whirl component at outlet velocity may be achieved by making the outlet blade angle θ, such that the absolute velocity at outlet V_2 is radial, as shown in Fig. 10.9. From the outlet velocity diagram then, it follows that

$$\tan \phi = \frac{V_2}{u_2}$$

but since $V_{w2} = 0$, then $V_2 = V_{f2}$

therefore, $$\tan \phi = \frac{V_{f2}}{u_2} \qquad (10.14)$$

The hydraulic efficiency is now determined.

$$\eta_h = \frac{V_{w1}\, u_1}{gH} \qquad (10.15)$$

10.5. WORKING PROPORTIONS OF A FRANCIS TURBINE

A Francis turbine runner is required to be designed to develop a known power P, when running at a known speed N rpm under a known head H. The probable values of η_h, η_o, n and ψ are assumed.

The hydraulic efficiency of the turbine may vary from 0.85 to 0.9 or more.

The overall efficiency of the turbine is always less than the hydraulic efficiency. Its value may range from 0.8 to 0.9 or more according to the size and installation of turbine.

Let n = ratio of width to diameter = $\frac{B}{D}$

and Ψ = the flow ratio = $\frac{V_{f1}}{\sqrt{2gH}}$

Ψ may range in value from 0.15 to 0.3

The high values of ψ and therefore of the flow component V_f will result in excessive losses in the form of velocity energy rejected from the wheel.

The high values of n bring us up against of the cardinal problems of low head turbine desing-the problem of providing a big enough outlet area for the water. It is of no use to give the runner a generous inlet area if we do not allow the water to escape freely. The difficulty can be partially overcome by the use of the mixed flow type of runner. The value of n ranges from 0.1 to 0.45.

The relationship between the runner speed and spouting velocity for the Francis turbine is not rigidly defined as for the Pelton wheel. In practice,

$K_u = \frac{u_1}{\sqrt{2gH}}$ is contained within the limits of 0.6 to 0.9.

10.6. DESIGN OF FRANCIS TURBINE RUNNER

(1) Calculate the required discharge for the given brake power or shaft power

$$P = \rho Q g H \eta_o$$

(2) Area of flow is calculated as

$$A_1 = (\pi D_1 - \text{no. of blades} \times \text{thickness}) B_1$$

$$= K_1 \pi D_1 B_1$$

Similarly, $A_2 = K_2 \pi D_2 B_2$

$$Q = \text{Area of flow x Velocity of flow}$$

$$= K_1 \pi D_1 B_1 V_{f1} = K_2 \pi D_2 B_2 V_{f2}$$

Normally it is assumed that

$$V_{f1} = V_{f2} \text{ and } K_1 = K_2 = K$$

Where K is a factor known as the vane thickness factor. Its value is less than unity, but usually it is taken as unity. In other words vane thickness is neglected. Then,

$$Q = \pi\ D_1\ B_1\ V_{f1}$$

$$= \pi\ n\ D_1^2\ V_{f1}$$

Since, $B_1 = n\,D_1$

The value of n varries from 0.1 to 0.45.

(3) Velocity of flow at inlet is calculated as

$$V_{f1} = \psi\ \sqrt{2\,g\,H}$$

The value of ψ varies from 0.15 to 0.3

Therefore, $\psi\ \sqrt{2\,g\,H} = \dfrac{Q}{\pi\,n\,D_1^2}$

From the above relation D_1 is obtained and using $B_1 = nD_1$, B_1 is found.

(4) The rim velocity u_1 can now be obtained

$$u_1 = \frac{\pi\,D_1\,N}{60} = K_u\ \sqrt{2gH}$$

Ku varies from 0.6 to 0.9

(5) The guide vane angle α, runner vane angle θ and runner vane angle ϕ at outlet are obtained from the velocity triangles

$$\tan\alpha = \frac{V_{f1}}{V_{w1}}$$

$$\tan\theta = \frac{V_{f1}}{V_{w1} - u_1}$$

$$\tan\phi = \frac{V_{f1}}{u_2}$$

(6) The hydraulic efficiency can be obtained as (assuming β = 90° for maximum efficiency).

$$\eta_h = \frac{V_{w1}\,u_1}{g\,H}$$

If blade friction losses, eddies etc are neglected, energy loss will be only in the form of velocity energy loss at the outlet and the energy utilised by the runner will be

$$g\,H = \frac{V_2^2}{2}$$

Therefore,

$$\eta_h = \frac{V_{w1}\,u_1}{g\,H} = \frac{g\,H - V_2^2/2}{g\,H} \qquad (10.16)$$

While designing, the condition of frictionless blades may be assumed, if not stated other wise.

(7) The runner diameter D_2 at outlet end varies from one third to two thrid of runner diameter D_1 at inlet and usually it is taken equal to $\frac{D_1}{2}$ then $B_2 = 2\,B_1$.

(8) The number of vanes should be either one more or one less than the number of guide vanes, in order to avoid setting up of periodic impulse.

10.7. AXIAL FLOW REACTION TURBINES

Axial flow reaction turbines are the trubines in which the flow is parallel to the axis of the shaft. There are two types of axial flow turbines, propeller (fixed blades) and Kaplan (adjustable moving blades). The Kaplan turbine has been named after victor Kaplan. The turbine works in a head range of 5 metres to 60 metres and in exceptional cases upto 80 m. The specific speed ranges 260 to 860 in S.I. units.

At many power plant sites, where large quantities of water were available at the low heads, dimensions of the Francis turbine started becoming too large and uneconomical, and another turbine named the propeller turbine was developed.

The propeller turbine has greater speed and power for a given diameter than do other types. For low heads or large power output the turbine speed would be too low to be economical unless the specific speed is as high as possible. The higher speed of propeller turbine results in a lower cost generator and some what smaller power house substracture and super structure.

The fixed blade propeller has a very high efficiency at a point near full load, but the efficiency drops off rapidly as the load decreases, until, at 40% of full load, the efficiency is only about 50%. The installation of the fixed blade propeller type runner should therefore, be limited to developments where the units may be operated at a point near maximum efficiency or not at all and where, under variable head, the load of the unit could be shifted to an output resulting in best efficiency.

Experiments by professor Kaplan with a model propeller turbine led him to the conclusion that the great disadvantage which it had of a very peaky efficiency curve could be overcome if the pitch of the blades could be altered during normal operation as the output changed. The variable pitch propeller turbine, more appropriately called the Kaplan turbine, has now reached a stage of evolution where, with few exceptions, it can be used with marked advantage over its predecessor, the fixed blade or propeller turbine.

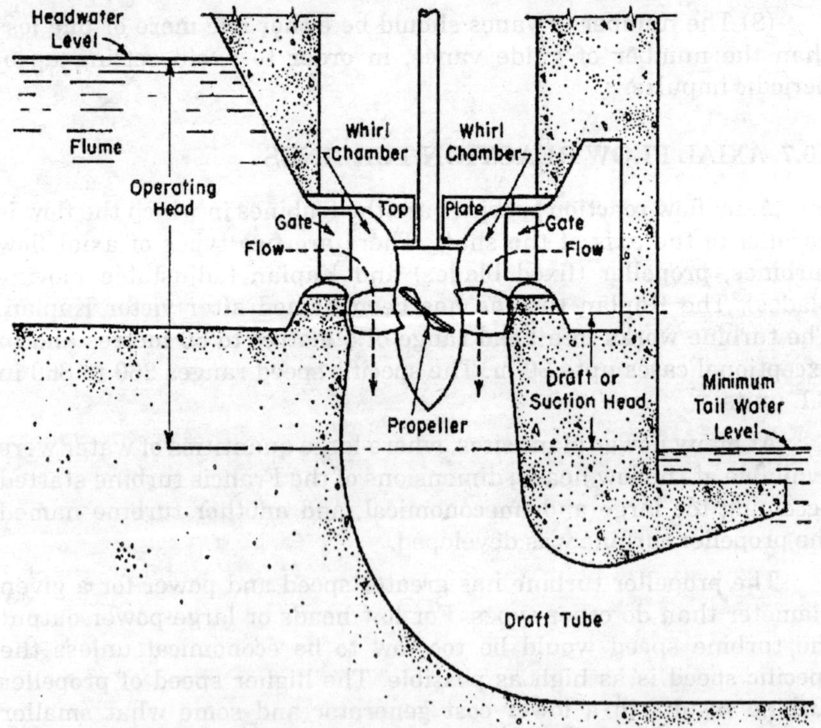

Fig. 10.10. Kaplan turbine for low heads (Ref. Skrotzki)

The Kaplan blades may be rotated to the most efficient angle by a hydraylic servomotor. A cam on the governor is used to cause the blade angle to change with the gate position so that high efficiency is always obtained at almost any percentage of full load. Photo shows a Kaplan turbine, with cutaway views of the blade servomotor and of the top plate. Fig. 10.10 shows a Kaplan turbine for low heads.

By reason of its high efficiency at all gate openings, the Kaplan turbine is particularly applicable to low head developments where conditions are such that the units must be operated at varying load and varying head. Capital cost and maintenance for such units are necessarily higher than for propeller units operated at point of maximum efficiency.

Consequently, for a low head development with fairly constant head and comprising a number of units, consideration should be given to installing propeller type runners for most of them and Kaplan type runners for only one or two units. Then the propeller units could be operated at point of maximum efficiency or not at all and the Kaplan units could take the necessary variation in load. Such an arrangement is particularly applicable to a large power system containing a multiplicity of units.

10.9. CHOICE BETWEEN PROPELLER AND FRANCIS TURBINES

In the area of Fig. 10.1 where the application of the propeller and Francis turbines overlap, the choice depends upon the relative cost and output. Because of its higher specific speed, a propeller turbine will run about twice as fast as a Francis turbine for the same head and output, or, the same speed and head, the propeller units would have four times the power. The large volume of water handled by a propeller turbine requires a high velocity, so that cavitation is a more of a problem than with a Francis tunner.

As indicated in Fig. 10.11, a propeller turbine has to be set 0.3 m to 4.6 m lower than a Francis unit.

Fig. 10.11. Hiehgt of turbine setting above tail water in relation to head. (Normal specific speed). Higher specific speed, altitude or temperature would require a lower setting.

The cost of additional excavation may largely offset the lower cost of the propeller turbine and generator because of its greater speed.

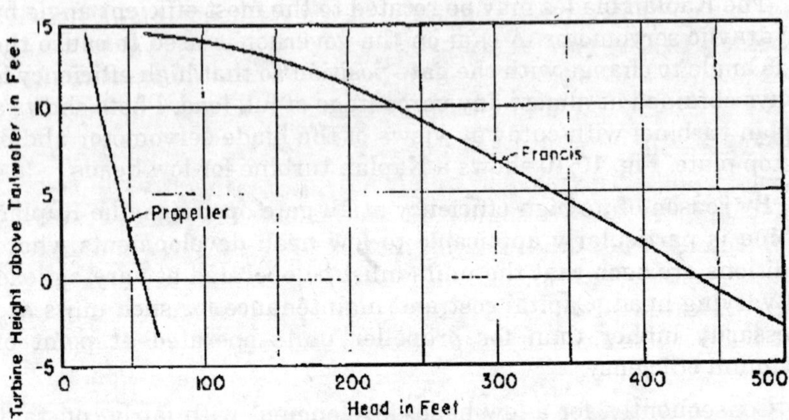

Fig. 10.11. Height of turbine setting above tail water in relation to head (Ref. Skrotzki)

10.10. CHOICE OF KAPLAN AND PROPELLER TURBINES

The operating experience at numerous power stations equipped with Kaplan turbines indicates that no difficulties exist in using this type of turbine upto a head of approximately 32 m for the largest outputs.

The present trend is towards the use of larger and fewer units of the Kaplan type, which is adopted for most low and medium head schemes except for the smallest outputs.

For a large scheme using Kaplan turbines it is of importance that limitations of maximum size and weight are known, since this is the limiting factor of the output that can be obtained from one unit. The largest and heaviest single part is the hub casting which cannot be split into parts to facilitate transport.

For low heads up to approximately 20 m, and where large water quantities are to be utilized, the advantages of Kaplan turbines are quite clear; but as the head increases the plant costs are greater relative to those associated with Francis turbines. This is primarily due to the proportionately large hub required for an increased head which, although an expensive item in itself, necessitates a large runner diameter than for a Francis turbine. Depending to a limited extent on the permissible depth of excavation, pitch circle diameter of the guide vanes of Francis and Kaplan turbines are about the same at 20 m head.

Kaplan units will be advantageous even for heads higher than 30 m, owing to the smaller change of peak efficiency over a head range compared with a Francis or propeller turbine. This generally leads to the use of Kaplan units for variable head schemes up to a present maximum of about 80 m.

Kaplan units are of particular advantage for run of river schemes where storage is small and the head water level must be maintained constant within narrow limits. With low flows the units can operate at best efficiency, and under flood conditions, when the tail water rises, the high overload capacity enables full output to be obtained without cavitation and without departure from the normal design for a reduction in head of as much as 10 per cent which may result from an increase in tailwater level.

10.11. KAPLAN AND PROPELLER TURBINES

The water supply part of the turbine consists of the turbine case, stay ring and guide blades. It should be noted that the designs of these component elements are similar to Francis trubine.

10.11.1. Scroll casing

The scroll casing or spherical casing has a thick steel lining secured in concrete by means of anchors and pullrods. Such lining and fastening are traced to the fact that when the turbine is in operation the casing walls take up large pulsating loads from water pressure, capable of shaking and damaging the lining of the casing.

The importance to adjustable blade turbines are the clearances (gaps) between blade tips and the casing. The less the gap, the less the leakage and the higher the turbine efficiency. To have a constant gap when the blade angle pit is changed, the turbine runner casing must be of spherical form. However, then difficulties may be encountered when mounting the assembled runner in its place. That is why the part of the casing on the top of the adjustable blade axis is made spherical, but not over its entire height, so that a minimum outlet diameter is not smaller than (0.98 to 0.96) of the turbine runner diameter. A lesser cross section causes a drop in the turbine power and efficiency. With a casing of this form remains constant at the adjustable blade axis. When the blade angle is increased the gap increases especially at the inlet end of the blade.

10.11.2. Stayring

The turbine stay ring is built up of number of vanes that are of elongated streamlines cross section. Stayring vanes may be anchored in the reinforced concrete; or to ensure a greater steel circular bands - the upper and lower. Individually mounted vanes are used only at comparatively low heads and large turbine diameters, and a stayring with bands is used at higher heads, and the use of bands is obligatory when the turbine case is made of steel or has a steel lining.

10.11.3. Wicket gates

The wicket gate consists of guide vanes secured in the lower ring and in the turbine cover by means of pivot pins, making it possible to set the vanes at an angle.

The characteristic parameter of the wicket gate are:

(1) The number of guide vanes: The wicket gate of large axial flow turbines operated at a low head comprising 32 vanes; and at higher heads there are 24, and sometimes 20 guide vanes;

(2) The height of the wicket gate: Sometimes use is made dimensionless quantities referred to the diameter; these quantities characterize to a greater extent the type of turbine. The relative height of the wicket gate is a ratio of height of the wicket gate to the turbine diameter. The relative height in the axial flow turbines falls within 0.45 to 0.35 (the smaller values pertain to greater heads);

(3) The diameter over the guide vane pivot pins: The relative diameter of the wicket gate is a ratio of the diameter over the guide vane pivot pins to the runner diameter.

For different turbines relative diameter varies within 1.16 to 1.25 (increases with the lesser number of guide vanes in the wicket gate).

The kind of the wicket gate predetermines the direction of water flow. When the stream surfaces are close to the planes normal to the turbine axis, such wicket gate is referred to as radial. When the stream surfaces are almost conical, the wicket gate is called a conical. When the surfaces are close to cylindrical, the wicket gate is called an axial.

10.11.4. Runner

The turbine runner differs in the following respects with respect to Francis turbline.

(1) The runner is of the axial flow type and water enters and leaves the runner blades parallel to the axis while the Francis turbines runner is mixed flow type (Fig. 10.12).

(2) The runner vanes resemble the blades of a propeller. They are cantilevers mounted on the runner hub and the outer edges are not connected to each other (Fig. 10.12).

(3) The number of blades varies between 4 and 8, considerably less than the Francis turbine, therefore, a lesser area is occupied by the blades and more area is available for flow. (Fig. 10.12).

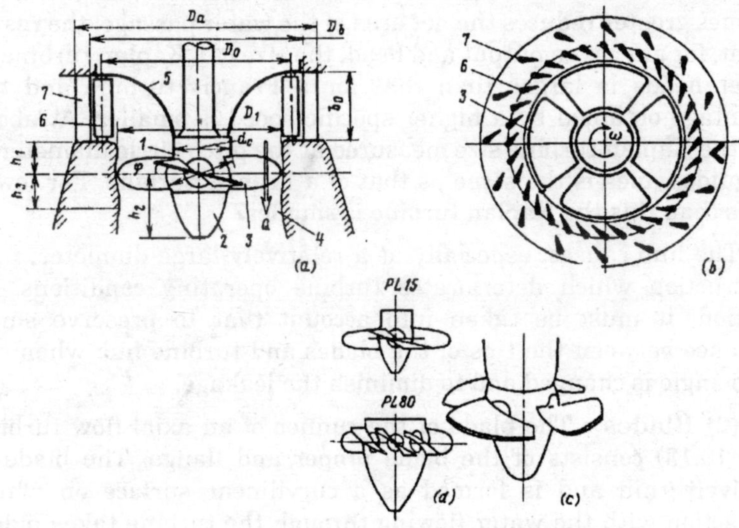

Fig. 10.12. Kaplan turbine (a) Cross sectional elevation; (b) Horizontal section through runner (c) Geneal appearance of runner; (d) Runner for various heads 1 - hub; 2- cones; 3- blades; 4-runner envelope; 5 - stationary fairing; 6 - wicket gates; 7-stay vanes (Ref. Ilyinykh)

The main components of the runner consists essentially of the hub, the blades, operating mechanism of blade, the hub cover and the nose piece.

(1) The hub : The hub is a hollow cylindrical piece with spherical midriff. The runner vanes are mounted on the specially provided recesses on the periphery of the hub; with their turnions projecting into the body of the hub. The operating mechanism is provided inside the body of the hub. The hub is generally spherical at the blade roots so that the clearance between the hub and blades is constant as the latter rotate. For low heads, however, the hub sometimes made cylindrical since then the varying clearance between the blades and hub has only a small effect on performance.

As the head increases, the loading on the blade increases, and a large hub in relation to the overall diameter of the blades is required in order to accommodate the blade bearings. This conforms with hydraulic requirements since a small hub diameter for low peripheral speed would necessitate an undesirable high degree of curvature of the blades adjacent to the hub in order to obtain the correct angle of attack.

The necessity of increasing the hub diameter as the head becomes greater reduces the net area of the water passage; the result is that, for any given output and head, the size of a Kaplan turbine at higher heads is larger than that for a Francis turbine and the advantage obtained by a higher specific speed is smaller. At about 20m the Kaplan turbine size measured at the pitch circle diameter of the guide vanes is the same as that of a Francis turbine. For lower heads than this the Kaplan turbine is smaller.

The hub causes, especially at a relatively large diameter, flow construction which deteriorates turbine operating conditions. In addition, it must be taken into account that to preserve small clearance between the tips of the blades and turbine hub when the blade angle is changed and to diminish the leakage.

(2) Blades : The blade of the runner of an axial flow turbine (Fig. 10.13) consists of the blade proper and flange. The blade is relatively thin and is formed as a curvilinear surface on which interaction with the water flowing through the turbine takes place, causing the conversion of the hydraulic energy into rotational energy of the runner and the shaft. The flange is used to secure the blade to the flange of the pivot with the aid of which the blade is secured in the bushing.

The cylindrical blade sections shown schematically in Fig. 10.13 give an idea of the form of the blades: the section located closer to the flamge, root sections, are thicker and steeper, the peripheral sections are thinner and flatter. Such a form is determined by the hydraulic conditioins under which water flows past the vane and also by strength. The load taken up by the vane is very large and distributed over the entire surface (for instance, in a turbine 9.5 m in diameter having 6 blades, at a head of 30 m the load on each blade amounts to 2700 kN). The blade is like a cantilever beam carrying a distributed load, and the bending moment, especially in the root sections, is very large, and this requires a corresponding section modules.

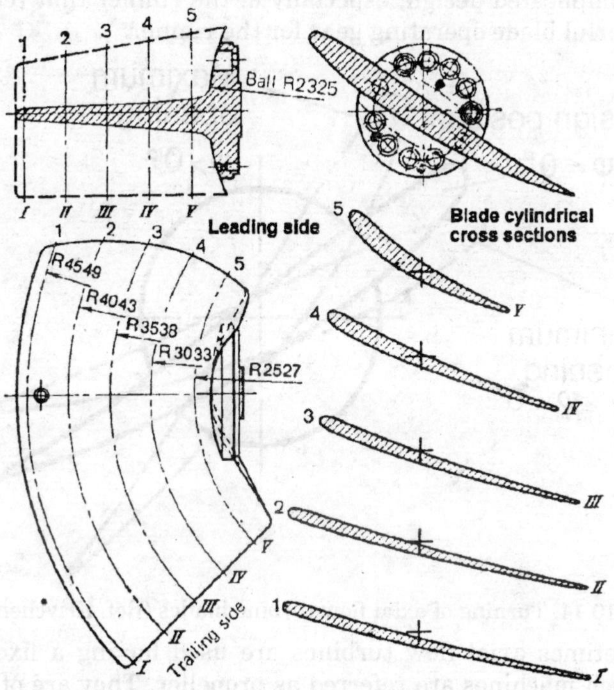

Fig. 10.13. Blade of axial flow turbine runner (Ref. Krivchenko)

The runners of adjustable blade Kaplan turbines permit the blades to change the blade angle, as illustrated in Fig. 10.14. Some design blade position is assumed as the reference point for the blade angle ϕ. In this position $\phi = 0$. When the blade angle ensure a maximum opening $\phi > 0°$, and when the blade angle is diminished $\phi < 0°$. Sometimes another reference point for the blade angle is used for instance, the plane perpendicular to the runner axis of rotation.

The blade angle is chosen in strict accordance with the opening of the guide blades and with account taken of the acting head. The relationship

$$\phi = f\,(\text{Opening of guide blades, H}) \qquad (10.17)$$

The so called combinator dependence, is set such that a maximum turbine efficiency is ensured under all operating conditions. It follows that adjustable blade turbines have a double regulation of power: regulation by changing the opening of the guide blades, and regulation by varying the runner blade angle. This gives great advantages from the power generating viewpoint, but leads to a more complicated design, especially of the runner that requires a very powerful blade operating gear for the runner.

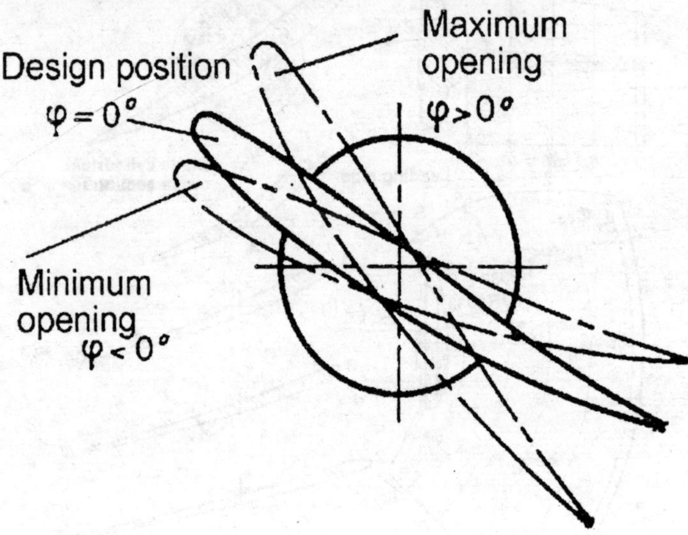

Fig. 10.14. Turning of axial flow turbine blades (Ref. Krivchenko)

Sometimes axial flow turbines are used having a fixed blade angle. Such machines are referred as propeller. They are of simpler desing, cost less, and have only one power regulating element - the guide blades, but their power generating indices are somewhat worse.

(3) Runner blade operating mechanism

Its function is to vary the runner blade angle to suit the instantaneous load conditioins in coordinatioin with the guide blade operating mechanism. This is done with the help of servomtor mechanism (Fig. 10.15)

Servomotor mechanism consists of a piston operating inside a cylinder and is under oil pressure on both sides. Pressurised oil is sup- plied to the servomotor through the hollow generator shaft from the supply head mounted on the excitor. The supply of the pressurised oil is regulated by the governing mechams.

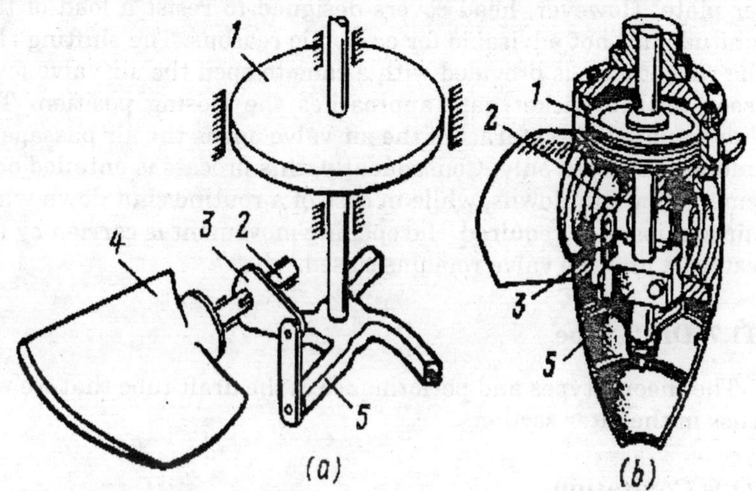

Fig. 10.15. Blade angle control mechanism (a) Schematic diagram; (b) Construction; 1- servomotor piston; 2 - blade axis; 3 - cam; 4-blade; 5-spider (Ref. Ilyinykh)

The governing of Kaplan turbines will be discussed in a separate chapter.

10.11.5. The turbine cover

The turbine cover the space on top of the runner; it accommodates the upper pivots of the guide vanes and supports the regulating ring. The cover houses the guide bearing of the turbine. The bearing structure of the generator thrust bearing often bears against the turbine cover. This bearing takes up the axial load constituting by the rotating parts of the unit. Safe turbine operation is ensured by making the cover not only of sufficient strength, but also rigid.

10.11.6. The air valve

The air valve mounted on the cover plate admits air through the air gap between the intermediatc and inner turbine cover into the guide channel when the turbine is shut down suddenly. That device is indispensable since in case of a sudden shutdown the water column issuing forth at the original velocity would, upon the discontinuation of discharge, separate from the cover plate, creating thereby a depression resulting in the excessive overloading of the cover plate. However, head covers designed to resist a load of this magnitude are not advisable for economic reasons. The shifting ring of the wicket gate is provided with a cane to open the air valve lever as soon as the wicket gate approaches the closing position. The spring actuated air cataract of the air valve opens the air passage at a sudden shutdown only. Consequently, this process is entailed only by emergency shutdowns, while in case of a routine shut down when no air admisiion is required, the opening movement is carried by the oil cataract and the valve remains closed.

10.11.7. Draft tube

The theory, types and performance of the draft tube that we will discuss in the later section.

10.11.8. Cavitation

A study of the operating condition of turbine without draft tube and cavitation is incomplete. Cavitation depends upon a critical value of the cavitation factor and suction height of the turbine above the tail race level. Cavition factor is a function of specific spped. Cavitation will be discussed in detail in a later chapter.

10.12. WORKING PROPORTIONS OF KAPLAN AND PROPELLER TURBINES

(1) Ratio of hub diameter to the outer diameter of the runner is established experimentally and depends upon the specific speed of the runner. With increasing value of specific speed, this ratio (d/D) decreases. The Table 10.1 gives the variation of (d/D), Head (H) and number of blades (Z) with a change in specific speed (N_s) of a Kaplan turbine.

Table 10.1. Tables Ns, d/D, Z and H

Ns	250	325	500	650	725	850
n=d/D	0.7	0.6	0.5	0.42	0.38	0.3
Z	10	7	5	4 to 5	4	3
H(m)	70	55	40	25	15	5

The value of n usually varies form 0.35 to 0.65.

(2) The discharge Q flowing through the turbine,

$$Q = \frac{\pi}{4}(D_2 - d^2)\ V_{f1}$$

(3) Assuming $V_{f1} = V_{f2}$

$$V_{f1} = \psi\ \sqrt{2\,g\,H}$$

The value of flow ratio ψ for Kaplan turbine is around 0.7.

(4) At any section of the blade, the radial distance from the axis at the inlet and the outlet is same.

Therefore,

$$u_1 = u_2 = \frac{\pi\,D\,N}{60}$$

and $$u_1 = K_u = \sqrt{2\,g\,H}$$

The speed ratio k_u varies from 0.9 to 2.1.

(5) The tangential velocity varies from section to section at the inlet and the outlet, therefore, the inlet and the outlet angles of the blades are changed at different section, greater at the tips and smaller at the hub (root). This gives to the twisted shape to the blades (Table 10.2).

(6) The expressions for the work done and the efficiencies of Kaplan turbine are same as those for Francis turbine.

Table 10.2. Comparison of turbines

Types of runner	α	θ	β	φ	Ku speed ratio	ψ flow ratio	No. of blades	m = b/D or n = d/D
Francis								
Slow	10° to 20°	60° to 90°	90°	15°	0.6 to 0.7	0.1	19 to 24	1/30
Medium	20° to 30°	90°	90°	15°	0.7 to 0.75	0.15 to 0.3	15 to 19	0.125 to 0.25
Fast	30° to 40°	90° to 135°	90°	15°	0.75 to 0.9	0.25 to 0.3	11 to 15	0.25 to 0.45
Propeller & Kaplan	45° to 70°	120° to 160°	90°	15°	0.9 to 2.1	Around 0.7	3 to 8	0.35 to 0.65

10.13. DRAFT TUBES

The energy output and cavitation index of hydroelectric turbines are considerably influenced by the design of their draft tubes. The design also determines the size and cost of the machine room of the hydroelectric station.

It has been impossible until recently to develop a theoretical method for determining dimensions of draft tubes because of their complex shape and the flow pattern after the rotor.

At present a large number of laboratories, research institutes, higher technical institutes and turbine manufactures are actively engaged in research on draft tubes. These investigations point to methods of improving the power indices and operating regimes of turbine. These investigations aid in analyzing the effect of variation of the size of the draft tubes from those recommended by the turbine manufactures.

10.13.1. The function of draft tubes

The draft tube is a conduit discharging water from the runner to the tail race, and has a twofold function :

(1) to recover as much as possible of the velocity energy (kinetic energy) of the water leaving the runner, thus increasing the dynamic draft head, and

(2) to utilize the vertical distance between the turbine exit and the tailwater level, called the static draft head.

10.13.2. Hydraulic characteristics of draft tubes

Refer Fig. 10.16 in which points 1, 2 and 3 have been considered at the runner entrance, runner exit and at the outlet end of the draft tube respectively.

The average velocity at the inlet of the draft tube, V_2 depends on mode of turbine operation and is determined with the help of outlet velocity triangle. Let us determine the absolute pressure at point 2 when the turbine in operation, $P_2/\rho g$. Applying Bernoulli's equation for point 2 and 3 (outlet from the draft tube).

$$\frac{P_2}{\rho g} + \frac{V_2^2}{2g} + Z_2 = \frac{P_3}{\rho g} + \frac{V_3^2}{2g} Z_3 + h_d \quad (10.18)$$

But $\quad \dfrac{P_3}{\rho g} = \dfrac{P_a}{\rho g} + h$

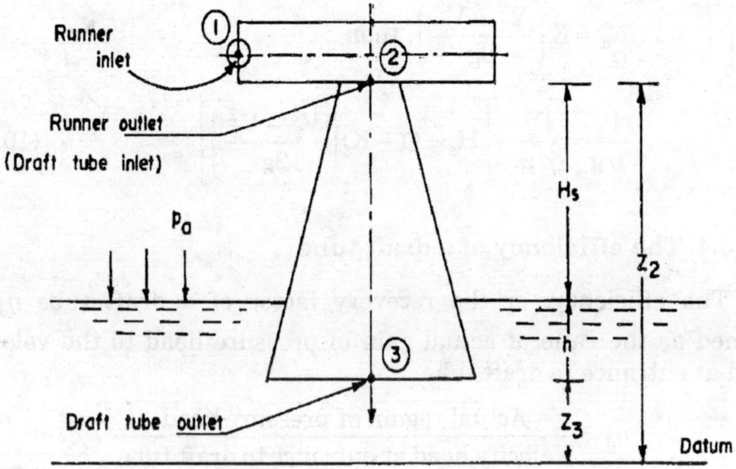

Fig. 10.16. Hydraulic characteristics of draft tube.

Where h is the depth below the level of the tail water level and h_d are the hydraulic losses in the draft tube.

Therefore,

$$\frac{P_2}{\rho g} + \frac{P_a}{\rho g} - (Z_2 - Z_3 - h) - \left(\frac{V_2^2 - V_3^2}{2g}\right) + h_d$$

Since $(Z_2 - Z_3 - h) = H_s$, the height of runner exit above tail water level.

Therefore,

$$\frac{P_2}{\rho g} + \frac{P_a}{\rho g} - H_s \left(\frac{V_2^2 - V_3^2}{2g}\right) + h_d \qquad (10.19)$$

The obtained results from equation (10.19) shows that the draft tube creates a lower pressure (vacuum) underneath the turbine runner, the vacuum being the sum of two parts : static fall of pressure H_s that is independent of the discharge and dynamic fall of pressure

$$\left[\left(\frac{V_2^2 - V_3^2}{2g}\right) - h_d\right]$$

that is specified by diffusion action of the draft tube and increasing with a rising flow rate. Generally h_d is expressed as

$$h_d = K\left(\frac{V_2^2 - V_3^2}{2g}\right), \text{ then}$$

$$\frac{P_2}{\rho g} = \frac{P_a}{\rho g} - \left[H_s + (1-K)\left(\frac{V_2^2 - V_3^2}{2g}\right)\right] \tag{10.20}$$

10.13.4. The efficiency of a draft tube

The efficiency, or the recovery factor of a draft tube η_d is defined as the ratio of actual gain of pressure head to the velocity head at entrance to draft tube.

$$\eta_d = \frac{\text{Actual regain of pressure head}}{\text{Velocity head at entrance to draft tube}}$$

$$= \frac{\frac{V_2^2}{2g} - \frac{V_3^2}{2g} - h_d}{\frac{V_2^2}{2g}}$$

$$= \frac{\left(\frac{V_2^2 - V_3^2}{2g}\right) - K\left(\frac{V_2^2 - V_3^2}{2g}\right)}{V_2^2/2g}$$

$$= \frac{(1-K)\left(\frac{V_2^2 - V_3^2}{2g}\right)}{V_2^2/2g} \tag{10.21}$$

For good draft tubes of a sufficiently great length η_d reaches 80 to 85 per cent.

10.14. TYPES OF DRAFT TUBES

Depending on the power output and the diameter of the runner the following types of draft tubes are used.

(1) Straight conical

(2) Elbow type and

(3) Bell mouth with or without cone.

Each of these types of draft tubes has its own particular advantages and disadvantages which make them highly suited in a particular situation.

10.15. CAVITATION

A Study of the operating ocnditions of turbine with a draft tube has shown that a lower pressure is created underneath the runner. In addition, as water flows past the blades of the runner, the pressure is more reduced on the back of the blades. It follows that in some part of the flow passage of hydraulic turbines the pressure may be rather low (a high vacuum). These conditions have their peculiarities which may cause the appearance of cavitation.

10.15.1. Definition of Cavitation

The term "Cavitation" basically refers to the ability of cold water to boil under low pressure.

Water contains a great number of finest solid and gaseous inclusions serving as weak points, the nuclei. The dissolved air contained in water also contributes to the formation of nuclei.

10.15.2 Causes of cavitation

When water pressure drops below the vapour pressure the intensive transition of the liquid into a gaseous state (steam) initiates at the nuclei boundary, and local discontinuities - cavities form mainly filled with water vapour. After cavities have formed, no further drop in pressure takes place since it is compensated for by a rapid increase in the volume of cavities. With an increase in pressure the cavities close (collapse) and the vapour condenses to liquid again. A cavity results and the surrounding liquid rushes in to fill it. The liquid moving from all direction collides at the centre of cavity, thus giving rise to very high local pressures (upto 1 GPa), which spreads just as quickly in an explosion like manner.

10.15.3. Effects on hydraulic machines

The cavitating flow has the following effects on hydraulic machines

(1) with a sufficiently developed cavitation the hydraulic losses increase resulting in a lower efficiency and lesser discharge.

(2) in the presence of cavitation in a hydraulic machine a characteristic noise is heard, and a higher than normal vibration appears;

(3) when a hydraulic machines is operated under conditions of cavitation, the material ultimately fails by fatigue, aided perhaps by corrosion, and so the surface becomes badly scored and pitted.

10.15.4. Cavitation Erosions

Cavitation erosions are mainly caused by the mechanical action of the cavitating flow which manifests itself in water hammers originating when cavities collapse on a streamlined surface or close to it. The peculiarity of this harmful action consists in that the frequency of the hydraulic hammers is very high. They cause fatigue phenomena in metals. The destruction takes place in the form of fatigue flaking, knockout of individual crystals and the surface of the affected metal becomes sponge shaped. The intensity of erosion is sometimes rather high and may reach an annual depth of 10 to 40 mm. This necessitates frequent repairs, change of working members and results in considerable operation costs for hydraulic machines. The wear of blades and other surfaces exposed to cavity flow also results in a supplementary reduction of efficiency and a correspo-nding drop in the power generated at a hydro electric power plants.

In addition to the above considered mechanical actions, cavity flow brings about chemical and electrical phenomena. The role of these factors has not been elucidated as yet, but it is clear that they contribute to a higher intensity of cavitation erosion.

10.15.5. Prevention of Cavitation

Prevention of cavitation can be done in the following way.

(1) Choosing suitable materials that can resist cavitation damage. Porous materials like cast iron are least resistive to damage, whereas materials of a dense, homogenous structure are better suited to resist cavitation.

(2) Specific speed greater than recommended values may result in cavitation. This may be avoided by setting the

runner at a lower elevation, resulting in increased excavation. Deeper tailrace excavation may be too expensive. Often it is more economical to repair, once a year, by welding, the damage done by cavitation than to prevent all cavitation.

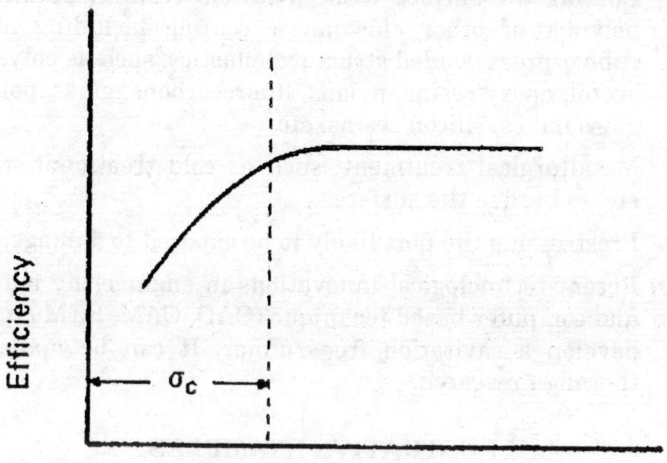

Fig. 10.17. Graph between efficiency v/s cavitation parameter (Ref. Massey)

(3) High velocities with sudden changes in velocity which result in lower pressure areas are conducive to cavitatioin. This can be avoided either by improved design which eliminates sudden changes in curvature and/or by lowering the setting of the runner with reference to tailwater, so that a vacuum sufficient to produce vaporization will not occur. In some cases it is necessary to set the runner below tailwater level. This is particularly true of propeller turbines.

(4) Shifting the area of damage by converting transient cavities to super cavities by pumping air or other means. Cavities by pumping air or other means. The air admitted raises the pressure existing in the region and forming of vapour bubbles is prevented. Air admitted forms large bubbles which travel downstream. The bubble do not collapse due to cushioning effect of air.

(5) A model of the prototype must be made and thoroughly tested in a laboratory.

(6) Giving fine finish to the surface or other surface treatments like welding cavitation resistant metal pieces or power on the surface likely to be exposed to cavitation damage.

(7) Coating the surface with cavitation resitant paintings of polymer or other elastomeric coating including natural rubber press bonded structural plastics, such as polyamide, acetol, epoxy resins, nylons, fluorocarbons, glass, polyester glass flakes, silicon resins, etc.

(8) Metallurgical treatment, such as cold treatment, rolling, etc. to harden the surface.

(9) Prestressing the part likely to be exposed to damage.

(10) Recent technological innovations in engineering materials and computer based technique (CAD, CAM, FEM etc.) is to develop a cavitation free runner. It can be made after thorough research.

ILLUSTRATIVE EXAMPLES

10.1 Water flows through a vertical Francis turbine at the rate of 15.5 m³/s and makes its runner to rotate at 428.4 rpm. The velocity and the pressure head at the inlet of the spiral casing are 8.5 m/s and 240 m respectively and the centre line of its inlet is 3 m above the tail race level. The diameter and the width of the runner at inlet are 2 m and 300 mm respectively. If the hydraulic and overall efficiencies of the turbine are 93% and 90% respectively, determine (a) output power, (b) guide vane angle and (c) the runner blade angle at inlet.

Solution

$$\text{Total head } H = \frac{P_1}{\rho g} + \frac{V_1^2}{2g} + Z_1$$

$$H = 240 + \frac{(8.5)^2}{2 \times 9.81} + 3 = 246.68\ m$$

$$P = \rho g Q H \eta_0 = 1000 \times 9.81 \times 15.5 \times 246.68 \times 0.9$$

$$= 33.76\ MW$$ Ans.

$$u_1 = \frac{\pi D N}{60} = \frac{\pi \times 2 \times 428.4}{60} = 44.862\ m/s$$

$$\eta_h = \frac{\omega_1\ u_1}{g\ H}$$

$$0.93 = \frac{V_{\omega 1} \times 44.862}{9.81 \times\ 246.68}$$

$$\therefore V_{\omega 1} = 50.166\ m/s$$

$$Q = \pi\ B_1\ D_1\ V_{f1} \quad 15.5 = \pi \times 0.3 \times 2 \times V_{f1}$$

$$V_{f1} = 8.223\ m/s$$

$$\tan\ \alpha = \frac{V_{f1}}{V_{\omega 1}} = \frac{8.223}{50.166} = 0.1639$$

$$\therefore \alpha = \tan^{-1}(0.1639) = 9.31°$$ Ans.

$$\tan\ \theta = \frac{V_{f1}}{V_{\omega 1} - u_1} = \frac{8.223}{50.166 - 44.862} = \frac{8.223}{5.304} = 1.55$$

$$\therefore \theta = \tan^{-1}(1.55) = 57.17°$$ Ans.

10.2 An inward flow reaction turbine of Francis type has a runner 600 mm diameter and 50 mm width at the outer rim, the inner diameter being 390 mm. The wheel blade angles at the inlet and outlet are 95° (inside angle of the velocity triangle 85°) and 14°respectively. The velocity of flow is uniform throughout the wheel and 8 percent of circumferential area of the runner is blocked by the blade thickness. The head on the turbine is 54 m, the hydraulic efficiency is 88 percent, the overall efficiency is 81 percent and the discharge at the outlet is radial. Determine

(a) the speed in rpm, and

(b) the power of the turbine

Solution : For velocity triangles-see text

$$u_1 = \frac{\pi\ D_1\ N}{60} = \frac{\pi \times 0.6 \times N}{60} = 0.0314\ N$$

$$u_2 = \frac{\pi D_2 N}{60} = \frac{\pi \times 0.39 \times N}{60} = 0.02\, N$$

$$\tan \theta = \frac{V_{f1}}{u_1 - V_{\omega 1}}$$

and $$\tan \varphi = \frac{V_{f2}}{u_2}$$

Given $V_{f1} = V_{f2}$

Therefore,

$$\frac{\tan \theta}{\tan \varphi} = \frac{u_2}{u_1 - V_{\omega 1}} \tag{1}$$

$$\eta_h = \frac{V_{\omega 1}\, u_1}{g\, H} = \frac{V_{\omega 1} \times 0.0314\, N}{9.81 \times 54}$$

$$\therefore V_{\omega 1} = \frac{0.88 \times 9.81 \times 54}{0.0314\, N} = \frac{14846.2}{N}$$

Free equation (1)

$$\frac{\tan 85^\circ}{\tan 14^\circ} = \frac{0.02\, N}{0.0314\, N - \dfrac{14846.2}{N}}$$

$$1.419\, N^2 = 680251.4$$

$$N^2 = 479387.87$$ Ans.

$$\therefore N = 692.378\ rpm$$

Therefore,

$$u_2 = 0.02\, N = 0.02 \times 692.378 = 13.84\ m/s$$

$$V_{f1} = V_{f2} = u_2 \tan \varphi = 13.84 \tan 14^\circ = 13.84 \times 0.2493 = 3.45\ m/s$$

$$Q = K\, \pi\, B_1\, D_1\, V_{f1} = 0.92\pi \times 0.05 \times 0.6 \times 3.45 = 0.3\ m^3/s$$

$$P = \rho\, g\, Q\, H\, \eta_0 = 1000 \times 9.81 \times 0.3 \times 54 \times 0.81 = 128.72\ kW$$ Ans.

10.3 A propeller, turbine operates at 23 MW generator under an effective head of 39 m. Assuming generator efficiency of 93%, speed ratio 2, flow ratio as 0.6, diameter of boss as 0.35 times the external diameter of runner, overall turbine efficiency of 90% and hydraulic efficiency of 94%, determine.

(a) The tip diameter of the runner.

(b) Speed and specific speed of the runner

(c) Guide vane angle and runner outlet angle, assuming swirl at the runner exit.

Solution:

$$u_1 = K_u \sqrt{2\,g\,H}$$

$$= 2\sqrt{2 \times 9.81 \times 39} = 55.324\ m/s$$

$$V_{f1} = \psi \sqrt{2\,g\,H}$$

$$= 0.6\sqrt{2 \times 9.81 \times 39} = 16.597\ m/s$$

$$P = \rho\, g\, Q\, H\, \eta_0\, \eta_g$$

$$23 \times 10^6 = 1000 \times 9.81 \times Q \times 39 \times 0.9 \times 0.93$$

$$Q = 71.82387\ m^3/s$$

$$Q = \frac{\pi}{4}\left(D^2 - d_b^2\right)V_{f1}$$

$$71.82387 = \frac{\pi}{4}\left[D^2 - (0.35\,D)^2\right] \times 16.597$$

$$D = 2.506\ m$$ Ans.

$$u_1 = \frac{\pi\, D\, N}{60}$$

$$55.324 = \frac{\pi \times 2.506 \times N}{60}$$

$$N = 421.66\ rpm$$ Ans.

$$N_s = \frac{N\sqrt{P}}{H^{5/4}} = \frac{421.66\sqrt{23000}}{(39)^{5/4}} = 680.384$$ Ans.

$$\eta_h = \frac{V_{\omega 1}\, u_1}{g\, H}$$

$$0.94 = \frac{V_{\omega 1} \times 55.324}{9.81 \times 39}$$

$$\therefore V_{\omega 1} = 6.5\ m/s$$

$$\tan\alpha = \frac{V_{f1}}{V_{\omega 1}} = \frac{16.597}{6.5} = 2.55338$$

$$\therefore \alpha = \tan^{-1}(2.55338) = 68.61^\circ$$ Ans.

$$\tan\varphi = \frac{V_{f1}}{u_2} = \frac{V_{f1}}{u_1} = \frac{16.597}{55.324} = 0.2999$$

$$\therefore \varphi = \tan^{-1}(0.2999) = 16.699^\circ$$ Ans.

10.4 A Kaplan turbine develops 10 MW under a net available head of 18 m. Assuming a speed ratio of 2, flow ratio 0.7, the hydraulic efficiency 93% and overall efficiency 82% and the diameter of the boss equal to one third of the runner, determine (a) the diameter of the runner and its speed and (b) the inlet blade angle at the edge of the blade.

Solution

$$u_1 = K_u \sqrt{2\,g\,H} = 2\sqrt{2 \times 9.81 \times 18}$$

$$= 37.59\ m/s$$

$$V_{f1} = \psi\sqrt{2\,g\,H}$$

$$= 0.7\sqrt{2 \times 9.81 \times 18} = 13.155\ m/s$$

$$P = \rho\,g\,Q\,H\,\eta_0$$

$$10 \times 10^6 = 1000 \times 9.81 \times 18 \times 0.82$$

$$Q = 69.06\ m^3/s$$

$$Q = \frac{\pi}{4}\left(D^2 - d_b^2\right)V_{f1}$$

$$69.06 = \frac{\pi}{4}D^2\left(1 - \frac{1}{9}\right) \times 13.55$$

$$\therefore D = 2.742\ m$$ Ans.

$$u_1 = \frac{\pi D N}{60}$$

$$37.59 = \frac{60 \times 2.742 \times N}{60}$$

$$\therefore N = 262 \ rpm$$ Ans.

$$\eta_h = \frac{V_{\omega 1} \, u_1}{g \, H}$$

$$0.93 = \frac{V_{\omega 1} \times 37.59}{9.81 \times 18}$$

$$\therefore V_{\omega 1} = 4.369 \ m/s$$

$$\tan\theta = \frac{V_{f1}}{u_1 - V_{\omega}}$$

$$\tan\theta = \frac{13.155}{37.59 - 4.369} = \frac{13.155}{33.221}$$

$$= 0.39598$$

$$\theta = \tan^{-1}(0.39598) = 21.6^{\circ}$$ Ans.

$$\therefore 180 - \theta = 180 - 21.6^{\circ} = 158.4^{\circ}$$ Ans.

10.5 An axial flow water turbine has a rotor 7.5 m overall diameter, with a boss of 2.25 m diameter. The blade tips at the upper inlet sides are inclined at $\tan^{-1} 2/5$ to the tangential direction and at the lower outlet side $\tan^{-1} 1/3$. If the turbine is to work under a pressure difference of 8 m, at an efficiency of 90 percent, what will be its power output, speed and specific speed.

Solution: For Figures - see text

From inlet velocity triangle

$$\frac{2}{5} = \frac{V_{f1}}{u_1 - V_{\omega 1}} \tag{1}$$

and from outlet velocity triangle

$$\frac{1}{3} = \frac{V_{f2}}{u_2} \tag{2}$$

For axial flow water turbine

$$V_{f1} = V_{f2} \quad \text{and } u_1 = u_2$$

Solving (1) and (2)

$$V_{f1} = 2\,V_{\omega 1} \qquad (3)$$

and $u_1 = 3\,V_{f1}$ (4)

$$\eta_h = \frac{V_{\omega 1}\, u_1}{g\, H}$$

$$0.9 = \frac{\left(V_{f1}/2\right)\left(3\,V_{f1}\right)}{9.81 \times 8}$$

$$V_{f1}^2 = \frac{0.9 \times 9.81 \times 8 \times 2}{3} = 47.088$$

$$\therefore V_{f1} = 6.862\ m/s$$

From (4), $u_1 = 3 \times 6.862 = 20.586\ m/s$

$$Q = \frac{\pi}{4}\left(D^2 - d_b^2\right)V_{f1}$$

$$= \frac{\pi}{4}\left[(7.5)^2 - (2.25)^2\right] \times 6.862$$

$$= 276\ m^3/s$$

$$\eta_0 = \eta_h \times 1 = 0.9 \times 1 = 0.9$$

In the Kaplan turbine mechanical losses are negligible therefore, the given efficiency is overall efficiency

$$P = \rho\, g\, Q\, H\, \eta_0$$

$$= 1000 \times 9.81 \times 276 \times 8 \times 0.9$$

$$= 19494.432\ kW$$

Ans.

$$u_1 = \frac{\pi\, D\, N}{60}$$

$$20.586 = \frac{\pi \times 7.5 \times N}{60}$$

$$N = 52.42\ rpm$$

Ans.

$$N_s = \frac{N\sqrt{P}}{H^{5/4}} = \frac{52.42\sqrt{19494.432}}{(8)^{5/4}}$$

$= 544$ Ans.

10.6 An axial flow reaction turbine of the propeller type has a speed of 30 rpm. The mean diameter of the runner blades is 5.5 m and the effective axial area of flow is 27.9 m². The inlet guide blade angle and the outlet runner blade angles are 150° and 20° respectively to the direction of motion of the wheel periphery. The velocity of flow is constant throughout. Calculate the quantity of water required per second, the theoretical power of the turbine, the hydraulic efficiency and the specific speed.

Solution

$$u_1 = \frac{\pi D N}{60} = \frac{60 \times 5.5 \times 30}{60} = 8.65\ m/s = u_2$$

$$\tan 30° = \frac{V_{f1}}{V_{\omega 1}} \qquad (1)$$

$$\tan 20° = \frac{V_{f2}}{u_2}$$

$$\therefore V_{f2} = 8.65 \times 0.3639 = 3.148\ m/s = V_{f1}$$

From equation (1)

$$V_{\omega 1} = \frac{V_{f1}}{\tan 30°} = \frac{3.148}{0.577} = 5.45\ m/s$$

$$\frac{V_{\omega 1}\, u_1}{g} = H - \frac{V_2^2}{2g} = H - \frac{V_{f2}^2}{2g}$$

$$4.8 = H - 0.505$$

$\therefore H = 4.8 + 0.505 = 5.31\ m$ Ans.

$Q = A\, V_{f1} = 27.9 \times 3.148 = 87.73\ m^3/s$ Ans.

$$P = \rho\, g\, Q\, H$$

$$= 1000 \times 9.81 \times 87.73 \times 5.31$$

$$= 4570.43\ kW$$ Ans.

$$\eta_h = \frac{V_{\omega 1}\, u_1}{g\, H} = \frac{5.45 \times 8.65}{9.81 \times 5.31} = 0.905 = 90.5\%$$ Ans.

$$N_s = \frac{N\sqrt{P}}{H^{5/4}} = \frac{30\sqrt{4570.43}}{(5.31)^{5/4}} = 251.61$$ Ans.

10.7 In a projected low head hydroelectric scheme 283 m³/s of water are available under a head of 3.66 m. Alternative schemes to use Francis turbines having a specific speed of 400 or Kaplan turbine with a specific speed of 686 are investigated. The normal running speed is to be 50 rpm in both the schemes.

Compare the two proposals in so far as the numbers of machines are concerned and estimate the power to be developed by each machine. The units in either installations are to be equal power and the efficiency of each type may be assumed to be 90%.

Solution

$$P = \rho\, g\, Q\, H\, \eta_0$$

$$= 1000 \times 9.81 \times 283 \times 3.66 \times 0.9 = 9144.9\ kW$$

First alternative – Francis turbine

$$N = \frac{N\sqrt{P}}{H^{5/4}}$$

$$400 = \frac{50\sqrt{P}}{(3.66)^{5/4}}$$

$$\therefore P = 1640.15\ kW$$

$$\text{Number of Francis turbine used} = \frac{9144.9}{1640.15} = 5.57 \approx 6$$

6 Francis turbines are used Ans.

Second alternative – Kaplan turbine

$$N_s = \frac{N\sqrt{P}}{H^{5/4}}$$

$$686 = \frac{50\sqrt{P}}{(3.66)^{5.4}}$$

$$\therefore P = 4824.04\ kW$$

Number of Kaplan turbine used

$$= \frac{9144.9}{4824.04} = 1.896 \approx 2$$

2 Kaplan turbines are used. Ans.

10.8. An inward flow reaction turbine discharges radially and the velocity of flow is constant and equal to the velocity of discharge from the turbine. Show that the hydraulic efficiency can be expressed by

$$\eta_h = \frac{1}{\dfrac{\tan^2\alpha}{1 + \dfrac{2}{\left[1 - \dfrac{\tan\alpha}{\tan\theta}\right]}}}$$

If vanes are radially at inlet, also show that

$$\eta_h = \frac{2}{2 + \tan^2\alpha}$$

Solution:

Given

$$V_{w2} = 0$$

$$V_{f1} = V_{f2} = V_2$$

From the inlet velocity triangle (see text)

$$\tan\alpha = \frac{V_{f1}}{V_{w1}}$$

$$\therefore V_{f1} = V_{w1}\tan\alpha \qquad (1)$$

$$\text{and } \tan\theta = \frac{V_{f1}}{V_{w1} - u_1}$$

$$V_{w1} - u_1 = \frac{V_{f1}}{\tan\theta}$$

$$= \frac{V_{w1}\tan\alpha}{\tan\theta} \qquad \text{[From (1)]}$$

$$u_1 = V_{w1} - \frac{V_{w1}\tan\alpha}{\tan\theta}$$

$$= V_{w1} = \left[1 - \frac{\tan\alpha}{\tan\theta}\right] \qquad (2)$$

We know that [Energy balance at runner]

$$H = \frac{V_{w1}\,u_1}{g} + \frac{V_2^2}{2g}$$

$$= \frac{V_{w1}\,u_1}{g} + \frac{V_{f1}^2}{2g}$$

Substituting the values of V_{f1} and u_1 from (1) and (2),

$$H = \frac{V_{w1}{}^2}{g}\left(1 - \frac{\tan\alpha}{\tan\theta}\right) + \frac{V_{w1}^2\tan^2\alpha}{2g}$$

$$= \frac{V_{w1}^2}{g}\left[1 - \frac{\tan\alpha}{\tan\theta} + \frac{\tan^2\alpha}{2}\right]$$

Hydraulic efficiency is given by

$$\eta_h = \frac{V_{w1}\,u_1}{g\,H}$$

$$= \frac{V_{w1}^2\left(1 - \dfrac{\tan\alpha}{\tan\theta}\right)}{g\,\dfrac{V_{w1}^2}{g}\left(1 - \dfrac{\tan\alpha}{\tan\theta} + \dfrac{\tan^2\alpha}{2}\right)}$$

$$= \frac{\left(1 - \dfrac{\tan\alpha}{\tan\theta}\right)}{\left(1 - \dfrac{\tan\alpha}{\tan\theta} + \dfrac{\tan^2\alpha}{2}\right)}$$

$$= \frac{1}{1 + \dfrac{\dfrac{\tan^2 \alpha}{2}}{1 - \dfrac{\tan \alpha}{\tan \theta}}}$$ Proved.

If vanes are radial at inlet, then $\theta = 90°$.

$$\therefore \eta_h = \frac{1}{1 + \dfrac{\dfrac{\tan^2 \alpha}{2}}{1 - \dfrac{\tan \alpha}{\infty}}} = \frac{1}{1 + \dfrac{\tan^2 \alpha}{2}}$$

$$= \frac{2}{2 + \tan^2 \alpha}$$ Proved.

10.9. In a reaction turbine with radial vanes at the inlet, the velocity of flow at outlet is K times that at inlet. Prove that

Hydraulic efficiency $\eta_h = \dfrac{2}{2 + k^2 \tan^2 \alpha}$

and $u_1 = \dfrac{\sqrt{2\,g\,H}}{2 + k^2 \tan^2 \alpha}$

Solution:

Given

$$V_{f2} = k\, V_{f1} = V_2$$

$$u_1 = V_{w1}$$

From the inlet velocity triangle (see text)

$$\tan \alpha = \frac{V_{f1}}{V_{w1}}$$

$$V_{f1} = V_{w1} \tan \alpha$$

$$k\, V_{f1} = k\, V_{w1} \tan \alpha$$

$$V_2 = k\, u_1 \tan \alpha \qquad (1)$$

We know that [Energy balance at runner]

$$H = \frac{V_{w1} u_1}{g} + \frac{V_2^2}{2g}$$

$$gH = V_{w1} u_1 + \frac{V_2^2}{2g}$$

$$gH = \frac{V_2^2}{2} + u_1^2 \qquad (Given\ u_1 = V_{w1})$$

$$= \frac{k^2 u_1^2 \tan^2 \alpha}{2} + u_1^2 \qquad \text{from (1)}$$

$$= u_1^2 \left(\frac{k^2 \tan^2 \alpha}{2} + 1 \right) \qquad (2)$$

$$u_1^2 = \frac{2\, g\, H}{2 + k^2 \tan^2 \alpha}$$

$$u_1 = \sqrt{\frac{2\, g\, H}{2 + k^2 \tan^2 \alpha}} \qquad \text{Proved.}$$

$$\eta_h = \frac{V_{w1} u_1}{gH} = \frac{u_1^2}{gH} = \frac{u_1^2}{u_1^2 \left(\frac{k^2 \tan^2 \alpha}{2} \right)} \qquad \text{[From (2)]}$$

$$= \frac{1}{1 + \frac{k^2 \tan^2 \alpha}{2}}$$

$$= \frac{2}{2 + k^2 \tan^2 \alpha} \qquad \text{Proved.}$$

10.10. A Kaplan turbine develops 2.8 MW under a net head of 5m. It is provided with a draft tube having a inlet diameter of 3m and set 1.2 m above the tailrace water level. A vacuum gauge connected to the draft tube indicates a reading of 4.2 m of water. Assuming the efficiency of the draft tube as 80%, determine the efficiency of the turbine

Solution : The expression obtained for the draft tube is

The expression obtained for the draft tube is

$$\frac{P_2}{\rho g} = -\left[H_s + \frac{V_2^2 - V_3^2}{2g}\right] + h_d$$

Neglecting the head loss in the draft tube, we have

$$\frac{P_2}{\rho g} = -\left[H_s + \frac{V_2^2 - V_3^2}{2g}\right]$$

$$-4.2 = -\left[1.2 + \frac{V_2^2 - V_3^2}{2g}\right]$$

$$\therefore \quad \frac{V_2^2 - V_3^2}{2g} = 3 \qquad (1)$$

The expression for draft tube efficiency is

$$\eta_d = \frac{\left(V_2^2 - V_3^2\right)/2g}{V_2^2/2g}$$

$$0.8 = \frac{3}{\dfrac{V_2^2}{2g}}$$

$$\frac{V_2^2}{2g} = \frac{3}{0.8}$$

$$\therefore \; V_2^2 = 73.575 \qquad \therefore V_2 = 8.5775\ m/s$$

The discharge through the draft tube (or from the runner).

$$Q = \frac{\pi}{4}(3)^2 \times 8.5775 = 60.63\ m^3/s$$

$$P = \rho\, Q\, g\, H\, \eta_0$$

$$2.8 \times 10^6 = 10^3 \times 60.63 \times 9.81 \times 5 \times \eta_0$$

$$\eta_0 = 0.9415 = 94.15\%$$

Ans.

10.11. A Kaplan turbine develops 2000 kW under a head of 6m. The turbine efficiency is 85%. The inlet diameter of conical draft tube is 3m and the recovery factor of draft tube is 90%. Find the actual regain of pressure head and the outlet diameter of the draft tube. Neglect losses in the draft tube.

Solution

$$P = \rho\, g\, Q\, H\, \eta_0$$

$$2000 \times 10^3 = 1000 \times 9.81 \times Q \times 6 \times 0.85$$

$$Q = 39.9752\ m^3/s$$

$$Q = \frac{\pi}{4} D_2^2\, V_2$$

$$39.9752 = \frac{\pi}{4}(3)^2\, V_2$$

$$V_2 = 5.655\ m/s$$

We know that

$$\eta_d = \frac{\text{Actual regain of pressure head}}{\text{Velocity head at entrance to the draft tube}}$$

$\therefore$ Actual regain of pressure head

$$= 0.9 \times \frac{V_2^2}{2g}$$

$$= 0.9 \times \frac{(5.655)^2}{2 \times 9.81} = 1.46692\ m$$

Neglects head losses in the draft tube.

$$\therefore\ \frac{V_2^2 - V_3^2}{2g} = 1.46692$$

or, $$V_2^2 - V_3^2 = 2 \times 9.81 \times 1.46692 = 28.7811$$

or, $$V_3^2 = 31.979 - 28.7811$$

or, $$V_3^2 = 3.1979$$

$$V_3 = 1.78827\ m/s$$

$$39.9752 = \frac{\pi}{4} D_3^2 \times 1.78827$$

$$(1.4)\ D_3^2 = 39.9752$$

$$D_3^2 = 28.462$$

$$D_3 = 5{,}33498\ m$$ **Ans.**

EXERCISES

A. THEORY

10.1. Describe the choice of turbine with reference to (a) High specific speed (b) High speed (c) High discharge.

10.2. Describe the importance of speed ratio K_u with reference to choice of turbine.

10.3. Draw a sectional arrangement of Francis turbine. Name the essential component parts of turbine.

10.4. Explain the functon of Scroll casing (Spiral casing)

10.5. Explain the function of speed rings.

10.6. What are the differences between Francis turbine and Kaplan turbine? Are the propeller and Kaplan turbines same?

10.7. Explain how the runner of a Francis turbine can be developed to that of a Kaplan type.

10.8. Describe the special features of a Kaplan turbine.

10.9. Describe in brief the choice between propeller and Francis turbine.

10.10. Describe briefly the choice between Kaplan and propeller turbines.

10.11. Explain the runner blade control mechanism of Kaplan turbine.

10.12. Explain why the air valve mounted on the cover plate of Kaplan turbine?

10.13. Explain the theory and function of a draft tube. Drive an expression for draft tube theory.

10.14. Sketch and describe the different types of draft tube.

10.15. Describe hydraulic characteristics of draft tubes.

10.16. Describe power generating characteristics of draft tubes.

10.17. Explain the power generating effect of draft tubes.

10.18. Explain the relation between kinetic energy leaving the runner with reference to specific speed for draft tubes.

10.19. Explain the operating conditions of draft tube other than optimum.

10.20. Draw a neat sketch of straight conical draft tube and explain.

10.21. Draw a neat sketch of elbow type draft tube and explain its parts.

10.22. Draw a neat sketch of elbow type draft tube and show the recommended dimensions.

10.23. Draw different types of bell mouth shaped draft tubes.

10.24. What are the advantages of the bell mouth shaped draft tubes?

10.25. Define cavition.

10.26. What are the causes of cavitation?

10.27. What are the effect of cavitation on hydraulic machines?

10.28. What do you mean by cavitation erosions?

10.29. What are the characteristics types of cavitation for hydraulic machines?

10.30. Describe stages of cavitation.

10.31. Define cavitation coefficient and permissible suction lift.

10.32. Describe the methods for preventing cavitation.

10.33. What is meant by cavitation? Develop an expression for cavitation number.

10.34. What is Thomas' cavitation factor? Explain its practical utility.

10.35. On what factors does the cavitation in water turbines depend? Describe some methods to avoid cavitation in water turbines.

10.36. What are the causes of cavitation? How will you prevent cavitation in hydraulic turbines.

10.37. Prove the relation for cavitation

$$H_{s\,max} = H_b - \sigma_c H$$

10.38. Prove the relation for draft tube

$$\frac{P_2}{\rho g} = \frac{P_a}{\rho g} - \left[H_s + (1-K)\left(\frac{V_2^2 - V_3^2}{2g}\right)\right]$$

10.39. Describe Deriaz turbine or diagonal turbine.

$$\sigma_c = \frac{H_a - H_{min} - H_s}{H}$$

10.40. What are the advantages of Deriaz turbine over the other turbines.

10.41. Name the constructional requirement of Deriaz turbine.

10.42. What are the advantages of runner pitch variation of Deriaz turbine?

10.43. Describe bulb turbines. Why this turbine is preferred over Kaplan turbine for small hydel plants.

10.44. Describe pit turbine for mini/micro hydel plants.

10.45. Describe starflo turbine for mini/micro hydel plants.

10.46. What are the advantages of pit turbines over the vertical Kaplan turbines for small hydel plants.

B. UNSOLVED PROBLEMS

10.1. The guide blade angles of an inward flow reaction turbine make 20° with the tangent at the entry. The moving vane angle at entry is 120°. The external diameter of the runner is 450 mm. The width at entry is 60 mm and at exit is 94 mm. Calculate the exit angle of the runenr so that, with no shock loss at entry, the discharge is radial. If the supply head in the casing is 18 m, loss in guides and runner is 1.5 m and there is atmospheric pressure at exit from the runner, calculate the speed in rpm and hydraulic efficiency. If the mechanical efficiency is 90%, calculate the output power. Neglect vane thickness. [594 rpm, 91.65%, 536.98 kW]

10.2. A Francis turbine with an overall efficiency of 75% is required to produce 203 kW. It is working under a head of 7.62 m. The peripheral velocity is 0.26 $\sqrt{2gH}$ and the radial velocity of flow at inelt is 0.96 $\sqrt{2gH}$. The wheel runs at 150 rpm and the hydraulic losses in the turbine are 22% of the available energy. Assuming radial discharge, determines (a) The guide blade angle. (b) The wheel vane angle at inlet (c) Diameter of the wheel at inlet and (d) Width of the wheel at inlet. [37°29′, 44° 4′, 0.405 m, 0.242m]

10.3. A vertical shaft inward flow reaction turbine runs at 215 rpm and uses 10.2 m^3/s of water when the net head is 24.4 m. The hydraulic efficiency is 90%. The inlet angle of the runner vane is 112° measured from the direction of the runner speed. Water enters the runner without shock with a velocity of flow of 7.3 m/s and enters the draft tube, without whirl, with a velocity of 6.1 m/s. The discharge velocity from the draft tube is 2.4 m/s. The mean height of the runner entry surface and entrance to the draft tube are 1.5 m and 1.2 m respectively above the tail race level. Find: (a) Diameter of the runner entry surface. (b) The pressure heads at the entrance to runner and entrance to draft tube, assuming that the loss of head due to friction is 0.9 m in the runner and 0.6 m in the draft tube. [1.4426m, 10.58m, – 2.2m]

10.4. A Francis turbine develops 500 kW at an overall efficiency of 85% when working under a static head of 5m, the draft tube being cylindrical and of diameter 2 m. What increase in power and efficiency of turbine would be there, head, speed and discharge remaining the same, if a tapered draft tube having an outlet diameter of 3 m and efficiency of conversion of 90% were substituted for the cylindrical one? [62.94 kW]

10.5. A Kaplan turbine runner is to be designed to develop 10000 kW. The net available head is 5.5 m. Assume that the speed ratio is 2.09 and the flow ratio is 0.68, and the overall efficiency is 60%. The diameter of the boss is of one third of the diameter of the runner. Find the diameter of the runner, its speed and specific speed. [2.639 m, 52.37 rpm, 612.83]

10.6. A propeller turbine runner has an outer diameter of 4.5 m and inner diameter of 2m and develops 28 MW, when running at 140 rpm under a head of 22 m. Hydraulic efficiency is 94% and overall efficiency is 85%. Find the discharge through the runner and the guide blade angle at inlet. [152.633 m^3/s, 62° 47′].

10.7. A turbine develops 5070 kW under net head of 12.2 m and at an overall efficiency of 82%. If a new source of supply is used and the net head increased to 18.3 m, assuming the overall efficiency to remain constant find: (a) the flow in m^3/s which will be required (b) the shaft power which will be developed (c) percentage increase in speed [63.27 m^3/s 9314.2 kW, 22.5%]

10.8. A Francis turbine has been manufactured to develop 11 kW at a head of 81 m and speed 375 rpm. The mean atmospheric pressure at the site is 1.01043 bar and vapour pressure 0.02943 bar. Calculate the maximum permissible height of the runner above the tail water level to ensure cavitation free operation. The critical cavitation factor

$$\sigma_c = 3.17 \times 10^{-8} N_s^2$$

Where N_s is the specific speed ot turbine [9.6178 m]

10.9. The guide blade angle of an inward flow reaction turbine make 20° with the tangent at the entry. The runner vane angle at entry is 120°. The outer diameter of the runner is 450 mm and the inner diameter is 300 mm. The width at entry is 60 mm. If the supply head in the casing is 18 m, loss of head in guides and runner is 1.5 m and there is atmospheric pressure at exit from runner, calculate the speed in rpm and the hydraulic efficiency. What is the output power, if the mechanical efficiency is 95%? Take the velocity of flow as contant and the discharge is radial. [578.3 rpm, 86.9%, 50.7 kW]

10.10. An inward flow reaction turbine has a runner of 450 mm diameter and 50 mm wide at the outer diameter. The inner diameter is 300 mm and the width here is 80 mm. The vanes occupy 8 per cent of the periphery. The guide vane angle is 24°, the inlet runner vane angle is 95° (vanes inclined forward to the direction of rotation) and the outlet angle is 30°.

If the hydraulic friction losses amount to 12 per cent and mechanical friction to 6 per cent of the supply head, calculate:

(a) the speed for no shock, at inlet when the pressure in the outer casing is 54 m above the pressure at discharge from the runner, (b) the output power in kW [984.1 rpm, 1752.72 kW]

10.11. In an inward flow turbine developing 200 kW under a head of 30 m. The guide and outlet runner vane angles are 20° and 25°, the inlet diameter is 1.5 times the outlet diameter, the ratio of the outlet area from the runner to that from the guides is 1.333, the outlet pressure is atmospheric, and final discharge is radial. Assume that the loss in the guides is 10 per cent of the velocity head generated therein, and the loss in the runner vanes is 20 per cent of the outlet relative velocity head, determine the outlet area from the guides, the pressure head at inlet to the wheel and the flow in m^3/s.

[0.1687 m^2, 0.78447 m^3/s, 14.916 m]

10.12. A mixed flow vertical reaction wheel works under a net head of 40 m and produces a shaft output of 4.5 MW at an overall efficiency of 82%. The shaft speed is 280 rpm and hydraulic efficiency is 90%. The inlet to the runner is 1.5 m above the tail race level and the pressure at the inlet is 1.4 bar gauge. The corresponding figures at the runner outlet are 1.2 m and 0.143 bar vacuum. There is no whril in the draft tube and the water enters at 5.4 m/s and leaves at 3 m/s. The runner external diameter is 1.2 m and the velocity of flow is 6 m/s. Assume velocity of flow is constant through the runner. Determine (a) the blade angle at inlet to the runner; (b) the exit diameter of the draft tube and (c) the head losses in the guide blades, runner and draft. [50°15', 2.3 m, h_{lg} = 1.4 m, h_{lr} = 1.874 m, h_{ld} = 0.767 m]

10.13. A Kaplan turbine develops 2000 kW under a head of 6 m. The turbine is set 2.5 m above the tail water level. A vacuum gauge inserted at the turbine outlet records a suction head of 3.1 m. If the turbine efficiency is 85%, what will be the efficiency of draft tube having a inlet diameter of 3m. [36.8%]

11

Governing of Hydraulic Turbines

11.1. INTRODUCTION

Hydraulic turbine governors are designed to regulate the speed of the unit within a desired range by increasing or decreasing the amount of water supplied to the turbine runner in order to maintain a balance between power input and power demand, acting when a change in power demand causes a fluctuation in speed.

The governor operates the gate mechanism of a reaction turbine, adjusting the opening of the wicket gates. On a Kaplan unit the governor also changes the blade angle of the propeller to correspond with the gate position. The governor operates the pressure regulator on a Francis turbine according to the rate at which the gate close. On an impulse wheel the governor controls the position of the needles and jet deflectors.

When the turbine supplies power to an electric system, the output depends on the gate or nozzle opening. The turbine slows down when the load increases and speeds up when the load decreases. If a circuit breaker opens so that the turbine output suddenly drops from full load to zero, the turbine will reach run away speed, almost twice normal, if the water input stays constant.

Generally no two turbine installations are exactly alike; so each governor has to adjust for its application. The natural change of turbine speed with load determines how much the governor has to move the jets or gates to bring the speed back to normal. Adjustments are made to proportion the action of each part of the governor to assure that the gates are moved as quickly and smoothly as possible to the proper opening. When turbines are supplied by long penstocks, the governor timing may have to be increased to prevent water hammer. On large power systems the spinning momentum or flywheel effect of all the rotating equipment connected is so large that sudden load changes result in only small variation in speed. Slowing down the governor timing on high head units does not seriously affect system speed control.

Almost all modern hydraulic turbines are directly coupled to alternators which must run constantly at the design speeds. The speed of an alternator is fixed by the number of pair of poles and the required frequency. For various reasons the frequency cannot be allowed to vary. The speed must, therefore, remain constant when the load changes, hence the necessity of speed control. For uniform voltage and frequency when the load on a turbine changes, the governor must adjust its water supply to meet the change in demand.

11.2. THE MAIN ELEMENTS OF GOVERNOR

The main elements of the governor are the following :

1. The speed responsive element, most frequently rotating flyballs with governor sleeve and a governor lever rotating around an adjustable fulcrum.
2. The control valve to supply fluid under pressure to the servomoter in order to actuate control mechanism.
3. The servomotor, a fluid pressure operated piston or pistons which move the turbine control mechanism.
4. The restoring or relay mechanism to hold the servomotor in the correct (or dead beat) position when the turbine output and load demand are equalized (It can be direct mechanical drive or electrical).
5. The fluid pressure supply required for the action of the servomotor.

11.3. ACTION OF A GOVERNOR

Fundamentally the action of a governor is as follows :

A change in speed causes a change in the position of the rotating flyballs. This change in position moves the distributing or control valve so that oil under pressure is admitted into the regulating cylinder, or cylinders, causing them to move the servomotor pistion connected to the gate mechanism of the turbine.

This movement of the gates is transmitted back to the distributing valve through the restoring mechanism, bringing the distributor valve back to its neutral, or dead beat, position after the gates have been moved sufficiently to compensate for the change in load.

11.4. GOVERNING OF IMPULSE TURBINE

The quantity of water coming out from the nozzle and from striking the buckets may be regulated in one of the following methods :

1. Spear regulation.
2. Deflector regulation
3. Combined spear and deflector control.

The modern turbines are provided with double regulation which means that combined spear and deflector control. Double regulation means regulation of speed and pressure. The speed is regulated by spear and the pressure is regulated by deflector arrangement.

Spear regulation is satisfactory when a relatively large penstock feeds a small turbine and the fluctuation of load is small. With the sudden fall in load, the turbine nozzle has to be closed suddenly which may create water hammer in the penstock.

Deflector control is adopted when supply of water is constant but the load fluctuates. The spear position can be adjusted by hand. As the nozzle has always a constant opening, it involves considerable wastage of water and can be used only when supply of water is abundant.

The deflector is generally a plate connected to the governor by means of levers. When it is required to deflect the jet, the plate can be brought in between the nozzle and buckets, thereby diverting the water away from the runner and directing into the tail race. The position of the deflector can be adjusted according to the discharge required.

The deflector deflects jet from the runner within a very short period so that no further energy is imparted to the latter. The deflector engages until the spear has been adjusted to a new position of the equilibrium. The closing of the spear can thus be retarded to avoid undue pressure rise, whilst only the short time which the deflector requires to act need be considered in determining speed rise and flywheel momentum. The flywheel momentum should however, be sufficient to deal with a sudden increase in load.

11.4.1. Combined spear and deflector control

When turbine is running at its normal speed, the positions of piston (in relay cylinder or servomotor), control valve (relay valve) and the balls of centrifugal governor and lever arm will be in their normal position. The oil supply by the gear pump remains in the middle portion i.e. space D. The gears of the pump go on rotating without developing further pressure and flow back to sump because ports or valves A and B are closed (Fig. 11.1).

When the load on turbine decreases, its speed increases. Due to this, the balls moves outward resulting in an upward movement of the sleeve. As the sleeve moves up, the left hand end of the main lever is raised, which causes the bell crank lever to move downward and simultaneously pushes the piston of the control valve down in its cylinder. The downward motion of the bell crank lever brings the deflector in front of the jet thereby diverting a portion of the jet away from the buckets.

The downward movement of control valve keeps port A still closed but opens port B. In this position of the double piston of relay valve, pressurized oil from the sump arrives the pipe 1 into space D and is supplied to the servomotor through port B. Oil returns from port A to the space C in relay valve and is sent back to the sump via pipe 2. Oil from space E in the relay valve also goes to the sump via pipe 3 (Fig. 11.1).

Owing to the arrival of more pressurized oil in space D of the servomotor, the piston moves towards right and that moves the spear into the nozzle reducing the area of the nozzle as well as reducing the area of flow of water. This reduces the discharge through the nozzle as desired and the normal turbine speed is restored. The pressurised oil of chamber D is prevented to flow through pipe 2 and 3 with the help of special constructed piston of relay valve because two semi circular projections are provided on both the pistons of relay valve.

The forward motion of the spear is accompanied by roller moving down on the cam and slowly the jet deflector starts rising. By the time spear has occupied its final forward position corresponding

to the modified discharge the deflector has moved up and does not interfere with the nozzle. The discharge is thus reduced, speed becomes normal and main lever, balls etc. occupy their normal position.

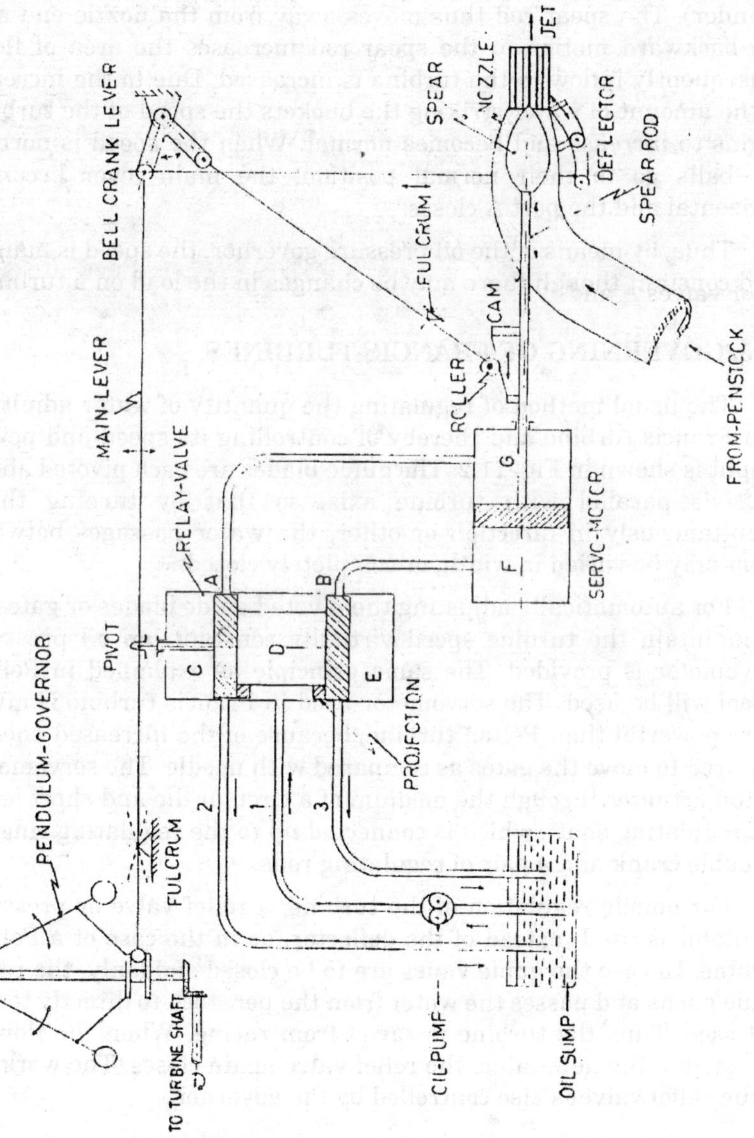

Fig. 11.1. Pelton wheel governing.

In case the load on turbine increases thus causing the speed to fall. The balls move down. The upward motion of main lever opens the port A but keeps the port B closed. The oil under pressure is delivered to the right side of the piston of the servomotor (relay cylinder). The spear rod thus moves away from the nozzle end and the backward motion of the spear rod increases the area of flow, consequently inflow to the turbine is increased. Due to the increase in the amount of water striking the buckets the speed of the turbine begins to increase and becomes normal. When the speed is normal the balls go to their normal position, the main lever becomes horizontal and the port A closes.

Thus, by means of the oil pressure governor, the speed is mainta ined constant though there may be changes in the load on a turbine.

11.5. GOVERNING OF FRANCIS TURBINES

The usual method of regulating the quantity of water admitted to a Francis turbine and thereby of controlling its speed and power output is shown in Fig. 11.2. The guide blades are each pivoted about an axis parallel with turbine axis, so that by turning them simultaneously in direction or other, the water passages between them may be varied in width, or completely closed.

For automatically adjusting the pivoted guide blades or gates so as maintain the turbine speed virtually constant, an oil pressure servomotor is provided. The same principle as explained in Pelton wheel will be used. The servomotor used in Francis turbine is much more powerful than Pelton turbine, because of the increased energy required to move the gates as compared with needle. The servomotor piston actuates through the medium of a circular die and short lever the regulating shaft, which is connected up to the regulating ring by a double crank and a pair of regulating rods.

For double regulation of the turbine, a relief valve or pressure regulator is used instead of the deflector, as in the case of a Pelton turbine. In case the guide vanes are to be closed suddenly, the relief value opens and passes the water from the penstock to directly to the tail race. Thus the turbine is saved from racing. When the flow is adjusted to the new value, the relief valve again closes. The working of the relief valve is also controlled by the governor.

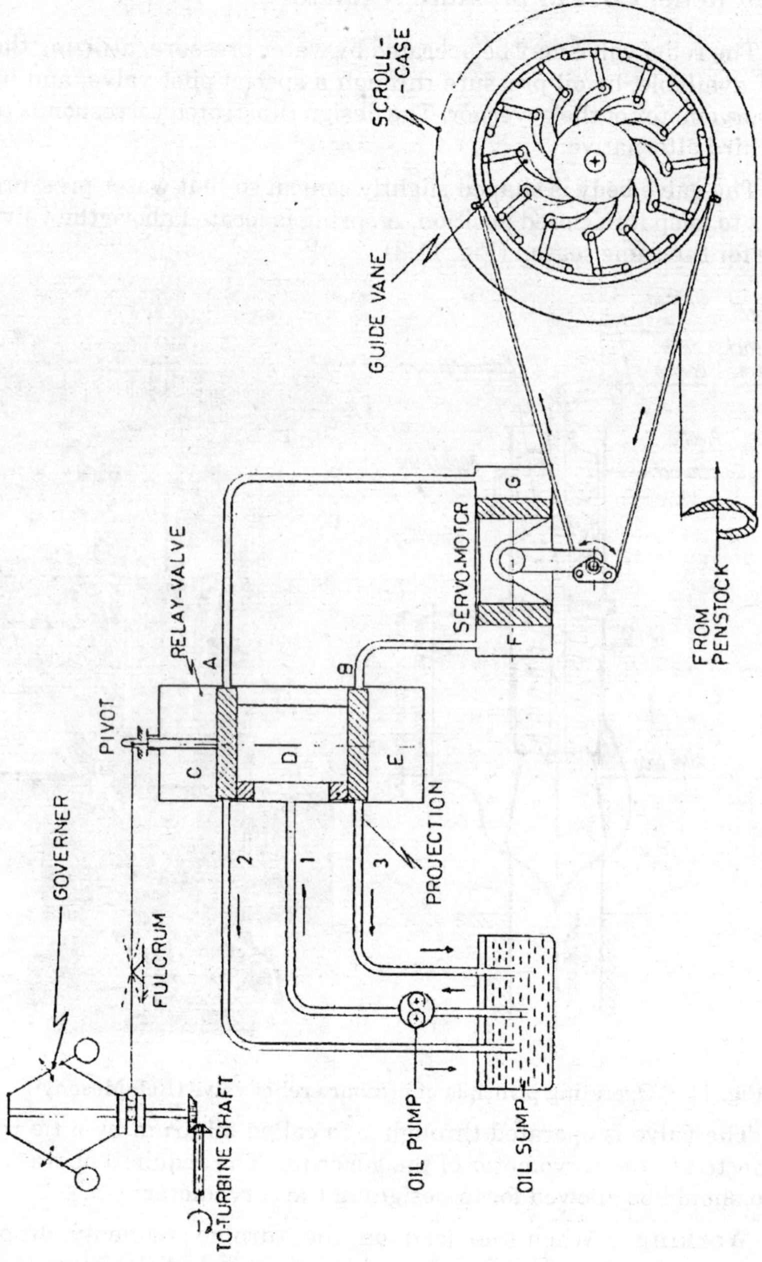

Fig. 11.2. Fracnis turbine governing.

11.5.1. Relief valve or pressure regulator

The relief valve may be operated by water pressure, utilizing the head available, by oil pressure through a special pilot valve, and by the servomotor of the governor. The design illustrated corresponds to the third alternative.

The valve body is shaped slightly conical so that water pressure tends to keep it in closed position. A spring is located above the valve body for the same reason (Fig. 11.3).

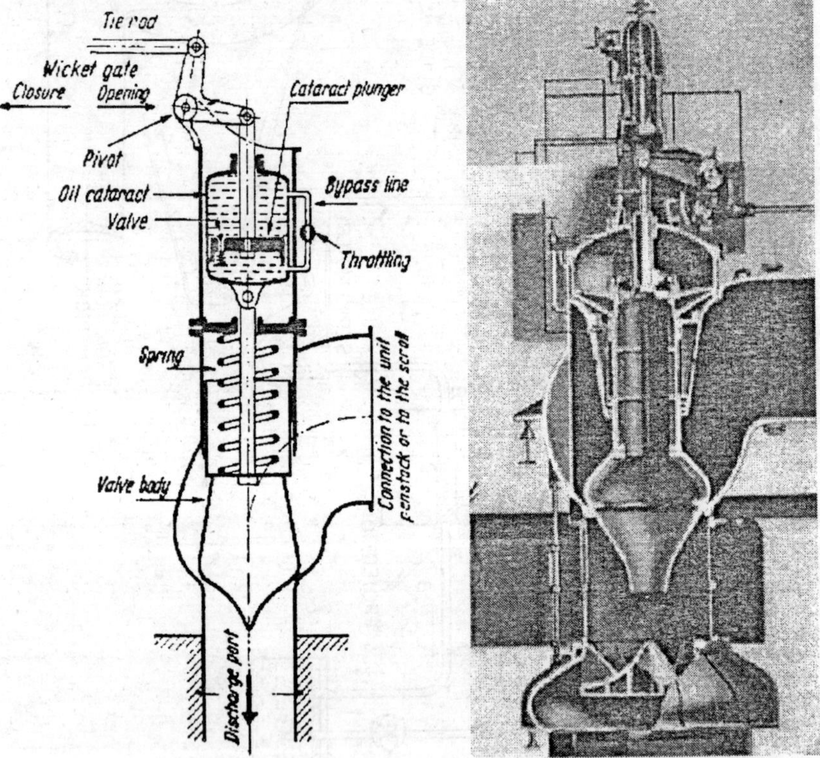

Fig. 11.3. Operating principle of pressure relief valve (Ref. Mosonyi)

The valve is operated through a so called oil drum by a tie rod connected to the servomotor of the governor. The required operating force should be allowed for in designing the servomotor.

Working : When the load on the turbine suddenly drops, governor, immediately closes the guide vanes and at the same time lifts the plunger of the oil drum. Since there is oil in the upper and underside of the plunger, and it is not possible to shift some of the oil from upperside to the underside quickly through a by-pass pipe, the plunger moves up as a whole alongwith the valve and water passes through the discharge port directly into the tail race.

Immediately, after this the plunger of the drum that remains in a fixed position, while the combined force of the spring and water pressure shifts the valve slowly downward, and oil returns into the space below the underside of the piston through the by-pass pipe providing communication between the two sides. The rate of closing can be controlled as desired by setting the throttle adjustment installed in the by-pass. The speed of closure should be adjusted so as to prevent, on the one hand, a pressure rise exceeding the specified limit and, on the other, an impermissible water loss that would arise if the closure were too slow.

In case of slow gate closure, i.e. during normal governor operation, the plunger of the oil drum is raised slowly and oil is forced through the by-pass to the underside of the plunger without lifting the valve.

When the load on the turbine is suddenly increased, the governor acts to open the guide vanes and exerts, at the same time, a downward force on the plunger. The valve itself resting already on its seat cannot move further downward and must be uncoupled from the plunger to ensure free movement for the latter. The by-pass is insufficient for this purpose and a one way valve is mounted therefore into the plunger, which permits the oil to escape directly to the upper side whenever the plunger is moved downward. If the discharge capacity of this valve is made large enough, the governor capacity required for the opening stroke can be reduced appreciably.

Under normal operating conditions, when the gates are closed at a slow rate, the pressure relief valve remains closed. Moderate pressure rises occurring on such occasions should be allowed for in the design of the penstock.

The application of relief valves is in general not justified unless the head is higher than 50m. For lower heads it is simpler and more economical to design the penstock for the water hammer pressures, especially if the pressure rise does not exceed 50% of the normal load.

Pressure relief valves are usually required to cause the minimum possible waste of water. They should be designed therefore, to close immediately and automatically after having been opened synchronously with gate closure. The relief valve should furthermore close slowly so that the resulting pressure rise remain below a specified value. An oil dashpot may be used for this purpose. Water hammer pressures of smaller intensity following any gate movement during speed regulation under normal operating conditions may be correlated to those considered in designing the relief valve.

The time of closure of the relief valve is approximately 20 to 30 (60) seconds.

When the load on the turbine is increased suddenly, i.e. in case of rapid opening, the relief valve has no function and is consequently designed to remain unaffected by a drop in pressure. Relief valves have therefore no limiting effect on the pressure drop occurring in the penstock.

No pressure relief valves are necessary with Pelton wheels, because they are protected by the jet deflectors at times of sudden closure and thus maintaining unchanged flow conditions in the penstock. In the second phase slow closure is affected by the nozzle so that no excessive overpressures can result.

11.6. GOVERNING OF KAPLAN TURBINE

The blade angle of the Kaplan turbine is regulated by a servomotor operated by oil pressure, as in Fig. 11.4.

The servomotor is a double acting piston in a cylinder at the coupling between the generator and turbine shafts. The piston is connected by a regulating rod in the hollow shaft of the turbine with the blade mechanism in the hub of the propeller, so that as the piston moves upward the crosshead linked to the gate levers decreases the angle of the blades, while a downward movement of the piston increases the blade angle.

The blade servomotor is actuated by oil pressure above or below the piston. Two pipes, one inside the other, extended through the hollow shaft of the generator from the servomotor to the oil supply head mounted on top of the generator. The pipes rotate with the shaft, while the oil supply head is stationary. The inner pipe connects to the space under the piston, while the outer pipe is connected to the top of the servomotor. The inner pipe moves up and down with the piston, transmitting the blade position through the compensating rod to the floating lever of the blade control valve.

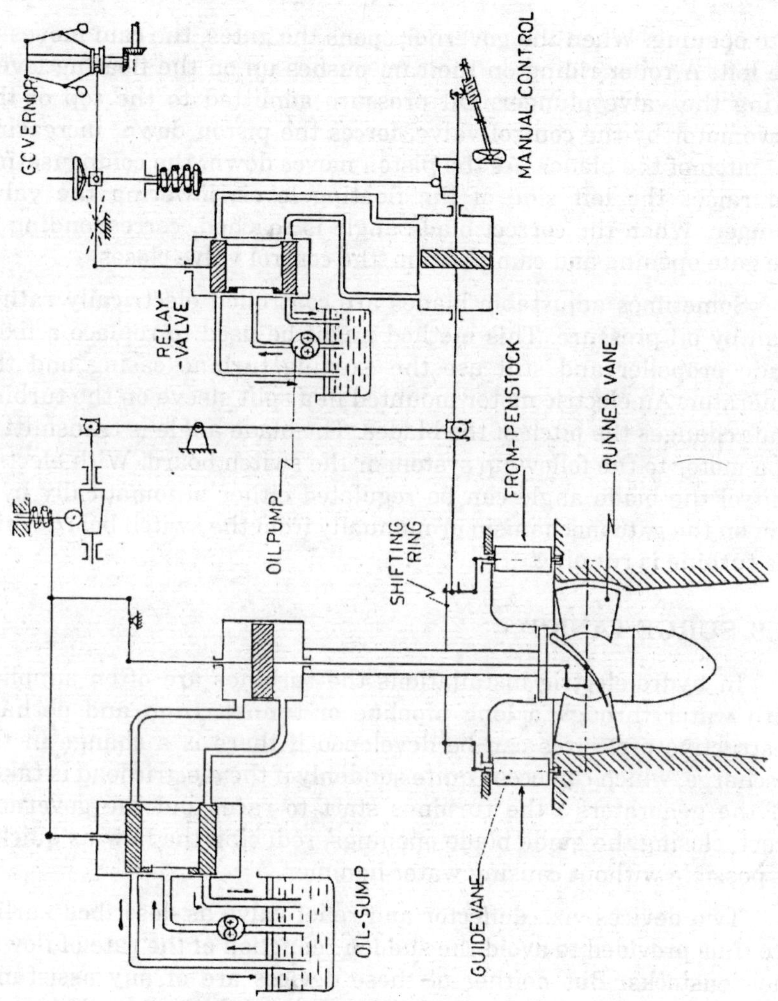

Fig. 11.4. Governing of Kaplan turbine

The blade control valve has a four way valve similar to that used in the governor for regulating gate opening. It is shown in the closed position in the diagram. When the valve plunger moves upward, it connects the inner pipe in the generator shaft to the discharge pipe of the oil pressure system and the outer pipe to the oil pressure. A downward movement of the valve plunger reverse these connections. The cam on top of the control valve connects to the gate mechanism. The shape of this cam maintains the proper blade position for every gate opening. When the governor opens the gates, the cam moves to the left. A roller riding on the cam pushes up on the floating lever, lifting the valve plunger. Oil pressure admitted to the top of the servomotor by the control valve, forces the piston down, increasing the pitch of the blades. As the piston moves down, the compensating rod raises the left side of the floating lever, lowering the valve plunger. When the correct blade angle is reached, corresponding to the gate opening and cam position, the control valve closes.

Sometimes adjustable blades are controlled electrically rather than by oil pressure. This method might be used to replace a fixed blade propeller and still use the existing turbine casing and the generator. An electric motor mounted in a split sleeve on the turbine shaft changes the pitch of the blades. The blade angle is transmitted by a motor to the follow up system or the switch board. With electric control the blade angle can be regulated either automatically by a cam on the gate mechanism or manually from the switch board while the turbine is running.

11.8. SURGE TANKS

In hydro-electric installations the turbines are often supplied with water through a long pipeline or tunnel. High and perhaps destructive pressures can be developed if there is a change in the discharge, which can occur quite suddenly if the electric load is taken off the generators : the turbines start to race, and the governors react, closing the guide blade openings, reducing the flow as quickly as possible without causing water hammer.

Two devices viz., deflector and relief valve as described earlier are thus provided to avoid the sudden reduction of the rate of flow in the penstocks. But neither of these devices are of any assistance when the load on the generator increases and the turbine is in need of more water. Thus in order to fulfil both the above noted requirements, in addition to the above noted devices certain other devices such as surge tank and forebay are usually employed. Surge tanks are employed in the case of high and medium head hydroelectric plants where the penstock is very long, and forebays are suitable for medium and low head hydroelectric plants where the length of the penstock is very short.

A surge tank is an open chamber or a cylindrical open topped storage reservoir, which is connected to the penstock at a point as close as possible to the turbine (Fig. 11.5).

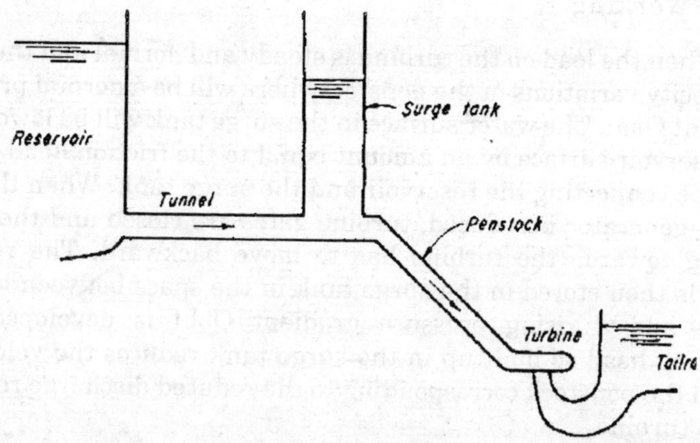

Fig. 11.5. A surge tank on hydro electric scheme.

11.8.1. The main functions

The main functions of a surge tank are :

(1) A surge tank stores or provide water. The water in the pipeline is, therefore, accelerated or decelerated slowly, and the amplitude of the pressure fluctuations in the system is reduced.

(2) A surge tank improves the regulating characteristics of a hydraulic turbine. Because of the surge tank, the length of the power conduit to be used for determining the water starting time is up to the surge tank rather than up to the upstream reservoir. The water starting time of a hydro-electric plant is therefore reduced, thus improving the regulating characteristics of the power plant.

(3) It reduces the amplitude of pressure fluctuations by reflecting the incoming pressure waves. For example, the water hammer waves produced in a penstock by load changes on a turbine are reflected back at the surge tank. Thus, the penstock length to be used in the waterhammer turbine and the upstream reservoir. Due to this reduction in the penstock length, the pressure rise or drop is less than if the surge tank were not provided. In addition, if a surge tank were not present at the juction of the penstock and the tunnel, then the tunnel would have to be designed to withstand the water hammer pressure.

11.8.2. Working

When the load on the turbine is steady and normal and there are no velocity variations in the penstock there will be a normal pressure gradient Oaa_1. The water surface in the surge tank will be lower than the reservoir surface by an amount equal to the friction head loss in the pipe connecting the reservoir and the surge tank. When the load on the generator is reduced, turbine gates are closed and the water moving towards the turbine has to move backward. The rejected water is then stored in the surge tank in the space between levels *a* and *b* and a rising pressure gradient Obb_1 is developed. The retarding head so built up in the surge tank reduces the velocity of flow in the penstock corresponding to the reduced discharge required by the turbine.

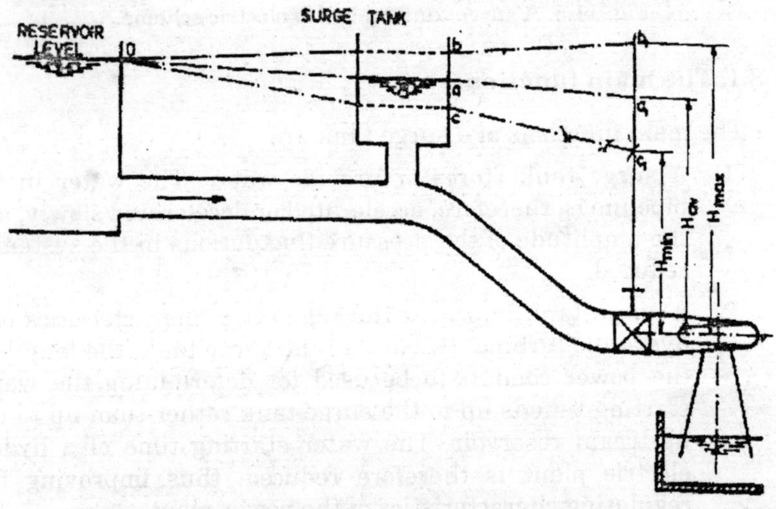

Fig. 11.6. Working of surge tank.

When the load on the generator increases, the governor opens the turbine gates to increase the rate of flow entering the runner. The increased demand of water by the turbine is partly met by the water stored between levels *a* and *c* in the surge tank. As such the water level in the surge tank falls and a falling pressure gradient Occ_1 is developed. In other words, the surgetank develops an accelerating head which increases the velocity of flow in the penstock to a valve corresponding to the increased discharge required by the turbine (Fig. 11.6).

11.8.3. Types of surge tank

Depending upon its configuration, a surge tank may be classified as simple, orifice, differential, one way, or closed (Fig. 11.7).

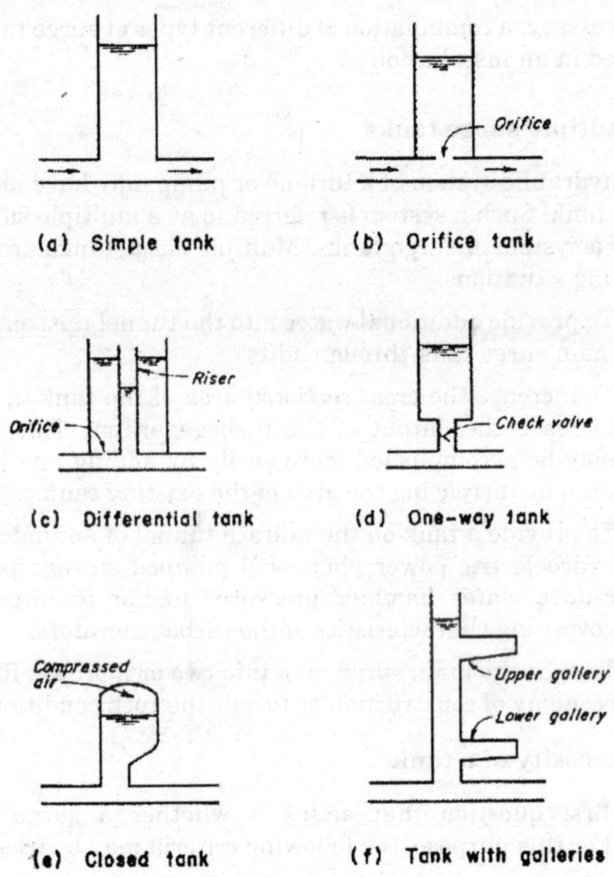

Fig. 11.7. Types of surge tanks.

A simple surge tank is an open shaft or stand pipe connected to the pipeline. If the entrance to the surge tank is restricted by means of an orifice, it is called an orifice tank. An orifice tank having a riser is termed differential. In a one way surge tank, the liquid flows from the tank into the pipeline only when the pressure in the pipeline drops below the liquid level in the surge tank. Following the transient state conditions, the tank is usually filled from the pipeline. If the top of the tank is closed and there is compressed air between the water surface and the top of the tank, then the tank is called a closed surge tank, a tank with air cushion, etc. Depending upon the requirement that the tank must fulfill, a simple tank may have upper or lower gallaries.

If necessary, a combination of different types of surge tanks may be provided in an installation.

11.8.4. Multiple surge tanks

The hydraulic system of a turbine or pump may have more than one surge tank. Such a system is referred to as a multiple surge tank system or a system of surge tanks. Multiple surge tanks are used in the following situations.

(1) To provide additional water into the tunnel upstream of the main surge tank through adits.

(2) To increase the cross sectional area of the tank in order to increase the output of the turbogenerator. This increase may be accomplished more easily by adding another tank than by increasing the area of the existing tank.

(3) To provide a tank on the tailrace tunnel of an underground hydroelectric power plant or a pumped storage project to reduce water hammer pressures and/or to improve the governing characteristics of the turbogenerators.

(4) To split the main surge tank into two or more shafts for the economy of construction or to suit the rock conditions.

11.8.5. Necessity of a tank

The first question that arises is whether a surge tank is required. For this purpose, the following criteria may be used:

(1) Surge tanks should be provided where the resulting reduction in the waterhammer pressure provides a more economical penstock-surgetank installation.

(2) A surge tank should be provided if the maximum speed rise following rejection of the maximum turbine output cannot be reduced to less than 60% of the rated speed by other practical methods, such as increasing the generator inertia or penstock diameter or decreasing the effective wicket gate closing time. The speed rise will be computed assuming one

unit operating alone if there is more than one unit on the penstock.

(3) As a rough rule of thumb, the provision of a surge tank should be investigaed if —

$$\Sigma \frac{L_i V_1}{H_n} > 3 \text{ to } 5$$

in which $\sum L_iV_i$ is computed from the intake to the turbine, and H_n is the minimum net head. In general, a surge tank should be preferred to a pressure regulator although the latter may be more economical for high head plants.

11.8.6. Location

A surge tank should be located as near to the turbine as the local topography permits.

11.8.7. Size of tank

The cross sectional area of a tank is determined to satisfy the following criteria :

(1) The tank is stable

(2) The tank does not overflow following load rejection under an overflow weir is provided.

(3) The tank does not drain (i.e., the water level does not fall to the tunnel crown) following maximum possible load acceptance with the reservoir at its minimum level.

The minimum cross sectional area required for stability has been a matter of great discussion. Jaeger proposed a safety factor n such that the area of the surge tank should be n times the Thoma area, A_{th}, with n>1. As a rough guide for periliminary design, n may be taken as 1.5 for a simple tank, and 1.25 for an orifice and a differential tank.

ILLUSTRATIVE EXAMPLES

11.1. A single jet Pelton wheel uses 840 liter/s of water when the level in the reservoir is 255m, above the centre line of the needle controlled nozzle. The efficiency of power transmission through pipeline and nozzle is 88%.

The head is increased to 270m, by raising the waterlevel in reservoir, and the needle in the nozzle is adjusted so that the power of the jet remains the same as before. Assuming that the velocity coefficient of the nozzle is 0.97 in both cases and that the loss of head in the pipeline is proportional in the pipeline is proportional to Q^2 where Q is the quantity of water used in liter/s, find

(a) the increase in efficiency of power transmission. (b) the reduction in the quantity of water used.

Solution

$$\eta_t = \frac{H_g - h_f}{H_g} = \frac{H}{H_g}$$

$$0.88 = \frac{H}{255}$$

$$\therefore H = 224.4\ m$$

and $h_f = H_g - H = 255 - 224.4 = 30.6m$

$$P = \rho\, Q\, g\, H$$

$$= 1000 \times 840 \times 10^{-3} \times 9.81 \times 224.4$$

$$= 1849.14\ kW$$

$$h_f \propto Q^2$$

$$h_f = K\, Q^2$$

$$\therefore K = \frac{h_f}{Q^2} = \frac{30.6}{(0.84)^2} = 43.3673$$

When gross head is increased to 270 m then

$$h_{f1} = K\,(Q_1)^2$$

$$\therefore H_1 = 270 - K\,(Q_1)^2$$

$$= 270 - 43.3673\, Q_1^{\,2}$$

Therefore,

$$P = \rho\, g\, Q_1\, H_1$$

$$1849.14 = 9.81\, Q_1 \times \left(270 - 43.36\, Q_1^{\,2}\right)$$

$$188.4954 = 270\, Q_1 - 43.36\, Q_1^{\,3}$$

$$\therefore Q_1^{\,3} - 6.2269\, Q_1 + 4.3472 = 0$$

Solving we get $Q_1 = 0.77\ m^3/s$

$\therefore$ Reduction in discharge

$$= 0.84 - 0.77 = 0.07\ m^3/s = 70\ liters/s$$

(a) $$h_{f1} = K\, Q_1^{\,2} = 43.36\,(0.77)^2 = 25.7081\, m$$

$$H_{f1} = H_{g1} - h_{f1} = 270 - 25.708 = 244.2918\ m$$

$$\therefore \eta_{t1} = \frac{H_{g1} - h_{f1}}{H_{g1}} = \frac{H_1}{H_{g1}} = \frac{244.2918}{270} = 0.90478 = 90.478\%$$

Increase in transmission efficiency

$$= \frac{0.90478 - 0.88}{0.88} = 0.0281 = 2.81\%$$

(b) Reduction in discharge

$$= \frac{0.84 - 0.77}{0.84} = 0.083 = 8.3\%$$ Ans.

11.2. A Pelton wheel with a needle controlled nozzle develops 700 kW when the total head is 208 m and the jet diameter is 100 mm. The loss of head due to friction in the pipeline and nozzle is 92 Q^2 (metre) where *Q* is the quantity flowing in m^3/s.

Assuming that the total head and runner efficiency remain constant, determine the percentage reduction in the quantity of water supplied when the power is reduced to 625 kW, (a) by alteration of position of the needle in the nozzle and (b) by partial closure of a value on the pipeline, the position of the needle (and therefore, the diameter of the jet) being unchanged. Take cv =1. Also obtain the loss of head across the valve in case(b).

Solution :

$$H_1 = H_{g1} - h_{f1} = 208 - 92\, Q_1^{\,2}$$

$$Q_1 = A\, V_1 = \frac{\pi}{4}(0.1)^2\, V_1$$

$$\therefore V_1 = 127.3\, Q_1\ m/s$$

$$V_1 = \sqrt{2\, g\, H}$$

$$V_1^2 = 2\, g\, H_1$$

$$(127.3\, Q_1)^2 = 2 \times 9.81 \times (208 - 92)$$

$$825.9576\, Q_1^2 = 208 - 92\, Q_1^2$$

$$917.9576\, Q_1^2 = 208$$

$$Q_1^2 = 0.2265$$

$$\therefore Q_1 = 0.476\ m^3/s$$

$$V_1 = 127.3 \times 0.476 = 60.596\ m/s$$

$$H = 208 - 92 \times 0.2265 = 208 - 20.838 = 187.162\ m$$

$$P_1 = \rho\, g\, Q_1\, H_1\, \eta$$

$$700 \times 1000 = 1000 \times 9.81 \times 0.476 \times 187.162 \times \eta_r$$

$$\therefore \eta_r = 0.8 = 80\%$$

Second case

Power is reduced from 700 kW to 625 kW

Therefore,

$$P_2 = \rho\, g\, Q_2\, H_2\, \eta_r$$

$$625 \times 1000 = 1000 \times 9.81 \times Q_2 \times \left(208 - 92\, Q_2^2\right)$$

$$208\, Q_2 - 92\, Q_2^2 = 89.6381$$

$$92\, Q_2^{\,3} - 208\, Q_2 + 79.6381 = 0$$

Solving we get $Q_2 = 0.4275\ m^3/s$

Therefore reduction in discharge

$$= \frac{0.4761 - 0.4275}{0.4761}$$

$$= 0.1020 = 10.20\%$$ Ans.

(b) Head loss across needle is h metre

Therefore,

$$H_3 = 208 - 92\, Q_3^2 - h$$

$$Q_3 = A\, V_3$$

$$Q_3 = \frac{\pi}{4}(0.1)^2\, V_3$$

$$16211.3893\, Q_3^2 = 2. \times 9.81 \times H_3$$

Therefore,

$$P_2 = \rho g\, Q_3\, H_3\, \eta_r$$

$$625 \times 1000 = 1000 \times 9.81 \times Q_3 \times 826.2685\, Q_3^2 \times 0.8$$

$$625 = 6484.555\, Q_3^3$$

$$Q_3^3 = 0.09638$$

$$\therefore Q_3 = 0.4584\, m^3/s$$

Percentage reduction in discharge

$$= \frac{0.4761 - 0.4584}{0.4761} = 0.371 = 3.71\%$$ Ans.

From equation (1)

$$826.2685\, Q_3^2 = 208 - 92\, Q_3^2 - h$$

$$918.2\, Q_3^2 = 208 - h$$

$$192.941 = 208 - h$$

$$\therefore h = 208 - 192.941 = 15.058\, m$$ Ans.

11.3 A Kaplan turbine develops 2000 kW under a head of 6 m. The turbine is set 2.5 m above the tail race level. A vacuum gauge inserted at the turbine outlet records a suction head of 3.1 m. If the turbine efficiency is 85%, what would be the efficiency of draft tube having inlet diameter of 3 m?

Also find the reading of the suction gauge if power developed is reduced to 1000 kW, the head and speed remaining constant.

Solution

$$P = \rho g Q H \eta_0$$

$$2000 \times 10^3 = 1000 \times 9.81 \times Q \times 6 \times 0.85$$

$$Q = 39.9752\, m^2/s$$

$$Q = \frac{\pi}{4} D_2^2 V_2$$

$$39.9752 = \frac{\pi}{4}(3)^2 V_2$$

$$V_2 = 5.655\, m/s$$

Applying Bernoullis theorem between points 2 and 3.

$$\frac{P_2}{\rho g} + \frac{V_2^2}{2g} + Z_z = \frac{p_3}{\rho g} + Z_3 + \frac{V_3^3}{2g}$$

$$-3.1 + \frac{(5.655)^2}{2 \times 9.81} + 2.5 = 0 + 0 + \frac{V_3^2}{2g}$$

$$-3.1 + 1.63 + 2.5 = \frac{V_3^2}{2g}$$

$$\frac{V_3^2}{2g} = 1.0311 \quad \therefore V_3^2 = 20.210826$$

$$\therefore V_3 = 4.4956\, m/s$$

$$\eta_d = \frac{(1-K)\left(V_2^2 - V_3^2\right)}{V_2^2}$$

[Because losses are not given, therefore, K = 0]

$$\eta_d = \frac{V_2^2 - V_3^2}{V_2^2} = \frac{31.979 - 20.21}{31.979} = \frac{11.769}{31.979} = 0.368 = 36.8\%$$

Case (b)

$P = 1000\, kW$, Head and speed same

Therefore, $P = \rho\, g\, Q\, H\, \eta_0$

$$1000 \times 10^3 = 1000 \times 9.81 \times Q \times 6 \times 0.85$$

$$\therefore Q = 19.9876\ m^3/s$$

$$Q = \frac{\pi}{4} D_2^2\, V_2$$

$$19.9876 = \frac{\pi}{4}(3)^2\, V_2 \quad \therefore V_2^2 = 7.9958$$

$$\therefore V_2 = 2.8277\ m/s$$

$$\eta_d = \frac{V_2^2 - V_3^2}{V_2^2} = 1 - \left[\frac{V_3}{V_2}\right]^2$$

$$0.368 = 1 - \frac{V_3^2}{7.9958}$$

$$V_3^2 = 5.0533 \quad \therefore V_3 = 2.24796\ m/s$$

Applying Bernoullis theorem between points (2) and (3)

$$\frac{p_2}{\rho\, g} + Z_2 + \frac{V_2^2}{2g} = 0 + 0 + \frac{V_3^2}{2\,g}$$

$$\frac{p_2}{\rho\, g} = \frac{V_3^2 - V_2^2}{2\,g} - Z_2 = \frac{5.0533 - 7.9958}{2 \times 9.81} - 2.5$$

$$= -\,0.1498 - 2.5 = -2.65\ m$$

The reading of the suction gauge = 2.65 m (vacuum) Ans.

11.4 An axial flow Kaplan water turbine with 10 m blade tip diameter and 3.33m boss diameter is to run at 60 rpm developing 22 MW under 6.67 m head. Assuming a hydraulic efficiency of 90 per cent and neglect mechanical losses. What must be the angles between the plane of rotation and the tangents to the blades at entry and exit to the rotor, both at tips and at the boss ? State all the assumptions made. If half the power is required at the same head and speed, by how much must the blades be twisted by the governor so that the efficiency is preserved?

Solution

Assumptions

(1) Velocities with in the machine do in fact conform with the blade direction, which, particularly with machine having such a small number of blades, is known to be wide of the truth.

(2) Velocity of flow is uniform through the machine under maximum efficiency, design point, conditions and that discharge is axial, the whirl having been removed by the blades.

Velocity of blade at tip,

$$u_1 = u_w = \frac{\pi D N}{60} = \frac{\pi \times 10 \times 60}{60} = 31.41\, m/s$$

$$\text{Velocity of blade at boss} = \frac{\pi\, d_b\, N}{60} = \frac{\pi \times 3.33 \times 60}{60} = 10.46\, m/s$$

$$P = \rho\, g\, Q\, H\, \eta_0$$

$$22 \times 10^6 = 1000 \times 9.81 \times Q \times 6.67 \times 0.9$$

$$Q = 373.58\, m^3/s$$

$$Q = \frac{\pi}{4}\left(D^2 - d_b^2\right) V_{f1}$$

$$373.58 = \frac{\pi}{4}(100 - 11.0889)\, V_{f1}$$

$$373.58 = 69.83\, V_{f1}$$

$$V_{f1} = 5.3498\, m/s$$

$$\eta_h = \frac{V_{w1}\, u_1}{g\, H}$$

$$0.9 = \frac{V_{w1}\, u_1}{9.81 \times 6.67}$$

$$V_{w1}\, u_1 = 9.81 \times 6.67 \times 0.9 = 58.889 \qquad (1)$$

$$\text{The velocity of whirl at tip inlet} = \frac{58.889}{u_1} = \frac{58.889}{31.41} = 1.87\, m/s$$

The velocity of whirl at boss inlet

$$= \frac{58.889}{u_2} = \frac{58.889}{10.467} = 5.629\ m/s$$

Now at any radius r

$$u_1 = 2\pi r \tag{2}$$

$$V_{w1} \times 2\pi r = 58.889 \qquad \text{[From (1)]}$$

$$V_{w1} = \frac{58.889}{2\pi r} = \frac{9.37}{r} \tag{3}$$

From inlet velocity triangle

$$\tan\theta = \frac{V_{f1}}{u_1 - V_{w1}}$$

$$V_{w1} = u_1 - \frac{V_{f1}}{\tan\theta} = 2\pi r - \frac{V_{f1}}{\tan\theta}$$

and $\quad V_{f1} = (u_1 - V_{w1})\ \tan\theta$

$$= \left(2\pi r - \frac{9.37}{r}\right)\tan\theta \tag{4}$$

At tip radius, $r = 5\ m$

$$V_{f1} = 29.5419\tan\theta \tag{5}$$

At mean radius $r = 3.33\ m$

$$V_{f1} = 18.119\tan\theta \tag{6}$$

At hub radius $r = \frac{3.33}{2} = 1.665\ m$

$$V_{f1} = 4.834\tan\theta \tag{7}$$

At the design point $V_{f1} = V_{f2} = 5.3498\ m/s.$

Approximate solution is obtained at mean radius with the assumption that the velocity of flow has the mean value there.

At mean radius, $\quad u_{av} = \frac{u_1 + u_2}{2} = \frac{31.41 + 10.46}{2} = 20.92$

$$V_w = \frac{9.37}{r} = \frac{9.37}{3.33} = 2.8138$$

Therefore,

$$\tan\theta = \frac{V_f}{u_{av} - V_w} = \frac{5.3498}{20.92 - 2.8138}$$

$$= \frac{5.3498}{18.105} = 0.2954$$

$$\theta = \tan^{-1}(0.2954) = 16.46° \qquad \text{Ans.}$$

As per question, the flow is half

i.e. 5.3498/2 = 2.6749 m/s

From equation (6)

$$\frac{V_f}{2} = 18.119 \tan\theta$$

$$2.6749 = 18.119 \tan\theta$$

$$\tan\theta = \frac{2.6749}{18.119} = 0.1476$$

$$\therefore\ \theta = \tan^{-1}(0.1476) = 8.397° \qquad \text{Ans.}$$

$$\text{Angle of twist} = 16.46 - 8.397 = 8.071° \qquad \text{Ans.}$$

EXERCISES

A. THEORY

11.1. What are the main elements of governor?

11.2. Explain the action of governor.

11.3. Describe spear rod control governing of Pelton wheel turbine. What are its limitation?

11.4. Why the spear rod control governing in the Pelton wheel cannot be used efficiently?

11.5. Why the modern Pelton wheel turbine are provided with double regulation?

11.6. Why we are using an additional governing method along with spear rod?

11.7. Draw a heat sketch of governing method of Pelton wheel showing both the methods.

11.8. Draw a neat sketch of governing of Francis turbine.

11.9. Describe the governing mechanism of Francis turbine with a neat sketch.

11.10. Explain the operating principle of pressure relief value.

11.11. What is an oil pressure governor? Draw a line sketch and describe its principal components.

11.12. How is speed of a reaction turbine regulated? What is the disadvantage in this method of governing? How is it rectified in Kaplan turbine governing mechanism?

11.13. Describe the governing mechanism of Kaplan turbine with a neat sketch.

11.14. Describe latest development of governing system.

11.15. Write the main functions of surge tank.

11.16. Explain the working of surge tank.

11.17. Classify the surge tank with reference to its configuration.

11.18. Why we use multiple surge tanks?

11.19. What is necessity of a surge tank.

11.20. Why surge tank is provide in hydroelectric power plant?

11.21. Why forebay is provided in hydroelectric power plant?

11.22. Differentiate between surge tank and forebay.

B. UNSOLVED PROBLEMS

11.1. A Pelton wheel works under a head of 300 m and a full load of 7.5 MW with an overall efficiency of 80%. There is a sudden drop in the load to 50% of the full load. If the turbine is governed for the constant speed, calculate the diameter of the jet at the part load. Take coefficient of velocity is equal to 0.98 [0.168 m]

11.2. The output power of a Pelton wheel is 6 MW. If the coefficient of velocity of the nozzle is 0.97, the relative velocity of the water to the bucket at exit is 0.95 times the relative velocity at inlet, the bucket speed is 0.46 times the jet speed and the deflecting angle of the buckets is 165°, calculate the hydraulic efficiency of the turbine. If the flow through the nozzle is reduced by 10% by closing a throttle value, the jet needle position being unaltered and the wheel speed remaining constant, calculate the new output power. Assume that the mechanical efficiency of the turbine remains a constent fraction of the hydraulic efficiency. [89.6%, 5.4 MW]

11.3. A Pelton turbine installed by the Tata Hydroelectric Co. Ltd. at Khapoli near Bombay develops 12570 kW when working under a head of 513m at 300 rpm. The nozzle diameter is 200 mm. Assuming that the losses in pipeline due to friction amount to kQ^2 where Q is the discharge in m^3/s and K is constant and is always unity. Calculated the discharge rate when power produced is reduced to 8820 kW (a) by operating the needle in the nozzle, (b) by closing the valve provided in mains.

Which of the two methods gives less head loss? What is the efficiency of the turbine under normal operating conditions? Assume that this efficiency and the gross head remain constant. Assume low value of c_v say 0.84 [$\eta_{turbine}$ = 96%, Q_1 = 1.83 m^3/s, Q_2 = 2.39 m^3/s, h_{II} = 3.35 m, h_{III} = 96 m. I method gives less head loss].

12

Performance of Hydraulic Turbines

12.1. INTRODUCTION

Dimensional analysis is one of the most important mathematical tools in the study of fluid mechanics and fluid machines. The main advantage of a dimensional analysis of a problem is that it reduces the number of variables in the problem by combining dimensional variables to form nondimensional parameters. Dimensional analysis is particularly helpful in experimental work because it provides a guide to those things that significantly influence the phenomena; thus it indicates the direction in which experimental work should go. A chapter an "dimensional analysis and similitude" given in Author's Fundamental of Fluid Mechanics will be useful for this purpose.

When experimental testing of a full size prototype is either impossible or prohibitively expensive, the only feasible way of attacking the problem is through model testing in the laboratory. If we are to predict the prototype behaviour from measurements on the model, it is obvious that we can not run just any test on any model. The model flow and the prototype flow must be related by known scaling laws. The laws of similitude make it possible to predict the performance of the prototype. The analytical prediction of the performance characteristics of turbomachines is complex, and often the characteristics are determined experimentally with the help of models.

For preliminary design purposes, it has been assumed that in general a turbine will work under varying conditions of head, speed, gate opening or the spear and output. As such uniformity rarely prevails in practice; The variations in the conditions for working may be as follows :

(1) The head on the turbine may change, and with it the output, the speed being correspondingly adjusted so that no sensible change in efficiency occurs, the gate opening remaining fixed.

(2) The head and the speed may remain steady, the output being varied by the movements of the gates or the spear (needle). These are the normal operating conditions for most of the turbines.

(3) The head and speed may vary. Such variations are common particularly in low head units. It may, however, be stated that although the speed is permitted to vary within vary narrow limits, the head may vary by even 50% or more.

(4) The speed may be allowed to vary by adjusting the load on the turbine, the head and gate opening remaining constant. These conditions can be developed only for laboratory turbines or those in the test plant and are otherwise uncommon.

(5) With gate fully open, and with no load on the turbine, the speed will be much higher than the normal speed. In these exceptional conditions, the term runaway speed is used.

In order to predict the behavior of a turbine working under varying conditions and the facilitate comparison between the performances of the turbines of the same type but having different outputs and speeds and working under different heads, it is often convenient to express the test results in terms of certain unit quantities. Similarly the turbines of different types certain specific quantities are also required to be considered.

12.2. PERFORMANCE UNDER UNIT HEAD

When comparing the performances of turbines of different outputs and speeds working under different heads, it is often convenient to calculate what the outputs would be if the heads were reduced to unity or one meter the speed being adjusted so that the efficiency remains unaffected. The efficiency of a given turbine will remain unchanged if the velocity triangles under the working head and under unit head should be geometrically similar so that shock at entry is avoided.

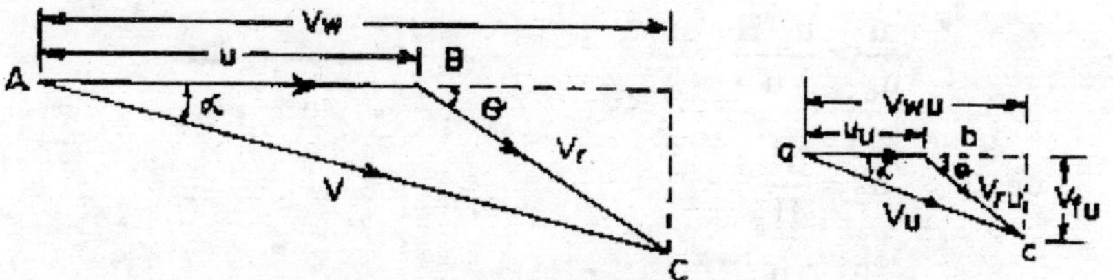

Fig. 12.1. Inlet velocity triangles of head H and unit head.

Fig. 12.1 shows two inlet velocity triangles for a Francis turbine, under its working head H and under unit head. ABC represents the or, velocity triangle and abc represent the velocity triangle for the same turbine under unit head, the subscript u being used to denote unit conditions.

Triangles ABC and abc are similar if their corresponding sides are parallel. It then follows that —

$$\frac{V_w}{u} = \frac{V_{wu}}{u_u} \tag{12.1}$$

and

$$\frac{V_f}{u} = \frac{V_{fu}}{u_u} \tag{12.2}$$

From equations (12.1) and (12.2)

$$\frac{V_{wu}}{V_w} = \frac{V_{fu}}{V_f} = \frac{u_u}{u} \tag{12.3}$$

The efficiency is unchanged, therefore,

$$\eta_n = \eta_{hu}$$

$$\frac{V_w\, u}{g\,H} = \frac{V_{wu}\, u_u}{g\,H_u}$$

or $$\frac{V_w\, u}{H} = V_{wu}\, u_u$$

or $$\frac{V_w}{V_{wu}} = \frac{u_u\, H}{u} \tag{12.4}$$

Substituting the value of $\frac{V_w}{V_{wu}}$ from equation (12.1) into equation (12.4) we get —

$$\frac{u}{u_u} = \frac{u_u H}{u}$$

or $$u_2^u = \frac{u^2}{H}$$

or $$u_u = \frac{u}{\sqrt{H}} \quad (12.5)$$

From equation (12.3), we get

$$\frac{V_{fu}}{V_f} = \frac{u_u}{u} = \frac{1}{\sqrt{H}}$$

or $$V_{fu} = \frac{V_f}{\sqrt{H}} \quad (12.6)$$

Since $$u = \frac{\pi DN}{60} \text{ and } u_u = \frac{\pi D_u N_u}{60}$$

or, $$\frac{u_u}{u} = \frac{N_u}{N}$$

or, $$\frac{N_u}{N} = \frac{1}{\sqrt{H}} \quad \text{[from equation (12.5)]}$$

or, $$N_u = \frac{N}{\sqrt{H}} \quad (12.7)$$

Where N_u represents the speed of the turbine under unit head.

Further it can be shown that both the speed ratio K_u and the flow ratio ψ remain constant under the head H and the unit head

$$Q = \pi B D V_f$$

and $$Q_u = \pi B D V_{fu}$$

Hence, $$\frac{Q_u}{Q} = \frac{V_{fu}}{V_f} = \frac{1}{\sqrt{H}} \quad \text{[From equation (12.6)}$$

Or, $$Q_u = \frac{Q}{\sqrt{H}} \quad (12.8)$$

Where Q_u represents the unit discharge. The power developed

$$P = \rho Q g H \eta_o$$

$$P_u = \rho Q_u g H_u \eta_{ou}$$

$$\text{or,} \quad \frac{P_u}{P} = \frac{Q_u}{QH} = \frac{1}{\sqrt{H}} \times \frac{1}{H} = \frac{1}{H^{3/2}}$$

$$\text{or,} \quad P_u = \frac{P}{H^{3/2}} \tag{12.9}$$

Where P_u represents the unit power. Although these identities have been deduced from the velocity triangles relating to a Francis turbine, they are applicable to all types of turbine, and we may therefore, say in general of terms:

(1) To reduce the performance of a turbine working under head H to conditions of unit head, all velocities must be reduced in the ratio $\frac{1}{\sqrt{H}}$.

(2) The unit speed Nu of a turbine is the working speed N divided by $\sqrt{H}$.

(3) Under unit conditions a turbine has the same speed ratio K_u and the same flow ratio ψ as it has under its working head H.

(4) The unit power P_u of a turbine, that is, its output under unit head, is found by dividing its normal of output P by $H\sqrt{H}$.

(5) A turbine working under a fixed ɓ .te setting and fixed values of K_u and ψ and $N/\sqrt{H}$ behaves exactly like an orifice is so far as the relationship between head, velocity and discharge is concerned.

The advantage of obtaining the performance at unit head is that from the known values of discharges, speed N and power P of a turbine working under head H, the values Q_1, N_1, and P_1 corresponding to some other head H_1, may be computed as indicated below:

Since Q_u, N_u and P_u will be same for both the heads, we have

$$Q_u = \frac{Q}{\sqrt{H}} = \frac{Q_1}{\sqrt{H_1}}; \therefore \; Q_1 = Q\sqrt{\frac{H_1}{H}}$$

$$N_u = \frac{N}{\sqrt{H}} = \frac{N_1}{\sqrt{H_1}}; \therefore \; N_1 = N\sqrt{\frac{H_1}{H}}$$

$$P_u = \frac{P}{H^{3/2}} = \frac{P_1}{H_1^{3/2}}; \therefore \; P_1 = P\left(\frac{H_1}{H}\right)^{3/2}$$

The above derived expressions are based on the assumption that the efficiencies remain constant at all the heads, which is however not correct. In practice it has been found that the efficiency varies with head which will cause the scale effect.

12.2 PERFORMANCE UNDER SPECIFIC CONDITIONS

The performance of turbine under unit head facilitate the comparison of the turbines of the same type. However, the turbines of different types may be compared by considering an imaginary turbine called specific turbine. The specific turbine is an imaginary turbine which is identical in shape, geometrical proportion, blade angles, gate setting etc.; with the actual turbine but reduced to such a size that it will develop one kilowatt power under unit head. The suffix **s** is used to denote the symbols expressing the performance of the specific turbine. The two runners now under comparison are shown in Fig 12.2.

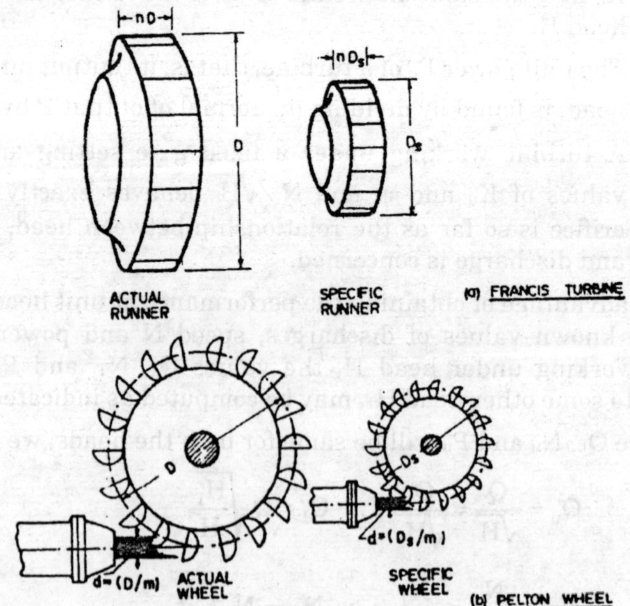

Fig. 12.2. Actual and specific runners.

Remembering that the actual turbine under unit conditions and the specific turbine both work under a head of 1 metre, it is at once evident that corresponding velocities in the two are identical, i.e. $u_s = u_u$, $V_{fs} = V_{fu}$, etc.; for the velocity triangle abc (Fig 12.1) serves for the specific turbine as well as for the actual turbine under unit conditions. Consequently,

$$\frac{P_u}{P_s} = \frac{\text{Power of actual turbine unit head}}{\text{Power of specific turbine}}$$

$$= \frac{\rho_u\, Q_u\, g\, H_u}{\rho_s\, Q_s\, g\, H_s}$$

$$= \frac{Q_u \times 1}{Q_s \times 1} = \frac{Q_u}{Q_s} \qquad (12.10)$$

But

$$\frac{Q_u}{Q_s} = \frac{\pi\, D_u\, (n\, D_u)\, V_{fu}}{\pi\, D_s\, (n\, D_s)\, V_{fs}}$$

$$= \frac{D_u^2}{D_s^2}$$

Because $V_{fu} = V_{fs}$

Therefore, frome equation (12.10)

$$\frac{P_u}{P_s} = \frac{D_u^2}{D_s^2}$$

Since $P_s = 1\,\text{kW}$, we have

$$D_s = \frac{D_u}{\sqrt{P_u}} \qquad (12.11)$$

From specific condition

$$u_u = u_s$$

$$\frac{\pi\, D_u\, N_u}{60} = \frac{\pi\, D_s N_S}{60}$$

$$N_s = N_u \left(\frac{D_u}{D_s}\right)$$

$$= N_u \sqrt{P_u} \qquad \text{[From equation (12.11)]}$$

$$= \frac{N}{\sqrt{H}} \sqrt{\frac{P}{H\sqrt{H}}} = \frac{N\sqrt{P}}{H^{5/4}} \qquad (12.12)$$

$$\therefore N_s = \frac{N\sqrt{P}}{H^{5/4}}$$

This value of N_s = speed of the specific runner, is termed the specific speed.

12.2.1 Specific speed

The significant aspect of the specific speed formula (Equation 12.12) is that it is independent of the dimensions or size both of the actual turbine and of the specific turbine; that is to say, all turbines of the same geometrical shape, working under the same values of K_u and ψ, and therefore, having the same efficiency, will have the same specific speed. We therefore, say that N_s represents the specific speed of the prototype or actual turbine, as well as of the specific turbine, and we use the term homologous to define such similarly proportioned machines. Unless otherwise stated, specific speeds refer to the performance of the turbine at the value of Ku giving maximum efficiency at normal full gate opening.

Here is a general definition : The specific speed of any turbine is the speed in revolutions per minute of a turbine geometrically similar to the actual turbine, but of such a size that under corresponding conditions it will develop 1 kW under unit head.

As the specific speed is not a pure number, its value depends upon the system of units employed.

In practice the conception of specific speed is of the utmost utility- the mere knowledge of the numerical value of N_s for a given runner conveys quite a definite notion of the shape or proportions of the runner, and moreover permits its working performance to be predicted. In the following paragraph it will be shown how information relating to specific speeds and specific runners can be used for roughly determining the main dimensions of turbines for given duties.

A non dimensional specific speed or non dimensional shape number K_n can readily be developed from the specific speed N_s, its value then being independent of the system of units. This pure number is represented by

$$K_n = \frac{N\sqrt{P/\rho}}{(gH)^{5/4}} \tag{12.13}$$

Where, N is in rps, P is in watts and ρ is the density of water.

The specific speed and the shape number (non dimensional specific speed) are connected thus-

$$K_n = \frac{N_s}{1042} \text{ (SI Units)} \tag{12.14a}$$

$$K_n = \frac{N}{1213} \text{ (MKS Units)} \tag{12.14b}$$

12.3 PERFORMANCE CHARACTERISTIC CURVES

The turbines are generally designed to work at particular values of P, Q, H, N and η_o which are known as the designed conditions. But often the turbines are required to work at conditions different from those for which they have been designed. Therefore, it is essential to determine the exact behaviour of the turbines under the varying conditions by carrying out tests either on the actual turbines or on their small scale models. The results of these tests are usually graphically represented and the resulting curves are known as characteristic curves. For the sake of convenience the characteristics curves are plotted in terms of unit quantities. These characteristic curves are of the following three types:

(1) Constant head characteristic curves (Main characteristic curves)

(2) Constant speed characteristic curves (Operating characteristic curves)

(3) Constant efficiency curves (Isoefficiency curves or Muschel curves).

12.3.1 Main characteristic curves

In order to obtain these curves the tests are performed on the turbine by maintaining a constant head and constant gate opening and the speed is varied by changing the load on the turbine. A series of values of N are thus obtained and corresponding to each value of

N, discharge Q and the output power P are measured. A series of such tests are performed by varying the gate opening, the head being maintained constant at the previous value. From the data of the tests the values of Q_u, P_u, N_u and η_o are computed for each gate opening. Then with Nu as abscissa the values of Q_u, Pu and n for each gate opening are plotted. The curves thus obtained for impulse turbine (Pelton wheel turbine) and the reaction turbines (Francis and Kaplan) for four different gate openings are shown in Fig 12.3 to 12.5.

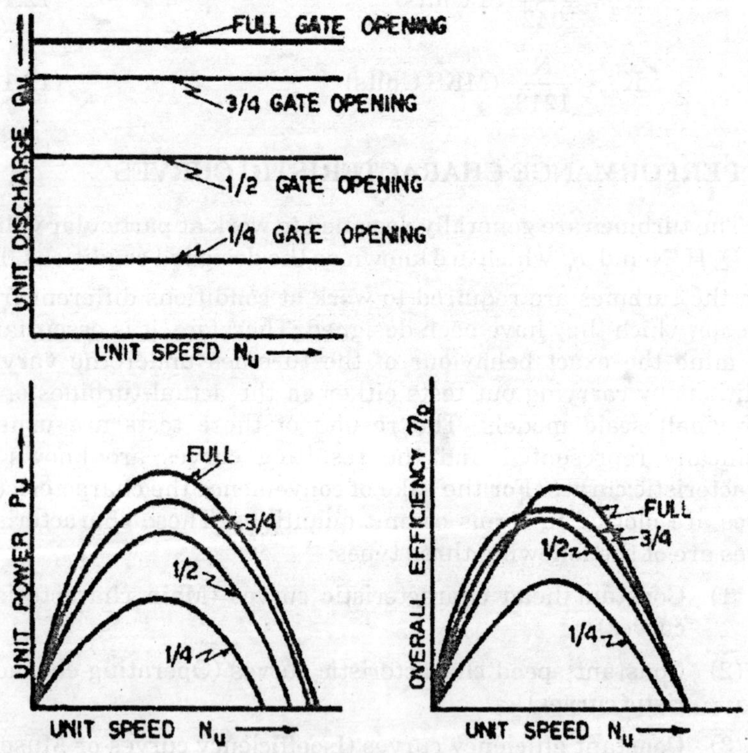

Fig. 12.3. Main characteristic curves for Pelton turbines.

For Pelton wheels since Q_u depends only on the gate opening and is independent of N_u, the Q_u v/s N_u plots are horizontal straight lines. However, for low specific speed Francis turbines Q_u v/s N_u are droping curves thereby indicating that as the speed increases the discharge through the turbine decreases. This is so because in these turbines a centrifugal head is developed which retards the flow.

Since the centrifugal head increases with the speed, the flow through the turbine is reduced as the speed increases. On the other hand, for high specific speed Francis turbines as well as Kaplan turbines, since the flow is axial there is no such centrifugal head developed which may cause the retardation of the flow.

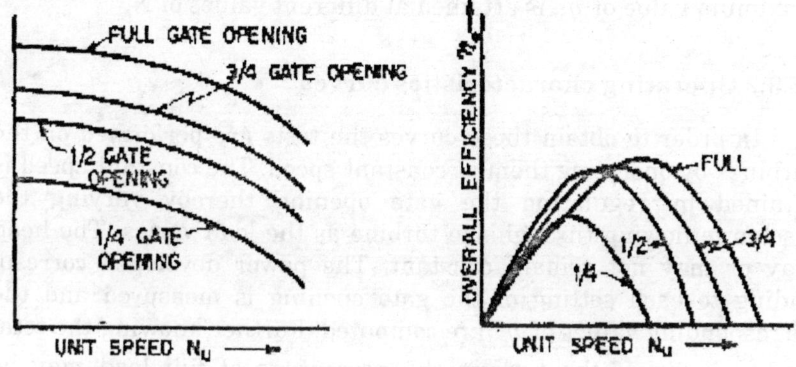

Fig. 12.4. Main characteristic curve for Francis turbine.

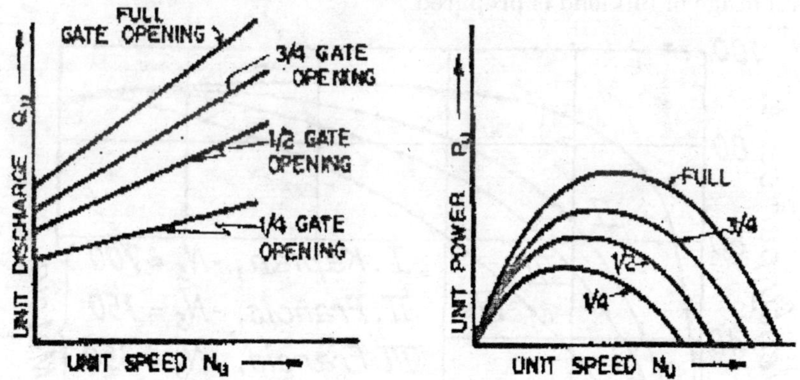

Fig. 12.5. Main characteristic curves for Kaplan turbine.

The curves of P_u v/s N_u and η_o v/s N_u are parabolic in shape for the different turbines as shown in Figures 12.3 to 12.5. It will be observed that for a Pelton wheel, for each gate opening the maximum value of η_o is attained at almost the same value of N_u which corresponds to $K_u = 0.46$. Under ideal conditions η_o v/s N_u curves would have a parabolic form, but owing to the increased windage and friction at high speeds, the maximum or runaway speed is not quite twice the working speed corresponding to maximum efficiency. The reason why the gross efficiency falls off if the value of Ku departs sensibly from 0.45 is that at other values of K_u the water leaves the buckets with on undue amount of unexpended velocity energy, which is wasted in the tail race.

In the case of reaction turbines for each gate opening the maximum value of η_o is attained at different values of N.

12.3.2 Operating characteristics curves

In order to obtain these curves the tests are performed on the turbines by operating them at constant speed. The constant speed is attained by regulating the gate opening thereby varying the discharge flowing through the turbine as the load varies. The head may or may not remain constant. The power developed corresponding to each setting of the gate opening is measured and the corresponding values of η_o are computed. Further knowing the total load capacity of the turbine the percentage of full load may be computed from the measured power, and a plot of efficiency vesus percentage of full load is prepared.

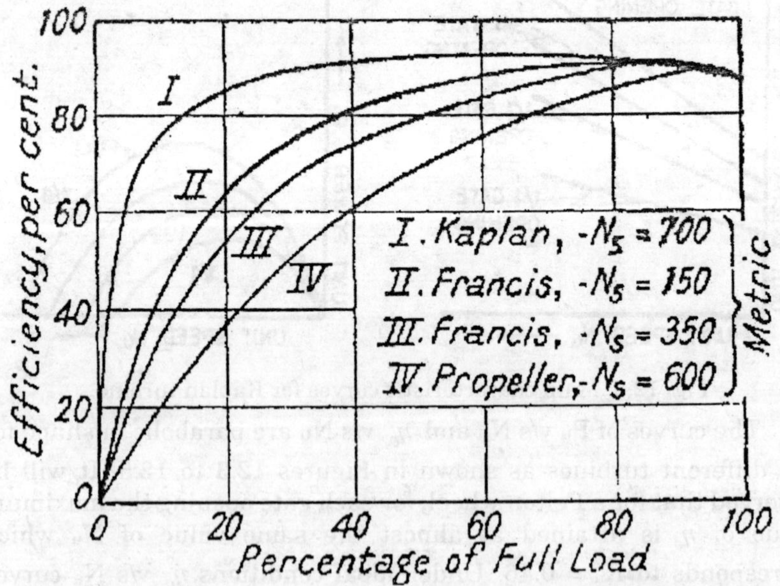

Fig. 12.6. Constant speed characteristics compared (Ref. Addison)

Typical characteristics for axial flow turbines are represented by the curves I and IV in Fig. 12.6, where they may be contrasted with the corresponding curves II and III for Francis turbines. The superiority of the curve I for the Kaplan turbine, which rivals or even excels the best performance of the Pelton wheel, as regards sustained high efficiency even at small gate openings.

The inlet and outlet velocity diagrams for Kaplan runner working under the specific conditions at full load and the part load are different. Although the corresponding change in the flow through the turbine does affect the shape of the diagrams, yet as the blade angles themselves are simultaneously adjusted, the losses inevitable in a Francis turbine at part loads are now very largely avoided.

On the other hand, the constant pitch propeller turbine at part load suffers losses due to eddying and to velocity energy rejected that are even more serious than in the Francis turbine, as is shown clearly in Fig. 12.6 (IV). It is thus essentially a machine for working continously at full load.

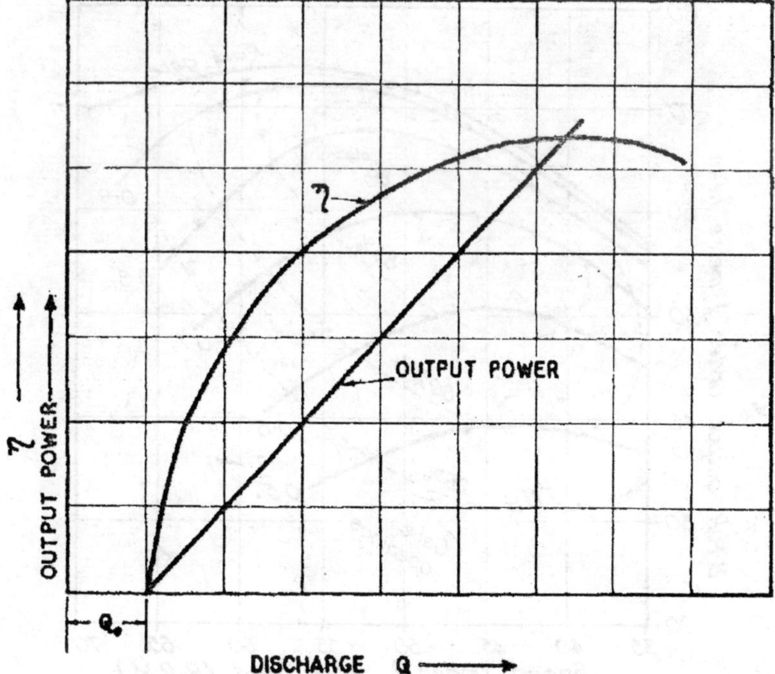

Fig. 12.7. η_0 and P v/s Q curves.

In addition to the above graph, P versus Q and η_o versus Q graphs are plotted from a certain minimum discharge Q_o Which is required to initiate the motion of the turbine runner from its state of rest. Further since Q is directly proportional to discharge if head is constant, P versus Q graphs is a straight line. On the other hand η_o versus Q graphs is curvilinear and as shown in Fig. 12.7 the overall efficiency increases with discharge and remains more or less constant beyond a particular value of discharge.

12.3.3. Isoefficiency or Muschel curves

These curves are plotted from data which can be obtained from the constant speed and constant head curves. The object of obtaining this curves is to determine the zone of constant efficiency so that we can always run the turbine with maximum efficiency.

This curve (Fig. 12.8) also give a good idea about the performance of the turbines at various efficiencies.

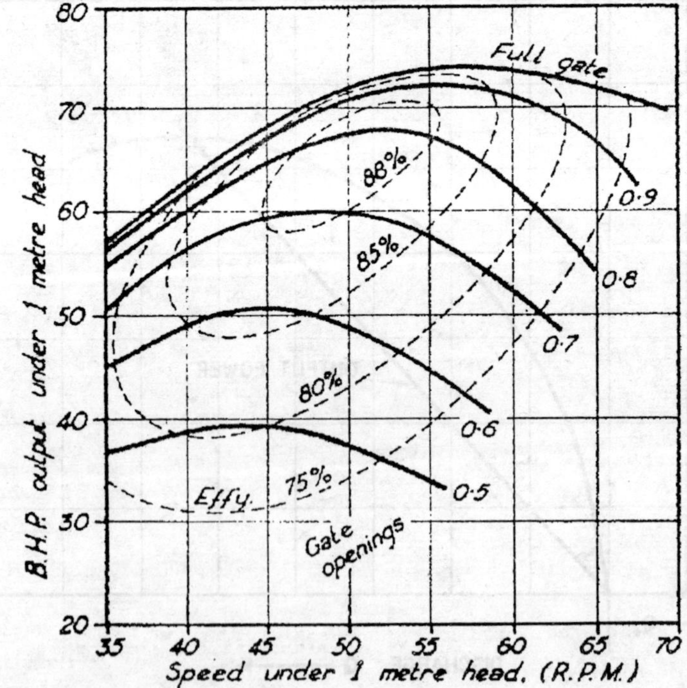

Fig. 12.8. Muschel curve.

If the peak points of various efficiency curves are joined by a smooth curves. The best performance curve for the turbine under a given head is obtained. Peak point on a constant efficiency curve is a point at which the tangent to the curve is horizontal. For a particular head and gate opening, highest efficiency is obtained when the curve for that opening cut the best performance curve values of the speed and the discharge corresponding to that point can be noted from the plot. If the speed is known intersection of vertical line at the given speed and the best performance. Curve provides the values of the discharge and the maximum efficiency.

12.4 MODEL TESTING OF TURBINES

In order to have an idea about the performance of actual turbine in advance, a small scale model of the turbine which is geometrically similar to the actual (Prototype) turbine is first prepared.

In accordance with the definition two turbines of one type, but of different size meet the following conditions :

(1) all turbine corresponding angles are a like :

$$\delta_{11} = \delta_{12}\ , \delta_{21} = \delta_{22}, \delta_{i1} = \delta_{i2} \qquad (12.15)$$

(2) the ratio of corresponding dimensions is constant :

$$\frac{D_{11}}{D_{12}} = \frac{D_{21}}{D_{22}} = \frac{B_{01}}{B_{02}} = --- \qquad (12.16)$$

The mode of operation of turbines of one type will be similar if the geometrical similarity of velocity parallelogram or velocity triangles at correspoints in the liquid passage remains valid. It follows that when operating conditions are similar.

(1) all corresponding velocities be similarly directed, i.e. all corresponding angles are equal.

(2) the ratio of all velocities at corresponding points must be constant:

On the basis of the above definitions we stipulate the regularities governing the relationship between the parameters of turbines of one type under conditions of similar modes of operation.

Model testing of turbines assists in obtaining a perfect design for the actual turbines, as well as in the development of the new types of turbines with higher specific speed and better efficiency.

The various variables involved in this case are discharge Q, head H, speed of rotation of runner N, runner diameter D, output power P, density ρ and viscosity μ of the flowing fluid. It may however of turbomachine generally shaft work gH is used as one of the variables instead of head H. By adopting the usual methods of dimensional analysis these variables may be grouped into following dimensionless parameters.

$$\frac{Q}{ND^3}, \frac{gH}{N^2D^2}, \frac{\mu}{\rho N D^2}, \frac{P}{\rho g H N D^3}$$

The above non dimensional number permits us to find the speed ratio, discharge ratio and power ratio.

Speed ratio, $$\frac{N_1}{N_2} = \frac{D_2}{D_1}\sqrt{\frac{H_1}{H_2}} \qquad (12.17)$$

Discharge ratio, $$\frac{Q_1}{Q_2} = \left(\frac{D_2}{D_1}\right)^2 \sqrt{\frac{H_1}{H_2}} \qquad (12.18)$$

Power ratio, $$\frac{P_1}{P_2} = \left(\frac{D_2}{D_1}\right)^2 \frac{H_1}{H_2}\sqrt{\frac{H_1}{H_2}} \qquad (12.19)$$

Formulae (12.17), (12.18) and (12.19) are called the laws of similarity which are widely used. In deriving the formulae we assumed that the efficiency ratio is unity.

The specific speed N_s and cavitation number σ are also to be considered for the modal analysis.

Thus for the complete similarily to exist between the modeal and the prototype turbines, the following conditions may be required to be satisfied. In most of the cases $g_m = g_p$, $\rho_m = \rho_p$ and $\eta_m = \eta_p$.

$$\left(\frac{Q}{ND^3}\right)_m = \left(\frac{Q}{ND^3}\right)_p \qquad (12.20)$$

$$\left(\frac{H}{N^2D^2}\right)_m = \left(\frac{H}{N^2D^2}\right)_p \qquad (12.21)$$

$$\left(\frac{P}{HND^2}\right)_m = \left(\frac{P}{HND^3}\right)_p \qquad (12.22)$$

$$(N_s)_m = (N_s)_p$$

$$\left(\frac{N\sqrt{P}}{H^{5/4}}\right)_m = \left(\frac{N\sqrt{P}}{H^{5/4}}\right)_P \quad (12.23)$$

and $\sigma_m = \sigma_P$

$$\left(\frac{H_a - H_v - H_s}{H}\right)_m = \left(\frac{H_a - H_v - H_s}{H}\right)_P \quad (12.24)$$

ILLUSTRATIVE EXAMPLES

12.1. A turbine is to operate under a head of 25m at 200 rpm. The discharge is 9 m3/s. If the turbine efficiency is 90 percent, make calculations for the power generated, specific speed of the turbine, and performance under a head of 20 m. Also state the type of turbine.

Solution

$$P = \rho g Q H \eta_0 = 1000 \times 9.81 \times 9 \times 25 \times 0.9$$

$$= 1986.525\ kW \quad \text{Ans.}$$

$$N_s = \frac{N\sqrt{P}}{H^{5/4}} = \frac{200\sqrt{1986.525}}{(25)^{5/4}} = 159.46 \quad \text{Ans.}$$

Since it lies between 50 to 250 the turbine is a Francis turbine.Ans.

$$N_u = \frac{N_1}{\sqrt{H_1}} = \frac{N_2}{\sqrt{H_2}}$$

$$\frac{200}{\sqrt{25}} = \frac{N_2}{\sqrt{20}} \quad \therefore \quad N_2 = 178.88\ rpm \quad \text{Ans.}$$

$$Q_u = \frac{Q_1}{\sqrt{H_1}} = \frac{Q_2}{\sqrt{H_2}}$$

$$\frac{9}{\sqrt{25}} = \frac{Q_2}{20} \quad \therefore Q_2 = 8.05\ m^3/s \quad \text{Ans.}$$

$$P_u = \frac{P_1}{H^{3/2}} = \frac{P_2}{H_2^{3/2}}$$

$$= \frac{1986.525}{(25)^{3/2}} = \frac{P_2}{(20)^{3/2}} \qquad \therefore P_2 = 1421.44\ kW \qquad \text{Ans.}$$

12.2. A model of Francis turbine one fifth of full size develops 3 kW under a head of 1.8 m. Find the power and speed of the full sized turbine operating under a head of 5.8 m. assume the efficiency of the model and full size to be the same. Assume speed of the model as 360 rpm. Find also the ratio of quantities of water flowing through the runner.

Solution

We know that

$$\frac{D_m N_m}{\sqrt{H_m}} = \frac{D_p N_p}{\sqrt{H_p}}$$

$$\therefore N_p = N_m \times \frac{D_m}{D_p} \times \sqrt{\frac{H_p}{H_m}}$$

$$= 360 \times \frac{1}{5} \times \sqrt{\frac{5.8}{1.8}} = 129.244\ rpm \qquad \text{Ans.}$$

We know that

$$\frac{P_p}{D_p^3 N_p H_p} = \frac{P_m}{D_m^3 N_m H_m}$$

$$\therefore P_p = P_m \times \left(\frac{D_p}{D_m}\right)^3 \times \frac{N_p}{N_m} \times \frac{H_p}{H_m}$$

$$= 3 \times (5)^3 \times \frac{129.244}{360} \times \frac{5.8}{1.8} = 433.805\ kW \qquad \text{Ans.}$$

We know that

$$\frac{Q_m}{D_m^3 N_m} = \frac{Q_p}{D_p^3 N_p}$$

$$\therefore \frac{Q_m}{Q_p} = \frac{N_m}{N_p} \times \left(\frac{D_m}{D_p}\right)^3 = \frac{360}{129.244} \times \left(\frac{1}{5}\right)^3 = 0.022283$$

$$\therefore \frac{Q_p}{Q_m} = 44.8764$$ Ans.

12.3. A 1/4 scale turbine model is tested under a head of 12m. The full scale turbine is required to work under a head of 36 m and to run at 428 rpm. At what speed must the model be run and if it develops 100 kW and uses 1 m³/s of water at this speed, what power will be obtained from the full scale turbine assuming that its efficiency is 3% better than that of the model? State the type of runner used in this turbine.

Solution

$$P_m = \rho\, g\, Q_m\, H_m\, \eta_{om}$$

$$100 \times 10^3 = 1000 \times 9.81 \times 1 \times 12 \times \eta_{om}$$

$$\eta_{om} = 0.84947 = 84.947\%$$

$$\eta_{op} = 0.84947 \times (1 + 0.03)$$

$$= 0.87947 = 87.495\%$$

We know that

$$\frac{D_m\, N_m}{\sqrt{H_m}} = \frac{D_p\, N_p}{\sqrt{H_p}}$$

$$N_m = N_p \times \frac{D_p}{D_m} \times \sqrt{\frac{H_m}{H_p}}$$

$$= 428 \times 4 \times \sqrt{\frac{12}{36}} = 988.42366\ rpm$$ Ans.

We know that

$$\frac{Q_p}{D_p^2\sqrt{H_p}} = \frac{Q_m}{D_m^2\sqrt{H_m}}$$

$$\therefore Q_p = Q_m \times \left(\frac{D_p}{D_m}\right)^2 \times \sqrt{\frac{H_p}{H_m}}$$

$$= 1 \times (4)^2 \times \sqrt{\frac{36}{12}} = 27.7128\ m^3/s$$

$$P_p = \rho\, g\, Q_p\, H_p\, \eta_{op} = 1000 \times 9.81 \times 27.7128 \times 36 \times 0.87495$$

$$= 8563.18154\ kW \quad = 8.5632\ MW \qquad \text{Ans.}$$

$$N_S = \frac{N_p \sqrt{P_p}}{H_p^{5/4}} = \frac{428\sqrt{8563.18154}}{(36)^{5/4}} = 449.14$$

Since the specific speed lies between 250 and 860 the turbine is a Kaplan turbine.

12.4 A Kaplan turbine with a 7m diameter propeller develops 26.45 MW at 75 rpm under a head of 12 m. (a) At what speed should a 0.7 m diameter model be run under a head of 6m for homologous conditions? (b) What flow rate should exist for model, assuming an efficiency of 93% for both prototype and model? (c) What power will be developed by the model? (d) What is the non dimensional speed?

Solution

$$P_p = \rho\, g\, Q_p\, H_p\, \eta_{op}$$

$$26.45 \times 10^6 = 1000 \times 9.81 \times Q_p \times 12 \times 0.93$$

$$Q_p = 241.5975\ m^3/s$$

We know that

$$\frac{Q_m}{D_m^2\sqrt{H_m}} = \frac{Q_p}{D_p^2\sqrt{H_p}}$$

$$\therefore Q_m = Q_p \times \left(\frac{D_m}{D_p}\right)^2 \times \sqrt{\frac{H_m}{H_p}}$$

$$= 241.5975\left(\frac{0.7}{7}\right)^2\sqrt{\frac{6}{12}} = 1.70835\ m^3/s \qquad \text{Ans.}$$

We know that

$$\frac{D_m N_m}{\sqrt{H_m}} = \frac{D_p N_p}{\sqrt{H_p}}$$

$$\therefore N_m = N_p \times \frac{D_p}{D_m} \times \sqrt{\frac{H_m}{H_p}} = 75 \times \frac{7}{0.7} \times \sqrt{\frac{6}{12}}$$

$$= 530.33\ rpm \qquad \text{Ans.}$$

We know that

$$\frac{P_m}{D_m^2\ H_m^{3/2}} = \frac{P_p}{D_p^2\ H_p^{3/2}}$$

$$\therefore P_m = P_p \times \left(\frac{D_m}{D_p}\right)^2 \times \left(\frac{H_m}{H_p}\right)^{3/2}$$

$$= 26.45 \times \left(\frac{0.7}{7}\right)^2 \times \left(\frac{6}{12}\right)^{3/2} = 0.0935\ MW = 93.5\ kW \qquad \text{Ans.}$$

$$K_n = \frac{N_p \sqrt{P_p / \rho}}{(g\ H_p)^{5/4}} = \frac{\frac{75}{60}\sqrt{\frac{26.45 \times 10^6}{10^3}}}{(9.81 \times 12)^{5/4}} = 0.524276\ rev. \qquad \text{Ans.}$$

12.5. **A Kaplan turbine is to develop 20 MW when running at 240 rpm under a net head of 45 m. In order to predict its performance, a model to a scale of 1/7 is tested under a net head of 20 m. At what speed should the model run, and what power would it develop? If the overall efficiency of the model were found to be 84%, at the best operating point, what quantity of water would the model and the full scale turbine require.**

Solution

$$P_p = \rho\, g\, Q_p\, H_p\, \eta_{op}$$

$$20 \times 10^6 = 1000 \times 9.81 \times Q_p \times 45 \times 0.84$$

$$Q_p = 53.9348\ m^3/s \qquad \text{Ans.}$$

We know that

$$\frac{D_m\ N_m}{\sqrt{H_m}} = \frac{D_p\ N_p}{\sqrt{H_p}}$$

$$\therefore N_m = N_p \frac{D_p}{D_m} \times \sqrt{\frac{H_m}{H_p}} = 240 \times 7 \times \sqrt{\frac{20}{45}}$$

$$= 1120 \ rpm \qquad \text{Ans.}$$

We know that

$$\frac{Q_m}{N_m D_m^3} = \frac{Q_p}{N_p D_p^3}$$

$$\therefore Q_m = Q_p \times \frac{N_m}{N_p} \times \left(\frac{D_m}{D_p}\right)^3$$

$$= 53.9348 \times \frac{1120}{240} \times \left(\frac{1}{7}\right)^3 = 0.7338 \ m^3/s \qquad \text{Ans.}$$

We know that

$$\frac{P_m}{D_m^2 H_m^{3/2}} = \frac{P_p}{D_p^2 H_p^{3/2}}$$

$$\therefore P_m = P_p \times \left(\frac{H_m}{H_p}\right)^{3/2} \times \left(\frac{D_m}{D_p}\right)^2 = 20 \times \left(\frac{20}{45}\right)^{3/2} \times \left(\frac{1}{7}\right)^2$$

$$= 0.120937 \ MW$$

$$= 120.937 \ kW \qquad \text{Ans.}$$

12.6 **A model turbine of scale ratio 1/8 was tested under a head of 15m in a laboratory where it was possible to vary the draft head. The model just began to cavitate when the runner was located 6m above the tail water level. The speed of the model was 480 rpm and its output 14.7 kW. The actual turbine had to work under the head of 45 m. The atmospheric pressure was 760 mm of mercury and the vapour pressure corresponding to water temperature was 0.33 m of water head in sight as well as in laboratory. Find : (a) Speed of the turbine (b) Power of the turbine and (c) The maximum safe setting of the turbine above the tail race.**

Solution

We know that

$$\frac{D_p N_p}{\sqrt{H_p}} = \frac{D_m N_m}{\sqrt{H_m}}$$

$$\therefore N_p = N_m \times \frac{D_m}{D_p} = \sqrt{\frac{H_p}{H_m}}$$

$$= 480 \times \frac{1}{8} \times \sqrt{\frac{45}{15}} = 103.923\ rpm$$ Ans.

We know that

$$\frac{P_p}{D_p^2 H_p^{3/2}} = \frac{P_m}{D_m^2 H_m^{3/2}}$$

$$\therefore P_p = P_m \times \left(\frac{D_p}{D_m}\right)^2 \times \left(\frac{H_p}{H_m}\right)^{3/2}$$

$$P_p = 14.7 \times (8)^2 \times \left(\frac{45}{12}\right)^{3/2} = 4888.54\ kW$$ Ans.

We know that

$$\sigma_{cm} = \sigma_{cp}$$

$$\frac{H_{atm} - H_{vm}}{H_m} = \frac{H_{atm} - H_{vp} - H_{sp}}{H_p}$$

$$\frac{10.33 - 0.33 - 6}{15} = \frac{10.33 - 0.33 - H_{sp}}{45}$$

$$0.2667 = \frac{10 - H_{sp}}{45}\ 10 - H_{sp} = 12$$

$$\therefore H_{sp} = -2 \text{ m of Water}$$ Ans.

12.7. A model of Francis turbine one fifth of full size, develops 3 kW at 300 rpm under a head of 1.75 m. Find the speed and power of full size turbine operating under a head of 6 m, if

(a) the efficiency of the model and the full size turbine are same,

(b) the efficiency of the model turbine is 76% and the scale effect is considered.

Solution

(a) We know that

$$\left(\frac{H}{D^2 N^2}\right) = \left(\frac{H}{D^2 N^2}\right)_m$$

$$N_p^2 = \left(\frac{{}^pH_p}{H_m}\right) \times \left(\frac{D_m}{D_p}\right)^2 N_m^2$$

$$= \frac{6}{1.75} \times \frac{1}{25} \times (300)^2$$

$$= 12342.857 = 111.0984 \; rpm$$ Ans.

$$\left(\frac{P}{D^2 H^{1/2}}\right) = \left(\frac{P}{D^2 H^{3/2}}\right)_m$$

$$P_p = \left(\frac{D_p}{D_m}\right)^2 \left(\frac{H_p}{H_m}\right)^{3/2} P_m$$

$$= 25 \times \left(\frac{6}{1.75}\right)^{1.5} \times 3 = 476.136 \; kW$$ Ans.

(b) We know that

$$\frac{1-\eta_{op}}{1-\eta_{om}} = \left(\frac{D_m}{D_p}\right)^{0.2}$$

$$\frac{1-\eta_{op}}{1-0.76} = \left(\frac{1}{5}\right)^{0.2}$$

$$\therefore \eta_{op} = 0.826 = 82.6\%$$ Ans.

We know that

$$\frac{D_p N_p}{\sqrt{\eta_{op} H_p}} = \frac{D_m N_m}{\sqrt{\eta_{om} H_m}}$$

$$N_p = \frac{D_m}{D_p} \times \sqrt{\frac{\eta_{op} H_p}{\eta_{om} H_m}} \times N_m$$

$$= \frac{1}{5} \times \sqrt{\frac{0.826}{0.76} \times \frac{6}{1.75}} \times 300 = 115.822 \ rpm$$ Ans.

We know that

$$\frac{P_p}{(\eta_{op} \ H_p)^{3/2} \ D_p^2} = \frac{P_m}{(\eta_{om} \ H_m)^{3/2} \ D_m^2}$$

$$P_p = \left(\frac{D_p}{D_m}\right)^2 \left(\frac{\eta_{op} \ H_p}{\eta_{om} \ H_m}\right)^{1.5} P_m$$

$$= 25 \times \left(\frac{0.826}{0.76} \times \frac{6}{1.75}\right)^{1.5} \times 3 = 539.4867 \ kW$$ Ans.

12.8. **The following figures relate to a test on a water turbine working under its designed head of 8.4 m.**

Pu = Unit power	8.9	9.3	9.57	9.57	9.4	9.1
Nu = Unit speed	56	65	75	84	93	102
m . = mass flow rate	3590	3540	3470	3390	3300	3200

Plot a curve of overall efficiency (η_0) against unit speed (Nu). Find the turbine speed at maximum efficiency and hence thespecific speed of the turbine. If the head is changed to 9.9 m estimate the power developedand efficiency at a turbine speed of 250 rpm.

Solution

To plot the required curve, overall efficiency must be worked out corresponding to all the given unit speeds.

$$P_u = \frac{P}{H^{3/2}}$$

$$\therefore = H^{3/2} \times P_u = (8.4)^{1.5} \ P_u$$

$$= 24.3455 \ P_u$$

(1) $P = 24.3455 \times 8.9 = 216.67 \ kW$

(2) $P = 24.3455 \times 9.3 = 226.41 \ kW$

(3) $P = 24.3455 \times 9.57 = 232.98 \ kW$

(4) $P = 24.3455 \times 9.57 = 232.98\ kW$

(5) $P = 24.3455 \times 9.4 = 228.84\ kW$

(6) $P = 24.3455 \times 9.1 = 221.54\ kW$

We know that

$$P = \rho\, g\, Q\, H\, \eta_o = \dot{m}\, g\, H\, \eta_o$$

(1) $\eta_o = \dfrac{P}{g\, H\, \eta_o} = \dfrac{216.67 \times 1000}{9.81 \times 8.4 \times 3590}\ 0.7324 = 73.24\%$

(2) $\eta_o = \dfrac{226.41 \times 1000}{9.81 \times 8.4 \times 3540} = 0.7761 = 77.61\%$

(3) $\eta_o = \dfrac{232.98 \times 1000}{9.81 \times 8.4 \times 3470} = 0.8147 = 81.47\%$

(4) $\eta_o = \dfrac{232.98 \times 1000}{9.81 \times 8.4 \times 3390} = 0.8340 = 83.4\%$

(5) $\eta_o = \dfrac{228.84 \times 1000}{9.81 \times 8.4 \times 3300} = 0.8415 = 84.15\%$

(6) $\eta_o = \dfrac{221.54 \times 1000}{9.81 \times 8.4 \times 3200} = 0.8401 = 84.01\%$

From the above, we have maximum efficiency is 84.15%.

Plot a curve between η_o and N_u. From this curve, corresponding to the maximum efficiency, $N_u = 93$ and hence $P_u = 9.4$ from the data given in the table. Hence the speed and specific speed corresponding to maximum efficiency are

$$N = N_u \sqrt{H} = 93 \sqrt{8.4} = 269.5 = 270\ rpm$$

$$N_s = N_u \sqrt{P_u} = 93 \sqrt{9.4} = 285.13$$ Ans.

(b) For the same efficiency 84.15%, when H = 9.9 m and N = 250 rpm

We have,

$$P_u = \frac{P_1}{H^{3/2}} = \frac{P_2}{H_2^{3/2}}$$

$$\text{Therefore, } P_2 = P_1 \left(\frac{H_2}{H_1}\right)^{3/2}$$

$$= 228.84 \left(\frac{9.9}{8.4}\right)^{1.5} = 292.7\ kW$$

$$\dot{m} = \rho Q \ \therefore Q = 3300./1000 = 3.3\ m^3/s$$

$$Q_u = \frac{Q_1}{\sqrt{H_1}} = \frac{Q_2}{\sqrt{H_2}}$$

$$\text{Therefore, } Q_2 = Q_1 \left(\frac{H_2}{H_1}\right)^{0.5}$$

$$= 3.3 \left(\frac{9.9}{8.4}\right)^{0.5} = 3.58\ m^3/s$$

$$\therefore P = \rho\, Q\, g\, H\, \eta_o$$

$$292.7 \times 1000 = 1000 \times 3.58 \times 9.81 \times 9.9 \times \eta_o$$

$$\therefore \eta_o = 84.12\%$$ Ans.

EXERCISES

A. THEORY

12.1. Define unit quantities. What are the assumptions on which unit quantitites defined?

12.2. Explain (a) Unit speed (b) Unit discharge (c) Unit power of a hydraulic turbine. Derive expressions for each of them.

12.3. What are the limitations of unit quantities?

12.4. Define specific speed.

12.5. Prove that $N_s = N_u \sqrt{P_u}$

12.6. Define specific speed, unit speed and unit power as applied to hydraulic machines. Deduce expressions to indicate their values.

12.7. Show that the non dimensional specific speed (K_n) of a fluid machines is in general given by

$$K_n = \frac{N\sqrt{P}}{\sqrt{\rho}\,(gH)^{5/4}}$$

Where the symbols have their usual meanings. Classify the fluid machines on the basis of their non dimensional specific speed.

12.8. Define the terms 'unit speed' and 'unit power'. Show that the N_u P_u is the specific speed.

12.9. Define the terms unit speed, unit power and specific speed and derive an expression for the specific speed in terms of power, head and speed.

12.10. Derive an expression for the specific speed of a single jet pelton wheel

$$N_s = 494.24\sqrt{c_v\,\eta_0}\;K_u\left(\frac{D}{d}\right)$$

Where, c_v is the coefficient of velocity of nozzle, K_u is the speed ratio, η_o is overall efficiency and $\left(\frac{D}{d}\right)$ is the jet ratio.

12.11. Derive an expression for the specific speed of a single jet Pelton wheel taking

$$K_u = 0.46, c_v = 0.98 \quad \text{and} \quad \eta_o = 85\%$$

$$N_s = 207.5\left(\frac{D}{d}\right)$$

12.12. Derive an expression for the specific speed of a Francis turbine

$$N_s = 984.45\;K_u\sqrt{K\,n\,\psi\,\eta_o}$$

Where Ku = speed ratio, ψ = flow ratio, n = breadth ratio.

12.13. Derive an expression for the specific speed of a Kaplan or Propeller turbine

$$N_s = 494.24\,K_u\sqrt{\psi\,(1-n^2)\,\eta_0}$$

12.14. Derive an expression for the specific speed of a Kaplan or propeller turbine taking $\eta_0 = 90\%$ and n = d/D = 0.35

$$N_s = 439.22\,K_u\sqrt{\psi}$$

12.15. What are the characteristic curves of a hydraulic turbine? How are they useful to practical Engineer?

12.16. Draw constant head characteristic curves of hydraulic turbines and explain them clearly.

12.17. Draw constant speed characteristic curves of hydraulic turbines and explain them clearly.

12.18. Draw constant efficiency curves of hydraulic turbines and explain them clearly.

12.19. Draw main characteristics, operating characteristic and Muschel curves for Kaplan turbine and explain them clearly.

12.20. Draw main, operating and isoefficiency curves for Francis turbine and expalin them clearly.

12.21. Draw main, operating and isoefficiency curves for the Pelton turbine and explain them clearly.

B. UNSOLVED PROBLEMS

12.1 A turbine develops 5880 kW when running at a speed of 200 rpm and working under a head of 200 m with an overall efficiency of 80%. Determine unit speed, unit discharge and unit power [14.142, 0.2648, 2.078] .

12.2. A model of Francis turbine built one tenth size of actual turbine. When tested, the model produces 1.1 kW at a speed of 1000 rpm under a head of 15 m. Estimate the equivalent speed and power of full size machine when working under the head of 100m. [1893.457 kW, 258.2 rpm].

12.3 A Kaplan turbine model designed to one seventh scale is tested under a head of 20 m. The large unit develops 18.375 MW at a speed of 250 rpm, under a head of 50 m. The overall efficiency of the model, at its best operating point is 80%. Estimate :

(a) Speed at which the model is run,

(b) Power the model develops;

(c) Quantity of water required by model,

(d) Quantity of water required for the full scale turbine [1106.79 rpm, 94.869 kW, 0.6044 m³/s 46.82 m³/s].

12.4. A water turbine develops 150kW at 300 rpm under a head of 15m. A similar machine is to be designed to give 750 kW under a head of 18m, but otherwise under the same conditions. Find the scale ratio and speed for this machine, and hence its specific speed [1 : 95 : 1, 168.5, 124.8]

12.5. A quarter scale turbine model is tested under a head of 10.8 m. The full scale turbine is required to work under a head of 30 m and to run at 7.14 rev/s. At what speed must the model be run? If it develops 100 kW and uses 1.085 m^3 of water per second at this speed, what power will be obtained from the full scale turbine, its efficiency being 3% better than that of the model? What is the nondimensional specific speed of the full scale turbine? [17.14 rev/s, 7.66 MW, 0.513 rev.]

12.6. A turbine develops 3725 kW under a net head of 12.2 m and at an overlall efficiency of 0.82. If a new source of supply is used and the net head is increased to 18.3 m, assuming the overall efficiency remain the constant, find (a) flow in m^3/s which will be required, (b) the brake power which will be developed and (c) the % increse in speed. [46.6 m^3/s, 6835 kW, 82.6%]

12.7. A model is to be built for studying the performance of a turbine having a runner diameter of 1 m and maximum output 2240 kW under a head of 50 m. What is the appropriate diameter of the runner of the model if the corresponding power is 10 kW and the available head is 8m? Show that the corresponding model speeds will be about 50% greater than those of the turbine. [0.265 m]

12.8. A model propeller turbine rotor 300 mm in diameter develops 26kW at 7.5 m head and 1200 rpm. The prototype developed 7500 kW with 6m head. Find the diameter and speed of prototype. [6m, 53.5 rpm]

12.9. A hydraulic turbine develops 9000 kW under a head of 10m at a speed of 90 rpm and gives an efficiency of 92.7%. Calculate the water consumption and the specific speed. If a model 1/10 full size is constructed to operate under a head of 8m, what must be its speed, power and water consumption to run under the similar conditions to the prototype? How would the model efficiency compare with the prototype? [98.96 m^3/s, N_s = 480, N_m = 804.98 rpm, R_m = 64.4 kW, Q_m = 0.885 m^3/s]

12.10. The following data were obtained from the main characteristics of a Kaplan turbine of runner diameter 1 m, P_u = 30.695; Q_u = 108.6; N_u = 63.6. Estimate the runner diameter, the discharge and the speed of a similar runner working under a head of 30 m and developing 2 MW. Determine the specific speed of the runner [629.7 mm, 553.2 rpm, 352.36].

13

Centrifugal Pumps

13.1 INTRODUCTION

The main difference between a turbine and a rotodynamic pump is that a turbine takes in liquid energy and gives out mechanical energy. While a rotodynamic pump takes in mechanical energy and gives out liquid energy. A rotodynamic pump adds energy to liquid while a turbine extracts energy from liquid.

Depending upon the shape of the impeller and the delivery space, impeller or rotodynamic pumps are divided into:

(a) Centrifugal pumps or radial flow pumps

(b) Mixed flow pumps

(c) Propeller or axial flow pumps.

In rotodynamic pumps, increase in energy level is due to a combination of centrifugal energy, pressure energy and kinetic energy. The role played by each of these types of energy depends on whether the pump is a radial flow, mixed flow or axial flow type.

In a radial flow pump, the energy, transfer occurs mainly when the flow is in its radial path; in an axial flow pump, when the flow is in its axial direction : and in a mixed flow pump when the flow has both radial and axial components. The ranges of operation of each of these types of pumps, based on specific speed. But only the radial flow type pumps which are commonly called centrifugal pumps are discussed in detail in this chapter.

13.2 CLASSIFICATION OF CENTRIFUGAL PUMPS

Centrifugal pumps may be classified according to

1. Operating head
2. Specific speed
3. Types of casing
4. Number of impellers per shaft
5. Disposition or layout of shaft
6. Number of entrances to the impeller and
7. Relative direction of flow through impeller.

13.2.1 Operating head

In general centrifugal pumps are suitable for both low and high heads. The basic design lends itself to operating heads of up to about 150 m in a single stage, but a more general usual categorization is given in Table 13.1.

Table 13.1. Operating head ranges

S.No.	Head types	Ranges
1	Low-head	Maximum total head up to 20 m.
2	Medium-head	Maximum total head 15 to 60 m.
3	High-head	Maximum total head over 60 m.

13.2.2 Specific speed

The shape of a pump impeller is closely related to its speed of rotation (N) and the pumping head (H). These factors are combined into a number known as the specific speed of the pump, which provides a common base for comparing pumps. It is the speed at which a pump will deliver 1 m³/s at 1 m head and is calculated as follows

$$N_s = \frac{N\sqrt{Q}}{H_m^{3/4}}$$

Where, N is in rpm, Q is in m³/s and H_m is in m.

The specific speed is independent of the size of the pump and so it describes the shape of the pump rather than how big it is. But specific speed is not dimensionless. It is important to make sure that SI units are used so that range of specific speeds and pump types is as shown in Table 13.2.

Table 13.2. Specific speeds for different pumps

Pump type	Specific speed Ns	Comments
Centrifugal	10 to 17	High head - low discharge
Mixed flow	70 to 170	Medium head - medium discharge
Axial flow	above 110	Low head - large discharge

13.2.3 Types of casing

According to the type of casing provided, centrifugal pumps are classified into the two classes.

(1) Diffuser or turbine pump

(2) Volute pump.

(1) Diffuser or turbine pump

In a diffuser pump the impeller is surrounded by a series of guide vanes which provide gradually increasing area of flow, thus the liquid gradually loses its velocity and gains in pressure. From the vanes the flow takes place into the surrounding casing which may be circular placed concentrically with the impeller or volute shaped like the scroll casing of a reaction turbine. The guide vanes with their enclosed channels are also known as diffuser ring. This pump very much resembles with a reversed turbine and is thus known as a turbine pump (Fig. 13.1).

The diffuser pump is capable of converting as much as 75% of the kinetic energy of the liquid discharged from the impeller into pressure energy. However, these pumps will work with maximum efficiency only for one rate of discharge at a given impeller speed. This is so because the guide vanes will be correctly set or shaped for one rate of discharge only and for other discharges a loss of energy by shock or turbulenece will occur at the entrance to the guide vanes, thereby, resulting in a low efficiency ring is usually employed only in multistage pumps. Diffuser pumps are more costly as comparison to volute pumps.

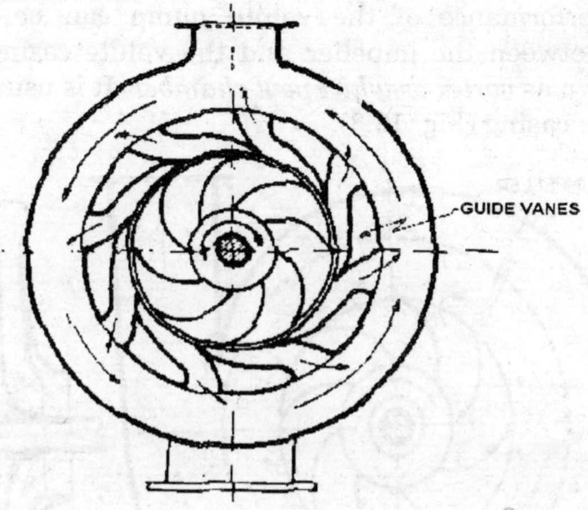

Fig. 13.1. Diffuser pump (single suction)

(2) Volute pump

In a volute pump, the impeller is surrounded by a spiral shaped casing known as volute chamber. The sectional area of the volute casing increases uniformly from the nose to the point of connection with the delivery pipe. The increasing area of flow provides to some extent a means for gradually reducing the velocity and by the time the liquid reached the delivery pipe the velocity becomes small and the pressure increases correspondingly (Fig 13.2). The increase in efficiency obtained by this method is not considerable because of eddy losses at exit periphery of the impeller and the entry to the delivery pipe.

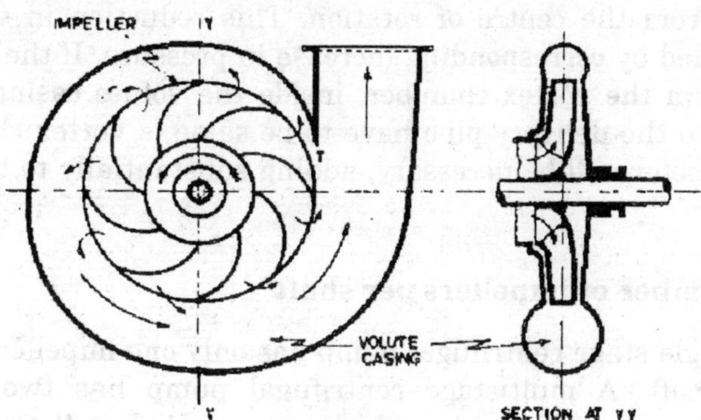

Fig. 13.2. Volute pump.

The performance of the volute pump can be improved by providing between the impeller and the volute casing an annular spaces known as *vortex or whirl pool chamber*. It is usually formed as a part of the casing (Fig. 13.3).

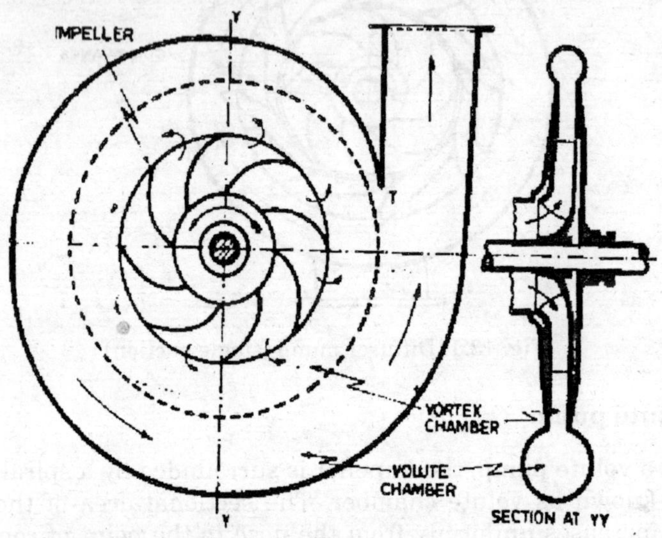

Fig. 13.3. Volute pump with vortex chamber.

Water from the impeller enters the vortex chamber with a whirling motion, and a single particle follows a spiraling path while passing through it. Since there is no work done on water while on this chamber, its energy must remain constant, thus the torque, which this water is capable of producing does not change. And if torque has to remain same, the velocity of whirl must decrease for increasing distance from the centre of rotation. This reduction in velocity is accompanied by corresponding increase in pressure. If the velocities at exit from the vortex chamber, inside the volute casing and the entrance to the delivery pipe have to be same, a vortex chamber of large diameter will be necessary, adding substantially to the cost of the unit.

13.2.4 Number of impellers per shaft

A single stage centrifugal pump has only one impeller mounted on the shaft. A multistage centrifugal pump has two or more impellers connected in series, which are mounted on the same shaft and are enclosed in the same casing (Table 13.3).

Table 13.3. Classification by geometry

Basic type	Stage(s)	Layout	Entry	Casing
Centrifugal	Single-stage	Horizontal	Single	Volute-split or end pull-out
	Single-stage	Horizontal	Double	Volute-split or end pull out
	Multi-stage	Horizontal	Axial (single or double)	Ring type for low to medium heads. Barrel type for high pressures
	Multi-stage (balanced)	Horizontal	Back-to-back	Volute (usually)
	Single-stage	Vertical	Single	Volute-split casing
	Single-stage	Vertical	Double	Volute-split casing
	Multi-stage	Vertical	Single	Volute
Mixed flow (helical)	Single	Horizontal	Single	Volute-split casing or end pull-out
	Single	Horizontal	Double	Volute-split casing
	Single	Vertical	Single	Volute-split casing
Mixed flow (diagonal)	Single	Horizontal	Single	Concentric
	Single	Vertical	Single	Concentric
	Multi	Usually vertical	Single	Concentric
Axial flow	Single	Horizontal	Single	Concentric
		Vertical	Single	Concentric
	Multi	Usually vertical	Single	Concentric

13.2.5 Disposition or layout of shaft

A geometric classification depends on whether the pump is designed for horizontal or vertical mounting (Table 13.3). For general purpose applications the choice is somewhat arbitrary. Horizontal pumps are the more common and have advantages from the point of view of ease of maintenance resistance to corrosion and abrasion and ease of mounting with minimum headroom needed. Vertical pumps have the advantage of occupying minimum floor space, ease of priming and reduction in required NPSH and generally flexibility to adapt to a particular duty with certain types of pumps the application may dictate mounting *eg*. submersible and immersible pumps are invariably of vertical type.

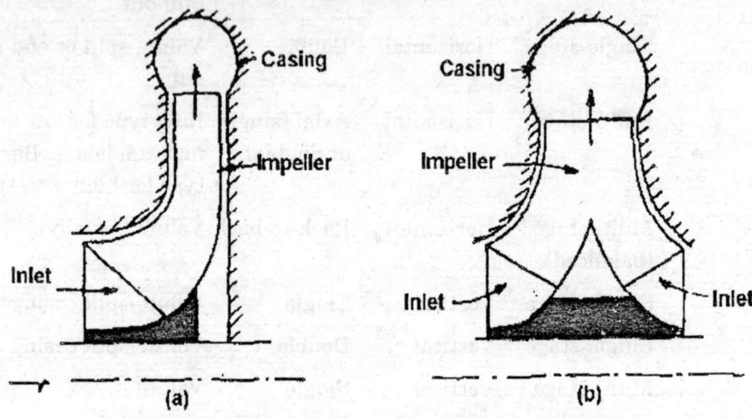

Fig. 13.4. Single and Double suction pumps.

13.2.6 Number of entrances to the impeller

Depending on the number of entrances to the impeller the centrifugal pumps may be classified as single suction pump and the double suction pump. In a single suction pump liquid is entering from a suction pipe on one side of the impeller. In a double suction pump liquid enters from both sides of the impeller (Table 13.3).

With a double entry pump the inflow is divided and taken to each side of a double sided impeller thus balancing the axial thrusts. Pumps with double entry are usually of large size, with higher specific speeds. Suction and delivery branches are commonly situated on one half of a split casing, the other half being removable for inspection of for dismantling the pump without the necessity of first disconnecting the suction and delivery pipes.

13.2.7 Relative direction of flow through impeller

On the basis of the direction of flow of the liquid through the impeller the pump may be classified as radial flow pump, mixed flow pump and axial flow pump.

Fig. 13.5 shows the basic design of radial, axial and mixed flow impellers.

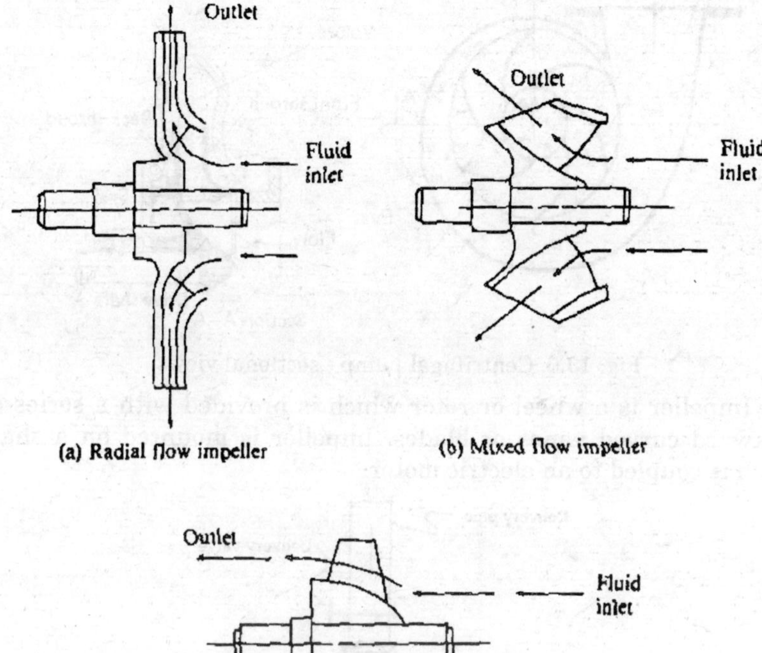

Fig. 13.5. Different direction of flow through impeller.

13.3 CENTRIFUGAL PUMP

A centrifugal pump-sectional views (Fig. 13.6) is just the reverse of a radial flow reaction turbine. In centrifugal pump, the water flows radially outward. The usual component parts of a centrifugal pump are (Fig 13.7).

1. Impeller
2. Casing
3. Suction pipe with foot valve and a striner.
4. Delivery pipe.

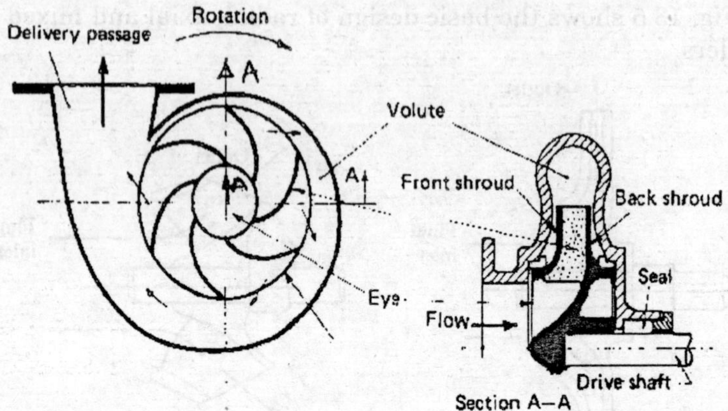

Fig. 13.6. Centrifugal pump - sectional views.

Impeller is a wheel or rotor which is provided with a series of backward curved vanes or blades. Impeller is mounted on a shaft which is coupled to an electric motor.

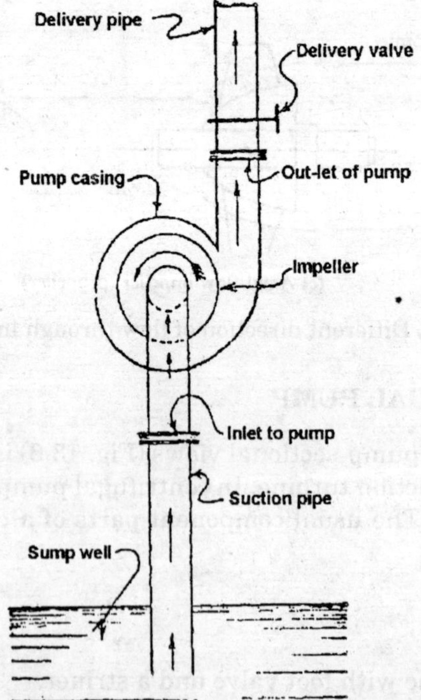

Fig. 13.7. Component Parts of A Centrifugal Pumps.

Casing is an airtight chamber which surrounds the impeller. It is similar to the casing of a reaction turbine. Reaction turbine casing are water tight.

Suction pipe with its upper end connected to the inlet of pump and its lower end dipping into a sump or suction well from which the liquid is to be pumped or lifted up. Water from inlet of pump goes to the eye of casing and from there it goes to the centre of the impeller. the lower end of suction pipe is fitted with a strainer which keeps out debris and allows pure liquid to be sucked up the suction pipe.

Delivery pipe which takes the water from the outlet of pump and delivers it to a point higher than the position of pump.

A delivery value on the delivery pipe, just near the outlet of pump to control the flow into the delivery pipe.

Prime movers which rotates the shaft of the impeller. Usually electric motor is provided for this purpose of driving the pump.

13.3.1 Types of vanes

The vanes on the impeller may be of the following types.

(i) Curved back vanes.

(ii) Radial vanes

(iii) Curved forward vanes.

The curved back vanes are the vanes which are curved back from the direction of rotation of the impeller, they thus present convex surface to the liquid which is impelled (or forced out) by them out of the impeller (Fig 13.8). Efficiency obtained is 85 to 90%. Centrifugal pumps are generally provided with this type of vane. Unless mentioned otherwise it must be assumed that the impeller has curved back vanes.

For this type of vane, $\theta < 90^\circ$ and $\varphi < 90^\circ$.

Radial vanes are those which have got their outlet tips in the radial direction. (Fig. 13.8b) They are less efficient as compared to curved back vanes. The efficiency obtained is about 80 to 85%. For this type of vane, $\theta < 90^\circ$ and $\varphi = 90^\circ$.

Curved forward vanes are the vanes which are curved forward in the direction of rotation of the impeller, thus, they present concave surface to the liquid which is impelled out by them. These vanes give very low efficiency. For this type of vane, $\theta < 90^\circ$ and $\varphi > 90^\circ$.

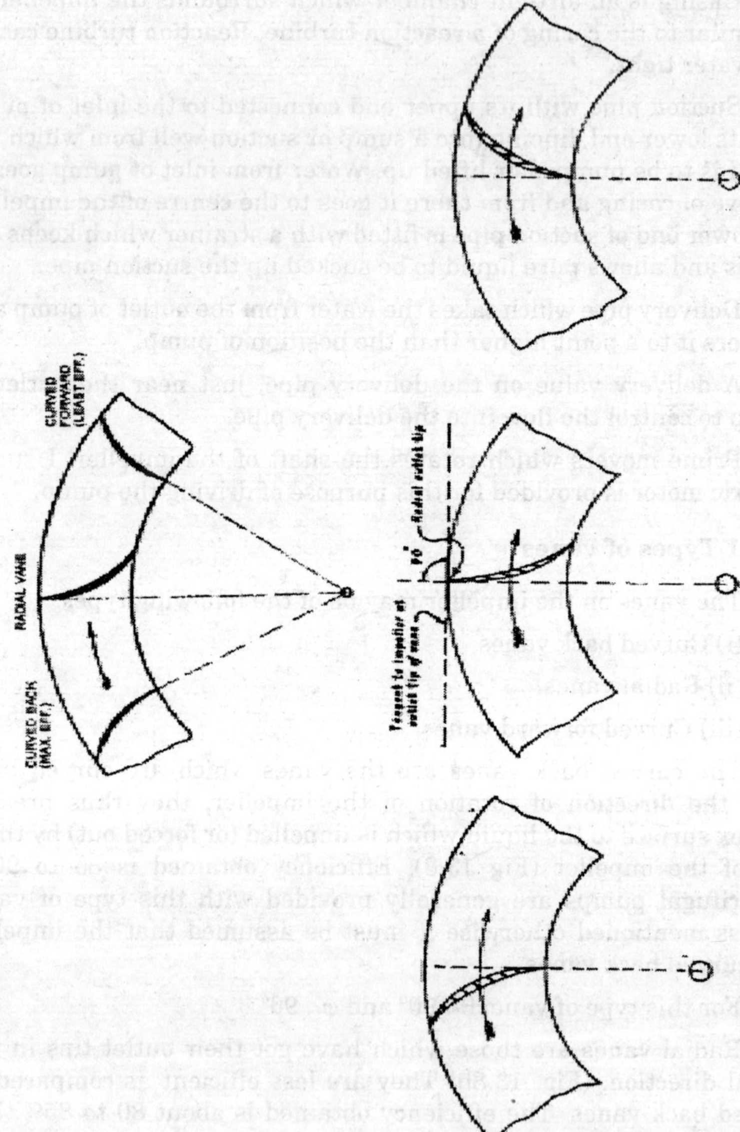

Fig. 13.8 (a) and (b). Types of vanes.

13.3.2 Work done and action of centrifugal pump

The expression for the work done by the impeller of a centrifugal pump on the liquid flowing it may be derived in the same way as for a turbine. The velocity triangles at inlet and outlet of the impeller are drawn in the same way as for a turbine.

For good efficiency, the following assumptions are usually made.

(1) The flow towards the impeller is radial (i.e. the absolute velocity at inlet is radial) or, there is no velocity of whirl. Hence $V_{w1} = 0$, $V_1 = V_{f1}$ and $\alpha = 90°$.

(2) Velocity of flow through the impeller is assumed to be constant. Hence, $V_{f1} = V_{f2}$. However, if in some problems sufficient data is given to find V_{f1} and V_{f2} separately. Such different values of V_{f1} and V_{f2} may be used.

The impeller of a centrifugal pump, showing one vane and two velocity triangles at the inlet and outlet tips of vane is given in Figs. 13.9(a) and 13.9 (b).

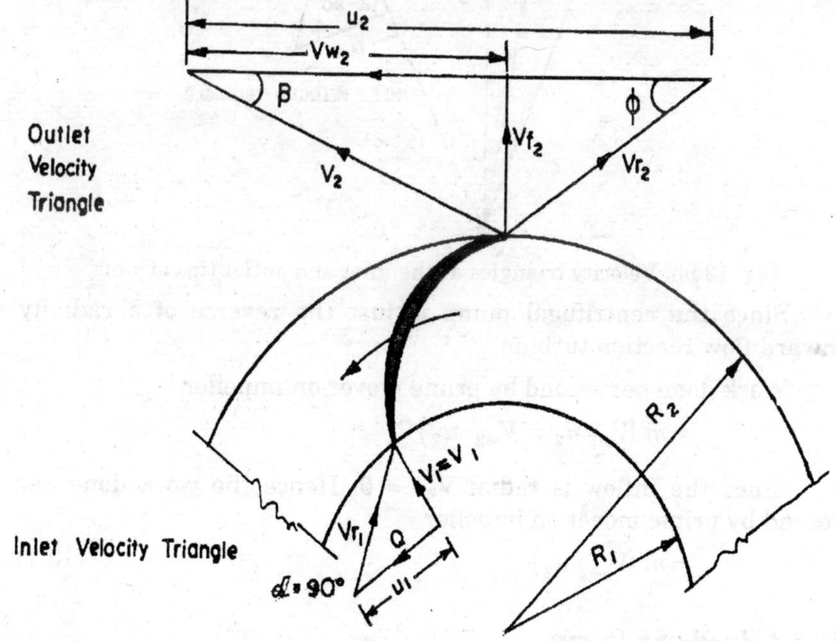

Fig. 13.9(a). Velocity triangles at the inlet and outlet tips of vane.

We know that for radial inward flow reaction turbine.

Work done by liquid on runner

$$= \dot{m}\left(V_{w1}\, u_1 \pm V_{w2}\; u_2\right)$$

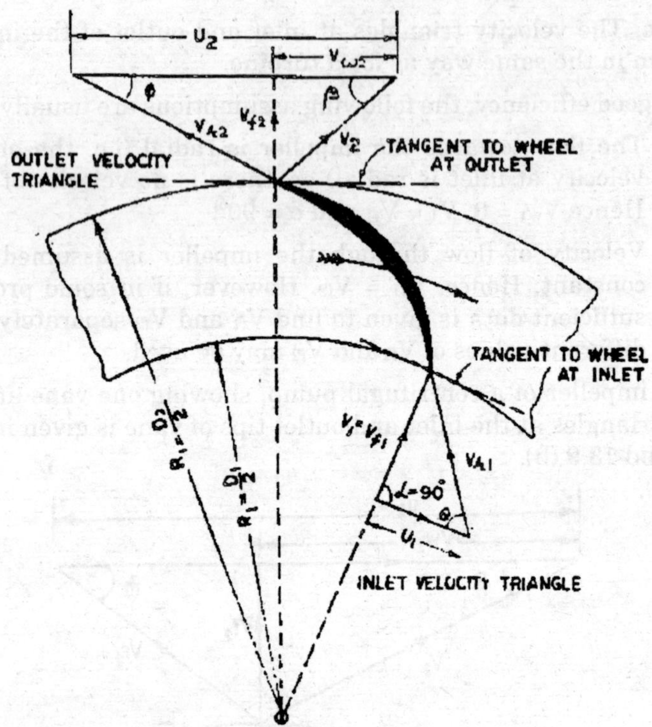

Fig. 13.9b. Velocity triangles at the inlet and outlet tips of vane.

Since the centrifugal pump is just the reverse of a radially inward flow reaction turbine.

Work done per second by prime mover on impeller

$$= \dot{m}\left(V_{w1}\, u_2 \pm V_{w2}\;\; u_2\right)$$

Since the inflow is radial V_{w1} = 0. Hence the work done per second by prime mover on impeller

$$= \dot{m}\left(V_{we2}\, u_2\right) \qquad (13.1)$$

13.3.3 Head of a Pump

The head of a centrifugal pump may be expressed in the following two ways.

1. Static head.
2. Manometric head (Total head or Gross head or Effective head or Euler head).

(1) Static head (H_s)

The sum of sunction head (h_s) plus delivery head (h_d) is known as static head).

$$H_s = h_s + h_d \tag{13.2}$$

(2) Manometric head (H_m)

Manometric head is defined as the head actually developed for its working. Manometric head is given by the following expressions:

(a) $$H_m = h_s + h_d + h_{fs} + h_{fd} + \frac{V_d^2}{2g} \tag{13.3}$$

$$= H_s + h_{fs} + h_{fd} + \frac{V_d^2}{2\,g} \tag{13.4}$$

(b) Manometric head developed by the pump is the difference of total heads between the outlet point B and the inlet point A of the pump. Therefore,

$$H_m = H_B - H_A$$

$$= \textit{Total head at delivery side} - \textit{Total head at suction side}$$

$$= \left[\frac{P_d}{\rho\,g} + \frac{V_d^2}{2g} + Z_d\right] - \left[\frac{p_s}{\rho\,g} + \frac{V_s^2}{2\,g} + Z_s\right] \tag{13.5}$$

$$= \left[\frac{p_d}{\rho\,g} + \frac{p_s}{\rho\,g}\right]$$

[If the suction gauges are set, level with pump axis or at same level ($Z_d = Z_s$)]

$$\therefore H_m = \left[\frac{p_d}{\rho\,g} - \frac{p_s}{\rho\,g}\right] \tag{13.6}$$

where, Pd/ρg is read on pressure gauge and ps/ρg is read on vaccum gauge i.e. ps/ρg is negative. Hence, H_m is the sum of the pressure head readings on the two gauges.

$$\therefore H_m = \frac{p_d}{\rho\,g} - \left[-\frac{p_s}{\rho\,g}\right]$$

$$= \frac{p_d}{\rho g} + \frac{p_s}{\rho g} \tag{13.7}$$

(c) Manometric head is slightly less than the head given (by prime mover i.e. **electric motor**) to the impeller. In other words,

$$H_m = \frac{V_{w2}\, u_2}{g} - \textit{hydraulic losses of head in the pump.} \tag{13.8}$$

Hydraulic losses occur in the impeller and in the diffuser.

(1) The kinetic energy leaving the diffuser is not normally recovered and must contribute to the total loss. **The loss in the head in the diffuser is**

$$(1 - \eta_d)\left[\frac{V_2^2}{2g} - \frac{V_B^2}{2g}\right]$$

where V_2 is the velocity with which liquid leaves the **impeller and V_B is the velocity at point B; this latter velocity is approximately equal to V_d.** This can also be expressed as some function of $V_2^2/2g$, say $(1 - k)\ V_2^2/2g$ or gain in pressure head $k\ V_2^2/2g$ where the value of k for volute casing is 0.4 and for the case of turbine pump, diffuser vanes may be from 0.67 to 0.75.

(2) Impeller head loss

(3) Exit head loss.

The expression $\left(V_2^2 - V_d^2\right)/2g$ may be written as

Velocity had converted = Pressure head actually recovered
+ Loss in conversion due to the widening of flow area.

Hence, $$\frac{V_2^2 - V_d^2}{2g} = k\,\frac{V_2^2}{2g} + \textit{loss due to expansion} \tag{13.9}$$

The above expressions for manometric head H_m will be found very useful in solving problems on centrifugal pumps.

13.3.4 Efficiencies of centrifugal pump

Centrifugal pump has the following kinds of efficiencies.

(a) Manometric efficiency, η_{mano}

(b) Mechanical efficiency, η_m

(c) Overall (or gross or total to total) efficiency, η_o

The expressions for these efficiencies are:

(a) Manometric efficiency

The manometric efficiency η_{mano} is defined as ratio of the manometric head developed by the pump to the head imparted by the impeller to the liquid. Thus

$$\eta_{mano} = \frac{H_m}{V_{w2}\, u_2 / g} = \frac{g\, H_m}{V_{w2}\, u_2} \tag{13.10}$$

$$= \frac{H_m}{H_m + Hydraulic\ losses\ in\ pump}$$

(b) Mechanical efficiency

The mechanical efficiency η_m is defined as the ratio of the power actually delivered by the impeller to the power supplied to the shaft by the prime mover or electric motor. Thus

$$\eta_m = \frac{Energy\ of\ impeller}{Energy\ given\ to\ the\ shaft}$$

$$= \frac{V_{w2}\, u_2\ /\ g}{\dfrac{V_{w2} . u_2}{g} + Mechanical\ losses\ in\ bearing\ etc.} \tag{13.11}$$

(c) Overall efficiency

The overall efficiency (Total to total or Gross) η_o of the pump is defined as the ratio of the power output from the pump to the power input from the prime mover or electric motor driving the pump. Thus

$$\eta_0 = \frac{\rho\, g\, Q\, H_m}{Power\ given\ to\ the\ shaft} \tag{13.12}$$

The overall efficiency is also equal to the product of all the efficiencies described above. Thus

$$\eta_0 = \eta_{mano} \times \eta_m \tag{13.13}$$

$$= \frac{H_m}{V_{w2}\, u_2 / g} \times \frac{(V_{w2}\, u_2)/\, g}{(V_{w2}\, u_2)/\, g + mechanical\ losses}$$

$$= \frac{H_m}{V_{w2}\, u_2 / g + Mechanical\ losses} \quad (13.14)$$

$$= \frac{\rho\, g\, Q\, H_m}{Power\ given\ to\ the\ shaft} \quad (13.15)$$

Which is same as equation (13.12)

Note : (1) Unless mentioned otherwise, by efficiency of centrifugal pump we mean the manometric efficiency or hydraulic efficiency, η_h.

(2) Shaft power of motor mean output power of motor.

13.3.5. Losses in centrifugal pump

The losses between the inlet point A and outlet point B of pump are given by the following relation (Fig. 13.10). Hydraulic losses in pump (i.e. between points A and B) = Losses between inlet A of pump and the inlet of impeller + losses in impeller + losses between outlet of impeller and the outlet point B of pump.

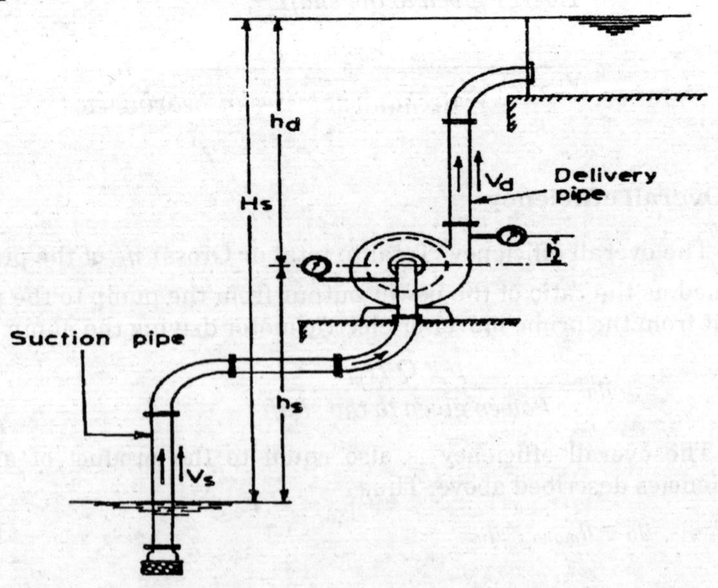

Fig. 13.10. Head on a centrifugal pump.

It should be noted that the losses between outlet of impeller and the outlet point B of the pump are due to the conversion of velocity head into pressure head : these losses due to expansion of flow area are relatively more important and are of more value as compared to other two types of losses. As already stated in the last article, these losses between outlet of impeller and the outlet of pump may be.

(i) In the volute casing only; in case of volute pump.

(ii) In diffuser blades and in the casing, in case of the turbine pump,

Fig. 13.11 shows how each of the power losses are subtracted from the initial input power. The rectangle OABC represents the total power input to the shaft while OADEFC is equivalent to the mechanical power loss. The impeller loss $\rho\ g\ Q_i\ h_i$ is next removed and is represented by rectangle EFGI. The next loss to be accounted for is the leakage loss $\rho\ g\ H_i\ q$ represented by rectangle DJKI, and finally the casing loss $\rho\ g\ h_c\,Q$ represented by rectangle MLGK is removed. This leaves us with rectangle JBLM, which represents the fluid power output or power developed by the pump $\rho g\ Q\ H$.

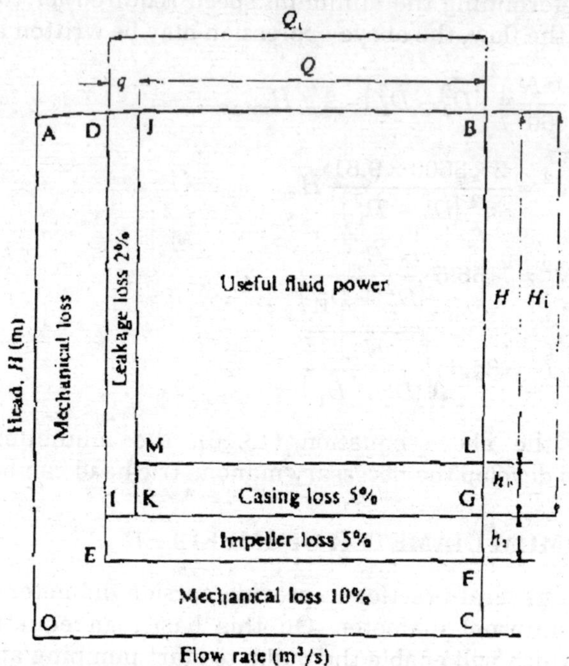

Fig. 13.11. Losses in a centrifugal pump.

13.4 MINIMUM STARTING SPEED

After priming, the speed of the pump is brought normal and then the delivery valve is opened. The head developed by the pump just before the delivery is opened is called *shut-off* head. If the impeller is rotating and the delivery valve is closed, the water is rotating in a forced vortex. Therefore, a centrifugal head or pressure head caused by the centrifugal force on the rotating water will be $(u_2^2 - u_1^2)/2g$. The flow will commence only when this head is at least equal to the external resistance H_m. Thus,

$$\frac{\left(u_s^2 - u_1^2\right)}{2g} \geq H_m$$

or,

$$\left(\frac{\pi D_2 N}{60}\right)^2 - \left(\frac{\pi D_1 N}{60}\right)^2 \geq 2g\, H_m$$

or,

$$\left(\frac{\pi N}{60}\right)^2 \left(D_2^2 - D_1^2\right) \geq 2\, g\, H_m$$

For determining the minimum speed required for the pump to commence the flow, the above expression may be written as

$$\left(\frac{\pi N}{60}\right)^2 \left(D_2^2 - D_1^2\right) = 2\, g\, H_m$$

$$N^2 = \frac{2 \times 3600 \times 9.81}{\pi^2 \left(D_2^2 - D_1^2\right)} H_m$$

$$N^2 = 7156.5 \frac{H_m}{\left(D_2^2 - D_{11}^2\right)} \qquad (13.16a)$$

$$\therefore N = 84.6 \sqrt{\frac{H_m}{\left(D_2^2 - D_1^2\right)}} \qquad (13.16b)$$

From the above equation (13.16), the minimum speed N required to develop the necessary manometric head can be obtained.

13.5 MINIMUM DIAMETER OF IMPELLER

It is a general practice to provide outside diameter of impeller twice the internal diameter. On this basis, an equation can be obtained which will enable the pump to start pumping at its normal speed. From the equation

$$D_2^2 - D_1^2 = 7156.5 \frac{H_m}{N^2}$$

or, $$D_2^2 - (0.5\, D_2)^2 = 7156.5 \frac{H_m}{N^2}$$

or, $$0.75\, D_2^2 = 7156.5 \frac{H_m}{N^2}$$

or, $$D_2^2 = \frac{7156.5}{0.75} \frac{H_m}{N^2}$$

$$D_2^2 = 9542 \frac{H_m}{N^2}$$

or, $$D_2 = 97.68 \frac{\sqrt{H_m}}{N} \tag{13.17}$$

13.6 SPECIFIC SPEED

The specific speed of a pump impeller will be the same as for the turbine runner. Thus

$$N_s = \frac{N\sqrt{P}}{H^{5/4}}$$

However, in the case of a pump there is no power output, infact the pump delivers a known volume of liquid per unit time. It would therefore, be better to express the specific speed of a pump impeller in terms of the volume of liquid delivered in liters per minute.

Now, $$P = \frac{Q}{10^3} \times \frac{10^3 \times 9.81 H\, S}{60}$$

Where

Q = Volume delivered in litres/min.

S = Specific gravity of liquid.

$$\therefore \sqrt{P} = 0.40 \times \sqrt{(S\, Q\, H)}$$

$$\therefore N_s = \frac{0.40\, N\sqrt{Q\, S}}{H^{3/4}}$$

In practice it is usual to omit 0.40 $\sqrt{S}$ and hence the specific speed of a pump impeller is expressed as follows

$$N_s = \frac{N\sqrt{Q}}{H_m^{3/4}} \tag{13.18}$$

The equation (13.18) can also be obtained by putting P α Q H_m in the specific speed expression of turbine

$$N_s = \frac{N\sqrt{Q\,H_m}}{H_m^{5/4}} = \frac{N\sqrt{Q}}{H_m^{3/4}} \tag{13.19}$$

The equation (13.18) may also be derived in the following manner.

Substituting the value of D_2 from equation (13.17) in equation (13.19) it becomes

Discharge $Q = K\,\pi\,B_2\,D_2\,V_{f2}$

or, $Q\ \alpha\ B_2\,D_2\,V_{f2}$

or, $Q\ \alpha\ D_2^2\,V_{f2}$ $\quad$ [*Since* $B_2\ \alpha\ D_2$]

Also, $V_{f1} = \psi\sqrt{2\,g\,H_m}$

or, $V_{f1}\ \alpha\ \sqrt{H_m}$

Thus $\dfrac{Q}{D_2^2\sqrt{H_m}} = constant$

Further

$$u_2 = \frac{\pi\,D_2\,N}{60}$$

or $D_2\ \alpha\ \dfrac{u_2}{N}$

Also $u_2 = K_u\sqrt{2\,g\,H_m}$

$u_2\ \alpha\ \sqrt{H_m}$

Thus $D_2\ \alpha\ \sqrt{H_m}\,/N$

or $$\frac{\sqrt{H_m}}{D_2 N} = constant \tag{13.20}$$

Substituting the value of D_2 from equation (13.20) in equation (13.19) it becomes

$$Q \; \alpha \; \frac{H_m^{3/2}}{N^2} \tag{13.21}$$

or, $$N^2 \; \alpha \; \frac{H_m^{3/2}}{Q}$$

or, $$N \; \alpha \; \frac{H_m^{3/4}}{\sqrt{Q}}$$

or, $$N = C \left[H_m^{3/4} / \sqrt{Q}\right]$$

Where C is a constant

$$\frac{N\sqrt{Q}}{H_m^{3/4}} = C \tag{13.22}$$

The specific speed of a centrifugal pump may be defined as the speed in rpm of a geometrically similar pump of such a size that under corresponding conditions it would deliver 1 liter of liquid per second against a head of 1 meter. It is represented by a N_s.

According to the definition of the specific speed, putting Q = 1 liters/s and H_m = 1 m in the equation (13.22), then C = N = N_s, thus

$$N_s = \frac{N\sqrt{Q}}{H_m^{3/4}}$$

The discharge Q is taken as 1 m³/s in the SI units.

13.7 AFFINITY LAWS FOR CENTRIFUGAL PUMPS

Most centrifugal pumps can be operated at different speeds to obtain varying capacities. Also, a given size pump casing can accommodate impeller of differing diameters. It is important to understand the manner in which capacity, head and power vary when either speed or impeller diameter is varied. These relationships, called affinity laws, are listed here.

(a) When speed varies

(i) Capacity varies directly with speed.

$$\frac{Q_1}{Q_2} = \frac{N_1}{N_2} \tag{13.23}$$

(ii) The total head capability varies with the square of the speed.

$$\frac{H_{m1}}{H_{m2}} = \left[\frac{N_1}{N_2}\right]^2 \tag{13.24}$$

(iii) The power required by the pump varies with the cube of the speed.

$$\frac{P_1}{P_2} = \left[\frac{N_1}{N_2}\right]^3 \tag{13.25}$$

(b) When impeller diameter varies

(i) Capacity varies directly with impeller diameter.

$$\frac{Q_1}{Q_2} = \frac{D_1}{D_2} \tag{13.26}$$

(ii) The total head varies with the square of the impeller diameter.

$$\frac{H_{m1}}{H_{m2}} = \left[\frac{D_1}{D_2}\right]^2 \tag{13.27}$$

(iii) The power required by the pump varies with the cube of the impeller.

$$\frac{P_1}{P_2} = \left[\frac{D_1}{D_2}\right]^3 \tag{13.28}$$

The relations $Q \alpha N, H \alpha N^2$, $P \alpha N^3$ allow the performance characteristics of a pump at one speed to be predicted from the results of experiments carried out at a different speed (Fig. 13.12).

They may thus also be used for correcting individual experimental values of Q, H and P obtained at speeds slightly different from the intended one. Since, however, they apply only when dynamic similarity is maintained, they do not describe all possible changes in performance with a change of speed. For example, Q does not necessarily change in proportion to N if H is held constant.

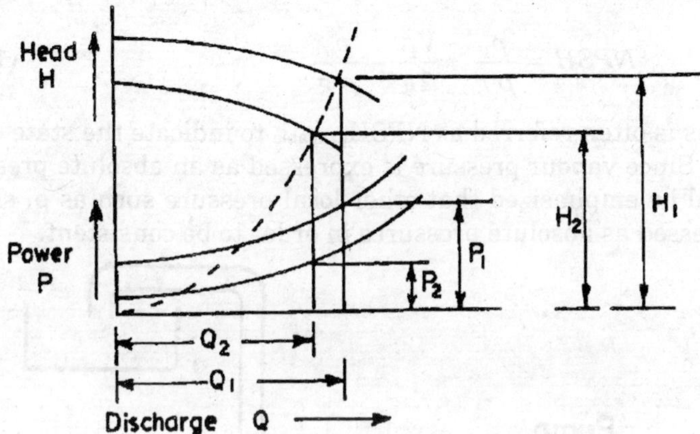

Fig. 13.12. Effect of change of specific speed in centrifugal pump.

When the range of speed is wide, the affinity laws are only approximately true because the modifying effect of Reynolds number is neglected. *It must also be remembered that cavitation can play havoc with dynamic similarly and so falsify the affinity laws.*

13.8 CAVITATION IN CENTRIFUGAL PUMPS

It is generally assumed, as a first approximation, that when the absolute pressure at some point within a liquid reaches the vapour pressure associated with the liquid temperature, cavities will form. These cavities will contain fluid vapour, free gas, or both. The phenomenon of cavity formation and collapse is known as cavitation.

Cavitation is likely to occur on the inlet side of a pump particularly if the pump is situated at a level well above the surface of the liquid in the supply reservoir. Cavitation, if severe enough, will reduce pump performance (additional mechanical energy losses), cause noisy operation, enhance vibration and erode metal from the impeller. For the sake of good efficiency and the prevention of damage to the impeller, cavitation should be avoided.

A cavitation number for centrifugal pumps may be derived in a manner similar to that for turbines.

The value of the total head at the centreline of the pump inlet (velocity head plus pressure head) minus the vapour pressure head of the liquid is called the net positive suction head (NPSH) for the pump.

$$NPSH = \frac{p_1}{\rho g} + \frac{V_1^2}{2g} - \frac{p_v}{\rho g} \tag{13.29}$$

This is often referred as $NPSH_{available}$ to indicate the state of the system. Since vapour pressure is expressed as an absolute pressure, it should be emphasized that other local pressure such as p_1 should be expressed as absolute pressures in order to be consistent.

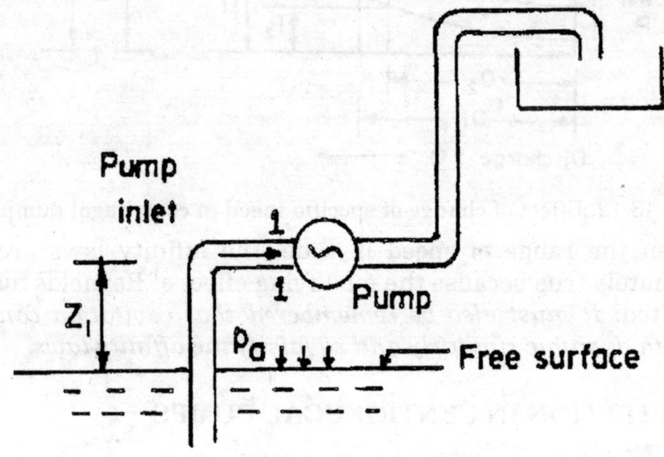

Fig. 13.13. Pump installation.

The energy equation for flow between the free surface of the lower sump in Fig. 13.13 and the pump inlet (section 1) in terms of absolute pressure is

$$\frac{p_a}{\rho g} + 0 + 0 = \frac{p_1}{\rho g} + \frac{V_1^2}{2g} + Z_1 + h_f \tag{13.30}$$

Where p_a is the atmospheric pressure and h_f is the head loss in the intake. From the above equation (13.30) and the definition of NPSH in equation (13.29).

$$NPSH = \frac{p_1}{\rho g} + \frac{V_1^2}{2g} - \frac{p_v}{\rho g}$$

$$= \frac{p_a}{\rho g} - Z_1 - h_f - \frac{p_v}{\rho g} \tag{13.31}$$

As the NPSH is decreased for a given pump and flow rate; a value will be reached for which cavitation is considered determintal, that is, the efficiency drops or the pump loses its prime. This point is called the minimum net positive suction head ($NPSH_{min}$) or $NPSH_{required}$ and is a function of the type of pump as well as the flow rate. The $NPSH_{required}$ must be measured experimentally for a particular pump, much as are the other pump characteristics. The NPSH increases with the flow rate, as indicated in Fig. 13.14 for a typical centrifugal pump. In the event that only a single value of NPSH is provided, this is the value at the rated flow conditions. If a pump is operated such that $NPSH_{available} > NPSH_{required}$, it will not cavitable determentally.

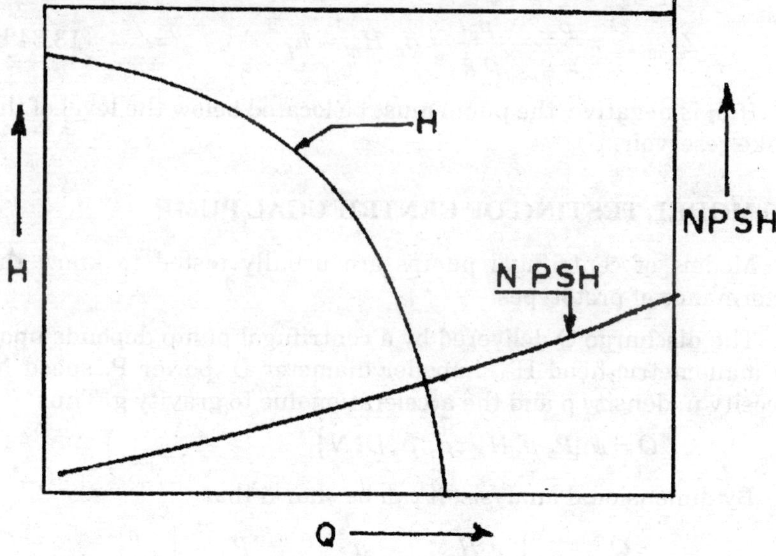

Fig. 13.14. Typical NPSH and Head Discharge Curves for a centrifugal pump.

The critical cavitation number σ_c is defined for a pump as the $NPSH_{required}$ divided by manometric head or total head.

$$\sigma_c = \frac{NPSH_{required}}{H_m} = \frac{\dfrac{p_a}{\rho g} - \dfrac{p_{\min}}{\rho g} - Z_I - h_f}{H_m} \tag{13.32}$$

The cavitation number σ for a pump is defined as the $NPSH_{available}$ divided by manometric head or total head.

$$\sigma = \frac{NPSH_{available}}{H_m} = \frac{\dfrac{p_a}{\rho g} - \dfrac{p_v}{\rho g} - Z_1 - h_f}{H_m} \qquad (13.33)$$

For cavitation not to occur $\sigma < \sigma_c$ *i.e.* $p_{min} > p_v$.

In order that σ should be as large as possible, Z_1 must be as small possible.

The maximum value of the static suction lift Z_1 to prevent cavitation is

$$Z_{1\,max} = \frac{p_a}{\rho g} - \frac{p_v}{\rho g} - NPSH_{required} - h_f \qquad (13.34a)$$

$$Z_{1\,max} = \frac{p_a}{\rho g} - \frac{p_v}{\rho g} - \sigma_c H_m - h_f \qquad (13.34\,b)$$

If Z_1 is negative, the pump must be located below the level of the intake reservoir.

13.9 MODEL TESTING OF CENTRIFUGAL PUMP

Models of centrifugal pumps are usually tested to know the performance of prototypes.

The discharge Q delivered by a centrifugal pump depends upon the manometric head H_m, impeller diameter D, power P, speed N, viscosity μ, density ρ and the acceleration due to gravity g. Thus

$$Q = \psi\,[P, \mu, H_m, g, \rho, D, N]$$

By dimensional analysis, it can be shown that

$$\frac{Q}{N D^3} = \psi\left[\frac{g H_m}{N^2 D^2}, \frac{\mu}{\rho N D^2}, \frac{P}{\rho N^3 D^5}\right]$$

Therefore,

$$\left[\frac{Q}{N D^3}\right]_m = \left[\frac{Q}{N D^3}\right]_p \qquad (13.35)$$

$$\left[\frac{g H_m}{N^2 D^2}\right]_m = \left[\frac{g H_m}{N^2 D^2}\right]_p \qquad (13.36)$$

$$\left[\frac{\mu}{\rho\, N\, D^2}\right]_m = \left[\frac{\mu}{\rho\, N^3\, D^5}\right]_p \tag{13.37}$$

$$\left[\frac{P}{\rho\, N^3\, D^5}\right]_m = \left[\frac{P}{\rho\, N^3\, D^5}\right]_p \tag{13.38}$$

The specific speed N_s and cavitation number are also to be considered for the mode analysis. Thus,

$$\left[\frac{N\sqrt{Q}}{H_m^{3/4}}\right]_m = \left[\frac{N\sqrt{Q}}{H_m^{3/4}}\right]_p \tag{13.39}$$

and $\sigma_m = \sigma_P$

$$\left[\frac{\frac{p_a}{\rho\, g} - \frac{p_v}{\rho\, g} - Z_1 - h_f}{H_m}\right]_m = \left[\frac{\frac{p_a}{\rho\, g} - \frac{p_v}{\rho\, g} - Z_1 - h_f}{H_m}\right]_p \tag{13.40}$$

13.10 MULTISTAGE CENTRIFUGAL PUMPS

A centrifugal pump is said to be multistage pump when more than one impeller are used. These impellers are arranged in series and parallel.

13.10.1 Pumps in series

When a head against which a pump has to work is too much, one pump will not prove useful, in such case a number of pumps are fixed on the same shaft in series.

The delivery pipe of the first pump (first stage) is coupled to the intake pipe of the second pump (second stage, etc.) as shown in Fig 13.15. The discharge on the Q from first pump passes to the inlet of the second pump, the discharge from second pump passes to the inlet of the third pump and so on till the discharge Q from the last pump and passes into the delivery pipe. While the liquid passes through each impeller, head H_m is impressed on it. Thus, if there are n impellers, the total head H_T developed by this multistage pump will be

$$H_T = n\ H_m \tag{13.41}$$

and $\quad Q_T = Q \qquad (13.42)$

Where n is the number of stages.

This rule (H_T = n H_m and Q_T = Q) makes it possible to plot the characteristic of the entire group of series connected pumps using characteristics of separate pumps.

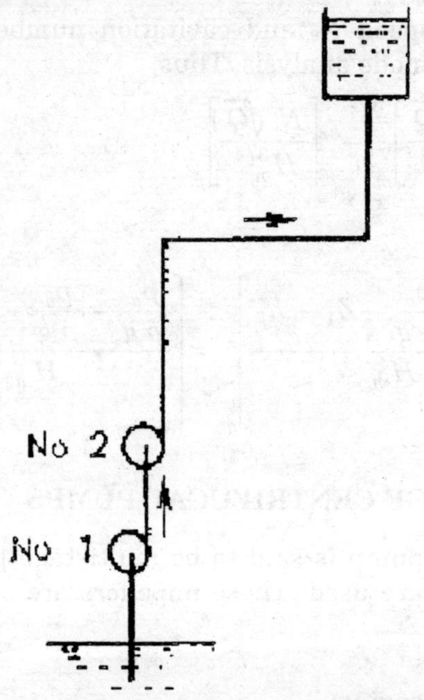

Fig. 13.15. Series connection of pumps.

Fig. 13.16 illustrates characteristic of two pumps H_1, η_2 and H_2, η_2 and total head curve H_{1+2} corresponding to series connection.

Accordingly, the average efficiency η_{av} for the entire group of pumps can be determined on the basis of capacity equation $\left(P = \dfrac{\rho g Q H}{\eta}\right)$

$$P = \frac{\rho g Q (H_1 + H_2)}{\eta_{av}}$$

$$= \frac{\rho g Q H_1}{\eta_1} + \frac{\rho g Q H_2}{\eta_2}$$

and

$$\eta_{av} = \frac{H_1 + H_2}{\dfrac{H_1}{\eta_1} + \dfrac{H_2}{\eta_2}} \quad (13.43)$$

If several identical pumps are connected in series, the efficiency does not change while total head becomes $H_T = n\ H_m$.

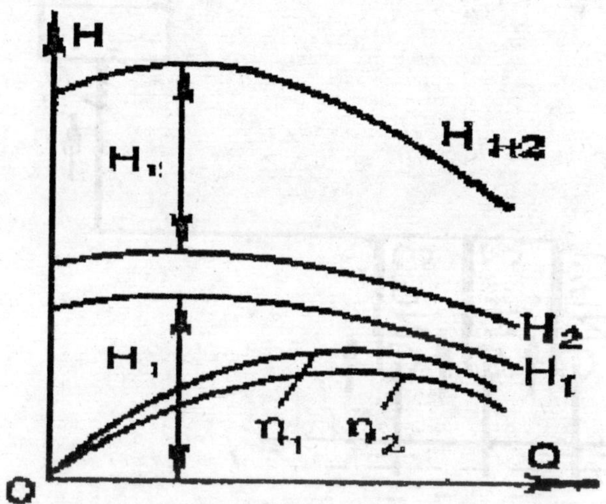

Fig. 13.16. Total characteristics of pumps at series connection.

Series connection of pumps is used when a high head is required, which cannot be attained by one pump, or when the pump unit is to have an adjustable head.

However, it should be borne in mind that housings and other parts of the pumps are designed for a particular pressure. The pump catalogues usually specify the maximum permissible pressure in the intake pipe, which determines the possibility of series connection of the pumps.

Note, that pumps of different heads can be run in series, but since the flow rates are identical the pumps should have close design deliveries i.e. they should have close dimensions.

13.10.2. Pumps in Parallel

When a high discharge is required to be pumped at a given small head H_m, the pump may be arranged in parallel. This can be done by mounting one impeller on each shaft (Fig. 13.17); the discharge from various delivery pipes will be connected in one common delivery pipe. If Q be the discharging capacity of one pump and n be the number of pumps arranged in parallel, we have

$$Q_T = n\,Q \tag{13.44}$$

and

$$H_T = H_m \tag{13.45}$$

Fig. 13.17. Parallel connection of pumps to common pipeline.

The above condition of pumps in parallel makes it possible to plot a total characteristic for the entire group of pumps operating in parallel.

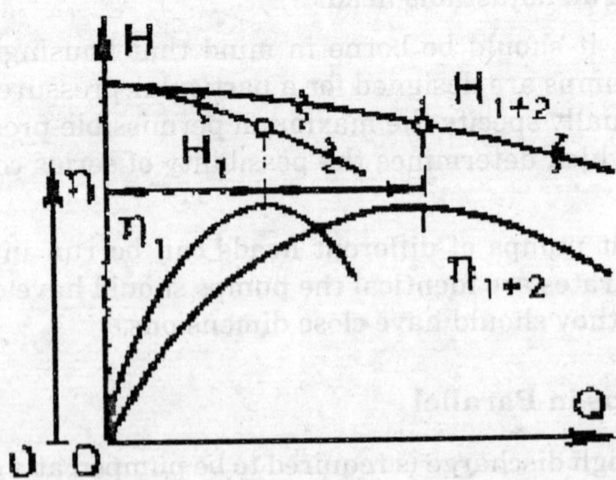

Fig. 13.18. Total characteristic of pumps at parallel connection.

Figure 13.18 illustrates a simple case when two similar pumps H_1 are connected in parallel. The total characteristic H_{1+2} is plotted by doubling the abscissas (deliveries) of one curve. The efficiency line η_{1+2} is shaped similarly. It should be noted, that the recommended area of employment for two pumps H_{1+2} is widened with respect to Q.

13.11 CHARACTERISTIC CURVES OF CENTRIFUGAL PUMPS

An important stage of pump unit designing is the choice of pumps, determination of their types, dimensions, basic parameters, required power of motors, as well as solution of problems involved in pump installation conditions which are greatly dependent on the permissible suction lift. In the course of operation, it is necessary to determine the most feasible operating modes of the pumps with respect to energy consumption, reliability, lingivity, operating conditions and costs. To solve all these problems, the designer should have fairly complete data on pump properties under different operating conditions. These data are represented in the form of pump characteristic curves.

The basic characteristics such as head, discharge, efficiency etc. depend on the shape and dimensions of the flow part (impeller) and operating conditions.

Specific relations of pump parameters versus operating conditions usually represented in the graphical from are called pump characteristics. The characteristics are given for a pump of a given type (definite shape and dimensions of the flow part) at a constant speed.

In order to predict the behaviour and performance of a pump under varying conditions, test are performed, and the results of the tests are plotted. The curves thus obtained are known as the characteristics curves of the pump. The following three types of characteristic curves are usually prepared for the centrifugal pumps:

(1) Main and operating characteristics.

(2) Constant efficiency or Muschel curves.

(3) Constant head and constant discharge curves.

13.11.1 Main and operating characteristics

In order to obtain the main characteristic curves of a pump it is operated at different speeds. For each speed the discharge Q is varied by means of a delivery valve and the different values of Q the corresponding values of Manometric head H_m shaft power P, and overall efficiency η_o, are measured or calculated. The same operation is repeated for different speeds of the pump. The H_m, P, η_o are plotted against Q for different speeds and plotted, so that three sets of curves are shown in Fig. 13.19. These curves are known as main characteristics of a pump. The main characteristics are useful in indicating the performance of a pump at different speeds.

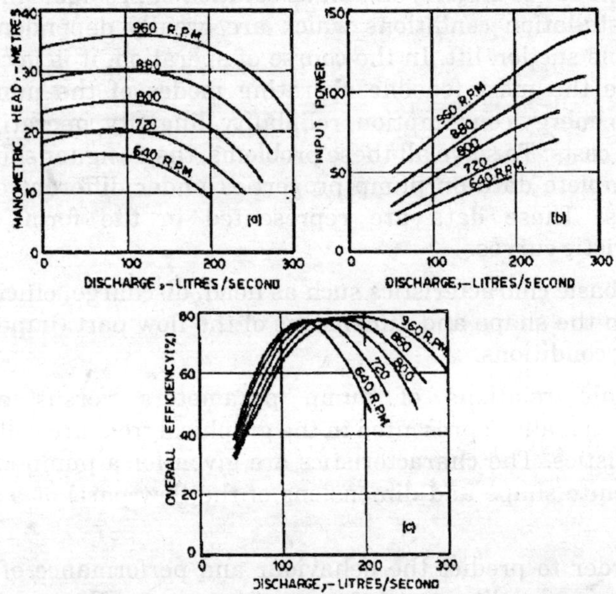

Fig. 13.19. Main characteristics of a centrifugal pump.

During the operation of a pump it is normally required to run at constant speeds, which is its designed speed (same as the speed of the driving motor). As such that particular set of main characteristics which corresponds to the design speed is mostly used in the operations of a pump and is, therefore, known as the operating characteristics. Typical characteristics curves for backward facing blades for constant speed is shown in Fig 13.20. The variation of discharge Q, with manometric head H_m and power P is shown in Fig 13.20.

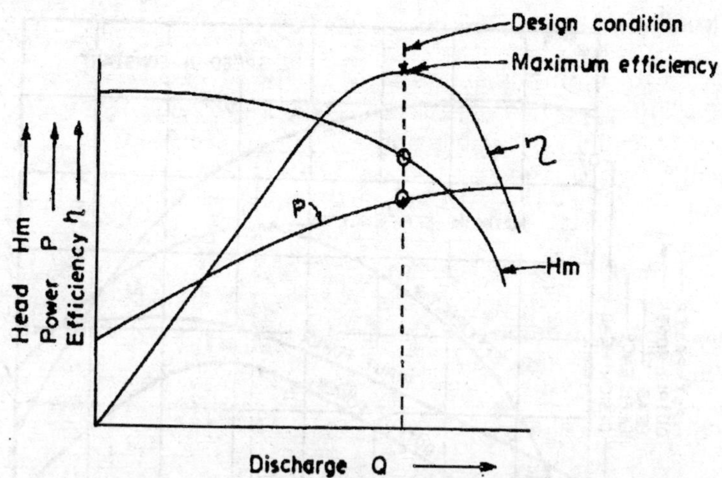

Fig. 13.20. Operating characteristics curve.

The specific speed of the centrifugal pump is always calculated at maximum efficiency. At maximum efficiency manometric head H_m and discharge Q obtained from the graph.

The head corresponding to zero or no discharge is known as the shut off head of the pump. From these characteristics it is possible to determine whether the pump will handle the necessary quantity of liquid against the desired head and what will happed if the head is increased or decreased. The P v/s Q curve will show what size of motor will be required to operate the pump at the required conditions and whether or not the motor will be overloaded under any other operating conditions.

13.11.2 Constant efficiency or Muschel curves

As shown in Fig 13.21 the iso-efficiency or constant efficiency curves may be obtained from H_m v/s Q and η_o v/s Q curves of Fig. 13.21. In order to plot the iso-efficiency curves, horizontal lines representing constant efficiencies are drawn on the η_o v/s Q curves. The points, at which these lines cut the efficiency curves at various speeds, are transferred to the corresponding H_m v/s Q curves. The points corresponding to the same efficiency are then joined by smooth curves which represent the iso-efficiency curves. From these curves the line of maximum efficiency as shown in Fig. 13.21 may be obtained. The iso-efficiency curves facilitate the direct determination of the range of operation of a pump with a particular efficiency.

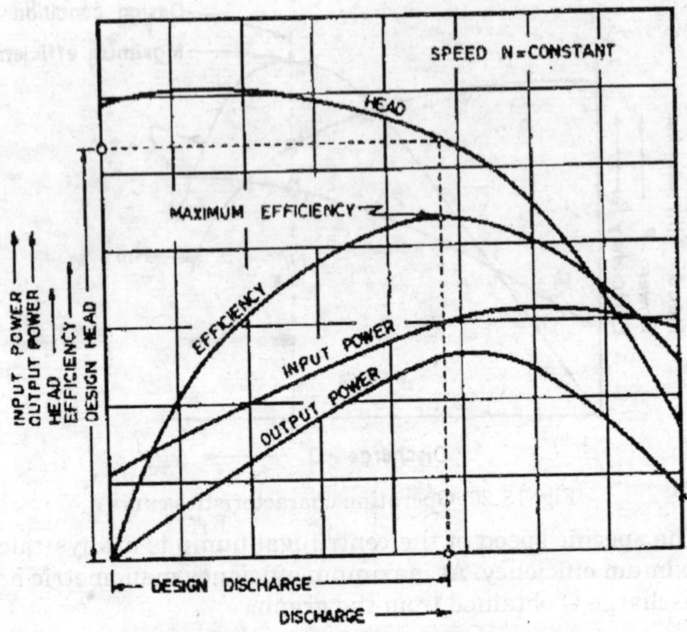

Fig. 13.21. Iso-efficiency or Muschel curves of a centrifugal pump.

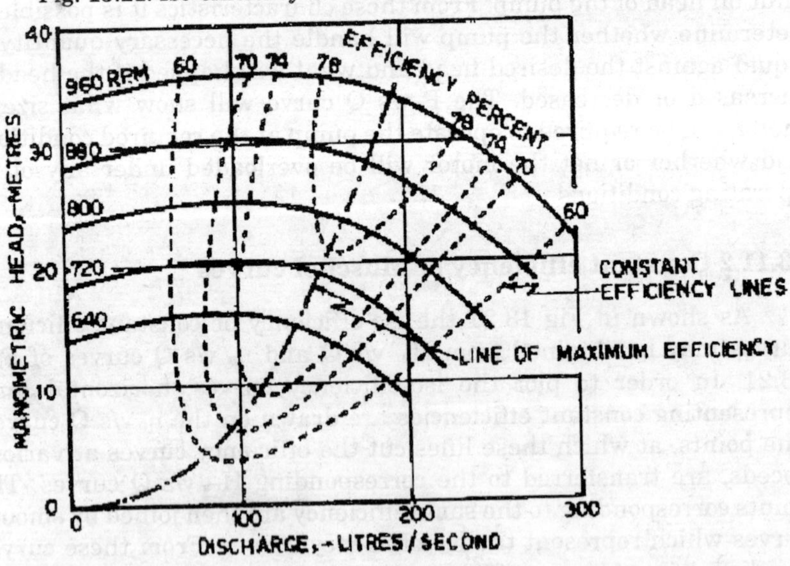

Fig. 13.22. Constant head and Constant discharge curves for variable speeds.

13.11.3 Constant head and Constant discharge curves

These curves are useful in determining the performance of a variable speed pump for which the speed constantly varies. In such cases if the head H_m is maintained constant then as the speed N varies the rate of flow Q will vary. As such a plot of Q v/s N can be prepared which can be used to determine the speeds required to discharge varying amounts of liquid at a constant pressure head. Similarly a plot of H_m v/s N and P v/s N can be prepared. The Q v/s N plot will be a straight line, H_m v/s N plot is a parabolic curve and P v/s N plot is a cubic profile (Fig 13.22).

13.12 PUMP SELECTION

Fig 13.23 shows one method for deciding what type of pump is suitable for a given services. Some general conclusions can be drawn from such a chart, but it should be emphasized that boundaries between zones are approximate. Two or more types of pumps may give satisfactory service under the same conditions. Such factors as cost, physical size, suction conditions, and the type of fluid may dictate a particular choice. In general.

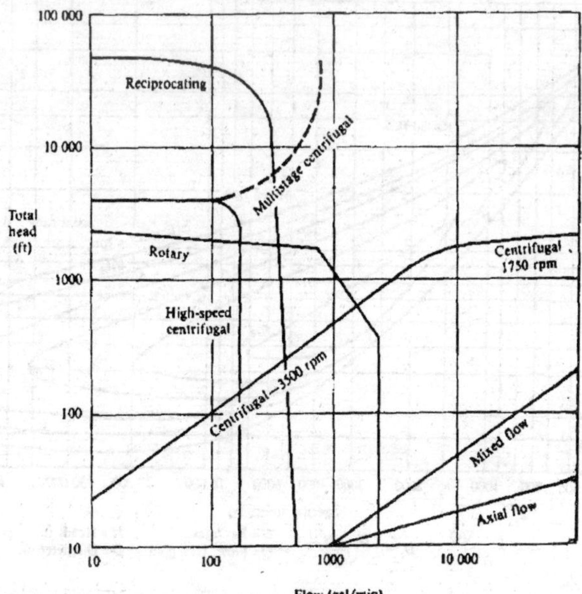

Fig. 13.23. Pump selection chart.

(1) Reciprocating pumps are used for flow rates up to about 500 gal/min (1892.5 liters/min) and from very low head to as high as 50000 ft (15240 m) of head.

(2) Centrifugal pumps are used over a wide range of conditions, mostly in high capacity, moderate head application.

(3) Single stage centrifugal pumps operating at 3500 rpm are economical at lower flow rates and moderate heads.

(4) Multistage pumps are desirable at high heads conditions.

(5) Rotary pumps (e.g., gear, vane, etc.) are used in application requiring moderate capacities and high heads or for fluids with high viscosities.

(6) Special high speed centrifugal pumps operating well above the 3500 rpm speed of standard electric motors are desirable for high heads and moderate capacities. Such pumps are sometimes driven by steam turbines or gas turbines.

(7) Mixed flow and axial flow pumps are used for very high flow rates and low heads.

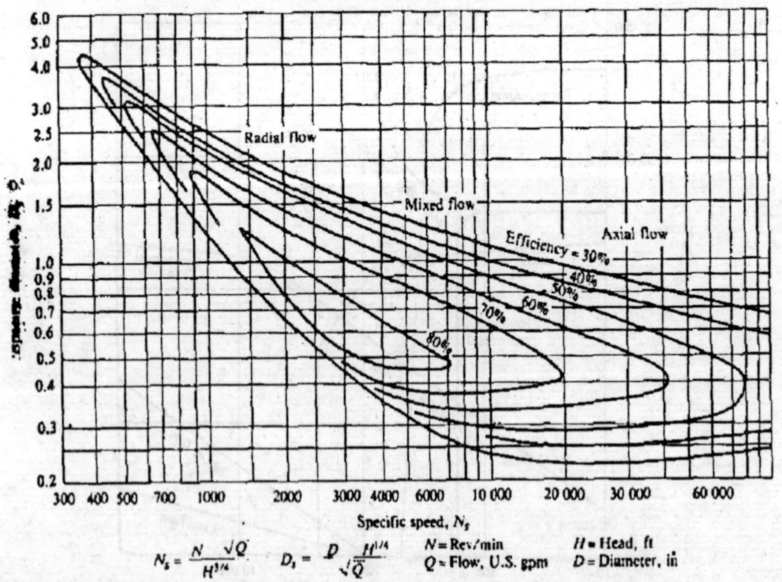

Fig. 13.24. Specific speed v/s specific diameter for centrifugal pump an aid to pump selection.

Another parameter that is useful in selecting the type of pump for a given application is the specific speed.

The specific speed is often combined with the specific diameter to produce a chart like that shown in Fig 13.24. The specific diameter is

$$D_s = \frac{D\,H^{1/4}}{\sqrt{D}}$$

where D_s is the impeller diameter in inches. Q in gal/min and H is in ft.

From Fig 13.24 we can see that radial flow centrifugal pumps are recommended for specific speeds from about 400 to 4000. Mixed flow pumps are used from 4000 to about 7000. Axial flow pumps are used from 7000 to over 60000.

ILLUSTRATIVE EXAMPLES

13.1. During a test on a centrifugal pump, the following readings were obtained.

Pressure gauge reading	= 1.32 bar
Vacuum gauge reading	= 300 mm of mercury
Effective height between gauge	= 0.45 m
Power of electric motor	= 22 kW
Discharge of pump	= 85 liters/s.
Diameter of delivery pipe	= 150 mm
Diameter of suction pipe	= 200 mm

Determine the overall efficiency of the pump.

Solution

$$\frac{P_s}{\rho\,g} = -\,300\ mm\ of\ mercury$$

$$= -\,0.3 \times 13.6 = -\,4.08\ m\ of\ water$$

$$\frac{P_d}{\rho\,g} = 1.32 \times \frac{100}{9.81} = 13.45\ m\ of\ water$$

$$Z_d = 0.45\ m,\ Z_s = 0$$

$$A_s = \frac{\pi}{4} D_s^2 = \frac{\pi}{4} (0.2)^2 = 0.0314\ m^2$$

$$A_d = \frac{\pi}{4} D_s^2 = \frac{\pi}{4} (0.15)^2 = 0.01767\ m^2$$

$$Q = A_s V_s = A_d V_d$$

$$V_s = \frac{Q}{A_s} = \frac{0.085}{0.0314} = 2.7056\ m/s$$

$$V_d = \frac{Q}{A_d} = \frac{0.085}{0.01767} = 4.81\ m/s$$

$$H_m = \left[\frac{P_d}{\rho g} + \frac{V_d^2}{2g} + Z_d\right] - \left[\frac{P_s}{\rho g} + \frac{V_s^2}{2g} + Z_s\right]$$

$$= \left[13.45 + \frac{(4.81)^2}{2 \times 9.81} + 0.46\right] - \left[-4.08 + \frac{(2.7056)^2}{2 \times 9.81} + 0\right]$$

$$= (13.45 + 1.1792 + 0.46) - (-4.08 + 0.373)$$

$$= 15.0892 - (-3.70689)$$

$$= 18.796\ m$$

$$P = \frac{\rho g Q H_m}{\eta_o}$$

$$\eta_o = \frac{\rho g Q H_m}{P} = \frac{1000 \times 9.81 \times 0.085 \times 18.796}{22 \times 1000}$$

$$= 0.7124 = 71.24\%$$ **Ans.**

13.2 A constant speed test of a centrifugal pump resulted in the law

$$H = 43.5 + 258\ Q - 3770\ Q^2$$

Where H_m is the total head generated in metre and Q is the rate of flow in m^3/s. The pump is to be used to deliver water through a pipe line 800 m long and 300 mm diameter, the static lift being 24.5 m. Neglecting velocity heads and taking a pipe friction coefficient $\lambda = 4\,f = 0.03$, calculate the operating head and discharge rate of the pump. Also estimate the power required to drive the pump assuming an overall efficiency of 70 percent for the particular Hm v/s Q point at which the pump operates.

Solution

$$h_f = \frac{\lambda\, l_d\, V_d^2}{2\, g\, d_d} = \frac{\lambda\, l_d\, Q^2}{12.1\, d_d^5} = \frac{0.03 \times 800 \times Q^2}{12.1 \times (0.3)^5}$$

$$= 816.24\ Q^2$$

$$H_m = H_{static} + h_f = 24.5\ \_+ 816.24\ Q^2$$

or, $43.5 + 258\,Q - 3777\,Q^2 = 24.5 + 816.24\ Q^2$

$$4586.24\,Q^2 - 258Q - 19 = 0$$

$$Q = 258 \pm \frac{\sqrt{(258)^2 - (4 \times 4586.24 \times (-19))}}{2 \times 4586.24}$$

Taking positive sign

$$Q = 0.09837\ m^3/s$$

$$\therefore\ H_m = 43.5 + 258\,(0.09837) - 3770\,(0.09837)^2$$

$$= 43.5 + 25.379 - 36.48$$

$$= 68.879 - 36.48 = 32.398\ m$$

$$P = \frac{\rho\, g\, Q\, H_m}{\eta_o} = \frac{1000 \times 9.81 \times 32.398 \times 0.0983}{0.7}$$

$$= 44.63\ kW$$ Ans.

Alter : This problem can be solved very easily with the help of graphical method.

13.3. A centrifugal pump delivers 0.1 m3/s of water at a rotational speed of 1200 rpm. The impeller vanes which lean back wards to the direction of rotation such that the vane tip angle is 50°. The impeller has an external diameter of 0.4 m, an internal diameter of 0.2 m and an axial width of 31.7 mm. Assuming that the diffuser efficiency is 55%, that the impeller head losses are 10% of the ideal head rise and that the diffuser exit is 0.15 m in diameter, estimate the manometric head and the hydraulic efficiency.

Solution

$$Q = \pi\ D_2\ B_2\ V_{f2}$$

$$0.1 = \pi \times 0.4 \times 0.0317 \times V_{f2}$$

$$\therefore V_{f2} = 2.51\ m/s$$

$$u_2 = \frac{\pi\ D_2\ N}{60} = \frac{\pi \times 0.4 \times 1200}{60} = 25.133\ m/s$$

From outlet velocity triangle

$$\tan \phi = \frac{V_{f2}}{u_2 - V_{w2}}$$

$$\tan 50^\circ = \frac{2.51}{25.133 - V_{w2}}$$

$$\therefore V_{w2} = 23.02686\ m/s$$

Total loss of head in the impeller (H_L)

= Impeller head loss + Loss of head in diffuser + Exit loss

$$=0.1\frac{V_{w2}\ u_2}{g} + (1-\eta_d)\left[\frac{V_2^2 - V_3^2}{2\,g}\right] + \frac{V_3^2}{2g}$$

$$V_2 = \sqrt{V_{w2}^2 + V_{f2}^2} = \sqrt{(23.02686)^2 + (2.51)^2} = 23.16325\ m/s$$

$$V_3 = \frac{Q}{A_d} = \frac{0.1}{\frac{\pi}{4}(0.15)^2} = \frac{0.1}{0.01767} = 5.658\ m/s$$

$$\therefore H_L = \frac{0.1 \times 23.02686 \times 25.133}{9.81} + (1.055)\left[\frac{(23.16325)^2 - (5.658)^2}{2 \times 9.81}\right]$$

$$+\frac{(5.658)^2}{2 \times 9.81}$$

$$= 5.89943 + 0.45 \times 25.7147 + 1.63$$

$$= 5.89943 + 11.57 + 16.3 = 19.1\ m$$

$$\therefore H_m = \frac{V_{w2}\ u_2}{g} - H_L = \frac{23.02686 \times 25.133}{9.81} - 19.1$$

$$= 58.99 - 19.1 = 39.89\ m \qquad \text{Ans.}$$

$$\eta_{mano} = \frac{g\,H_m}{V_{w2}\,u_2} = \frac{9.81 \times 39.89}{23.02686 \times 25.133} = 0.6762 = 67.62\% \qquad \text{Ans.}$$

13.4. The impeller of a centrifugal pump has an outer diameter of 250 mm and an effective outlet area of 17000 mm². The outlet blade angle is 32°. The diameters of suction and discharge openings are 150 mm and 125 mm respectively. At 24.2 rps and discharge 0.03 m³/s the pressure heads at suction and discharge openings were respectively 4.5 m below and 13.3 m above atmospheric pressure, the measurement points being at the same level. The shaft power was 7.76 kW. Water enters the impeller without shock or whirl. Assume that the true outlet whirl component is 70% of the ideal, determine the overall efficiency and the manometric efficiency based on the true whirl component.

Solution:

$$A_s = \frac{\pi}{4} D_s^2 = \frac{\pi}{4}(0.15)^2 = 0.01767\ m^2$$

$$A_d = \frac{\pi}{4} D_d^2 = \frac{\pi}{4}(0.125)^2 = 0.01227\ m^2$$

$$Q = A_s\,V_s = A_d\,V_d$$

$$\therefore V_s = \frac{Q}{A_s} = \frac{0.03}{0.01767} = 1.697\ m/s$$

and $$V_d = \frac{Q}{A_d} = \frac{0.3}{0.01227} = 2.445\ m/s$$

$$H_m = \left[\frac{P_d}{\rho g} + \frac{V_d^2}{2g} + Z_d\right] - \left[\frac{P_s}{\rho g} + \frac{V_s^2}{2g} + Z_s\right]$$

$$Z_s = Z_d, \quad \frac{P_d}{\rho g} = 13.3\ m, \quad \frac{P_s}{\rho g} = -4.5\ m, \quad \frac{V_d^2}{2g} = \frac{(2.445)^2}{2\,g} = 0.3\ m$$

and $$\frac{V_s^2}{2g} = \frac{(1.697)^2}{2\,g} = 0.1467\ m$$

$$\therefore H_m = [13.3 + 0.3] - [-4.5 + 0.1467] = 17.95\ m$$

$$u_2 = \frac{\pi D_2 N}{60} = \pi \times 0.25 \times 24.2 = 19\ m/s$$

$$Q = A\ V_{f2}$$

$$0.03 = 17000 \times 10^{-6}\ V_{f2}$$

$$\therefore V_{f2} = 1.7646\ m/s$$

From outlet velocity triangle

$$\tan 32^\circ = \frac{V_{f2}}{u_2 - V_{w2}}$$

$$0.6248 = \frac{V_{f2}}{u_2 - V_{w2}}$$

$$\therefore V_{w2} = 16.1759\ m/s$$

True whirl velocity $= 0.7 \times V_{w2} = 0.7 \times 16.1759 = 11.32\ m/s$

$$\eta_h = \eta_{mano} = \frac{g\ H_m}{V_{w2}\ u_2} = \frac{9.81 \times 17.95}{11.32 \times 19} = 0.818 = 81.8\% \qquad \text{Ans.}$$

$$P = \frac{\rho\ g\ Q\ H_m}{\eta_o}$$

$$\therefore \eta_0 = \frac{1000 \times 9.81 \times 0.03 \times 17.95}{7.76 \times 1000} = 0.68075 = 68.075\% \qquad \text{Ans.}$$

13.5. A centrifugal pumo delivers 12 m³/min at 1200 rpm with a manometric efficiency of 75%. The impeller is 300 mm diameter with a width at exit of 75 mm. The vane occupy 12% of the periphery and are swept backwards. making an angle of 40° with the tangent at the outer periphery. Calculate the fraction of the kinetic energy of discharge from the impeller which is subsequently recovered in the casing assuming no loss of head in the impeller. What is the manometric efficiency if recovery of energy in casing is nil?

Solution:

$$u_2 = \frac{\pi D_2 N}{60} = \frac{\pi \times 0.3 \times 1200}{60} = 18.85\ m/s$$

$$Q = K\ A\ V_{f2}$$

$$\frac{12}{60} = 0.88 \times \pi \times 0.3 \times 0.075 \times V_{f2}$$

$$\therefore V_{f2} = 3.215\ m/s$$

From outlet velocity triangles

$$\tan 40^\circ = \frac{V_{f2}}{u_2 - V_{w2}}$$

$$0.839 = \frac{3.215}{18.85 - V_{w2}}$$

$$\therefore V_{w2} = 15\ m/s$$

$$\eta_{mano} = \frac{g\,H_m}{V_{w2}\,u_2}$$

$$\therefore H_m = \frac{\eta_{man.}\,V_{w2}\,u_2}{g} = \frac{0.75 \times 15 \times 18.85}{9.81} = 21.6169\ m$$

$$H_m = \frac{V_{w2}\,u_2}{g} - \text{Hydraulic losses in pump}$$

$$21.6169 = \frac{15 \times 18.85}{9.81} - \text{Hydraulic losses in pump.}$$

$$\therefore \text{Hydraulic losses in pump} = 28.82 - 21.6169 = 7.2\ m$$

$$V_2 = \sqrt{(V_{f2})^2 + (V_{w2})^2} = \sqrt{(3.215)^2 + (15)^2}$$

$$= \sqrt{10.336 + 225} = \sqrt{235.336}$$

$$= 15.34\ m/s$$

$$\text{K.E. at outlet} = \frac{V_2^2}{2g} = \frac{(15.34)^2}{2 \times 9.81} \approx 12$$

$\therefore$ K.E. recovered or pressure rise in diffuser

= K.E. at outlet – Hydraulic losses in pump

$= 12 - 7.2 = 4.8$

Therefore, $\dfrac{\text{K.E. recovered}}{\text{K.E. at outlet}} = \dfrac{4.8}{12} = 0.4 = 40\%$ Ans.

$$H_m = \frac{V_{w2}\,u_2}{g} - K.E.\ at\ outlet = 28.82 - 12 = 16.82\ m$$

$$\eta_m = \frac{g\,H_m}{V_{w2}\,u_2} = \frac{9.81 \times 16.82}{15 \times 18.85}$$

$= 0.5835 = 58.35\%$ Ans.

13.6. A single stage centrifugal pump is to be used to pump water through a vertical distance of 30 m at the rate of 45 liters/s. Suction and delivery pipes will have a combined length of 36 m and a friction factor λ = 4f = 0.024. Both will be 150 mm diameter. Losses at valves etc. are estimated to total 2.4 times the velocity head in the pipes. The basic design of pump has a dimensionless specific speed of 0.074 rev, forward curved impeller blades with an outlet angle of 125° to the tangent and a width of impeller passages at outlet equal to one tenth of the diameter. The blades themselves occupy 5% of he circumference. If a manometric efficiency (neglecting whirl slip) of 75% may be expected, determine a suitable impeller diameter.

Solution :

Given datas

$$Q = 45 \times 10^{-3}\ m^3/s,\ K_n = 0.074\ rev,\ \eta_{man} = 0.75$$

$$l_s + l_d = 36\ m, \phi = 180 - 125 = 55°,\ B_2 = \frac{D_2}{10}$$

$$K = 0.95, \lambda = 4f = 0.024,\ D_s = 0.15\ m$$

$$Q = A\,V$$

$$45 \times 10^{-3} = \frac{\pi}{4}(0.15)^2 \times V$$

$$V = 2.546\ m/s = V_s = V_d$$

$$h_f = \frac{\lambda\,l\,V^2}{2\,g\,D} = \frac{0.024 \times 36 \times (2.546)^2}{2 \times 9.81 \times 0.15} = 1.9\ m$$

Head losses at valves etc.

$$h_{lv} = \frac{2.4\,V^2}{2\,g} = \frac{2.4 \times (2.546)^2}{2 \times 9.81} = 0.7929\,m$$

Manometric head

$$H_m = H_{static} + \sum losses$$

$$= H_{static} + h_f + h_{lv}$$

$$= 30 + 1.9 + 0.7929$$

$$= 32.6929\,m$$

$$= K_n = \frac{N\sqrt{Q}}{(g\,H)^{3/4}}$$

$$0.074 = \frac{N\sqrt{45 \times 10^{-3}}}{(9.81 \times 32.6929)^{3/4}}$$

$$0.074 = 2.799 \times 10^{-3}\,N$$

$$N = 26.437\,rps = 1586.238\,rpm$$

$$u_2 = \frac{\pi\,D\,N}{60} = \frac{\pi\,D \times 1586.238}{60} = 83.055\,D$$

$$V_{f2} = \frac{Q}{\pi\,D_2\,B_2\,K} = \frac{0.045}{\pi \times D \times \frac{D}{10} \times 0.95} = \frac{0.1507}{D^2}$$

$$\tan 55° = \frac{V_{f2}}{V_{w2} - u_2}$$

$$1.428 = \frac{\frac{0.1507}{D^2}}{V_{w2} - 83.055\,D}$$

$$V_{w2} - 83.055\,D = \frac{0.10552}{D^2}$$

$$V_{w2} = \frac{0.10552}{D^2} + 83.055\,D$$

$$\eta_{man} = \frac{g\,H_m}{V_{w2}\,u_2}$$

$$0.75 = \frac{9.81 \times 32.6929}{(83.055\,D)\left[\frac{0.10552}{D^2} + 83.055\,D\right]}$$

$$\frac{8.764}{D} + 6898.13\,D^2 = 427.623$$

$$\therefore 8.764 + 6898.13\,D^3 = 27.623D$$

$$\text{or,}\quad 6898.13\,D^3 - 427.623\,D + 8.764 = 0$$

$$\text{or,}\quad 787.098\,D^3 - 48.793\,D + 1 = 0$$

Solving by trial and error method

$D = 0.238\ m$ Ans.

13.7. Prove that the manometric efficiency of a centrifugal pump, when all the kinetic energy at exit from the runner is lost, is given by

$$\eta_{mano} = \frac{u_2^2 - V_{f2}^2 \operatorname{cosec}^2 \phi}{2\,u_2\,(u_2 - V_2 \cot \phi)}$$

State the assumptions made in the above analysis and explain why the angle φ is rarely less than 20°.

Solution

$$H_{mano} = \frac{V_{w2}\,u_2}{g} - \frac{V_2^2}{2\,g}$$

From the outlet velocity triangle

$$V_2^2 = V_{w2}^2 + V_{f2}^2 = (v_2 - V_{f2} \cot \phi)^2 + V_{f2}^2$$

and $V_{w2} = u_2 - V_{f2} \cot \phi$

Therefore,

$$H_{mano} = \frac{(u_2 - V_{f2} \cot \Phi)u_2}{g} - \frac{(u_2 - V_{f2} \cot \Phi)^2 + V_{f2}^2}{2\,g}$$

$$= \frac{2\,u_2^2 - 2\,u_2\,V_{f2} \cot \Phi - u_2^2 - V_{f2}^2 \cot^2 \Phi + 2\,u_2\,V_{f2} \cot \Phi - V_{f2}^2}{2g}$$

$$= \frac{v_2^2 - V_{f2}^2 \left(1 + \cot^2 \Phi\right)}{2g}$$

$$= \frac{u_2^2 - V_{f2}^2 \operatorname{cosec}^2 \Phi}{2g} \qquad (1)$$

$$\eta_{mano} = \frac{g\, H_m}{V_{w2}\, u_2}$$

$$= \frac{g \left(u_2^2 - V_{f2}^2 \operatorname{cosec}^2 \Phi\right)}{2\, g\, u_2 \left(u_2 - V_{f2} \cot \Phi\right)}$$

$$= \frac{u_2^2 - V_{f2}^2 \operatorname{cosec}^2 \Phi}{2\, u_2 \left(u_2 - V_{f2} \cot \Phi\right)} \qquad (2)$$

Proved

Now assuming a constant value proved for the flow ratio $\psi = 0.24$ and computing the value of u_2 in terms of H_m from equation (1) for different values of ϕ, it will be observed that as the value of ϕ varies from 90° to 20°, the value of η_{mano} increase from 0.47 to 0.73. It is obvious from equation (2) that a further decrease in the value of ϕ will result in the increase in the efficiency, but it is impracticable to have the values of ϕ less that 20°, since it will result in the long and narrow passages causing high frictional losses. As such in practice the value of ϕ is not reduced below 20°, but some other devices in the form of guide vanes etc are adopted which will convert a part of the velocity energy into the useful pressure energy, thereby increasing the efficiency of the pump.

13.8. Prove that the pressure rise in the impeller of a centrifugal pump is given by

$$\frac{V_{f1}^2 + u_2^2 - V_{f2}^2 \operatorname{cosec}^2 \phi}{2\,g}$$

The symbols has the usual meanings.

Solution

Applying Bernoullis theorem at the inlet and outlet of the impeller and neglecting losses from inlet to outlet.

$$\frac{p_1}{\rho g} + \frac{V_1^2}{2g} + Z_1 = \frac{p_2}{\rho g} + \frac{V_2^2}{2g} + Z_2 - \frac{V_{w2}\,u_2}{g}$$

If $\quad Z_1 = Z_2$, then

$$\frac{p_1}{\rho g} + \frac{V_1^2}{2g} = \frac{p_2}{\rho g} + \frac{V_2^2}{2g} - \frac{V_{w2}\,u_2}{g}$$

or, $\quad \dfrac{p_2 - p_1}{\rho g}$ = pressure rise in the impeller

$$= \frac{V_1^2}{2g} - \frac{V_2^2}{2g} + \frac{V_{w2}\,u_2}{g} \qquad (1)$$

From inlet velocity triangle

$$V_1 = V_{f1} \qquad (2)$$

From the outlet velocity triangle

$$\tan\varphi = \frac{V_{f2}}{u_2 - V_{w2}}$$

$$V_{w2} = u_2 - \frac{V_{f2}}{\tan\varphi} = u_2 - V_{f2}\cot\varphi \qquad (3)$$

Also, $V_2^2 = V_{f2}^2 + V_{w2}^2$

$$= V_{f2}^2 + (u_2 - V_{f2}\cot\varphi)^2$$

$$= V_{f2}^2 + u_2^2 + V_{f2}^2 \cot\varphi - 2\,u_2 \cot\varphi$$

$$= V_{f2}^2\left(1 + \cot^2\varphi\right) + u_2^2 - 2\,u_2\,V_{f2}\cot\varphi$$

$$= V_{f2}^2\,\mathrm{cosec}^2\varphi + u_2^2 - 2\,u_2\,V_{f2}\cot\varphi \qquad (4)$$

Substituting the values of V_1, V_{w2}, V_2^2 from equation (2), (3) and (4) in equation (1), we get

Pressure rise

$$= \frac{V_{f1}^2}{2g} - \frac{1}{2g}\left[V_{f2}^2\,\mathrm{cosec}^2\varphi + u_2^2\ \hat{u}\ 2u_2\,V_{f2}\cot\varphi\right] + \frac{1}{g}\left(u_2 - V_{f2}\cot\varphi\right)u_2$$

$$= \frac{1}{2g}\left[V_{f1}^2 - V_{f2}^2\,\mathrm{cosec}^2\varphi - u_2^2 + 2\,u_2\,V_{f2}\cot\varphi + 2\,u_2^2 - 2\,u_2\,V_{f2}\cot\varphi\right]$$

$$= \frac{1}{2g}\left[V_{f1}^2 - V_{f2}^2 \operatorname{cosec}^2 \varphi + u_2^2\right]$$

$$= \frac{V_{f1}^2 + u_2^2 - V_{f1}^2 \operatorname{cosec}^2 \varphi}{2g}$$ Proved.

13.9. A centrifugal pump running at 1450 rpm has the characteristic as given below

Q (liters/s)	11.3	16.9	22.6	28.3	34	39.6	45.2
H_m (meters)	25.8	25	24.1	23.2	21.4	18.9	15.8
η_o (percent)	65	70	73	74	72	69	62

Draw the operating characteristic of the pump and determine its specific speed (rpm m³/s, m). If the pump is driven by a motor, determine the power of the motor when operating at maximum efficiency.

Solution:

From the plot (Fig. 13.25), maximum efficiency $\eta_o = 74.1\%$ and the corresponding $H_m = 23.2\ m$ and $Q = 28.4\ litres/s = 0.0284\ m^3/s$

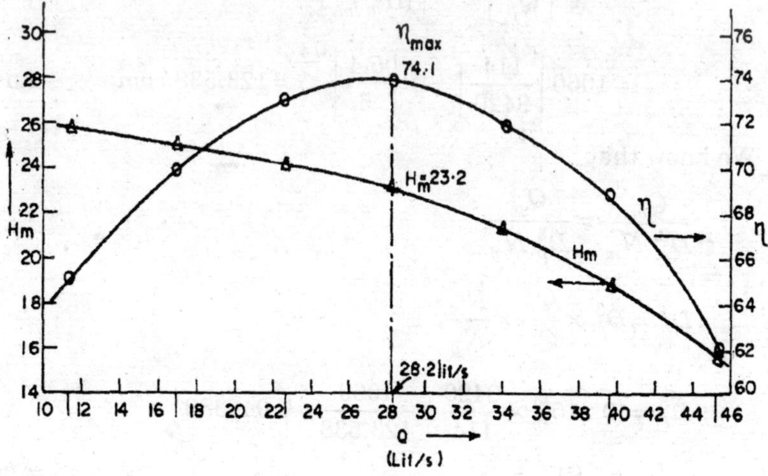

Fig. 13.25. Characteristic curve. P. 13.9.

$$\therefore N_s = \frac{N\sqrt{Q}}{H_m^{3/4}} = 23.116$$ Ans.

$$P = \rho\, g\, Q\, H_m / \eta_o = 1000 \times 9.81 \times 0.0284 \times 23.2/0.741$$

$$= 8.7228\ kW \qquad \text{Ans.}$$

13.10. A centrifugal pump model, having impeller diameter 215 mm. When tested at a speed of 1060 rpm, gave its best efficiency while discharging 114 liters/s at a head of 5 m, the shaft power 6.8 kW. The prototype is required to deliver 9420 liters/s at a head of 5.4 m. Find :

(a) The speed of prototype.

(b) The impeller diameter of prototype.

(c) Shaft power required to drive the prototype.

Solution

$$(N_s)_{model} = (N_s)_{prototype}$$

$$\frac{N_m \sqrt{Q_m}}{(H_m)_m^{3/4}} = \frac{N \sqrt{Q_p}}{(H_m)_P^{3/4}}$$

$$\therefore N_p = N_m \left[\frac{Q_m}{Q_p}\right]^{1/2} \times \left[\frac{(H_m)_p}{(H_m)_m}\right]^{3/4}$$

$$= 1060 \left[\frac{114}{9420}\right]^{1/2} \times \left[\frac{5.4}{5}\right]^{3/4} = 123.538\ rpm \qquad \text{Ans.}$$

We know that

$$\frac{Q_m}{D_m^3 N_m} = \frac{Q_p}{D_p^3 N_p}$$

$$D_p^3 = D_m^3 \times \frac{Q_p}{Q_m} \times \frac{N_m}{N_p}$$

$$D_P^3 = (0.215)^3 \times \frac{9420}{114} \times \frac{1060}{123.538} = 7.046389$$

$$\therefore D_p = 1.917\ m \qquad \text{Ans.}$$

We know that

$$\frac{P_m}{D_m^5\,N_m^3} = \frac{P_p}{D_P^5\,N_p^3}$$

$$\therefore P_p = P_m \times \left[\frac{D_p}{D_m}\right]^5 \times \left[\frac{N_p}{N_m}\right]^3$$

$$= 6.8 \times \left[\frac{1.917}{0.215}\right]^5 \left[\frac{123.538}{1060}\right]^3 = 606.846\ kW \qquad \text{Ans.}$$

13.11. **A centrifugal pump is to be designed for a capacity of 5660 liters/s, a head of 132 m and a speed of 450 rpm. Model tests are to be made in a laboratory where the maximum flow available is 200 liters/s. A 200 kW motor is available, if the overall efficiency of model is assumed to be 89.2%, find : (a) The speed of the model. (b) The linear scale ratio.**

Solution

For model

$$P = \frac{\rho\, g\, Q\, H_m}{\eta_o}$$

$$220 \times 10^3 = \frac{1000 \times 9.81 \times 0.2 \times H_m}{0.892}$$

$$\therefore H_m = 100\ m$$

For Prototype

$$N_s = \frac{N_p\,\sqrt{Q_p}}{(H_m)_p^{3/4}}$$

$$= \frac{450\sqrt{5.66}}{(132)^{5/4}} = 27.49$$

We known that

$$(N_s)_m = (N_s)_p$$

$$\frac{N_m\,\sqrt{Q_m}}{(H_m)_m^{3/4}} = 27.49$$

$$\therefore N_m = \frac{27.49 \times (100)^{3/4}}{\sqrt{0.2}} = 1943.9\ rpm \qquad \text{Ans.}$$

We know that

$$\frac{\sqrt{(H_m)_m}}{D_m\ N_m} \quad \frac{\sqrt{(H_m)_p}}{D_p\ N_p}$$

$$\therefore \frac{D_m}{D_p} = \frac{N_p}{N_m}\sqrt{\frac{(H_m)_m}{(H_m)_p}} = \frac{450}{1943.9}\sqrt{\frac{100}{132}} = 0.20 = \frac{1}{5}$$

$$\therefore D_m : D_p = 1 : 5 \qquad \text{Ans.}$$

13.12. A large centrifugal pump is to have a dimensionless specific speed of 0.183 rev and is to discharge 2 m³ of liquid per second against a total head of 15 m. The kinematic viscosity of the liquid may very between 3 and 6 times that of water. Determine the range of speeds and test heads for a one quarter scale model investigation of the full size pump, the model using water.

Solution

For Prototype

$$K_n = \frac{N\sqrt{Q}}{(g\ H)^{3/4}}$$

$$0.183 = \frac{N\sqrt{2}}{(9.81 \times 15)^{3/4}}$$

$$\therefore N = 5.467\ rps$$

$$\therefore N_p = 328.025\ rpm$$

We know that

$$\left[\frac{v}{N\ D^2}\right]_{model} = \left[\frac{v}{N\ D^2}\right]_{prototype}$$

$$N_m = \frac{v_m}{v_p} \times \left[\frac{D_p}{D_m}\right]^2 \times N_p = \frac{1}{3} \times (4)^2 \times 5.467 = 29.157\ rps$$

For the ratio $= \frac{v_p}{v_m} = 6$

Then, $N_m = \frac{1}{6} \times (4)^2 \times 5.467 = 14.5786 \; rps$

$\therefore$ Range of speed $= 14.5786 \; to \; 29.157 \; rps$

We know that

$$\frac{H_m}{D_m^2 N_m^2} = \frac{H_p}{D_p^2 N_p^2}$$

Case (a) $N_m = 14:5786 \; rps$

$$\therefore H_m = \left[\frac{D_m}{D_p}\right]^2 \times \left[\frac{N_m}{N_p}\right]^2 \times H_p$$

$$= \left[\frac{1}{4}\right]^2 \times \left[\frac{14.5786}{5.467}\right]^2 \times 15 = 6.67 \; m$$

Case (b) $N_m = 29.157 \; rps$

$$\therefore H_m = \left[\frac{1}{4}\right]^2 \times \left[\frac{29.157}{5.467}\right]^2 \times 15 = 26.667 \; m \qquad \text{Ans.}$$

$\therefore$ Range of head $= 6.67 \; to \; 26.667 \; m$ **Ans.**

13.13. **A centrifugal pump, having four stages in parallel, delivers 12 m³/min of liquid against a head of 25 m, the diameter of the impeller being 220 mm and a speed 1700 rpm.**

A pump is to be made up with a number of identical stages in series, of similar construction to those in the first pump, to run at 1250 rpm, and to deliver 15 m³/min against on head of 250 m.

Find the diameter of the impeller and number of stages required.

Solution

$$N_s = \frac{N\sqrt{Q}}{H_m^{3/4}} = \frac{1700\sqrt{12/4 \times 60}}{(25)^{3/4}} = 34 \left(rpm, m^3/s, m\right)$$

Specific speed is same as this pump is of similar construction (i.e. geometrically and dynamically similar)

Therefore,

$$N_s = \frac{N\sqrt{Q}}{(H_m)^{3/4}}$$

$$34 = \frac{1250\sqrt{15/60}}{(H_m)^{3/4}}$$

$$\therefore H_m = (18.38)^{4./3} = 48.51m$$

Total head to be lifted by this pump is 250 m

Therefore number of stages required

$$= \frac{\textit{Total head to be lifted}}{\textit{Head by each pump}} = \frac{250}{48.514}$$

$= 5.15 = 6$ *stages* Ans.

Now $\sqrt{H_m} / DN$ is a constant which should be same for both types of pumps as they are hydraulically similar.

Therefore,

$$\frac{\sqrt{25}}{0.22 \times 1700} = \frac{\sqrt{48.514}}{D \times 1250}$$

$$0.01336 = \frac{5.572 \times 10^{-3}}{D}$$

$\therefore D = 0.417\ m = 417\ mm$ Ans.

13.14. A multistage boiler feed pump is required to pump water at 30 kg/s against a pressure difference of 26.5 bar when running at a speed of 2900 rpm. The relative density of the preheated feed water is 0.96.

If all the impellers are identical and the specific speed per stage is not to be less than 670 (rpm, liters/s, m), find :

(a) The least number of stages and the head H_m per stage.

(b) The diameter of the impeller, assuming the peripheral velocity as $0.96\sqrt{2gH_m}$

(c) The shaft power required to drive the pump, assuming an overall efficiency of 79%.

Solution

$$W = s\,\rho\,g\,Q$$

$$30 \times 9.81 = 0.96 \times 1000 \times 9.81 \times Q$$

$$Q = 31.85 \times 10^{-3}\ m^3/s = 31.85 \text{ liters/s}$$

$$H_T = 26.5\ bar = \frac{26.5 \times 100}{0.96 \times 9.81} = 281.388\ m$$

$$N_s = \frac{N\sqrt{Q}}{(H_m)^{3/4}}$$

$$(H_m)^{3/4} = \frac{N\sqrt{Q}}{N_s} = \frac{2900\sqrt{31.85}}{670} = 24.427$$

$$\therefore H_m = 70.876\ m$$

$$\therefore \textit{Number of stages} = \frac{H_T}{H_m} = \frac{281.388}{70.876} = 3.97 = 4$$ **Ans.**

$$u = 0.96\sqrt{2\,g\,H_m} = \frac{\pi\,D\,N}{60}$$

$$0.96\sqrt{2 \times 9.81 \times 70.876} = \frac{\pi\,D \times 2900}{60}$$

$$\therefore D = 0.235\ m = 235\ mm$$ **Ans.**

$$P = \frac{S\,\rho\,g\,Q\,H}{\eta_o} = \frac{0.96 \times 1000 \times 9.81 \times 31.85 \times 10^{-3} \times 281.388}{0.79}$$

$$= 106.838\ kW$$ **Ans.**

13.15. Cavitation tests were performed on a pump giving the following results Q = 0.05³/s, H = 37 m, barometric pressure 760 mm of mercury, ambient temperature 25°C. Cavitation begin when the total head at pump inlet was 4 m. Calculate the value of Thoma cavitation coefficient and NPSH. What would be the maximum height of this pump above water level if it is to operate at the same point on its characteristic in the ambient conditions of barometric pressure of 640 mm of mercury and temperature of 10°C?

Solution

$$H_s = 760 \text{ mm of mercury}$$

$$= \frac{760}{1000} \times 13.6 = 10.336 \text{ m of water}$$

$At\ 25°\ C, p_v = 0.03169\ bar$ (from steam table)

$$\therefore H_v = 0.03169 \times \frac{100}{9.81} = 0.323\ m$$

$$NPSH_{available} = H_a - H_v - Z_I - h_{fs}$$

$$= 10.336 - 0.323 - 4 \qquad |h_{fs} \text{ is not given}|$$

$$= 6.013\ m \text{ Ans.}$$

$$\sigma = \frac{NPSH_{available}}{H_m} = \frac{6.013}{37} = 0.1625 \qquad \text{Ans.}$$

$$\text{(b) } H_a = 640 \text{ mm of mercury} = \frac{640}{1000} \times 13.6 = 8.704\ m$$

$At\ 10°C, p_v = 0.012276\ bar$ [From steam table]

$$\therefore H_s = 0.012276 \times \frac{100}{9.81} = 0.125\ m$$

$$\sigma = \frac{H_a - H_v - Z_I - h_{fs}}{H}$$

$$0.1625 = \frac{8.704 - 0.125 - H_s}{37} \qquad |h_{fs} \text{ is not given}|$$

$$6.0125 = 8.5788 - H_s$$

$$H_s = 2.566\ m \qquad \text{Ans.}$$

13.16. A centrifugal pump of specific speed 0.683 (based on units rps, m³/s, m) has a critical Thoma number equal to 0.2, the proposed installation of the pump requires its centre line to be 5.2 m above the sump water level. The pump when running at 1450 rpm delivers 0.0637 m³/s. The losses in the suction pipe are estimated as 0.457 m of water. If the barometric pressure is 749 mm of mercury and the temperature of the water is 27°C for which the vapour pressure is 0.03567 bar, establish whether cavitation likely to occur.

Solution

$$H_a = 749\ mm\ of\ mercury$$

$$= \frac{749}{1000} \times 13.6 = 10.1864\ m\ of\ water$$

$$H_v = 0.03567 \times \frac{100}{9.81} = 0.3636\ m\ of\ water \qquad [At\ 27°\ C]$$

$$NPSH_{available} = H_a - H_v - Z_I - h_{fs}$$

$$= 10.1864 - 0.3636 - 5.2 - 0.457 = 4.16458\ m$$

$$N_s = \frac{N\sqrt{Q}}{H_m^{3/4}}$$

$$0.683 = \frac{1450\sqrt{0.0637}}{60 \times (H_m)^{3/4}}$$

$$H_m = (8.93)^{4/3} = 18.5276\ m$$

$$NPSH_{required} = \sigma_c \times H_m = 0.2 \times 18.5276 = 3.7055$$

Because $NPSH_{available} > NPSH_{required}$

Therefore no cavitation. Ans.

13.17. A single stage centrifugal pump runs at 500 rpm and delivers 300 m³/min of water against a head of 120 m. The pump impeller is 2 m in diameter and has a positive suction lift of 3 m. Laboratory tests are conducted on a model with 450 mm impeller and on a reduced head of 95 m. Calculate the speed, discharge and suction lift for the laboratory tests. Assume vapour pressure as 0.34 m of water and atmospheric pressure as 10.15 m of water.

Solution :

We know that

$$\left[\frac{H_m}{D^2 N^2}\right]_{model} = \left[\frac{H_m}{D^2 N^2}\right]_{prototype}$$

$$\frac{95}{(0.45)^2 \times N_m^2} = \frac{120}{(2)^2 \times (500)^2}$$

$$\therefore N_m = 1977.2367 \ rpm$$ Ans.

We know that

$$\frac{Q_m}{N_m D_m^3} = \frac{Q_p}{N_p D_p^3}$$

$$\frac{Q_m}{1977.2367 \times (0.45)^3} = \frac{300}{500 \times (2)^3}$$

$$\therefore Q_m = 13.5132 \ m^3/\text{min.}$$ Ans.

We know that

$$\sigma_m = \sigma_p$$

$$\left[\frac{H_a - H_v - Z_I}{H_m}\right]_m = \left[\frac{H_a - H_v - Z_I}{H_m}\right]$$

$$= \frac{10 - 0.34 - Z_I}{95} = \frac{10.15 - 0.34 - 3}{120}$$

$$= \frac{10 - 0.34 - Z_I}{95} = 0.05675$$

$$\therefore Z_I = 4.41875 \ m$$ Ans.

13.18. When a laboratory test was carried out on a pump, it was found that, for a pump total head of 36.5 m at a discharge of 48.5 litres/s, cavitation began when the sum of the static pressure plus the velocity head at inlet was reduced to 3.26 m. Barometric pressure was 750 mm of mercury and the vapour pressure of water 1.8 kPa. If the pump is to operate at a location where atmospheric pressure is reduced to 620 mm of mercury and the vapour pressure of water is 830 Pa, what is the value of the cavitation parameter when the pump develops the same total head and discharge? Is it necessary to reduce the height of the pump above the supply, and if so by how much?

Solution

$$H_a = 750\ mm\ of\ mercury = 10.2\ m\ of\ water$$

$$p_v = 1.8\ kPa = 0.018\ bar$$

$$\therefore H_v = \frac{0.018 \times 100}{9.81} = 0.1834\ m$$

$$\frac{p_1}{\rho g} + \frac{V_1^2}{2g} = 3.26\ m \qquad (Given)$$

We know that

$$NPSH = \frac{p_1}{\rho g} + \frac{V_1^2}{2g} - \frac{p_v}{\rho g} = \frac{p_a}{\rho g} - \frac{p_v}{\rho g} - Z_I - h_f$$

$$\therefore \frac{p_1}{\rho g} + \frac{V_1^2}{2g} = \frac{p_a}{\rho g} - (Z_1 + h_f)$$

$$\therefore (Z_1 + h_f) = 10.2 - 3.26 = 6.94\ m$$

$$NPSH = \frac{p_1}{\rho g} + \frac{V_1^2}{2g} - \frac{p_v}{\rho g} = 3.26 - 0.1834 = 3.0766\ m$$

$$\sigma = \frac{NPSH}{H_m} = \frac{3.0766}{36.5} = 0.0843$$ **Ans.**

(b) $H_a = 622\ mm\ of\ mercury$

$$= \frac{622}{1000} \times 13.6 = 8.4592\ m\ of\ water$$

$$p_v = 830\ Pa = 8.3 \times 10^{-3}\ bar$$

$$\therefore H_v = 8.3 \times 10^{-3} \times \frac{100}{9.81} = 0.0846 m$$

$$NPSH = H_a - H_v - (Z_1 + h_f)$$

$$3.0766 = 8.4592 - 0.0846 - (Z_1 + h_f)$$

$$\therefore (Z_1 + h_f) = 8.4592 - 3.1612 = 5.29799\ m$$

Height reduced $= (Z_1 + h_f)_1 - (Z_1 + h_f)_{11}$

$= 6.94 - 5.29799 = 1.642\ m$ Ans.

Since the flow rate is the same, $h_{f1} = h_{f2}$, and the pump must be lowered a distance $Z_1 - Z_2 = 1.642\ m$ at the new location.

13.19. A pump situated 4 m above sump water level begins to cavitate when delivering 0.56 m³/s against a total head of 5 m with a speed of 1200 rpm and power input of 30 kW. What will be the maximum speed at which the pump may be run before it cavitates when it is installed 2 m below the water level if it is to work at the same efficiency? What will then be the flow rate, power and head produced by the pump?

Assume that the atmospheric pressure is equivalent to a head of 10 m of water and the absolute water vapour pressure at the temperature 29°C of the sump water is equivalent to 0.408 m head of water.

Solution

$$\sigma_c = \frac{H_a - H_{min} - Z_1 - h_{fs}}{H_m}$$

$$= \frac{10 - 0.408 - 4}{5} = 1.11828$$

We know that $\sigma_{c1} = \sigma_{c11}$

$$1.11828 = \frac{H_a - H_{min} - Z_1 - h_{fs}}{H_m}$$

$$= \frac{10 - 0.408 - (-2)}{H_m} = \frac{11.592}{H_m}$$

$$\therefore H_m = \frac{11.592}{1.11828} = 10.3658\ m$$ Ans.

$$P = \frac{\rho\, g\, Q\, H_m}{\eta_0}$$

$$30 \times 10^3 = \frac{1000 \times 9.81 \times 0.56 \times 5}{\eta_0}$$

$$\eta_0 = 0.9156 = 91.56\%$$

We know that

$$\frac{N_1}{\sqrt{H_1}} = \frac{N_2}{\sqrt{H_2}}$$

$$\frac{1200}{\sqrt{5}} = \frac{N_2}{\sqrt{10.3658}}$$

$$\therefore N_2 = 1200 \times \sqrt{\frac{10.3658}{5}}$$

$$= 1200 \times \sqrt{2.07316} = 1727.816\ rpm$$

and

$$\frac{Q_1}{\sqrt{H_1}} = \frac{Q_2}{\sqrt{H_2}}$$

$$\therefore Q_2 = Q_1 \times \sqrt{\frac{H_2}{H_1}} = 0.56 \times \sqrt{\frac{10.3658}{5}}$$

$$= 0.56 \times 1.4398 = 0.8063\ m^3/s$$ **Ans.**

$$p_2 = p_1 \left[\frac{H_2}{H_1}\right]^{3/2} = 30 \times \left[\frac{10.3658}{5}\right]^{3/2} = 89.55\ kW$$ **Ans.**

13.20. An experiment of single stage centrifugal pump was found to operate, free from cavitation, under a minimum net positive suction head of 4.2 m when pumping water at the rate of 140 liters/s against a head of 39 m. The impeller diameter was 450 mm and speed 1200 rpm.

A geometrically similar impeller is required to discharge 5660 liters/s against a head of 120 m. Assuming a water temperature of 20°C and allowing 0.35 m for hydraulic losses, at what speed and suction head should the prototype operate? Assume barometric pressure as 760 mm of mercury

Solution:

At 20°C, $H_v = 0.2384\ m$ [*From steam table*]

We know that

$$\sigma_{model} = \sigma_{prototype}$$

$$\left[\frac{NPSH}{H_m}\right]_m = \left[\frac{NPSH}{H_m}\right]_p$$

$$\frac{4.2}{39} = \frac{(NPSH)_p}{(H_m)_p}$$

$$0.107669 = \frac{(NPSH)_p}{120}$$

$$NPSH_p = 0.10769 \times 120 = 12.923$$

$$NPSH_p = H_a - H_v - Z_I - h_f = 10.336 - 0.2384 - Z_1 - 0.35$$

$$\therefore Z_1 = 9.747 - 12.923 = -3.175\ m$$ Ans.

We know that

$$(N_s)_m = (N_s)_p$$

$$\frac{N_m\sqrt{Q_m}}{(H_m)_m^{3/4}} = \frac{N_p\sqrt{Q_p}}{(H_m)_p^{3/4}}$$

$$\therefore N_p = 1200 \times \left[\frac{120}{39}\right]^{3/4} \times \left[\frac{0.14}{5.66}\right]^{1/2}$$

$$= 438.45\ rpm$$ Ans.

and $\dfrac{Q_m}{D_m^3\,N_m} = \dfrac{Q_p}{D_p^3\,N_p}$

$$\therefore D_p^3 = D_m^3 \times \frac{Q_p}{Q_m} \times \frac{N_m}{N_p} = (0.45)^3 \times \frac{5.66}{0.14} \times \frac{1200}{438.45} = 10.0828$$

$$\therefore D_p = (10.0828)^{1/3} = 2.16\ m$$ Ans.

EXERCISES

A. Theory

13.1. What is the main difference between turbine and rotodynamic pump?

13.2. Classify the rotodynamic pump with reference to shape of the impeller and delivery space and draw neat sketch.

13.3. Explain with the help of neat sketches the working of a single stage centrifugal pump.

13.4. Define (a) Static head, (b) Manometric head with reference to centrifugal pump.

13.5. Define (a) Manometric efficiency, (b) Overall efficiency, (c) mechanical efficiency for centrifugal pump.

13.6. Explain hydraulic losses in centrifugal pump.

13.7. Define static and manometric head of a centrifugal pump. State the. different types of head losses which may occur in pump installation.

13.8. What are the different efficiencies of a centrifugal pump?

13.9. Define specific speed of a centrifugal pump. Derive an expression for the same.

13.10. How does the specific speed of a centrifugal pump differ from that of a turbine.

13.11. How will you obtain an expression for the minimum speed for starting a centrifugal pump?

13.12. Explain affinity laws for centrifugal pumps.

13.13. Define Thoma number and explain its uses in conjuction with cavitation characteristics of rotodynamic pumps.

13.14. Distinguish between available and required NPSH and state briefly how they are determined.

13.15. Draw a typical curve between NPSH, head and discharge for a centrifugal pump.

13.16. What is cavitation and what are its causes ? How will you prevent the cavitation in hydraulic machines.

13.17. What is the difference between single stage and multi stage pumps?

13.18. Describe model testing of centrifugal pump ?

13.19. Describe multistage pump with the help of sketches

(a) pumps in parallel

(b) pumps in series.

13.20. Draw and explain the performance characteristics of pump.

13.21. Draw and discuss the operating characteristics of a centrifugal pump.

13.22. What is priming ? Why is it necessary?

13.23. How will you select the pump for a particular application ?

Draw the figure for deciding what type of pump is suitable for a given service.

13.24. Draw the pump selection chart and explain the conclusions drawn from the chart.

13.25. Draw the pump selection chart between specific speed and specific diameter for centrifugal pump.

13.26. Prove that for centrifugal pump

$$V_{f2} = u_2 \tan \phi$$

at the maximum efficiency condition.

13.27. Show that the difference of pressure of pressure head between the inlet of the impeller and the casing of a centrifugal pump, neglecting friction is given by

$$\frac{u_2^2 + v_{f1}^2 - V_{f2}^2 + K\,V_2^2}{2g}$$

where k ($V_2^2/2g$) is the fraction of the kinetic energy in the water leaving the impeller which is converted into pressure head in the casing.

13.28. Prove that, in general for a centrifugal pump running at speed N and giving a discharge Q, the manometric head is expressed in the form.

$$H_m = A\,N^2 + B\,N\,Q + C\,Q^2$$

Where A, B and C are constants.

B. Unsolved problems

13.1. A centrifugal pump impeller is 250 mm external diameter and the water passage is 32 mm wide at exit. The circumference is reduced by 12 percent on account of vane thickness. The impeller vanes are inclined at 140 to the forward tangent at exit. Manometric efficiency 83%, rpm =1000, Q= 2.86 m³/min, Calculate the conversion efficiency of the diffuser ring. Assume no losses in the impeller. (59.4%).

13.2. The impeller of a centrifugal pump is 300 mm outside diameter and 150 mm insider diameter. The impeller vane angles are 30° and 25° at the inner and outer peripheries respectively and the speed is 1450 rpm. Assume radial flow through the impeller, calculate the work done by the impeller per Neuton of water, the velocity flow being constant. [20.11 mN/N of water].

13.3. The diameter of the impeller of a pump is 1.2 m and its peripheral speed is 9 m/s. Water enters radially and is discharged from the impeller with a velocity whose radial component is 1.5 m/s. The vanes are curved backwards at exit and make an angle of 30° with the periphery. If the pump discharges 3.4 m³/min, what will be the turning moment on the shaft? [217 Nm]

13.4. If the static lift for a centrifugal pump is h meters, the speed of rotation N rev/min. and the external diameter of the impeller D meters, deduce the expression $N = 83.5\sqrt{n}/D$ for the speed at which pumping commences, assuming only rotation of water in the impeller, at the "no flow" condition. Such a pump delivers 1.27 m³/min. at 1200 rpm the Impeller of diameter is 350 mm and breadth at outlet flanges is 272 kN/m². Taking manometric efficiency at 63%, calculate the impeller exit blade angle. [32°202]

13.5. A centrifugal pump is to discharge 110 liters/s at a speed of 1450 rpm against a head of 23 m. The impeller diameter is 250 mm, its width 50 mm and the manometric efficiency is 75%, determine the vane angle at the outer periphery of the impeller. [42°12].

13.6. A centrifugal pump, having an overall efficiency of 75%, delivers 1820 liters of water per minute to a height of 18 m through a pipe 100 mm diameter and 90 m long. If L = 4f = 0.048 Calculate power to drive the pump. [20.44 kW]

13.7. At the design speed of 1000 rpm, a centrifugal pump is to deliver water against a head of 14.5 m. The vanes are curved backwards to an angle of 30° with the periphery. The impeller diameter is 300 mm, the outlet width 50 mm, and the hydraulic efficiency 95%, what is the discharge? [165 liters/s]

13.8. A test on a centrifugal pump having an impeller of 300 mm diameter show that it gave the best efficiency when running at 900 rpm and delivering 1.8 m^3/min against a head of 15.2 m. If an exactly similar design is required to deliver 3.6 m3/min against a head of 18.3 m, determine the diameter of the impeller and the speed in rpm [400 mm, 735 rpm].

13.9. A centrifugal pump with an impeller of 190 mm diameter gives at maximum efficiency a discharge of 3.9 m^3/min of fresh water at 1800 rpm against a head of 4.2 m. What should be the speed of rotation of a similar impeller of 380 mm diameter to give 54.5 m3/min of sea water of density 1025 kg/m3 and what pressure would it then generate? [3150 rpm. 515 kPa]

13.10 A centrifugal pump impeller has an outer diameter of 300 mm and an inner diameter of 150 mm. The pump runs at 1200 rpm. The impeller vanes are set at a blade angle of 30° at the outlet. If the velocity of flow is constant at 2 m/s.

Calculate

(a) The velocity and direction of water at outlet.

(b) The head developed if manometric efficiency is 85%.

(c) The blade angle at inlet.

[(a) 15.5 m/s, 7°242 (b) 25.12 m (c) 11°592]

13.11. A centrifugal pump discharges 7050 liters/minute against a total head of 26.5 m when running at 600 rpm. The outer diameter of impeller is 600 mm and inner diameter is 300 mm. The area of flow through the wheel is 615 cm^2. The vanes are set back at an angle of 45°. Water enters the wheel radially without shock.

Calculate:

(a) The manometric efficiency.

(b) Vane angle at inlet

(c) The head loss at inlet to impeller if the discharge is reduced by 50%, the speed remaining the same.

[82%, 11°24', 1.13m]

13.12. The centre line of the shaft of a centrifugal pump is at the water level in the sump. If angle at outlet is 30°, diameter at outlet is 450 mm, velocity of flow at outlet is 3 m/s and speed

1000 rpm, calculate the total head expected. Take the pressure head at impeller outlet to be 18 m, volute losses to be 0.5 $V_2^2/2g$ and assume radial entrance to impeller. [26.8 m]

13.13. Determine the least number of stages for a multistage pump to deliver 60 liters/s against a total head of 210 m when running at 1450 rpm if the specific speed (per stage) must be less than 670 (rpm, liters/s, m [5]).

13.14. A multistage centrifugal pump is required to lift 1.8 m3/min of water from a mine, the total head including friction 760 m. If the speed of the pump is 2900 rpm, find the least number of stages if the specific speed per stage is not to be less than 620, (rpm, liters/s, m). [10]

13.15. A centrifugal pump, having 4 stages in parallel delivers 10800 liters/min. of liquid against a head of 24.3 m, the diameter of the impellers being 225 mm, and the speed 1700 rpm. A pump is to be made up of a number of identical stages in series, of similar construction to these in the first pump, to run at 1250 rpm and to deliver 14400 liters/min against a head of 246 m. Find the diameter of the impellers and the number of stages required [436 mm, 5 stages]

13.16. A centrifugal pump with an impeller of 18.75 cm diameter delivers, at maximum efficiency 0.064 m3/s against a head of 4.2 m when running at 1800 rpm. What should be the speed of rotation of a dynamically similar impeller of 37.5 cm diameter required to deliver 0.9 m^3/s of sea water of density 1028 kg/m3, and what head would it then generate. [52 m, 3180 rpm].

13.17. A centrifugal pump when tested at a constant speed gave the following results.

Discharge (liters/min.)	0	225	450	675	900	1125
Manometric Head (m)	12	11.7	11.1	9.6	7.5	3.6
Efficiency (%)	0	46	66	75	70	48

Two such pumps are to be installed to pump against a static head of 6 m and the resistance of the pipeline is given by $Q^2/10^4 \times 67$.

Estimate the discharge in liters/min. and the power required to drive the pump when.

(a) only one pump is running and

(b) two pumps are running in parallel. [For (a) Q = 922 liters/min, H_m = 7.2 m, η = 69%, 1.575 kW (b) Q = 1418 liters/min, H_m = 9.3 m, η = 75%, 2.87 kW]

13.18. **When running at 1600 rpm the characteristics of a Centrifugal pump are as follows :**

H (m)	38.4	39.6	39.9	39	36.9	33.2
η (%)	0	65	74	78	80	79
Q (m^3/s)	0	2.8	5.6	8.5	11.3	14.2

Plot the curves of head efficiency expected for 1300 rpm. If the impeller diameter is 0.3 m, give the hydraulic efficiency for zero delivery. [0.6]

13.19. **A centrifugal pump is intended to lift 42.5 m^3/min against a head of 4.9 m. A model is made and delivers 1.7 m^3/min against 4.9 m head when making 1450 rpm the efficiency the being a maximum. Find the specific speed and estimate the speed and diameter of the impeller of the large pump if the diameter of the impeller of the model pump is 200 mm. [290 rpm, 1 m].**